Metal Oxide Nanostructures Chemistry

Metal Oxide Nanostructures Chemistry

SYNTHESIS FROM AQUEOUS SOLUTIONS

SECOND EDITION

Jean-Pierre Jolivet

OXFORD
UNIVERSITY PRESS

OXFORD
UNIVERSITY PRESS

Oxford University Press is a department of the University of Oxford. It furthers the University's objective of excellence in research, scholarship, and education by publishing worldwide. Oxford is a registered trade mark of Oxford University Press in the UK and certain other countries.

Published in the United States of America by Oxford University Press
198 Madison Avenue, New York, NY 10016, United States of America.

Translation of the French language edition of:
De la solution à l'oxyde—2e edition
De Jean-Pierre Jolivet

Library of Congress Cataloging-in-Publication Data
Names: Jolivet, Jean-Pierre, author.
Title: Metal oxide nanostructures chemistry : synthesis from aqueous solutions / Jean-Pierre Jolivet.
Other titles: De la solution a l'oxyde. Chimie aqueuse des cations metalliques, synthese de nanostructures. English
Description: 2nd edition. | New York, NY : Oxford University Press, 2019. | Translation of the French language edition of: De la solution a l'oxyde, 2e edition : chimie aqueuse des cations metalliques, synthese de nanostructures (2015). | Includes bibliographical references and index. Identifiers:
LCCN 2018045235 | ISBN 9780190928117
Subjects: LCSH: Oxides—Surfaces. | Precipitation (Chemistry) | Metallic oxides. | Nanostructured materials.
Classification: LCC QD181.O1 J6513 2019 | DDC 546/.72159—dc23
LC record available at https://lccn.loc.gov/2018045235

Printed on Demand

CONTENTS

PREFACE

The aim of this book is to present an overview of metal oxide nanoparticle fabrication in aqueous solution. Indeed, nanomaterials generate considerable interest because of their specific behaviors and properties, which have made their use increasingly diversified. Among these nanomaterials, metal oxides hold a special place because of the diversity of their properties.

The synthesis of oxide nanoparticles in allowing the control of their crystal structure, shape, size, and size distribution to optimize their physical properties has long been one of the ultimate goals of colloidal chemistry. This goal is still part of an intense effort, as shown by the huge number of articles published on this subject even to the present day. Of the large variety of synthesis pathways, precipitation in aqueous solution from metal salts is the most common and versatile method for obtaining oxide nanoparticles by sustainable green chemistry corresponding to an ecodesign of materials. Indeed, the technique does not require the use of solvents or ligands that present unwanted effects, even toxic ones. The environmental interest in the technique may consequently be added to the technological interest it has inspired.

This book is intended for students, researchers, and engineers interested in the precipitation of nano-oxides and their surface physical chemistry, not only in the field of materials but also in geochemistry and mineralogy—two fields that are directly and deeply associated with the aqueous chemistry of metal cations.

This book is in fact the second completely revised and extended edition of *Metal Oxide Chemistry and Synthesis* published in 2000 by John Wiley and Sons. The basic concepts of the chemistry of cations in solution are still valid, but after almost 20 years, immense progress has been achieved: the precipitation of most of the elements has been widely studied; the growth of particles by ordered aggregation (oriented attachment) today represents experimental evidence; theoretical modeling of oxide surface energy and the dynamics of nucleation and growth have been developed; and so on. It therefore became necessary to update the presentation of knowledge in this field, and it is for this reason that I rewrote this book, based on the course I was teaching at the University Pierre and Marie Curie, Paris, France. In the new edition of the book, the first five chapters include the major concepts: specificity of nanostructures, mechanisms of hydroxylation and condensation of cations in solution, structural and kinetic aspects of precipitation, and surface reactivity of oxides. The last three chapters are devoted to the chemistry of some technologically and environmentally important elements, such as aluminum, iron, titanium, manganese, and zirconium.

I warmly thank Philippe Belleville, Corinne Chanéac, David Chiche, Anne Duchateau, Cédric Froidefond, Julien Hernandez, Hanno Kamp, Magali Koelsh, Stéphane Lemonnier, Micaela Nazaraly, Thierry Pagès, Céline Pérégo, Agnès Pottier, David Portehault, Tamar Saison, and Lionel Vayssières, who completed their PhDs under my (co)direction. Their work added considerably to this book.

I am greatly indebted to Elisabeth Tronc, Marc Henry, Corinne Chanéac, Sophie Cassaignon, Olivier Durupthy, and all post–PhD students for the work we achieved together. I hope that in this book they find my expression of gratitude and friendship.

I would also like to thank all my colleagues from the French and foreign academic communities as well as from industry, who, through diverse collaborations or during simple exchanges, contributed closely and even furthered this work. I particularly want to thank Benjamin Gilbert (Lawrence Berkeley National Laboratory, USA), Alain Manceau (University of Grenoble, France), and Claudine Noguera (University P. & M. Curie, Paris, France) for their advice, suggestions, and exciting discussions.

Finally, I wish to express my gratitude and extend my friendship to Jacques Livage and Clément Sanchez for the high quality of their scientific work and for the happy years we spent together in the laboratory Chimie de la Matière Condensée de Paris.

ABBREVIATIONS

AO	atomic orbital
EPR	electron paramagnetic resonance
EXAFS	extended X-ray absorption fine structure
FESEM	field effect scanning electron microscopy
HRTEM	high-resolution transmission electron microscopy
IEP	isoelectric point
MO	molecular orbital
NMR	nuclear magnetic resonance
PZC	point of zero charge
SAXS	small angle X-ray scattering
SEM	scanning electron microscopy
TEM	transmission electron microscopy

Metal Oxide Nanostructures Chemistry

1 }

Nanomaterials

SPECIFICITIES OF PROPERTIES AND SYNTHESIS

The concept of material concerns matter in solid state that is endowed with usable properties for practical applications. It is indeed in the solid state that matter exhibits the highest mechanical strength and chemical inertness, providing solidity and sustainability because the solid is based on an extended stiff crystalline framework. It is also in the solid state that many properties exist, including optical, electrical, and magnetic properties, providing great technological progress. A typical example is electronics which owes its enormous development to doped silicon. A *material* may therefore be defined as a useful solid.

The properties of a solid depend directly on its chemical composition, crystalline and electronic structures, texture, as well as morphology and casting. This last point, which is often neglected, is illustrated by amorphous silica glass, which is used largely for its properties such as chemical inertness, mechanical strength, optical transparency, and low thermal and electrical conductivities. These various properties are highlighted through the many possibilities of casting and shaping: flat glass (optical transparency for glazing); hollow glass (chemical inertness and mechanical strength for bottling); short fibers (glass wool for heat insulation) and long fibers (optical fibers); massive pieces (insulators for electric power lines); and thin films (insulating layers for miniaturized electronics).

Metal oxides exhibit a wide range of exploitable properties useful for innumerable applications. Silica, SiO_2, as flat glass, has excellent optical properties, but other oxides such as $LiNbO_3$ and $KTiOPO_4$ exhibit interesting nonlinear optical properties, allowing changes in the wavelength of the transmitted light. Certain oxides are good electrical insulators (SiO_2), but others are true electronic conductors (VO_2, Na_xWO_3), ionic conductors (β-alumina $NaAl_{11}O_{17}$, NaSiCON $Na_3Zr_2PSi_2O_{12}$, yttria-stabilized cubic zirconia $Zr_{1-x}Y_xO_{2-x/2}$), and also superconductors (cuprates such as $YBa_2Cu_3O_{7-x}$ and $Bi_4Sr_3Ca_3Cu_4O_{16+x}$). Compounds such as $BaTiO_3$, $PbZr_{1-x}Ti_xO_3$, and $PbMg_{1/3}Nb_{2/3}O_3$ are ferroelectric solids used largely as miniaturized electronic components, whereas spinel ferrite γ-Fe_2O_3, barium hexaferrite $BaFe_{12}O_{19}$, and garnet $Y_3Fe_5O_{12}$ are more or less coercive ferrimagnetic solids used in magnetic recording or as permanent magnets. Alumina and aluminosilicates (clays and zeolites) are used in many areas as absorbents or catalysts because of their crystalline structure exhibiting interfoliar spaces or channels in which internal diffusion of ionic and molecular species can occur.

In addition to properties due to crystalline structure and chemical composition, some properties are specifically due to or depend on the size and shape of particles in the nanometric range (from 1 to some tens of nanometers). Today, these properties are exploited for multiple uses in many fields, for instance, that of materials: composites, membranes with controlled porosity, reinforcement and ignifugation of polymers, hydrophobic coatings, ferrofluids, and so on; the energy field: including photocells, photocatalytic self-cleaning glazings, and antireflective treatments of mirrors for power lasers; medicine: diagnostic, vectorization, imaging; and the environmental field: catalytic cleanup of Diesel exhaust gases, water nanofiltration, and so forth. Interest in ultradivided matter has recently revealed the concept of nanomaterial. The origin of its peculiarities is briefly studied in the next section.

1.1 Specific Properties of Nanoparticles

For several decades, ultrafine matter has attracted great scientific and technological interest, which has led to the development of *nanomaterial sciences*. A nanomaterial is defined with regard to the size of its individual components in the nanometric range and the specific properties it exhibits regarding those of the same matter in a bulky state. These specific properties may be a peculiar physics owing to the small volume of matter enclosed in a nanoparticle and/or a chemical reactivity enhanced by the huge surface area developed in nanoparticle assemblies up to some thousand $m^2.g^{-1}$.

1.1.1 VOLUME EFFECTS

Small volumes of solid matter exhibit properties that are different from those of massive state (micronic size or more) toward energy-matter interaction involved with light and electric and magnetic fields. The phenomenon occurs when at least one dimension of the object is of the order of magnitude of fundamental physical lengths—for instance, the mean free path of electrons in solid, the wavelength of light or of phonons in solid, as well as the size of magnetic domain. Owing to quantum confinement, specific effects which depend on the size and shape of nanoparticle or crystalline domains are resulting. Such effects, related to nanoscaled solids, have led to the development of various scientific and technical fields such as plasmonics, photonics, spintronics, and nanomagnetism, for which physical laws are now relatively well established [1]. Among the consequences of size effects, the various colors of semiconducting nanoparticle dispersions, such as cadmium and zinc chalcogenides, and of metal nanoparticles, such as gold, silver, and copper, are easily observable.

Cadmium and zinc chalcogenides, CdE, ZnE (E = S, Se, Te), are semiconductors whose optical properties (absorption, fluorescence) strongly depend on size effects because of changes in the electronic structure of the solid. Optical absorption corresponds to electron excitation from the highest occupied level in the valence band toward the lowest empty level in the conduction band (see § 1.1.3c). The difference in energy between these levels (gap) determines the wavelength of absorption and consequently the color of the material because the gap in chalcogenides corresponds to the energy of visible light (Fig. 1.1a). The effect is highlighted by the shift of the light absorption peak toward shorter wavelengths (blue shift) for CdSe nanoparticles of decreasing size (Fig. 1.1b). Fluorescence is the light emission that accompanies the return to the fundamental state of electrons excited beforehand by ultraviolet (UV) light. The color of fluorescence depends on the particle size for the same reason as for absorption (Fig. 1.1c). Beyond about 10 nanometers, the size effect on the electronic structure becomes insignificant.

Gold is a yellow metal, even though it is used as a very thin coating of some micrometers in thickness for gilding, whereas gold nanoparticle dispersions exhibit various colors ranging from red to green or blue. For metallic conductors, the size effect plays on the oscillations of the conduction electron cloud generated by light excitation. It results in a plasmon resonance that modifies the light absorption [2]. The absorbed energy depends on the wavelength according to the following equation [3]:

$$E(\lambda) = \frac{24\pi N a^3 \varepsilon_m^{3/2}}{\lambda}\left[\frac{\varepsilon_i}{(\varepsilon_r + 2\chi\varepsilon_m)^2 + \varepsilon_i^2}\right]$$

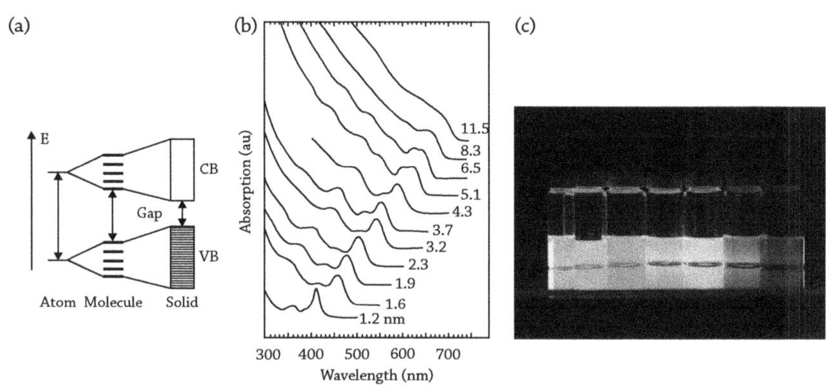

FIGURE 1.1 (a) Scheme of an electronic structure showing the demultiplication of energy levels with the size of the object and consequently with the number of involved atomic orbitals in bonds. It results in a change in the width of the forbidden band (gap) between the valence band (VB) and the conduction band (CB). (b) Absorption spectra of CdSe nanoparticles dispersed in hexane, with mean sizes from 1.2 to 11.5 nm (reproduced with permission from [42] Copyright 1993 American Chemical Society). (c) Light emission under UV irradiation of CdSe nanoparticles dispersed in hexane. The particle size varies from 1.5 nm (blue) to 7 nm (red) (courtesy of B. Dubertret, ESPCI, Paris, France).

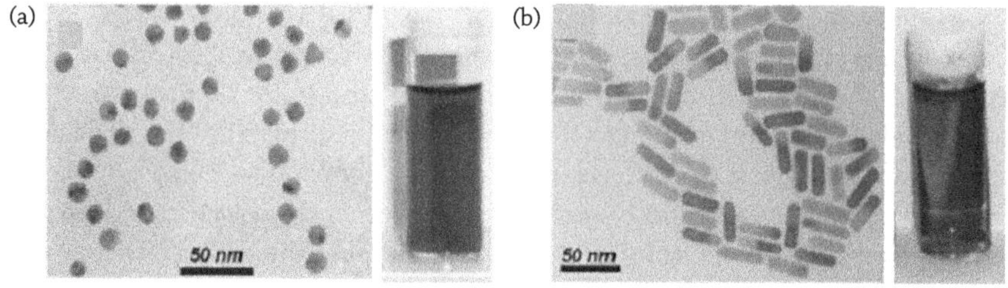

FIGURE 1.2 TEM images of aqueous dispersions of gold nanospheres (a) and nanorods (b) and colors of these dispersions (reproduced from Liz-Marzan L.M., Nanometals: formation and color, Materials Today 2004, 26–31 with permission of Elsevier).

where N represents the density of nanoparticles of radius a; ε_i and ε_r stand for imaginative and real parts of their dielectric constant; and ε_m is the dielectric constant of the surrounding medium. The χ parameter is relative to the particle shape. It is equal to 1 for a sphere and 17 for a spheroid with an aspect ratio of 5:1.

Since the maximum in light absorption is located along the whole range of visible radiations from 400 to 650 nm, the color of dispersions can range from blue to red, passing by yellow and green (Fig. 1.2). The same phenomenon occurs for other metals such as silver, gray in massive state, and copper [4]. Copper nanoparticles have been used since ancient times to color "ruby glass," and they are still used for decorative effects. Their main interest today is their role in developing new technologies such as plasmonics.

Another phenomenon arising from size effect is the magnetic relaxation that occurs in magnetic nanomaterials such as spinel iron oxide nanoparticles (γ-Fe_2O_3). This material is ferrimagnetic because the individual magnetic moments of Fe^{3+} ions in tetrahedral and octahedral occupied sites within the oxygen atoms framework are strongly coupled in opposite directions (see Chap. 7). The global magnetic moment is aligned around a direction called easy magnetization direction (here, the 111 axis of the cubic unit cell of maghemite). In fact, the magnetic moment of the nanoparticle thermally vibrates around the easy magnetization direction, and it can revert at the frequency $\tau = \tau_0 \exp E_B$. E_B, the energy barrier, is given by $E_B = KV / kT$ (Fig. 1.3). K is the magnetocrystalline anisotropy constant, V is the volume of the magnetic domain, and kT is the thermal excitation parameter.

In the bulky solid, the energy barrier is high, and the magnetization reversal does not occur unless the temperature is high enough. With nanosized particles, the energy barrier becomes close to kT, so that the magnetization can relax and reverse at room temperature (RT). The phenomenon is evidenced by Mössbauer spectroscopy of nanoparticles of different sizes dispersed in a stiff matrix made of a polymer in order to avoid interparticle magnetic coupling (Fig. 1.3d). The spectrum at RT of particles 30 nm in mean diameter is made of a sextuplet characteristic of a magnetically blocked system (the energy barrier is higher than thermal agitation). The spectrum of smaller particles is made of a paramagnetic doublet that is characteristic of a dynamical system. Nanoparticles 6 nm in mean diameter behave as paramagnetic ions with a magnetization that totally relaxes at RT because

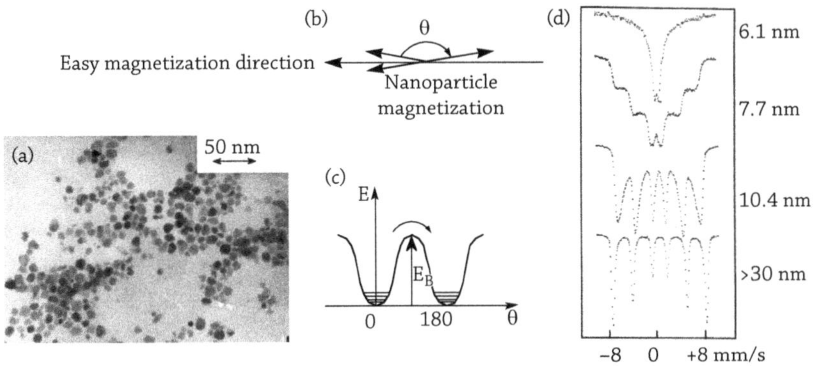

FIGURE 1.3 (a) TEM image of maghemite γ-Fe_2O_3 nanoparticles. (b) Fluctuations of the magnetic moment of a particle around an easy magnetization axis in the direction of the axis or in the opposite direction. (c) Scheme of the energy barrier E_B limiting the magnetization reversal (Θ = 0 or 180°). In the two energy sinks corresponding to Θ = 0 and 180°, vibrational energy levels exist, corresponding to oscillations of the magnetization in the direction of the axis or in the opposite direction.
(d) Mössbauer spectra at RT of dispersions of particles of different mean sizes from 6 to 30 nm.

the energy barrier is becoming much lower than the thermal agitation. That is the superparamagnetism phenomenon because the relaxing magnetic moment is that of the entire particle (the individual moments remain ferrimagnetically coupled) and is not that of an isolated ion [5].

Another interesting example of volume effect concerns the insertion of lithium into the rutile structure of TiO_2 following the reaction:

$$TiO_2 + xLi^+ + xe^- \Leftrightarrow Li_xTiO_2$$

The insertion of a Li^+ ion is balanced by the reduction of a Ti^{IV} ion to Ti^{III}. With micronic-sized particles, the reaction does not significantly occur (x = 0.03) because the compacity and stiffness of the crystalline framework do not allow the increase in size of the titanium ion (r_{Ti4+} = 0.60 Å, r_{Ti3+} = 0.67 Å). With TiO_2 nanoparticles that are around 10 nm in thickness and 200 nm in length synthesized by precipitation from aqueous acidic solutions of $TiCl_4$ (see Chap. 7), the reaction proceeds up to $x \approx 0.85$ at RT, according to different processes [6]: electrochemically (in a Swagelock-type cell, the sample being mixed with carbon in the presence of $LiPF_6$ to allow electrical conduction) (Fig. 1.4); chemically (with n-butyl lithium solubilized in hexane); and mechanically (by the grinding of TiO_2 and metal Li under argon atmosphere). The structural study shows that the reaction induces the expansion of the crystalline unit cell of around 10% for x = 0.6 (Fig. 1.4). For a higher reduction rate and consequently a more important intercalation stage, TiO_2 rutile transforms irreversibly into $LiTiO_2$ (with the rock-salt-type structure), which exhibits a reversible electrochemical capacity of 0.5 Li/Ti with good cyclability. Because of their very small size, the rutile titanium dioxide nanoparticles allow the intercalation of Li^+ ions and may form interesting electrode materials.

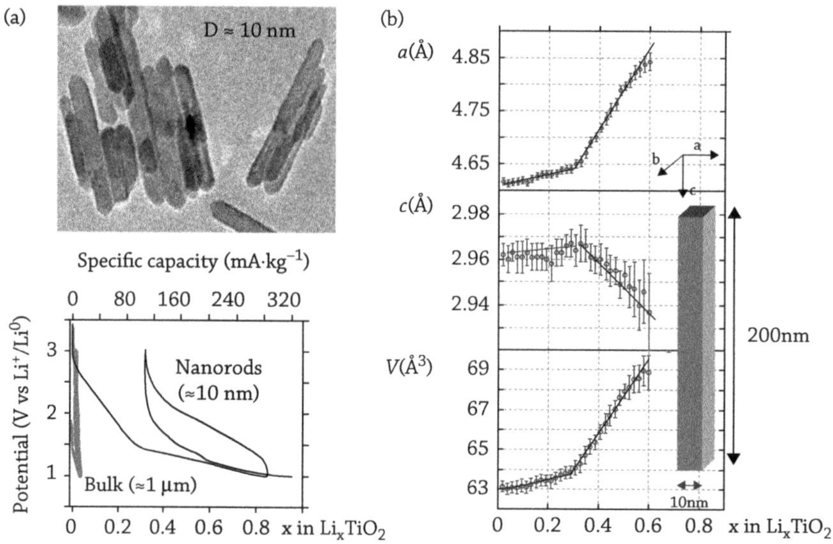

FIGURE 1.4 (a) TEM image of TiO_2 rutile nanorods and curves of galvanostatic cycling for micro- and nanoparticles; (b) changes in rutile unit-cell parameters and volume in nanorods during the electrochemical insertion of lithium. For $x > 0.6$, the rutile framework transforms into $LiTiO_2$ of rock-salt crystal type (reproduced from [6] with permission of Elsevier).

1.1.2 SURFACE EFFECTS

The decrease in size induces surface effects because the surface-to-volume ratio strongly increases when the particle size decreases. For spheres of radius *r*, the surface-to-volume ratio varies as $3/r$. For a metal oxide with a close compact crystalline structure, around 5% of the oxygen atoms of a sphere 20 nm in diameter are surface atoms, while for a sphere 15 nm in size, 15% of oxygen atoms are surface atoms; the proportion reaches 60% for spheres 2 nm in size. Consequently, surface effects may induce new and interesting properties. For instance, for densified nanomaterials, ultrahard nanophased metals are obtained when the crystal size decreases (Fig. 1.5), while, with nanoceramics, the material is "superplastic" in comparison to the solid formed of micronic or larger-sized grains [7]. Such unusual behavior is due to the increase in the amount of grain boundaries when the grain size decreases within the material. With regard to metals, the grain boundaries act as barriers limiting the sliding of atomic planes under mechanical strain and increasing hardness and resistance to the deformation of nanophased solids. In the case of ceramics, which are intrinsically hard and brittle materials, the grain boundaries allow the grains to move under stress; this produces a certain deformability, rendering the nanophased material less fragile than the classical ceramics.

The large specific area of nanometric systems gives them a high physicochemical reactivity. It results in the strong interaction of nanoparticles with their surroundings, giving rise to various phenomena involving adsorption: solvation; electrostatic

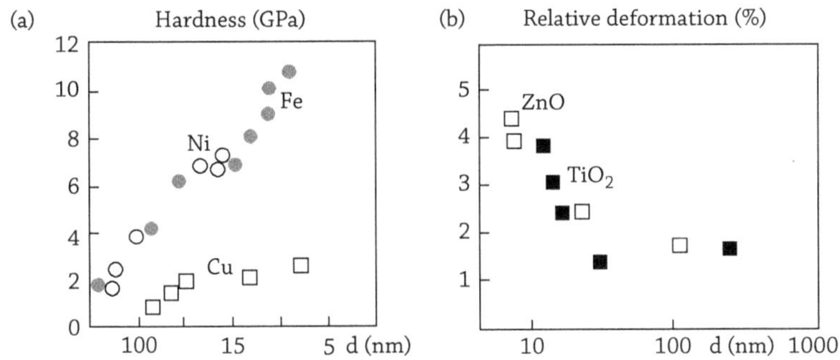

FIGURE 1.5 (a) Variation in hardness at RT of some metals as a function of grain size. Hardness is 2 to 5 times higher for nanophases than for coarse-grained samples. (b) Relative deformation of nanophased ZnO and TiO_2 samples as a function of grain size (adapted from [7]).

surface charge and dispersion of oxide nanoparticles in water; adsorption and transport of ionic and molecular species in natural waters; and biological effects and possible toxicity during interaction with cell or living organisms. Surface effects also influence the (electro)chemical reactivity of nanoparticles, as shown by heterogeneous catalysis [8]. For instance, the activity of palladium in pyrene hydrogenation increases by two orders of magnitude when the particle size decreases from 10 to 2 nm [9]. Nanometric magnesium oxide appears to be a better absorbent of acid gases such as SO_2 and CO_2 and chlorinated derivatives of benzene than the conventional compound with higher granulometry [10, 11]. Both the increase of surface area and the presence of defects and structural distortions contribute to the chemical activity of the catalyst. The greater activity of nanomaterials gives them good performance in environmental remediation [12]. Also, the high surface reactivity can be used in nanoparticle shape and size control during their synthesis (Chap. 4).

Size has a significant impact on diverse properties of the matter. It is therefore important to examine its influence on the main factors of the reactivity: thermodynamic stability and crystalline and electronic structure of the solid.

1.1.3 THE INFLUENCE OF SIZE EFFECT ON THERMODYNAMIC, STRUCTURE, AND REACTIVITY OF OXIDE NANOPARTICLES

Nanometric oxides are of obvious technological interest. They are also very important to the environment because of their ubiquity, abundance, and chemical and structural diversity [13, 14]. Some of them, especially iron and manganese oxides, give rise to important redox phenomena that regulate the cycle of many elements in the environment [15, 16]. Redox reactions between soluble Fe^{II} ions and ferric solid phases also have an important role in the mobilization and transformation of toxic heavy elements such as arsenic and lead [17, 18]. Understanding how these systems work and learning how particle size influences their reactivity is vital. This problem is complex and remains an open question despite the publication of numerous experimental studies.

Intuitively, one might predict that the reactivity of a solid increases with the decrease in particle size because of the higher exposed surface area and the larger proportion of catalytically active zones bearing specific coordinated sites and/or structural or geometric defects (steps, terraces, corners) that provide kinetic advantages. This prediction is effectively verified in many examples, but in some cases, it is observed that the reactivity of nanoparticles decreases as their size decreases. Consequently, it is interesting to examine the main consequences of the size effect on the thermodynamic, structural, and electronic characteristics of nano-oxides. To study comparable systems with various crystalline structures, the chosen examples concern iron oxides, but the exposed concepts are also applicable to many other nanostructured systems.

a) Thermodynamic Stability

Because of the increase in surface area with the decrease of particle size from the micron to nanometer range, the surface energy component (ΔG_{surf}, always positive) becomes competitive with the bulk component (ΔG_{bulk}, always negative) of the free enthalpy of formation ($\Delta G = \Delta G_{bulk} + \Delta G_{surf}$) of the ultradivided system (§ 5.4.4). This shows that the ultradivided material is intrinsically metastable compared to the bulk material. If transformations are kinetically possible (by Ostwald ripening and/or by aggregation in aqueous dispersions (§ 4.2.3)), such a system should spontaneously evolve to reduce the amount of interfaces. In the case of polymorphic systems such as iron, manganese, titanium, and aluminum oxides, an important consequence is the change in the relative stabilities of the polymorphs. Indeed, in bulk state, the formation enthalpies (ΔH) of the polymorphs are in general not substantially different and their entropies (ΔS) are similar. As a result, the metastability ($\Delta G = \Delta H + T\Delta S$) of the less stable phases is not very high. It follows that at the nanometric scale a metastable bulk phase can become the most stable phase among the different polymorphs (Fig. 1.6) [19–24].

The phase stability is also related to aggregation and sintering at high temperature, as is evidenced by dispersing metastable nanoparticles in a chemically inert rigid matrix such as silica glass or a polymer. Under these conditions, the growth of the nanoparticles (the decrease in surface area), as well as their structural transformation, does not occur, even at high temperature. That is the case, for instance, for anatase TiO_2 nanoparticles whose transformation into rutile occurs at around 300°C in powdered samples. The transformation is shifted up to 800°C when the nanoparticles are dispersed and confined in a silica glass [25]. Powdered maghemite, γ-Fe_2O_3, nanoparticles transform into hematite, α-Fe_2O_3, at around 400°C, but dispersed in a silica glass, the transformation does not occur below 1400°C, when the silica matrix crystallizes into cristobalite, allowing the aggregation of iron oxide nanoparticles [26]. At around 1200°C, when the silica matrix begins to soften, allowing diffusion and limited aggregation of maghemite nanoparticles, a rare phase of iron oxide, ε-Fe_2O_3, is obtained (see Chap. 5) [27]. This phase, isotypical with κ-alumina (transition alumina), may be considered as an intermediate between maghemite (isotypical with γ- alumina) and hematite

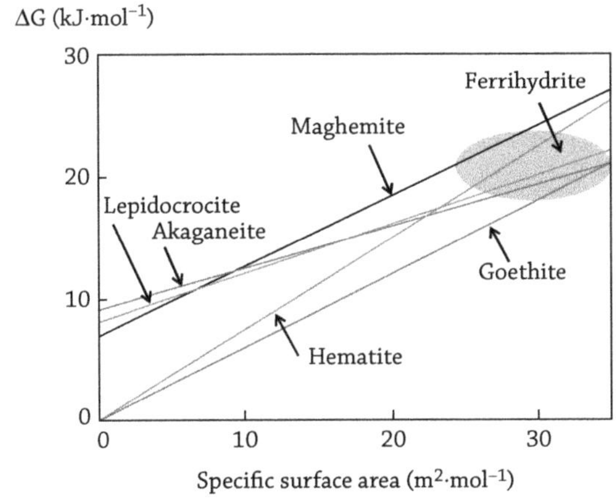

FIGURE 1.6 Free enthalpies of transformation of iron oxides and oxyhydroxides with hydrated surfaces into bulk hematite and liquid water at 298 K, calculated as a function of specific surface area (expressed in m^2 per mole of iron to take into account the stoichiometry of phases with different compositions). The slope of the lines represents the hydrated surface energy. The cross-points correspond to the changes in relative stabilities of the different phases (adapted from [20]).

(isotypical with corundum, α-Al_2O_3). Without a confinement effect, maghemite directly transforms into hematite at low temperature. So, as long as the system cannot dissipate the surface energy, the lowering of free energy does not occur and a potential transformation is blocked.

Surface hydration also has an important role because it significantly lowers the interfacial tension (§ 5.4.4). In addition, the particle solubility generally increases with decreasing size [28, 29]. It is one of the relevant parameters of reactivity, especially when biogeochemical phenomena are involved. Therefore, complex phase transformations can occur in an aqueous environment (Fig. 1.6). Typically, goethite, α-FeOOH, becomes thermodynamically stable against hematite, α-Fe_2O_3, for a size of spherical particle smaller than 60 nm, while lepidocrocite, γ-FeOOH, is more stable than maghemite, γ-Fe_2O_3, for a particle size smaller than 12 nm [20]. Similar phenomena occur with titania [22, 30] and manganese oxides, which exhibit various crystallochemistries [24] (see Chap. 7). Particle size therefore exerts a major role in the formation and stability of the different polymorphs, but kinetic factors may limit the transformations.

b) Crystal Structure and Structural Defects

Nanocrystals are often associated with crystalline disorder. Internal distortions are induced by strong surface reconstructions whose effects can spread to the core of a few nanometers in diameter particles [31]. These surface reconstructions may result from the small particle size, which requires a strong surface curvature and/or the juxtaposition of numerous facets and undercoordinated surface atoms that induce strong mechanical stresses [32].

Because of this surface disorder, the structure and chemical composition near the surface can differ from those in the bulk. This is well exemplified in maghemite, γ-Fe_2O_3: particles smaller than 11 nm exhibit an altered crystal structure with the creation of cation vacancies among tetrahedral sites of the surface [33]. These defects, which do not exist in larger particles, are revealed by their reactivity as being specific sites for adsorption and coordination of As^{III} [34]. Arsenic adsorption is an important environmental topic because its presence in large quantities in natural groundwaters poses serious public health problems in various countries throughout the world. Thus, arsenic adsorption by oxides could be an interesting way to avoid its dissemination. In this context, very small maghemite nanoparticles exhibit an unusual behavior. The amount of adsorbed arsenic increases from 0.02 to 1.8 mmol.g^{-1} when the particle size decreases from 300 to 11 nm [35]. However, the adsorbed amounts, normalized by the surface unit, are the same for 300- to 20-nm-sized particles (6 μmol/m^2 or 3.6 atoms/nm^2), indicating that the adsorption mechanism is the same in this particle size range. In contrast, for smaller particles, the adsorption capacity increases, and for particles 11 nm in mean diameter, the adsorbed arsenic amount is around three times larger (18 μmol/m^2 or 11 atoms/nm^2). The surface structural change leading to new adsorption sites and a significant decrease in surface energy [20, 36] helps explain this increase in adsorption capacity. Therefore, two adsorption stages are involved: (1) at low coverage rate, As ions are coordinated by the most reactive sites, which are the tetrahedral Fe^{III} vacancies at the center of hexagons made by sharing FeO_6 octahedra (Fig. 1.7); (2) After filling these sites, subsequent As adsorption occurs as tridentate complexes bounded to the classical sites available only with larger particles [37].

Similarly, an extended X-ray absorption fine structure (EXAFS) structural study of goethite nanoparticles showed that a decrease in particle size results in a distortion in the geometry of Fe^{III} surface sites [32]. This change disadvantages the cation adsorption: the adsorption rate of Hg^{II} ions on goethite particles of 5 nm is

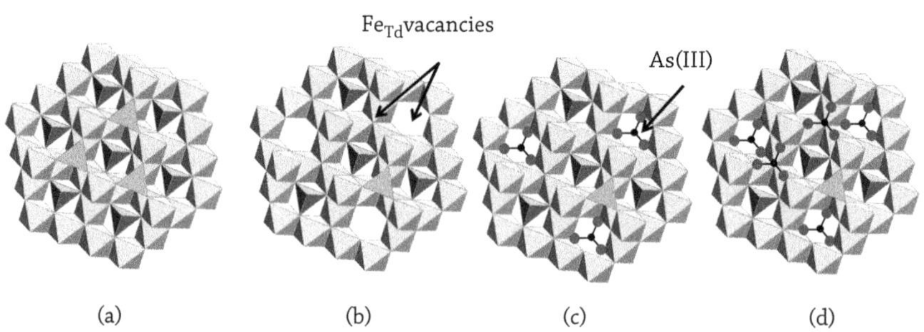

FIGURE 1.7 Scheme of adsorption sites for As^{III} on nanoparticles 6 nm in diameter based on structural information deduced from X-ray diffraction and As K-edge XAS analyses. (a) Idealized structure of {111} faces of maghemite, γ-Fe_2O_3 (octahedra and tetrahedra are FeO_6 and FeO_4, respectively). (b) Nanoparticle exhibiting Fe^{III} vacancies. (c) Filling of the most reactive sites by arsenic at low coverage rate. (d) Adsorption on lattice positions at higher surface coverage allowing a decrease of surface energy.

much lower than that for 25 or 75 nm [32]. In other cases, distortions of the surface geometry can promote adsorption. Thus, the presence of more numerous defects on hematite particle surfaces of decreasing size (88, 25, 7 nm) promotes the coordination of Cu^{2+} ions, which are themselves highly distorted by the Jahn–Teller effect [38]. Defects here are favorable to sites matching with the low symmetry of the cation environment. These effects are sometimes masked, however, because of particle aggregation, which decreases the accessible surface to adsorption.

c) Electronic Structure

Size effects also result in significant changes in the electronic structure of ultradivided solids compared to that of bulk materials [1]. The electronic structure of the solid may be roughly deduced by extrapolation from those of coordination complexes [39, 40]. The key factors of the reactivity of transition metal complexes or molecules are the energy gap between the highest occupied molecular orbital (HOMO) and the lowest unoccupied molecular orbital (LUMO), as well as the electron population of the HOMO. In solid matter, the multiplication of orbital levels due to the high quantity of involved orbitals leads to the concept of bands. By analogy, the HOMO and LUMO are, respectively, named valence band (VB) and conduction band (CB). We therefore gradually pass from a model with discrete molecular orbitals to a model of energy bands containing an infinity of crystal orbitals for the bulk solid, with a quasi-continuous variation (decrease) of the HOMO–LUMO gap (molecular orbitals) or the CB–VB gap (band model) (Figs. 1.1a and 1.8).

For isolators and semiconductors, the VB is fully occupied and the CB is empty at 0 K. These two bands are separated by a band gap (BG) or energy gap that is generally not as wide as the energy difference between HOMO and LUMO in the corresponding molecular unit (Fig. 1.8). For iron oxides and oxyhydroxides (hematite and goethite), the BG is around 2 eV. For an insulator such as SiO_2, the BG is nearly 9 eV. Magnetite, Fe_3O_4, with a BG around 0.1 eV, is a semimetal. The BG value represents the energy needed to excite an electron from the VB to

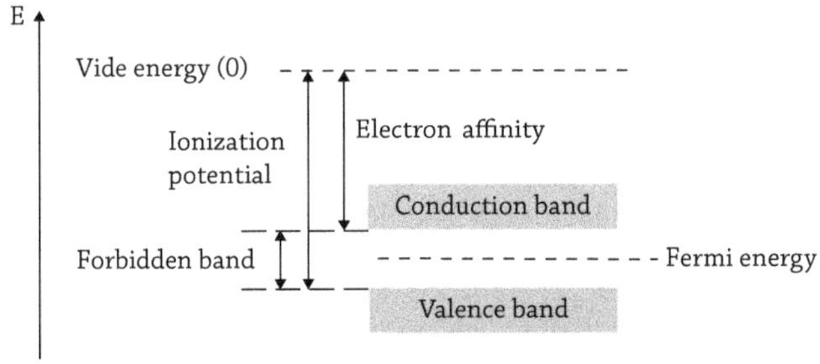

FIGURE 1.8 Scheme of the electronic structure of a solid showing valence band (VB) and conduction band (CB) separated by a forbidden band or band gap (BG). For isolators and semiconductors, the Fermi energy that represents the electronic chemical potential is located between the VB fully occupied and the CB empty at 0 K.

the CB where it may freely move outside the electrostatic interaction with a hole created in the VB (electron-hole pair). At room temperature, in iron oxides and hydroxides having a band gap around 2 eV, the thermal or light energy allows the excitation of a small number of electrons to the CB, providing a very low electric conductivity. The size of the interaction domain (bound state) of the lowest energy of the electron-hole pair (exciton) is characterized by the Bohr radius. At the nanoscale, the particle size reaches the order of the Bohr radius, so that an electron and a hole can never be sufficiently far away from each other in order that their interaction is negligible. Consequently, the lowest excitation energy of an electron is higher than the bulk BG. This is the effect of confinement, which involves the BG opening [41].

The phenomenon is well illustrated by the blue shift that occurs in the optical absorption of CdSe semiconductors of decreasing size from around 10 to 2 nm (Fig. 1.1) [42]. The BG width is usually determined experimentally by UV-visible optical spectroscopy, and the positions in energy of VB and CB are obtained from combined X-ray absorption and emission spectroscopies [43, 44]. For bulk solids, experimental data found in the literature are generally consistent with each other, while in the nanometer range, the data are more conflicting. For instance, in the case of TiO_2 in which there are no 3*d* electron exchange effects, conduction band states should be delocalized, and any true quantum confinement should be properly modeled by the electron-in-a-box model [41]. However, a lot of debate has arisen about whether there is quantum confinement in TiO_2 (e.g., [45, 46]). The apparent conclusion is that a confinement effect occurs only for a diameter less than 1 nm for TiO_2.

Similar discrepancies related to electronic structure also exist for hematite owing to the two competing contributions in energy transfer: the ligand to metal charge transfer (located at the top of the valence band) and the *d*-*d* ligand field intraband electronic transition (located at the bottom of conduction band). UV-Vis absorption and X-ray photoelectron spectroscopy (XPS) spectra show that the defected octahedral coordination of iron ions in small particle sizes induces an opening of the BG from 2.18 to 2.55 and 2.95 eV, with decreasing size from 120 to 38 and 7 nm, respectively [44]. However, an *ab initio* cluster model strategy consisting also of an extra electron in hematite indicates that the wavefunctions appear to extend out only to one or two neighbors [47]. Therefore, it is difficult to imagine how such a localized wavefunction should experience any confinement in a 7-nm-sized "box." Accordingly, no change in the BG width is observed by soft X-ray spectroscopy for bulk hematite and 30- to 8-nm-sized particles [43]. Theoretical and experimental works are obviously required to fully understand such *d*-electron semiconductors. Great care must also be taken regarding the purity of the studied phases. As an example, hematite nanoparticles synthesized by thermolysis of acidic Fe^{III} solutions are often contaminated by ferrihydrite or goethite nondetected by X-ray diffraction (XRD) but at which spectroscopies are very sensitive. The presence of such impurities could be an important cause of divergence in experimental data.

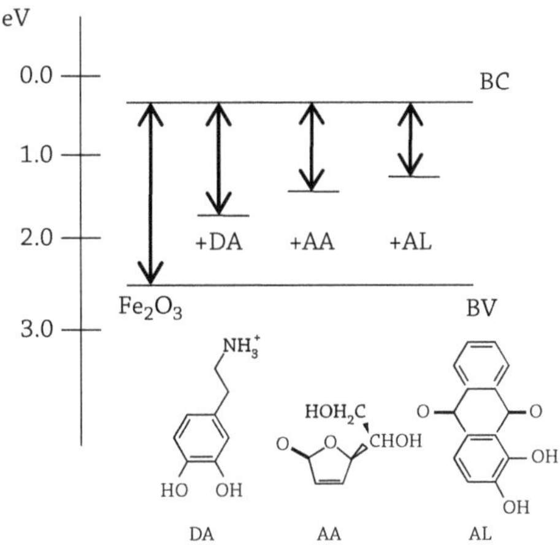

FIGURE 1.9 Scheme of the electronic structure modification of hematite by coordination of enediol ligands: dopamine (DA), ascorbic acid (AA), and alizarin (AL). Additional electronic states are introduced inside the band gap of the oxide, and less energy is required to excite electrons from the valence band (VB) into the conduction band (CB).

The surface disorder, which is even more important than the particles are small, may be due in part to the undercoordinated subsurface atoms. It may be more or less effectively lessened by coordinating strongly chelating ligands, such as enediol (Fig. 1.9), which can complete the coordination number of subsurface cations [48]. Thus, ligands can not only restructure the surface but also modify the electronic structure of the solid in forming surface charge-transfer complexes leading to a strong electronic coupling of ligand to surface. That introduces energy states inside the band gap, and the optical absorption of coated particle suspensions is thus shifted to low energy, creating the phenomenon of "sensitization." By using enediols as chelating ligands with 6-nm-sized hematite nanoparticles, the width of the band gap decreases from 2 to 1.40, 1.12, and 0.89 eV with alizarin, ascorbic acid, and dopamine, respectively [48]. A similar phenomenon is involved in the grafting of dyes on titania nanoparticles in photovoltaic cells [49].

1.2 Specificity and Requirements in the Fabrication Methods of Nanoparticles

Much progress has been made during the last several decades in developing highly advanced analytical techniques and tools that permit characterization of nanostructures and an understanding of their physical properties. Nevertheless, several prominent milestones in the progress of the nanosciences have resulted from the synthesis of well-defined nanoparticles, including the discovery of fullerenes [50] and carbon nanotubes [51, 52], the synthesis of well-defined quantum dots [54–56] and the shape control of CdSe nanocrystals [42], and the elaboration of multi-nanolayered metals for the giant magnetoresistance effect [56–58]. However, despite a vigorous expansion in the methods of nanoparticle synthesis, it is still difficult to generalize underlying physical or chemical principles behind arbitrary nanomaterials. Still lacking is a general mechanistic understanding of nanoparticle formation that might guide the development of new materials. Although the synthesis of nanoparticles with control over size, size distribution, and shape has been a major part of colloidal chemistry for decades, it remains an intensely studied topic, as is evident by the substantial body of literature on the subject. Nevertheless, the development of nanosciences and their technological applications in various fields such as optics, electronics, magnetism, catalysis, medical diagnostics, and imaging depends on the production of nanomaterials with strictly controlled size and size distribution, shape, chemical composition, as well as crystal structure, because these features greatly impact properties.

Ultradispersed solids in the form of nanoparticles are intrinsically thermodynamically metastable, largely because of the very high interfacial area. Surface area represents a positive contribution to the free enthalpy of the system (§ 1.1.3a). Consequently, a spontaneous evolution of a nanoparticle dispersion has to occur in order to increase nanoparticle size or affect the formation of nanostructured domains leading to a decrease in surface area, provided that the activation energies of the possible pathways are not too high. Consequently, it follows that

- an ultradispersed system with a high surface energy can only be kinetically stabilized;
- ultrafine powders cannot be synthesized by methods involving energies that exceed a threshold, but rather through methods of "soft chemistry" that maintain the formed particles in a metastable state; and
- additive and/or synthesis conditions that reduce the surface energy are needed to form nanoparticles stabilized against sintering, recrystallization, and aggregation.

Under these conditions, any solid matter such as metal oxides, chalcogenides, metals, or carbon can be obtained at the nanometric scale.

The synthesis methods for nanoparticles belong to two categories. The first one involves division of a massive solid into smaller portions. This "top-down" approach may involve mechanosynthesis (milling or attrition) and volatilization of a solid by laser ablation, for instance, followed by condensation of the gaseous components. The second category involves condensation of atoms or molecular species in a gaseous phase or in solution. This is the "bottom-up" approach in which the chemistry of metal complexes in solution holds an important place. This approach is far more popular in the synthesis of the nanoparticles of oxides, chalcogenides, and metals.

The liquid-phase chemistry is a powerful tool for convenient and reproducible syntheses of nanocrystals with a precise control of their size, shape, crystal structure, and composition. It also allows the dispersion of these nanocrystals in either an aqueous or a nonaqueous medium, modifying their surface by treatment with various species for various uses and also making it compatible with biological systems.

The precipitation in aqueous medium from metal salts (e.g., chlorides, nitrates, and sulfates) is the most widespread method used to form metal oxide nanoparticles. This topic is discussed in this book.

Metal-organic complexes (in which the cation is within an oxygenated environment of M-OC bonds) or organometallic complexes (in which the cation is bounded to carbon, nitrogen, or boron atoms) may be used through hydrolytic or nonhydrolytic processes to form oxide nanoparticles. However, cost and handling difficulties often have limited uses. Such precursors are nevertheless interesting with regard to the functionalization of oxide nanoparticles with various organic species and the formation of hybrid organic–inorganic systems. These types of nanomaterials are reviewed fully in several recent articles and reference books (see 59–61), and so they will not be considered here.

References

[1] Scholes G.D., Rumbles G. Excitons in nanoscale systems. *Nat. Mater.* 2006, 5, 683–696.

[2] Xia Y., Hallas N.J. Shape-controlled synthesis and surface plasmonic properties of metallic nanostructures. *MRS Bull.* 2005, 30, 338–343.

[3] Haes A.J., Haynes C.L., McFarland A.D., Schatz C.G., Van Duyne R.P., Zou S. Plasmonic materials for surface-enhanced sensing and spectroscopy. *MRS Bull.* 2005, 30, 368–375.

[4] Kerker M. The optics of colloidal silver: something old and something new. *J. Coll. Interface Sci.* 1985, 105, 297–314.

[5] Dormann J.L., Fiorani D., Tronc E. *Magnetic relaxation in fine particle systems.* Advances in Chemical Physics 1997, 98. New York: Wiley.

[6] Baudrin E., Cassaignon S., Koelsch M., Jolivet J.P., Dupont L., Tarascon J.M. Structural evolution during the reaction of Li with nano-sized rutile type TiO_2 at room temperature. *Electrochem. Comm.* 2007, 9, 337–342.

[7] Siegel R.W., Fougere G.E. Mechanical properties of nanophase materials, in nanophase materials, synthesis-properties-applications. In Hadjipanayis G.C. and Siegel R.W., Eds. NATO ASI Series, Series E: Applied Sciences, Vol. 260, 233–261, 1994.

[8] Li Y., Somorjai G.A. Nanoscale advances in catalysis and energy applications. Nano Lett. 2010, 10, 2289–2295.

[9] Wilcoxon J.P., Martino T., Klavetter E., Sylwester A.P. Synthesis and catalytic properties of metal and semiconductor nanoclusters, in nanophase materials, synthesis-properties-applications. In Hadjipanayis G.C. and Siegel R.W., Eds. NATO ASI Series, Series E: Applied Sciences, Vol. 260, 771–780, 1994.

[10] Li Y.X., Li H., Klabunde K.J. Destructive adsorption of chlorinated benzene on ultrafine (nanoscale) particles of magnesium oxide and calcium oxide, in nanophase materials, synthesis-properties-applications. In Hadjipanayis G.C. and Siegel R.W., Eds. NATO ASI Series, Series E: Applied Sciences, Vol. 260, 793–796, 1994.

[11] Stark J.V., Klabunde K.J. Nanoscale magnesium oxide used to absorb SO_2 and CO_2, in nanophase materials, synthesis-properties-applications. In Hadjipanayis G.C. and Siegel R.W., Eds. NATO ASI Series, Series E: Applied Sciences, Vol. 260, 707–800, 1994.

[12] Khin M.M., Nair A.S., Babu V.J., Murugan R., Ramakrishna S. A review on nanomaterials for environmental remediation. *Energy Environ. Sci.* 2012, 5, 8075–8109.

[13] Hochella Jr M.F. There's plenty of room at the bottom: Nanoscience in geochemistry. *Geochim. Cosmochim Acta* 2002, 66, 735–743.

[14] Hochella Jr M.F., Lower S.K., Maurice P.A., Penn R.L., Sahai N., Sparks D.L., Twining B.S. Nanominerals, mineral nanoparticles and Earth systems. *Science* 2008, 319, 1631–1635.

[15] Hem D.J. Reactions of metal ions at surfaces of hydrous iron oxide. *Geochim. Cosmochim Acta* 1977, 41, 527–538.

[16] Stumm W., Sulzberger B. The cycling of iron in natural environments: Considerations based on laboratory studies of heterogeneous redox processes. *Geochim. Cosmochim. Acta* 1992, 56, 3233–3257.

[17] Charlet L., Morin G., Rose J. Wang Y. Auffan M., Burnol A., Fernandez-Martinez A. Reactivity at (nano)particle-water interfaces, redox processes, and arsenic transport in the environment. *C. R. Geoscience* 2011, 343, 123–139.

[18] Dong D. M., Nelson Y. M., Lion L. W., Shuler M. L., Ghiorse W. C. Adsorption of Pb and Cd onto metal oxides and organic material in natural surface coatings as determined by selective extractions: new evidence for the importance of Mn and Fe oxides. *Water Res.* 2000, 34, 427–436.

[19] Navrotsky A. Energetics of nanoparticle oxides: interplay between surface energy and polymorphism. *Geochem. Trans.* 2003, 4, 34–37.

[20] Navrotsky A., Mazeina L., Majzlan J. Size-driven structural and thermodynamic complexity in iron oxides. *Science* 2008, 319, 1635–1638.

[21] Laberty C., Navrotsky A. Energetics of stable and metastable low-temperature iron oxides and oxyhydroxides. *Geochim. Cosmochim. Acta* 1998, 62, 2905–2913.

[22] Zhang H., Banfield J.F. Thermodynamic analysis of phase stability of nanocrystalline titania. *J. Mater. Chem.* 1998, 8, 2073–2076.

[23] Fritsch S., Post J.E., Navrotsky A. Energetics of low-temperature polymorphs of manganese dioxide and oxyhydroxide. *Geochim. Cosmochim. Acta* 1997, 61, 2613–2616.

[24] Birkner N., Navrotsky A. Thermodynamics of manganese oxides: Effects of particle size and hydration on oxidation-reduction equilibria among hausmannite, bixbyite and pyrolusite. *Amer. Miner.* 2012, 97, 1291–1298.

[25] Pottier A., Cassaignon S., Chanéac C., Villain F., Tronc E., Jolivet J.P. Size tailoring of TiO_2 anatase nanoparticles in aqueous medium and synthesis of nanocomposites. Characterization by Raman spectroscopy. *J. Mater. Chem.* 2003, 13, 877–882.

[26] Chanéac C., Tronc E., Jolivet J.P. Magnetic iron oxide-silica nanocomposites. Synthesis and characterization. *J. Mater. Chem.* 1996, 6, 1905–1911.

[27] Tronc E., Chanéac C., Jolivet J.P. Structural and magnetic characterization of ε-Fe_2O_3 J. Solid State Chem. 1998, 139, 93–104.

[28] Alexander G.B. The effect of particle size on the solubility of amorphous silica in water. *J. Phys. Chem.* 1957, 61, 1563–1564.

[29] Stumm W., Morgan J.J. *Aquatic Chemistry: Chemical Equilibria and Rates in Natural Waters*. New York: Wiley, 1996.

[30] Gribb A.A., Banfield J.F. Particle size effects on transformation kinetics and phase stability in nanocrystalline TiO_2. *Amer. Miner.* 1997, 82, 717–728.

[31] Gilbert B., Huang F., Zhang H., Waychunas G.A., Banfield J.F. Nanoparticles: strained and stiff. *Science* 2004, 305, 651–654.

[32] Waychunas G.A., Kim C.S., Banfield J.F. Nanoparticulate iron oxide minerals in soils and sediments: unique properties and contaminant scavenging mechanisms. *J. Nanoparticle Res.* 2005, 7, 409–433.

[33] Brice-Profeta S., Arrio M.-A., Tronc E., Menguy N., Letard I., Cartier dit Moulin C., Noguès M., Chanéac C., Jolivet J.-P., Sainctavit Ph. Magnetic order in γ-Fe_2O_3 nanoparticles: a XMCD study. *J. Mag. Mag. Mater.* 2005, 288, 354–365.
[34] Auffan M., Rose J., Proux O., Borschneck D., Masion A., Chaurand P., Hazemann J.L., Chanéac C., Jolivet J.P., Wiesner M.R., Van Geen A., Bottero J.Y. Enhanced adsorption of arsenic onto maghemite nanoparticles: As(III) as a probe of the surface structure and heterogeneity. *Langmuir* 2008, 24, 3215–3222.
[35] Yean S., Cong L., Yavuz C.T., Mayo J.T., Yu W.W., Kan A.T., Colvin V.L., Tomson M.B. Effect of magnetite particle size on adsorption and desorption of arsenite and arsenate. *J. Mater. Res.* 2005, 20, 3255–3264.
[36] Jolivet J.P., Froidefond C., Pottier A., Chanéac C., Cassaignon S., Tronc E., Euzen P. Size tailoring of oxide nanoparticles by precipitation in aqueous medium. A semi-quantitative modelling. *J. Mater. Chem.* 2004, 14, 3281–3288.
[37] Morin G., Wang Y., Ona-Nguema G., Juillot F., Calas G., Menguy N., Aubry N., Bargar J.R., Brown Jr., G.E. EXAFS and HRTEM evidence for As(III)-containing surface precipitates on nanocrystalline magnetite: implications for As sequestration. *Langmuir* 2009, 25, 9119–9128.
[38] Madden A.S., Hochella Jr. M. F., Luxton T. P. Insights for size-dependent reactivity of hematite nanomineral surfaces through Cu^{2+} sorption. *Geochim. Cosmochim. Acta* 2006, 70, 4095–4104.
[39] Burdett J.K. *Chemical Bonding in Solids*. New York: Oxford University Press, 1995.
[40] Sherman D.M. The electronic structures of manganese oxides minerals. *Amer. Miner.* 1984, 69, 788–799.
[41] Gilbert B., Banfield J.F. Molecular-scale processes involving nanoparticulate minerals in biogeochemical systems. *Rev. Miner. Geochem.* 2005, 59, 109–155.
[42] Murray C. B., Norris D. J., Bawendi M. G. Synthesis and characterization of nearly monodisperse CdE (E = S, Se, Te) semiconductor nanocrystallites. *J. Am. Chem. Soc.* 1993, 115, 8706–8715.
[43] Gilbert B., Frandsen C., Maxey E.R., Sherman D.M. Band-gap measurements of bulk and nanoscale hematite by soft X-ray spectroscopy. *Phys. Rev. B* 2009, 79, 035108.
[44] Chernyshova I. V., Ponnurangam S., Somasundaran P. On the origin of an unusual dependence of (bio)chemical reactivity of ferric hydroxides on nanoparticle size. *Phys. Chem. Chem. Phys.* 2010, 12, 14045–14056.
[45] Monticone S., Tufeu R., Kanaev A.V., Scolan E., Sanchez C. Quantum size effect in TiO_2 nanoparticles: does it exist? *Appl. Surf. Sci.* 2000, 162–163, 565–570.
[46] Satoh N., Nakashima T., Kamikura K., Yamamoto K. Quantum size effect in TiO_2 nanoparticles prepared by finely controlled metal assembly on dendrimer templates. *Nature Nanotechnology* 2008, 3, 106–111.
[47] Rosso K.M., Smith D.M.A., Dupuis M. An ab initio model of electron transport in hematite (α-Fe_2O_3) basal planes. *J. Chem. Phys.* 2003, 118, 6455–6466.
[48] Rajh T., Chen L.X., Lukas K., Liu T., Thurnauer M.C., Tiede D.M. Surface restructuring of nanoparticles: an efficient route for ligand-metal oxide crosstalk. *J. Phys. Chem. B* 2002, 106, 10543–10552.
[49] Grätzel M. *Energy Resources Through Photochemistry and Catalysis*. New York: Academic Press, 1983.
[50] Kroto H.W., Heath J.R., O'Brien S.C., Curl R.F., Smalley R.E. C60: Buckminsterfullerene. *Nature* 1985, 318, 162–163.

[51] Ijima S. Helical microtubes of graphitic carbon. *Nature* 1991, 364, 56–58.
[52] Ijima S., Ichihashi T. Single-shell carbon nanotubes of 1 nm diameter. *Nature* 1993, 363, 603–605.
[53] Berry C.R. Structure and optical absorption of AgI microcrystals. *Phys. Rev.* 1967, 161, 848–851.
[54] Efros A.A., Efros A.L. Interband absorption of light in a semiconductor sphere. *Sov. Phys. Semicond.* 1982, 16, 772–775.
[55] Rossetti R., Nakahara S., Brus L.E. Quantum size effects in the redox potentials, resonance Raman spectra, and electronic spectra of CdS crystallites in aqueous solution. *J. Chem. Phys.* 1983, 79, 1086–1088.
[56] Grünberg P., Schreiber R., Pang Y., Brodsky M. B., Sowers H. Layered magnetic structures: evidence for antiferromagnetic coupling of Fe layers across Cr interlayers. *Phys. Rev. Lett.* 1986, 57, 2442–2445.
[57] Baibich M.N., Broto J.M., Fert A., Nguyen Van Dau F., Petroff F., Eitenne P., Creuzet G., Friederich A., Chazelas J. Giant magnetoresistance of (001)Fe/(001)Cr magnetic superlattices. *Phys. Rev. Lett.* 1988, 61, 2472–2475.
[58] Fert A., Morel R., Barthélémy A., Cros V., Duvail J.-L., George J.-M., Parent F., Petroff F., Vouille C. Magnétorésistance géante dans les nanostructures magnétiques, *J. Phys. IV France* 7, C6, 1997, 151–161.
[59] Brinker C.J., Scherer G.W. *Sol-gel Science. The Physics and Chemistry of Sol-gel Processing.* San Diego: Academic Press, 1990.
[60] Gomez-Romero P., Sanchez C., Eds. *Functional Hybrid Materials.* Weinheim, Germany: Wiley-VCH, 2003.
[61] Sanchez C., Julian B., Belleville P., Popall M. Applications of hybrid organic-inorganic nanocomposites. *J. Mater. Chem.* 2005, 15, 3559–3592.

2 }

Water and Metal Cations in Solution

Water has an exceptional ability to dissolve minerals. It is safe and chemically stable, and it remains liquid over a wide temperature range. Thus, it is the best solvent and reaction medium for both laboratory and industrial purposes. Water is able to dissolve ionic and ionocovalent solids because of the high polarity of the molecule (dipole moment $\mu = 1.84$ Debye) as well as the high dielectric constant of the liquid ($\varepsilon = 78.5$ at 25°C).

This high polarity allows water to exhibit a strong solvating power: that is, the ability to fix onto ions as a result of electrical dipolar interactions. Water is also an ionizing liquid able to polarize an ionocovalent molecule. For example, the solvolysis phenomenon increases the polarization of the HCl molecule in aqueous solution. Finally, owing to the high dielectric constant of the liquid, water is a dissociating solvent that can decrease the electrostatic forces between solvated cations and anions, allowing their dispersion as $H^{+}{}_{solvated}$ and $Cl^{-}{}_{solvated}$ through the liquid. (The attractive force F between charges q and q' separated by the distance r is given by Coulomb's law, $F = qq'/\varepsilon r^2$.) These characteristics are rarely found together in common liquids. The dipole moment of the ethanol molecule ($\mu = 1.69$ Debye) is close to that of water, but the dielectric constant of ethanol is much lower ($\varepsilon = 24.3$). Ethanol is a good solvating liquid, but a poor dissociating one; consequently, it is considered a bad solvent of ionic compounds.

Dissolution in water of an ionic solid such as sodium chloride is limited to dipolar interactions with Na^+ and Cl^- ions and their dispersion in the liquid as solvated ions, regardless of the pH of the solution. Cations with higher charge, especially cations of transition metals, retain a fixed number of water molecules, thereby forming a true coordination complex $[M(OH_2)_N]^{z+}$ with a well-defined geometry. In addition to the dipolar interactions, water molecules behave as true ligands because they are Lewis bases exerting an electron σ-donor effect on the empty orbitals of the cation. In these complexes, the coordinated water molecules may be more or less stable. Depending on their polarization state in the complex, some water molecules spontaneously lose protons and are transformed into hydroxo (OH^-) or oxo (O^{2-}) ligands. Cations may thus behave as acids in solution, exchanging protons between the solvated complex and the solvent water molecules. Since the resulting hydroxo ligand within the coordination sphere is at the origin of condensation phenomena in solution and of precipitation, it is of primary importance to understand the coordination sphere of the cations under various conditions, including acidity, concentration, and temperature.

2.1 Water as Solvent, Physicochemistry of the Liquid

2.1.1 ELECTRONIC STRUCTURE OF THE WATER MOLECULE

The water molecule has C_{2v} symmetry. Its molecular orbitals (MOs) are described by linear combinations of atomic orbitals (AOs) of oxygen (2*s*, 2*p*) and hydrogen (1*s*) atoms. The MOs have one of the symmetries of the four irreducible representations of the C_{2v} group (a_1, a_2, b_1, b_2). Expressions for the four lowest energy MOs have been calculated as follows [1]:

$$1a_1 \quad 0.85(2s)+0.13(2p_z)+0.81(1s_a+1s_b) \quad E=-36\text{ eV}$$

$$1b_2 \quad 0.54(2p_y)+0.78(1s_a-1s_b) \quad E=-19\text{ eV}$$

$$2a_1 \quad 0.46(2s)-0.83(2p_z)-0.33(1s_a+1s_b) \quad E=-14\text{ eV}$$

$$1b_1 \quad (2p_x) \quad E=-12\text{ eV}$$

Each of the occupied MOs describes two electrons, whereas the higher-energy antibonding MOs ($3a_1$, $2b_2$) are empty (Fig. 2.1). The photoelectronic spectrum of

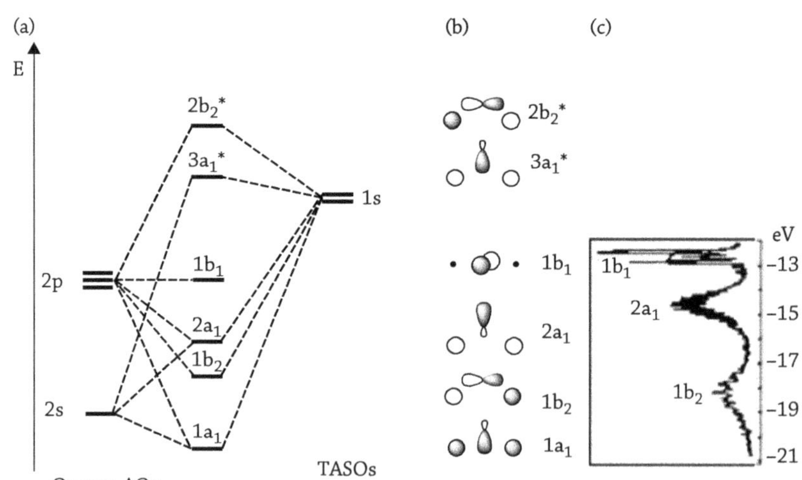

FIGURE 2.1 (a) Conventional molecular orbital energy diagram for the water molecule. (b) Representation of orbital overlappings. (c) Photoelectron spectrum of water molecule obtained by UV excitation with the He(I) line (21.2 eV) (adapted from [2]).

water shows the bonding character of the $1b_2$ and $2a_1$ MOs through a fine vibrational structure and the strictly nonbonding character of the $1b_1$ MO [2].

The main properties of the water molecule may be deduced from its electronic structure:

- The high polarity that explains the solvation of ions results from the electronegativity (χ) difference between oxygen and hydrogen. Oxygen ($\chi_O = 3.5$) attracts much greater electronic charge than hydrogen ($\chi_H = 2.1$). The high coefficients of oxygen AOs in the expressions of occupied MOs indicate that the electronic density is shifted toward the oxygen atom inside the molecule. The electronic charge carried by each atom (partial charge) may be calculated from the above-mentioned expressions for the MOs. Performing such calculations is a complex task because it requires knowledge of the overlapping integrals for each occupied MO. Moreover, the results are extremely variable, depending on the method of calculation used. For convenience, we choose to use the model of partial charges based on the principle of electronegativity equalization (§ 2.4). Although the model is relative and approximate, it provides realistic values of the charges carried by atoms within a combination.

According to Eq [A.10] (§ 2.4), the average electronegativity χ of the water molecule is

$$\chi = \frac{2\sqrt{\chi_H^*} + \sqrt{\chi_O^*}}{\left(2/\sqrt{\chi_H^*}\right) + \left(1/\sqrt{\chi_O^*}\right)} = 2.49$$

and, from Eq [A.8]:

$$\delta(H) = \frac{\chi - \chi_H^*}{1.36\sqrt{\chi_H^*}} = +0.2 \qquad \delta(O) = \frac{\chi - \chi_O^*}{1.36\sqrt{\chi_O^*}} = -0.4$$

The high polarity of hydrogen and oxygen atoms accounts for the high dipole moment of the molecule. It is important to note, however, that the relationship between the ionic character of bonds and the dipole moment of a molecule is not trivial. The ionic moments of bonds are only one contribution to the molecular dipole moment. The size difference between atoms as well as the dipole moments of free electron pairs must be taken into account. Evaluating the effective charges based solely on the molecular dipole moments is therefore subject to question.

- The Lewis base character of the molecule is due to electrons described by $2a_1$ MO. This MO is essentially built from hybridization of $2s$ and $2p_z$ AOs of oxygen. It is strongly delocalized outside the molecule, as indicated by the highly negative contributions in the combination. Thus, the water molecule behaves as a ligand donor of a σ pair, which explains its ionizing properties and its ability to coordinate with Lewis acids (complexing properties). The nonbonding b_1 MO, entirely localized on oxygen (Fig. 2.1), exhibits a very low π-donor character and does not allow formation of a bond.

2.1.2 STRUCTURE OF LIQUID WATER

Water possesses distinct properties when compared to similar compounds such as H_2S, H_2Se, H_2Te. Its boiling and melting points are unusually high, the volume change at melting is negative, and the maximum density is 4°C; the diffusion of protons and hydroxyl ions in water is particularly fast compared to that of other ions. These peculiarities result from the association of water molecules with hydrogen bonds. Such a type of interaction occurs when a hydrogen atom, bounded to an electronegative atom (F, O, N), is able to associate with another electronegative atom, as illustrated in the following scheme:

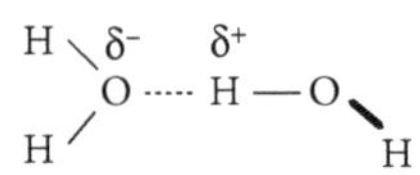

Within the H–bond of around 20 to 40 kJ.mol^{-1} in energy, the hydrogen atom may occupy two positions of minimum energy between the two oxygen atoms. By tunneling across a small energy barrier, the hydrogen atom is able to shift between the two positions.

Under usual conditions, water crystallizes as hexagonal ice in which strongly directional H–bonds allow oxygen atoms to be quasi-symmetrically tetrahedrally surrounded by hydrogen atoms (Fig. 2.2). During the melting of ice, the breakdown of some H–bonds reduces the geometrical constraints between the water molecules. They become somewhat mobile and move closer to each other, thereby inducing an increase in density of up to 4°C. At higher temperatures, thermal agitation prevails and causes the dilatation of the liquid with a decreasing density. Molecular dynamic calculations [3, 4] show that at RT, the proportion of isolated molecules within the liquid is very low (Fig. 2.2). The large majority of molecules form polygonal assemblies through 2 to 3 H–bonds with a tetrahedral local

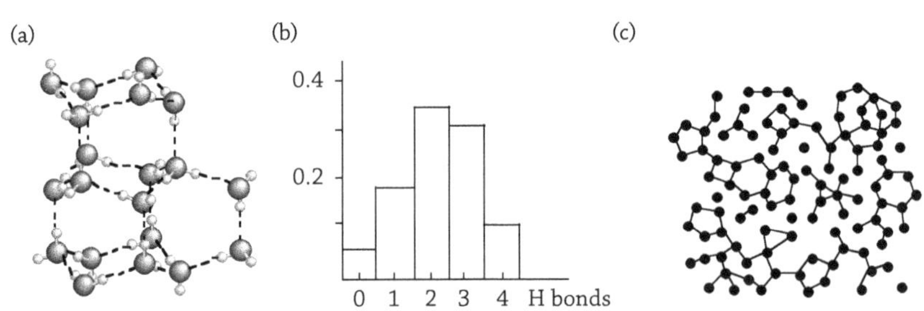

FIGURE 2.2 (a) Structure of hexagonal ice (bond lengths O···H···O 2.76 Å, O-H 1 Å, O···H 1.75 Å, angles H-O-H 109.5°). (b) Fractions of water molecules involved in various numbers of H-bonds in liquid at 10°C. (c) Scheme of H–bond network in liquid water (water molecules are represented by points) [3].

geometry. These assemblies have a short lifespan of around 10^{-10} s, but they can re-form permanently [5]. Therefore, the liquid, not being topologically related to hexagonal ice, forms a disordered tridimensional network of H–bonds with statistical connections. This allows rapid transfer of protons and hydroxyl ions [4, 6, 7].

The ionizing and dissociating power of liquid water provokes an appreciable self-dissociation of molecules:

$$2\,H_2O \Leftrightarrow H_3O^+_{\text{solvated}} + HO^-_{\text{solvated}} \qquad K_e = |H_3O^+||HO^-| = 10^{-14} \text{ at } 25°C$$

Hence, the water molecule is amphoteric, which gives the liquid the character of a protic solvent. Its acid–base properties are linked to the exchange of proton between water and donors (acids) or acceptors (bases).

The structure of the oxonium H_3O^+ ion has been investigated using ^{1}H-NMR with crystallized strong acid monohydrates $HClO_4{\cdot}H_2O$, $HNO_3{\cdot}H_2O$, and $H_2SO_4{\cdot}H_2O$ [8, 9]. The results of this investigation accord with a plane or slightly pyramidal geometry forming H-O-H angles of 120° or 118°, respectively, with H-H distances of 1.72 Å.

Solvation of oxonium ion involves H–bonds in forming $[H_3O{\cdot}nH_2O]^+$ entities. The number of water molecules forming the "coordination sphere" is dependent on whether the ion is in solution or in a solid network. The di-aquo ion $[H_2O\text{-}H\text{-}OH_2]^+$ ($n = 1$) has been identified by using X-ray and neutron diffractions by crystals of $HBr{\cdot}2H_2O$ and, in some complexes, trans $(Co(en)_2Cl_2)Cl{\cdot}2H_2O{\cdot}HCl$ [10]. The O···O distances in the di-aquo ion are about 2.40 Å, the hydrogen atom being approximately equidistant from the oxygen atoms.

Structural information on solutions from X-ray and neutron diffraction confirms the presence of four water molecules in the immediate surroundings of H_3O^+ ions ($n = 4$) [11, 12]. Neutron diffraction indicates a tetrahedral configuration of the solvated oxonium ion in molar aqueous solutions of HCl or HBr [13], while the analysis of radial function distribution obtained from X-ray diffraction by concentrated HCl solutions reveals a pyramidal geometry [14]. Three solvating water molecules, together with the H_3O^+ ion, form three short H–bonds (O···O distance of 2.44 Å). The fourth water molecule, found much further away (O···O distance of 2.90 Å, similar to that observed in pure water), is located at the corner of a trigonal pyramid (Fig. 2.3). These two techniques yield different results

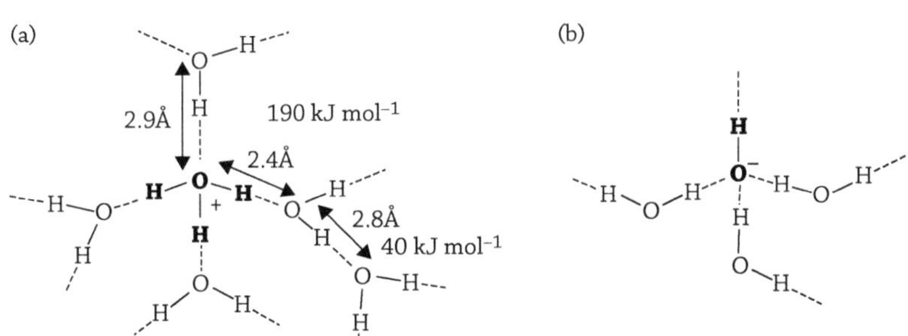

FIGURE 2.3 Possible structures of solvated ions: (a) proton $[H_9O_4]^+$, (b) hydroxyl ion $[H_7O_4]^-$.

because neutron diffraction gives only an average O···O distance [14]. The energy of short distances, calculated using the molecular orbital method, is about 190 kJ.mol^{-1}, and that of the long bond is almost zero. Consequently, one of the water molecules is in a weak interaction, and the solvated proton may be considered as the $[H_9O_4]^+$ molecular entity, which itself is solvated by other water molecules through weaker H–bonds, 20 to 40 kJ.mol^{-1} in energy.

The hydroxyl ion, HO^-, present in the structure of many monosolvated complexes $[HO\text{-}H\text{-}OH]^-$ (O···O distance 2.29 Å), exists in solution as the $[H_7O_4]^-$ molecular entity, which itself is solvated by other water molecules [10, 15] (Fig. 2.3). This ion is the strongest Brönstedt base in water. This is why self-dissociation of water is low and why the O^{2-} ion does not exist in aqueous solution. Therefore, oxides dissolve in water because the protonation of the O^{2-} ion occurs.

2.1.3 HYDRATION OF IONS AND STRUCTURE OF SOLUTIONS

The structure of solvated ions is deduced from data on spectroscopic techniques (e.g., extended X-ray absorption fine structure [EXAFS], X-ray absorption near edge structure [XANES], nuclear magnetic resonance [NMR], infrared, and Raman) and on radiation diffraction and scattering (X-ray, neutrons) [11, 12]. It has been shown that solvation corresponds to the association of cations with water molecules by oxygen atoms, whereas anions are associated with water molecules by hydrogen atoms (Fig. 2.4).

The number of water molecules involved in the solvation of a cation increases when its polarizing power increases, that is, when the charge z increases and the radius r of the cation decreases. The polarizing power of an ion is defined as the z/r^2 ratio. Water molecules are attracted and organized over several layers around ions; it is therefore important to distinguish the first shell of hydration in contact with the cation from the more distant other shells.

The dynamics of formation and the breaking of H–bonds between water molecules involved in the solvation layer is slower than that between molecules in the bulk of the liquid. This is shown by nonlinear femtosecond mid-infrared spectroscopy, which helps distinguishing between absorption by O-H bonds of solvation water molecules and that in the bulk [16]. The time constant of the

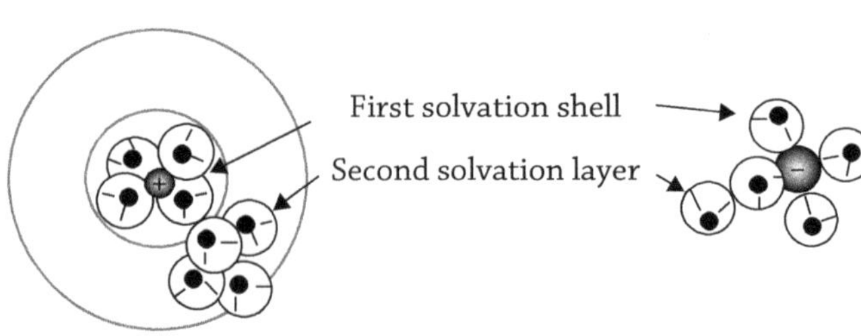

FIGURE 2.4 Scheme of the hydration of hydrophilic cations (left) and anions (right) in aqueous solution.

H-O···Y^- oscillator (12 ps for Cl^- and 25 ps for I^- in 1 mol.l^{-1} solution) is about 20 to 50 times higher than that (500 fs) of the H–O···H oscillator characteristic of liquid water. The large difference in dynamics of H–bonds results from the difference in structure inside these two zones. Within the bulk of the liquid, the disordered tridimensional network allows simultaneity of the formation and breaking of H–bonds. On the one hand, the energy cost of one H–bond break is low (the correlation time is short) because it is compensated by the simultaneous formation of a new H–bond in another place. On the other hand, inside the first solvation layer of an ion, eliminating one water molecule involves breaking the solvation structure with a high-energy cost because it cannot be as easily compensated for as that in the bulk liquid. That means that the important correlation time in the solvation layer is not solely the result of direct interaction between one ion and surrounding water. Rather, it is also linked to the mutual interaction of the water molecules inside the solvation layer where they have an ordered arrangement around the ion. The phenomenon thus appears to be essentially of an entropic nature. In addition, the fact that the correlation time is shorter in the solvation sphere of chloride ion than in that of bromide or iodine ions indicates that it may not only be determined by the bonding energy of water molecules in the solvation layer but also by the distortion frequency of the solvent cage and is smaller for Cl^- than for Br^- and I^-.

a) The First Hydration Sphere

Monovalent cations such as alkaline ions are surrounded by a first hydration layer that is poorly defined in both composition and geometry. Li^+ ion, the smallest and most polarizing in the series, is surrounded by four water molecules in tetrahedral symmetry. Larger and less polarizing ions (Na^+, K^+, Cs^+) have a first solvation layer made up of six to eight molecules (Table 2.1). The symmetry of the solvation sphere of these last cations cannot be clearly defined because of the rapid exchange of solvation molecules with those of solution [12]; their number varies

TABLE 2.1 } Ionic radius r (Å), usual coordination N, polarizing power z/r^2, hydration enthalpy ΔH (25°C, kJ.mol^{-1}), exchange rate constant k (s^{-1}) of water molecules in the first hydration layer of some metal cations (from 17).

Ions	r	N	z/r^a	$-\Delta H$	k
Li^+	0.59	4	2.87	522	5.10^8
Cs^+	1.70	8	0.35	285	5.10^9
Mg^{2+}	0.49	4	8.32	1940	10^5
Ba^{2+}	1.43	10	0.98	1320	$\approx 10^9$
Cr^{2+}	0.82	6	2.97	1930	3.10^8
Fe^{2+}	0.77	6	3.37	1956	3.10^6
Ni^{2+}	0.70	6	4.08	2120	4.10^4
Al^{3+}	0.53	6	10.68	4700	≈ 1
Fe^{3+}	0.65	6	7.10	4450	10^2
Cr^{3+}	0.63	6	7.56	4620	3.10^{-6}

with temperature, concentration, and the kind of anions present in the solution. These cations, having filled electronic layers, develop relatively weak interactions with water molecules and form an ionic bond.

Cations from *s* and *p* blocks in periodic classification, which are smaller and of higher charge than alkaline ions, develop stronger electrostatic interactions with water molecules and have strong hydration energies (Table 2.1). The geometry of these hydrated cations is well defined. The structural techniques (X-ray, neutron and electron diffraction, and various spectroscopies) [11, 12] indicate different symmetries: for instance, tetrahedral for tetraaquo $[Be(OH_2)_4]^{2+}$ ion, octahedral for hexaaquo $[M(OH_2)_6]^{2+}$ ions (M = Mg, Ca, Zn, Cd, Hg, Pb, etc.) and $[Al(OH_2)_6]^{3+}$, and square-based antiprismatic for $[Sr(OH_2)_8]^{2+}$. The coordination of these cations is directly related to their size. The di- and trivalent ions from the first transition series, which are of somewhat similar size, are all hexacoordinated. The molecular orbital model accounts for the bond system in the complex (Fig. 2.5). The interactions between the cation and the water molecules clearly result from their σ-donor ligand role. Hence, the bond exhibits an ionocovalent character, and the hydrated cations are true coordination complexes. Distortions of the coordination polyhedron are observed with Cr^{2+} and Cu^{2+} of d^4 and d^9 electron configurations, respectively, due to the Jahn–Teller effect. The lability of coordinated water molecules is variable (Table 2.1). Some hexaaquo complexes, strongly stabilized by the crystal field, are inert in substitution reactions. That is the case for $[Ni(OH_2)_6]^{2+}$ and $[Cr(OH_2)_6]^{3+}$ of d^8 ($t_{2g}^6e_g^2$) and d^3 ($t_{2g}^3e_g^0$) electron

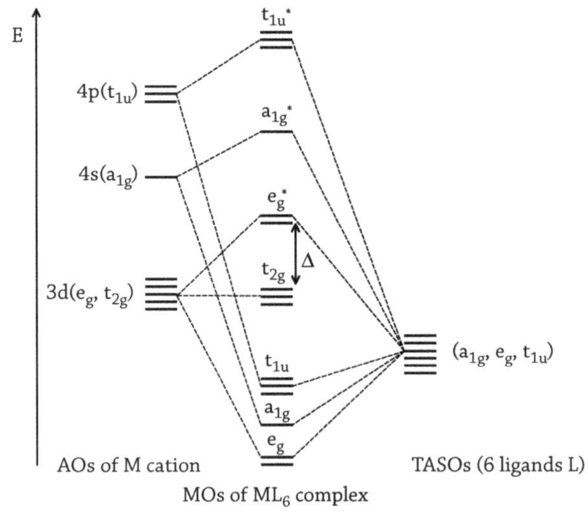

FIGURE 2.5 Molecular orbital energy diagram for ML_6 octahedral complexes of transition elements involving only σ-type metal–ligand interactions. The metal AOs and ligand terminal adapted in symmetry orbitals (TASOs) are defined by their symmetry labels. The TASOs are linear combinations of the $2a_1$ MO of water (Fig. 2.1). The crystal field parameter, Δ, represents the energy splitting of the cation *d* orbitals in the complex.

configurations, respectively, for which the formation of the transition state involves the largest number of changes in energy (§ 3.2.1).

The cations in a high oxidation state (+IV and more) are too polarizing to allow water molecules to exist in their coordination sphere, and they are more or less deprotonated. Elements from the *p* block (e.g., Si^{IV}, Sn^{IV}, Sb^{V}) form oxohydroxo anionic complexes. Transition elements (V^{V}, Cr^{VI}, Mo^{VI}, W^{VI}, Mn^{VII}) of very small size adopt the tetrahedral coordination as anionic tetraoxo complexes $(MO_4)^{(8-z)-}$ in which the electron transfer M←O occurs through a strongly covalent system of σ and π bonds (§ 2.2).

b) Outer Solvation Spheres

The second solvation sphere is more difficult to characterize because of the rapid exchange with the molecules within the solution. The aforementioned spectroscopic and diffraction techniques, as well as numerical simulations, indicate that, for a large majority of cations, 12 water molecules form the immediate surrounding of the solvated ion. The average O···O distance between the oxygen atoms of molecules in the first and second hydration spheres, being about 2.7 Å, is shorter than the average intermolecular O···O distance in pure water (of about 2.9 Å) [11, 12]. The number of water molecules attracted and oriented by a cation increases with its polarizing power.

c) Structure of Solutions

The presence of a cation in solution induces a local ordering of water molecules as a result of dipolar interactions propagating beyond the second solvation sphere. Four to five layers of water molecules are affected by the diffusion of ions within the solution. The hydration of an ion may be defined by the number of water molecules drained in its motion through the solution, while its coordination represents the number of water molecules in contact with it, in the first solvation sphere. The orientation and ordering within the hydration shell notably modify the structure of the solvent [18]. An LiCl solution is more viscous (1.146 cP at 20°C, 0.1 $mol.l^{-1}$) than pure water at the same temperature (1 cP), whereas a CsCl solution is less viscous (0.947 cP at 20°C, 0.1 $mol.l^{-1}$).

Ions of small size and high-charge density locally modify the original structure of water in breaking H–bonds and reorganizing many layers of solvent around them, in a structure that is entirely different from that of pure water. Such behavior is characteristic of ions such as Li^+, Na^+, Mg^{2+}, F^-, and HO^-. These ions are called type I structuring ions.

Ions of intermediate size and charge develop interactions with water that are strong enough to break the H–bonds locally, but they cannot immobilize numerous water molecules around them. Loss of the original water structure is not balanced by the restructuration of water molecules around the ion. This is the case for K^+, Cs^+, NH_4^+, NO_3^-, ClO_4^- ions called destructuring or breaking ions. Solvation of such ions is much weaker than that of structuring ions, which explains the decrease in viscosity of a Cs^+ solution compared to a Li^+ solution.

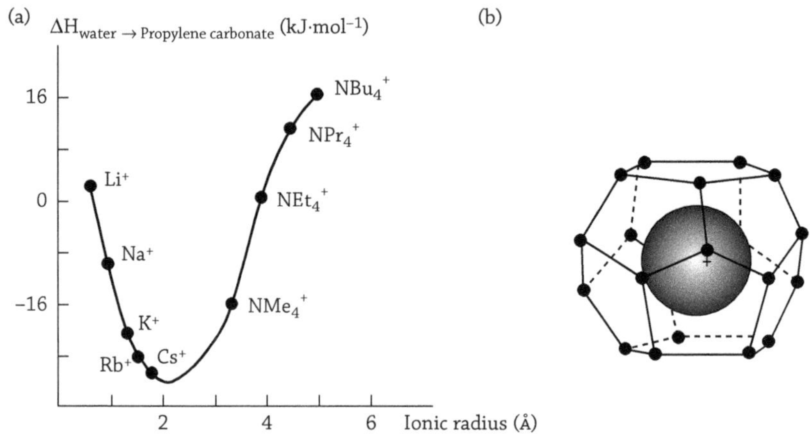

FIGURE 2.6 (a) Variation of transfer enthalpy of different-sized monovalent cations from water to propylene carbonate (reproduced with permission from [20] Copyright 1969 American Chemical Society). (b) Scheme of solvation of type II structuring ions. Water molecules are represented by points.

Weakly polarizing ions such as quaternary ammonium cations $[NR_4]^+$ ($R = CH_3$, C_2H_5, etc.) are not able to break the H–bond framework of liquid water, but they are incorporated into the liquid in considerably modifying its local architecture. The liquid forms polyhedral cages around these ions similar to those found in clathrates, the gas hydrates [6], or zeolites [19] (Fig. 2.6b). This kind of interaction is sometimes called hydrophobic interaction in contradistinction to the hydrophilic character of type I structuring ions. These ions are so-called type II structuring ions.

These three kinds of behavior are clearly distinguishable on the basis of variation of transfer enthalpy from water to propylene carbonate of monovalent cations of variable size [20] (Fig. 2.6a). The breaking ions, which develop weaker interactions with water, have lower transfer energy.

2.1.4 WATER UNDER HYDROTHERMAL CONDITIONS

The hydrothermal conditions for water correspond to temperatures and pressures higher than 100°C and 1 bar, respectively. These conditions are sometimes used in synthesis of materials to modify the chemistry of cations in aqueous solution. They favor the formation of metastable structures that are more complex and of lesser symmetry involving weaker enthalpy and entropy variations than is found in usual conditions [21, 22]. Hydrothermal conditions also are those associated with geological processes, during which many minerals have been formed. In the laboratory, such conditions are obtained by heating solutions in a closed container (autoclave) at temperatures ranging from 100°C to around 400°C. The synthesis of zeolites is a demonstrative example of the use of such conditions [23]. Many others are indicated in this book.

The influence of hydrothermal conditions on the speciation and chemistry of cations is still not fully understood, but it is highly probable that changes in the properties of the solvent play a fundamental role because they are largely modified by such conditions. The thermodynamic properties of water at temperatures up to 1000°C and pressures of several tens of kilobars are well known. Quantitative data are summarized in several reviews [21, 24, 25]. The three main points of these reviews are as follows:

- The dielectric constant decreases with increasing temperature and increases with increasing pressure [26] (Fig. 2.7a). Hydrothermal solutions are therefore characterized by low dielectric constants. Electrolytes, which are entirely dissociated in "normal" conditions, form mainly ion pairs or electrostatically weakly charged complexes. Inorganic salts are less soluble than they are in normal conditions, and water tends to behave like polar organic solvents.
- The viscosity decreases with increasing temperature [27], which leads to the higher mobility of dissolved species than is the case in normal conditions.
- The ionic product strongly increases with increasing temperature [28] (Fig. 2.7b). The variation, established from conductivity measurements, is given by the relation $\log K_e = -(3018/T) - 3.55$.

The increase in temperature enhances the dissociation of water that increases the concentrations of H^+ and OH^- ions and reduces the acidity scale of water, defined as $-\log K_e$. The strength of acids and bases is therefore notably modified when compared to normal conditions. In addition to the decrease in dielectric constant of the medium, the hydrothermal conditions may have a different influence on the behavior of acid–base couples of the type HA/A^- and HB^+/B, as that is the case in normal conditions for solvents such as water and alcohol [29].

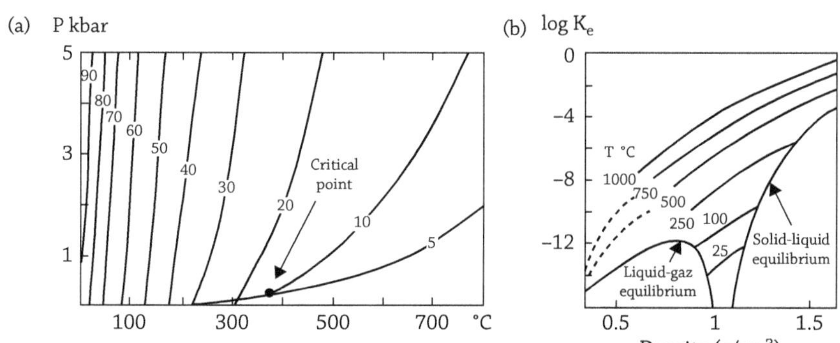

FIGURE 2.7 (a) Variation of the dielectric constant of water as a function of temperature and pressure (adapted from [26]). (b) Variation of ionic product K_e of water as a function of the temperature and density of the liquid (adapted from [28]).

For instance, the basicity of chloride ions is considerably strengthened in hydrothermal conditions: in water, at 500°C and 2 kbar, the constant of the equilibrium:

$$Cl^- + H_2O \Leftrightarrow HCl + HO^-$$

is nine orders of magnitude higher than in normal conditions (20°C, 1 bar) [26]. For this reason, quaternary ammonium $[R_4N]^+$ and amines R-NH_2, R_2-NH, and R_3-N (R stands for a hydrocarbon chain) may play a special role in the hydrothermal synthesis of compounds with zeolitic structures [23, 30]. These amines correspond to acid–base couples of type HB^+/B, which, like $[R_4N]^+$ cations, give rise to steric effects and a hydrophilic–hydrophobic balance tunable by the structure of the R group, as well as electrostatic interactions and H–bonds with various ionic or molecular species in solution.

On the one hand, under mild hydrothermal conditions, up to approximately 200°C in the presence of the liquid phase, the increase of pressure to as much as about 10 bars does not seem to have a significant effect on the obtained products. On the other hand, thermal effects that induce the lowering of kinetic barriers, provoke change in the physicochemical properties of the solvent in favoring electrostatic interactions and H–bond formation. Such interactions play a major role in molecular recognition [31] and are probably present when the surprising polyanionic structures formed in these conditions (§ 3.3.2) originate. The state of current knowledge is too tentative, however, to allow us to definitively interpret the mechanisms involved in these conditions.

2.2 Acidity and Cation Speciation

Based on knowledge of the electronic structures of the $[M(OH_2)_6]^{z+}$ complex (Fig. 2.5) and the water molecule (Fig. 2.1), one may observe that the overlap of valence orbitals of the cation and of the water molecules allows the electron transfer from the $2a_1$ MO of water to the empty or partially filled cationic orbitals. The electron transfer is more intense (the M-O bond is stronger) that the polarizing power of the cation high is, that is, the cation has a high charge and a small size.

The M←OH_2 electron transfer decreases the density in the $2a_1$ bonding MO in water, which weakens the O-H bond. The ionic character of the O-H bond in coordinated water molecules increases (see Table A1), and these molecules behave as stronger acids than the solvent molecules. They tend to deprotonate according to the equilibria:

$$\left[M\text{-}OH_2\right]^{z+} \Leftrightarrow \left[M\text{-}OH\right]^{(z-1)+} + H^+_{aq} \Leftrightarrow \left[M\text{-}O\right]^{(z-2)+} + 2\,H^+_{aq} \quad [2.1]$$

The acidity of the aquo-ion depends on the intensity of the σ electron transfer, and the cation may be coordinated with three kinds of ligands: aquo (H_2O), hydroxo (HO^-), and oxo (O^{2-}).

The elements of low formal charge (z = 1, 2), weakly polarizing toward oxygen, form cationic aquocomplexes in a large acidity range. Oxides of these elements are ionic and behave as bases in water:

$$Li_2O + 9\ H_2O \Leftrightarrow 2\left[Li(OH_2)_4\right]^+ + 2\ HO^-_{aq}$$

On the other side of the periodic classification, elements with a high formal charge (z = 5, 6, 7) strongly polarize oxygen and form anionic oxocomplexes. Oxides are covalent and acid in water:

$$z = 5 \quad N_2O_5 + H_2O \rightarrow 2\ NO_3^- + 2\ H^+_{aq}$$

$$z = 6 \quad SO_3 + H_2O \rightarrow SO_4^{2-} + 2\ H^+_{aq}$$

$$z = 7 \quad Cl_2O_7 + H_2O \rightarrow 2ClO_{4^-} + 2\ H^+_{aq}$$

Some metals have different oxidation states. Their oxides may therefore behave as bases and as acids. For chromium, for instance,

$$Cr^{III}\text{forms a basic oxide:}\quad Cr_2O_3 + 9\ H_2O + 6\ H^+_{aq} \rightarrow 2\left[Cr(OH_2)_6\right]^{3+}$$

$$Cr^{VI}\text{forms an acid oxide:}\quad CrO_3 + 2\ HO^-_{aq} \rightarrow \left[CrO_4\right]^{2-} + H_2O$$

The behavior of cations against hydrolysis is well known. It is described by thermodynamic equilibrium constants [31–36] that allow calculation of the cation speciation as a function of pH, concentration, and the eventual presence of various ligands [37] (Fig. 2.8).

The acidity of cations depends on their intrinsic characteristics such as formal charge (oxidation state), size, and electronegativity. As the equilibria in solution are sensitive to pH, it is possible to some extent to modify the nature of the coordination sphere in changing the acidity of the medium by adding a base or an acid, as shown by speciation diagrams (Fig. 2.8). Ferrous ion forms an hexaaquo complex over a wide range of acidity, and hydroxylation occurs from pH 7, while hydroxylation of ferric ion begins at pH 0. With the same formal charge as ferric ion and though a smaller size, Al^{3+} ion is far less acidic because of its lower electronegativity. Although bigger and less electronegative than ferric ion, the higher charged Zr^{4+} ion is far more acidic: the $[Zr(OH_2)_8]^{4+}$ complex practically does not exist in solution, even at pH 0. With Si^{4+} ion, the decrease in size and the increase in electronegativity enhance the polarizing power of the cation so much that water molecules no longer exist in the coordination sphere of silicon. Its more acidic form is the purely hydroxocomplex $Si(OH)_4$.

Following a more qualitative and general approach, hydroxylation of a cation in solution can be written as

$$\left[M(OH_2)_N\right]^{z+} \Leftrightarrow \left[M(OH)_h(OH_2)_{(N-h)}\right]^{(z-h)+} + hH^+_{aq}$$

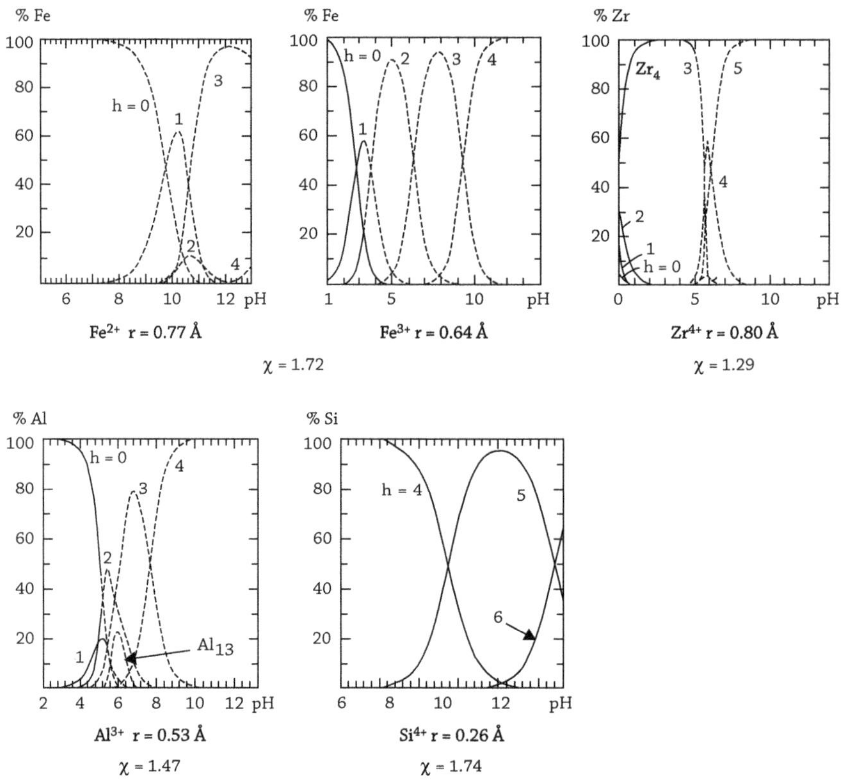

FIGURE 2.8 Speciation diagrams for very diluted solutions (10^{-5} mol.l^{-1}, except 10^{-2} mol.l^{-1} for Fe^{2+} and Zr^{4+}) of some cations of ionic radius r and electronegativity χ. h is the deprotonation degree of the complex $[M(OH_2)_N]^{z+}$. Al_{13} and Zr_4 are polycationic condensed forms corresponding to $h = 2.5$ and 2, respectively (§ 3.2). The dotted curves correspond to precipitation of a solid (adapted from [32]).

The hydroxylation rate, h, depends mainly on four parameters. Three of them intrinsically depend on the cation:oxidation state (the formal charge), size, and electronegativity. The fourth parameter, extrinsic, is the acidity of the medium. It may be independently tuned by adding a base or an acid. Temperature and concentration are of course important factors affecting the hydroxylation state of a cation in solution (§ 2.3), but the whole of thermodynamic data may be resumed on a schematic diagram exhibiting three domains indicating the type of ligand present in the coordination sphere of a cation as a function of its oxidation state and the pH of the solution [38, 39] (Fig. 2.9). In the aquo domain, all coordinated water molecules retain their protons, whereas in the oxo domain, the oxo ligands cannot be protonated. In the hydroxo domain, at least one hydroxo ligand exists in the coordination sphere.

Mn^{VII} ion exists as $[MnO_4]^-$ tetraoxo complex regardless of the pH. This species has no basic character because of the very low charge density on oxygen atoms,

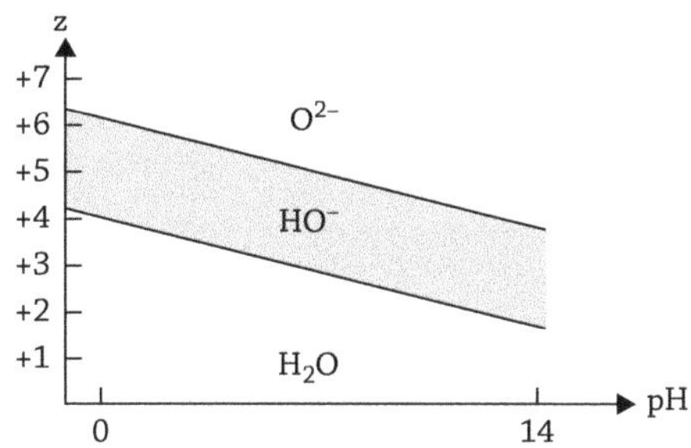

FIGURE 2.9 Diagram indicating the nature of ligands in the coordination sphere of a cation as a function of its formal charge, *z*, and the pH of the medium.

which are highly polarized by the extremely charged and small-sized cation. Thus, MnO_3OH is a strong acid in water, and aquo and hydroxo forms of the cation never exist in the usual acidity range [32]. At the same time, Mn^{II} ion, like many divalent cations, exhibits a low acid character. At pH < 6, it forms the hexaaquo $[Mn(OH_2)_6]^{2+}$ complex. Addition of a base leads to aquohydroxo complexes but never oxo species because the Mn^{II} cation is unable to polarize the oxygen atoms enough to break the OH bond to form the O^{2-} ligand.

Cr^{VI} ion exists at pH > 7 in the tetraoxo $[CrO_4]^{2-}$ form, but at pH < 7, protonation occurs because the polarization of oxygen by Cr^{VI} is less strong than it is with Mn^{VII}. $[CrO_3(OH)]^-$ species exist in solution around pH 4. In a more acidic medium, $CrO_2(OH)_2$ predominates in dilute solution, but additional protonation is not possible. On the other hand, cationic positively charged species are formed with Cr^{III} ion, such as the hexaaquo $[Cr(OH_2)_6]^{3+}$ complex at pH < 2. At around pH 4–5, the neutral aquohydroxo $Cr(OH)_3(OH_2)_3$ complex leads to hydroxide, which is soluble in alkaline medium as chromite, $[Cr(OH)_4(OH_2)_2]^-$. The oxo ligand, however, does not appear in the coordination sphere of Cr^{III} at high pH because polarization of oxygen is too low.

The increase in polarization of oxygen by atoms of increasing formal charge is well illustrated in the following series of tetrahedral species:

	$Si^{IV}(OH)_4$	$P^{V}O(OH)_3$	$S^{VI}O_2(OH)_2$	$Cl^{VII}O_3(OH)$
pK_1	9.8	2.1	(−1)	(−9)

As the charge of the cation increases, the number of oxo ligands increases at the expense of hydroxo ligands, which are of increasing acidity, as shown by the evolution of pKs. Oxo ligands have no basic character because they are bound to the central cation through a σ and π bonds system, and they are not protonable.

The cation's decreased size lowers its coordination number, *N*, and simultaneously increases its polarizing power. The effect appears clearly in the most acidic forms of some tetravalent elements:

	r (Å)	N	Most acidic form	
Ti^{4+}	0.68	6	aquohydroxo	$[Ti(OH)_2(OH_2)_4]^{2+}$
Si^{4+}	0.41	4	aquohydroxo	$[Si(OH)_3(OH_2)]^{+}$
C^{4+}	0.15	3	oxohydroxo	$[CO(OH)_2]^{0}$

Both charge and size work together since the size of the ion depends on its charge. This effect is illustrated in the following series of chlorine oxides [40]:

	HClO	$HClO_2$	$HClO_3$	$HClO_4$
z(Cl)	+1	+3	+5	+7
pK	7.3	2	0	(-9)

Specific characteristics of some cations merit discussion, as follows.

- Many divalent elements are hexacoordinated and exhibit a relatively weak acidity. Their hydroxylation starts at around pH 6.5–7. However, in the series Ni^{II}, Pd^{II}, Pt^{II}, the hydroxylation of palladium and platinum occurs at around pH 1 [32]. The acidity of these two elements is much higher than that of nickel. For the three cations with the d^8 electron configuration, the crystal field parameter of the aquo ligand (Fig. 2.5) notably increases for elements of the second and third transition series, which for Pd^{2+} and Pt^{2+} ions stabilize the diamagnetic configuration $(d_{xz},d_{yz})^4(d_{xy})^2(d_{z^2})^2(d_{x2-y2})^0$ with square plane coordination [41]. Therefore, despite the cation's increased size, the polarization of each aquo ligand is stronger in tetracoordinated $[M(OH_2)_4]^{2+}$ complexes of palladium and platinum than in the hexacoordinated $[M(OH_2)_6]^{2+}$ complex of nickel. This explains the difference in acidity.
- As a rule, the first coordination sphere of cations contains ligands involving the exchange of only one proton, typically forming aquohydroxo and hydroxo/oxo couples, depending on the polarizing power of the cation and the acidity of the medium. That is not always the case, however, especially with regard to transition cations; the reason is the shortening of bonds due to π overlap, a phenomenon observed for small cations of high charge ($z \geq 4$). A typical example is given by V^V (ionic radius 0.54 Å). The species $[V(OH_2)_6]^{5+}$ does not exist in aqueous solution, even in extremely acidic medium. Its hydroxylation rate is at least 4, and the main species is the *cis*-dioxovanadic complex, $[VO_2(OH_2)_4]^+$ [32, 33, 42]. The dioxoaquo configuration is favored compared to $[VO(OH)_2(OH_2)_3]^+$ or $[V(OH)_4(OH_2)_2]^+$ configurations because the cation, which is strongly polarizing, forms V–O bonds in which an important π character allows the lowering of its partial charge and consequently the ionicity of the bonds within the complex. The result is a strong distortion of the coordination sphere. This effect is less evident with V^{IV} of a lower oxidation state and larger size (ionic radius 0.59 Å) than V^V. V^{IV} forms the vanadyl complex $[VO(OH_2)_5]^{2+}$ containing only one short V–O bond (1.6 Å) [43–45]. The π character of the short

bonds strongly lowers the basicity of the oxo ligand and explains the stability as a monomer of vanadic and vanadyl species in solution (§ 3.3.2). The water molecules located in the *trans-configuration* of short V–O bonds are far from vanadium (V···O distance 2.4 Å in the vanadyl ion); this explains their high lability and the various coordinations of V^{IV} and V^{V} ions [46, 47]. Titanium IV (ionic radius 0.61 Å), which is slightly larger than V^{IV}, does not seem to form such oxo bonds. The strongly acidic $[Ti(OH)_2(OH_2)_4]^{2+}$ complex forms the $[Ti(OH)_3(OH_2)_3]^{+}$ and $[Ti(OH)_4(OH_2)_2]^{0}$ species, which are unstable as monomers [32].

- Of particular interest is the lowering of the coordination number of some cations of low formal charges ($z < 4$), which are soluble in basic medium as anionic tetrahydroxo forms (typically, Al^{3+}, Zn^{2+}) and as tetraoxo (Fe^{3+}; see Chap. 7). The lowering of the coordination as the pH increases may be explained by the decreased polarization of coordinated water molecules that occurs when the hydroxylation rate increases, decreasing the global charge of the complex. This effect may be correlated with the increased lability of the coordinated water molecules, that is, the decrease in their partial charge (Table 2.2).

In the noncharged species $Al(OH)_3(OH_2)_3$, the water molecules carry a partial charge that is practically equal to zero. It is therefore better to consider these molecules as part of the second solvation sphere rather than as ligands. The partial charge of water molecules in hexacoordinated aluminum complexes gradually decreases until it becomes negative when the hydroxylation rate increases (Table 2.3).

This situation is, of course, unrealistic because it would involve an electron transfer from the cation to the water molecule. In fact, ^{27}Al-NMR spectra indicate aluminum in a tetrahedral coordination in alkaline medium [50] (see Chap. 6); this confirms that in such conditions the water molecules have lost their nucleophilic character. Transition metal cations of high formal charge ($z > 4$), strongly polarizing and hexacoordinated in acidic medium, adopt in an

TABLE 2.2 } Exchange rate constant k (s^{-1}, 25°C) for water molecules in hydroxylated species of Fe^{III} (from [48, 49]).

	$[Fe(OH_2)_6]^{3+}$	$[Fe(OH)(OH_2)_5]^{2+}$	$[Fe(OH)_2(OH_2)_4]^{+}$
k	$1.7\ 10^2$	$4.5\ 10^5$	10^6

TABLE 2.3 } Partial charge of the water molecule in the coordination sphere of Al^{3+} as a function of the hydroxylation of the cation.

	$[Al(OH_2)_6]^{3+}$	$[Al(OH)(OH_2)_5]^{2+}$	$[Al(OH)_2(OH_2)_4]^{+}$	$[Al(OH)_3(OH_2)_3]^{0}$	$[Al(OH)_4(OH_2)_2]^{-}$
χ	2.754	2.675	2.588	2.487	2.373
$\delta(H_2O)$	+0.37	+0.26	+0.14	0	–0.16

alkaline medium the tetrahedral coordination for a different reason. Because of their small size and the presence of empty and low-energy *d* orbitals, the tetraoxo complexes are stabilized by the formation of a π–bond system, as discussed earlier.

2.3 Mechanisms of Hydroxylation and Redox Reactions in Solution

Because of the high lability of water molecules in most aquocomplexes (Tables 2.1 and 2.2), hydroxylation by alkalinization of these complexes may be viewed as an exchange of ligands by substitution:

$$\left[M(OH_2)_N\right]^{z+} + HO^-_{aq} \rightarrow \left[M(OH)(OH_2)_{N-z}\right]^{(z-1)+} + H_2O$$

However, the H–bond framework within the liquid allows a fast diffusion pathway for proton and hydroxyl ion, so that the reaction likely proceeds according to the hydroxyl's direct attack of the aquo ligand:

The same mechanism may occur for the hydroxylation of oxo species. Taking into account the strong basicity (nucleophilic character) of the oxo ligand, we observe that the reaction likely proceeds according to a direct attack by the proton or water itself:

$$\left[MO_{N'}\right]^{q-} + H^+_{aq} \rightarrow \left[MO_{N'-1}(OH)\right]^{(q-1)-} + H_2O$$

The kinetics of hydroxylation by alkalinization is extremely rapid [32] and is diffusion-controlled, which means the reaction rate is limited only by the reactants' rate of approach.

Hydrolysis is, strictly speaking, a reaction involving water as a reactant:

$$\left[M(OH_2)_N\right]^{z+} + hH_2O \rightarrow \left[M(OH)_h(OH_2)_{N-h}\right]^{(z-h)+} + hH^+_{aq}$$

The thermodynamic data of this reaction are [32]:

$\Delta H^\circ_{298} = (75.2 - 9.6\,z)$ kJ mol^{-1} $\Delta S^\circ_{298} = (-148.4 + 73.1\,z)$ J mol^{-1}, that gives:

$$\Delta G^\circ_{298} = (119.5 - 31.35\,z)\,\text{kJ mol}^{-1}.$$

The reaction is spontaneous ($\Delta G^\circ < 0$) for elements of charge $z \geq 4$. Therefore, at RT, tetravalent elements do not exist as purely aquocomplexes, even in strongly acidic medium. For elements of charge $z < 4$, ΔG° is negative only if the temperature is higher than 298 K. It is therefore necessary to heat the solution to trigger hydroxylation of the cation (forced hydrolysis or thermohydrolysis). Interestingly, this technique has been used to synthesize oxide nanoparticles with narrow-size distribution [51, 52]. Indeed, alkalinization of a solution by addition of a base inevitably forms local pH gradients leading to hydrolysis products that condense anarchically. In such conditions, the obtained solid is often amorphous and/or the particle size is often heterogeneous because of the overlapping nucleation and growth kinetic steps (§ 4.2). For instance, heating a solution to 60 – 100°C allows trivalent elements (Al, Fe, Cr), to achieve hydrolysis homogeneously under conditions that are close to thermodynamic equilibrium. In such a condition, the reaction between slowly formed hydroxylated precursors allows the decoupling of the precipitation kinetic steps, leading to narrow particle-size distribution. This book presents many examples of the usefulness of thermohydrolysis.

Redox reactions involving soluble metal complexes more often are dependent on acidity. Acidity, on the one hand, determines the nature of involved complexes (aquo, hydroxo, or oxo) and, on the other hand, acidity may be implicated in the mechanism of the electron transfer. The reaction between a reducer (Red 1) and an oxidant (Ox 2) may proceed according to an outer-sphere (OS) mechanism, in which the coordination spheres of the reactives are in close contact. The reaction may also proceed by an inner-sphere (IS) mechanism involving a substitution with formation of a bridging ligand, when at least one of the reactives carries labile ligands [17, 41, 53]. In these two cases, the electron exchange involves a variation in size of the cation, but in order for the transfer to take place, compression and expansion of the coordination sphere of the reducing and oxidizing species, respectively, have to occur to equalize their own energy levels. Such a breathing of complexes, thermically assisted, is characterized by the vibrational energy curves of the entire system (Fig. 2.10).

The electron is transferred within the transition state (OS or IS) when the sizes of the coordination spheres are in accordance—that is, when they have the same energy (that corresponds to the intersection of energy profiles). What role may acidity have in the redox transfer between soluble metal complexes?

As an example, let us consider the reduction of permanganate $[MnO_4]^-$ in aqueous solution producing the $[Mn(OH_2)_6]^{2+}$ complex. The half redox reaction involving manganese species has to be written as

$$\left[Mn^{VII}O_4\right]^- + 5e^- + 8H^+ + 2H_2O \rightarrow \left[Mn^{II}\left(OH_2\right)_6\right]^{2+}$$

accounting for the balance of charges and ligands. The protons involved in the reaction are required to maintain electrical neutrality. In fact, their chief role is to contribute to the expansion of the coordination sphere of Mn^{VII} (ionic radius 0.27 Å) by protonation of oxo ligands to produce *in fine* aquo ligands in the coordination sphere of Mn^{II} ion (ionic radius 0.67 Å). The reaction, involving the exchange of five electrons by manganese atoms, leads to a considerable increase in the size of the cation and therefore requires a strong acidic medium to force the protonation of permanganate in the transition state and finally to stabilize the aquo Mn^{II}

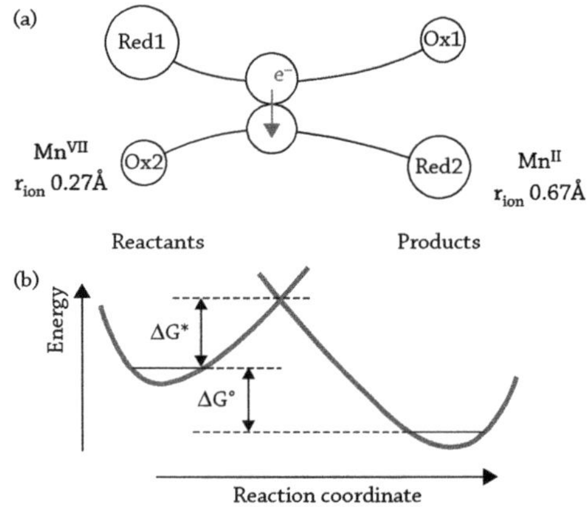

FIGURE 2.10 Scheme of a redox process Red1 + Ox2 → Ox1 + Red2 showing (a) the change in size of the reactants and products and (b) the corresponding profile of potential energy in the electron exchange in solution. The difference in energy between the intersection of curves and the initial state represents the activation energy ΔG^* of the process, whereas the difference between the final and initial states represents the free enthalpy ΔG° of the reaction.

complex. Of course, the electron transfer and the ligand protonation have to be concerted because of the weak basicity of the permanganate species. If the acidity is not high enough, the reaction stops. This happens more often after the exchange of three electrons by manganese atoms in producing an Mn^{IV} ion coordinated with hydroxo and aquo ligands according to the half reaction

$$\left[Mn^{VII}O_4\right]^- + 3e^- + 4H^+ + 2H_2O \rightarrow \left[Mn^{IV}(OH)_4(OH_2)_2\right]^0$$

The reaction yields the black oxide MnO_2, and subsequent manganese reduction does not occur even when there is excess reductant. Thus, the control of acidity in a redox system may allow control of the precipitation of a given solid phase. This book gives many examples concerning titanium and manganese oxides (Chap. 8). Here the coupling of acidity and redox transfer in aqueous medium is discussed using kinetic arguments. However, one may also observe that on thermodynamic potential-pH diagrams, the reducing power of most redox couples is at its maximum in the acidic medium, where the coordination sphere of the oxidant is more enlarged.

2.4 Evaluation of Partial Charges on Atoms in Combination

To estimate the likelihood of a reaction, it is useful to know the distribution of electron charge density within a chemical entity. It may be characterized by the partial charge carried by the individual atoms. Many attempts at these estimates have been made, but they have not been entirely successful largely because a charge carried by an atom is not a measurable quantity. Like the other methods, the partial charge model [54, 55] based on Sanderson's principle of electronegativity equalization [56] does not allow absolute charge estimation. However, its application to some problems of molecular reactivity in solution, such as the behavior of an element toward hydrolysis and condensation reactions, is interesting owing to its simplicity. It also allows, to some extent, compensation for the lack of experimental data. Used with caution (see the end of this section), it appears to be a useful teaching tool.

2.4.1 IONOCOVALENCY AND PARTIAL CHARGES

The chemical bond may be described by molecular orbitals (MOs) characterizing the electron status in the field of nuclei. In the simplest case of an A–B molecule, a bonding molecular orbital of the form

$$\Psi = a\psi_A + b\psi_B$$

indicates, within the coefficients a and b, the contribution of the ψ_A and ψ_B atomic orbitals. The coefficients characterize the ionicity of the bond, which is directly linked to the energy of the valence orbitals and to the electronegativities χ_A and χ_B of the neutral atoms. The most electronegative atom, A, for instance, attracts the bonding electrons and acquires a negative partial charge $\delta_{(A)}-$. The B atom, which is less electronegative, acquires a positive partial charge $\delta_{(B)}+$. For atoms of very different electronegativities, the bond becomes purely ionic. The MO degenerates into two AOs, one of which captures the two bonding electrons. The coefficients in the combination are $a = 1$ and $b = 0$, and the charges carried by the atoms are $\delta_{(A)} = -1$ and $\delta_{(B)} = +1$.

The formal charge of an atom corresponds to its oxidation state. It represents the charge of the isolated atom or a purely ionic combination. It may be used to establish the stoichiometry of a combination, on the base of a −2 charge for oxygen and +1 for hydrogen.

The partial charge of an atom in a combination is defined by the global atomic population. It includes the formal charge (oxidation state) and half of the overlapping charge due to the bonding with the ligands (Mulliken's population), which is assumed to be located on atoms [1]. In fact, the electrons "given" by the ligands have a weak probability of presence on the atom "acceptor." That means there is no redox phenomenon that will lower the formal charge of manganese in the $[MnO_4]^-$ complex from +7 to +0.66 [57]. Similarly, in the octahedral complex $[Ni(OH_2)_6]^{2+}$, magnetic measurements show a d^8 ion ($t_{2g}^6 e_g^2$), with a magnetic moment corresponding to two single electrons. The electronic spectrum of the complex corresponds to transitions between the terms of the d^8 configuration with the O_h symmetry. The partial covalency of the bond is reflected by parameters for the bonded Ni^{2+} ion that differ from those for the free ion, such as the spin-orbit coupling constant and the lowering of the interelectronic repulsion parameter (Raccah's parameter).

2.4.2 ELECTRONEGATIVITY

Electronegativity is the ability of an atom in a combination to attract electrons [58]. The concept is a flexible one, and many scales have been proposed to define it [59].

Absolute Electronegativity. The total energy of an isolated atom as a function of the number N of electrons around a nucleus of atomic number Z is shown in Figure 2.11.

Assuming that the variation is continuous and differentiable, Parr et al. [60] have shown that the slope $\partial E/\partial N$ is strictly equal to the chemical potential of electrons:

$$-\mu_e = -(\partial E/\partial N)_z \qquad \text{[A.1]}$$

The second-order development of the expression of energy may be written as [59]

$$E(N) \approx E(Z) + (N-Z)(\partial E/\partial N)_z + [(N-Z)^2/2!](\partial^2 E/\partial N^2)_z$$

The first ionization potential I and the electron affinity A are expressed as

$$I \approx E(Z-1) - E(Z) = -(\partial E/(\partial N)_z) + 1/2(\partial^2 E/\partial N^2)_z$$

$$A \approx E(Z) - E(Z+1) = -(\partial E/(\partial N)_z) - 1/2(\partial^2 E/\partial N^2)_z$$

that gives

$$(1/2)(I+A) \approx -(\partial E/\partial N)_z = -\mu_e = \chi \qquad \text{[A.2]}$$

The quantity $-(\partial E/\partial N)_z$ has been called "absolute electronegativity" [61]. This definition introduces electronegativity as an electron chemical potential similar to the thermodynamic chemical potential: in any system of atoms in a nonequilibrium state, an electronic density flux must occur from the high potential zone toward that of low potential. This gives an absolute estimation of electronegativity because both the ionization potential and the electron affinity refer to the *fundamental state* of the atom [62].

The hardness of an atom, as introduced by Pearson [61–63], represents the resistance of the electron chemical potential to the change of the electron number around the atom. It is expressed as

$$\eta_a = (1/2)(\partial \mu/(\partial N)_z = (1/2)(\partial^2 E/\partial N^2)_z = (1/2)(I-A) \qquad \text{[A.3]}$$

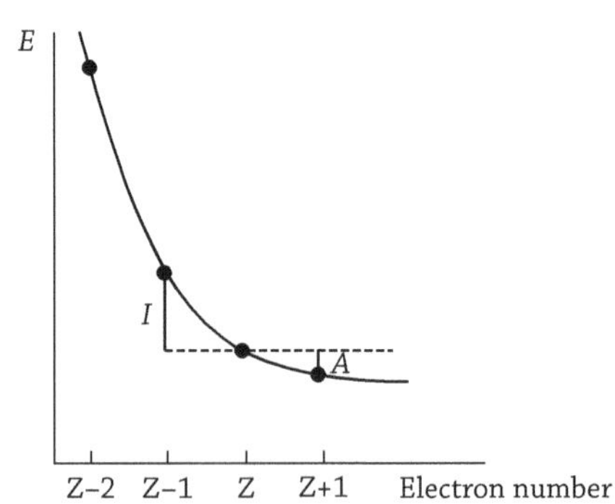

FIGURE 2.11 Total energy of an atom (atomic number Z) as a function of the number of electrons [59]. I and A are the first ionization potential and the electron affinity, respectively, of the neutral atom.

The hardness of an atom depends on its size. As the energy E is proportional to q^2/r (q is the electric charge, r the distance), hardness is then proportional to r^{-1}. This means that the more a system is spatially confined, the harder and less polarizable it is. So, the hardness represents the resistance of the electronic chemical potential to a change in the electron number around the atom.

According to the definition (Eq. [A.2]), the absolute electronegativity of an atom depends on its charge:

$$\chi_a = -\mu_e = -\mu_e^{\circ} + \eta \Delta N \qquad \text{[A.4]}$$

$-\mu_e^{\circ} (= \chi_a^{\circ})$ is relative to the neutral atom (standard state). The term $\eta \Delta N$ represents the deviation from the standard state owing to the charge variation ΔN.

Strictly speaking, Eq. [A.4] concerns isolated atoms. When atoms approach together, they undergo an external potential because of the proximity of other nuclei and electrons. The absolute electronegativity indicates only the direction of the electron density flux between atoms and characterizes the reactivity of *free atoms* [62]. These considerations apply to molecules as well.

The absolute electronegativity is, however, compatible with the MO model. Indeed, the first ionization potential of a diamagnetic molecule is the energy of the highest occupied molecular orbital (HOMO). The electron affinity is the energy of the lowest unoccupied molecular orbital (LUMO) [62]. The electronic chemical potential, $\mu_e = -\chi_a$, is, then, the average energy of these orbitals, and the hardness, η, is equal to half of the band gap. The donor or acceptor character of a ligand toward a cation may therefore be deduced by comparing their electronegativities.

Many other empirical evaluations of electronegativity have been proposed [59]. Allred and Rochow's definition [64] is based on the electrostatic attraction force exerted on electrons by the nucleus $Z_{eff}/r^2 \cdot Z_{eff}$ is the effective nuclear charge calculated from Slater's rules, and r is Pauling's covalent radius. Therefore, the size and shape of the *atom in combination* are taken into account in Allred–Rochow's electronegativity. The absolute electronegativity characterizes *isolated atoms* and their tendency to react (electronic chemical potential), but it does not take into account the change in size and shape of electron density or the external potential to which the electrons are subjected when the atoms combine.

2.4.3 PARTIAL CHARGES MODEL

The partial charges model is based on the idea that, in a combination, the atoms exchange electron density and acquire a partial charge until their electronegativities are equal. Because electronegativity, when defined as a potential, is a function of the charge of atom (Eq. [A.4]), it is first necessary to establish such variation and then to evaluate the common electronegativity in the combination.

Variation of Electronegativity with the Charge of an Atom. Let us consider neutral atoms in a combination. The absolute terms χ_a and η_a of the free atoms are subjected to a perturbation $\Delta\chi$ and $\Delta\eta$ owing to changes in their size and shape stemming from a covalency effect. The local electronic potential on each neutral i atom is then expressed by $(\chi_{ai} + \Delta\chi)$. The electron transfers due to the formation of bonds lead to partial charges on the atoms in creating an electrostatic potential. This electrostatic potential comprises a term $(\eta_{ai} + \Delta\eta)\delta_i$ that results in the charge δ_i on the atom i and a term $\Sigma(\delta_j / R_{ij})$ due to the presence of charges δ_j carried by the nearby atoms j at distances R_{ij}. This electrostatic potential is added to the local electronic potential, so that the electronegativity of combinated atoms is then given by [66]

$$\chi_i = \chi_{ai} + \Delta\chi + (\eta_{ai} + \Delta\eta)\delta_i + \sum\nolimits_{j \neq i, j=1}^{n} (\delta_j / R_{ij}) \qquad \text{[A.5]}$$

This last expression requires knowledge of perturbations to which electronegativities and hardnesses are subjected as a result of the covalency effect in the combination. But as the terms $\Delta\chi$ and $\Delta\eta$ are unknown, we write

$$\chi^* = \chi_a + \Delta\chi \qquad \eta^* = \eta_a + \Delta\eta$$

χ^* is Allred–Rochow's electronegativity, calculated for any element. Hardness and electronegativity are related, and since η is proportional to the inverse of a distance, the choice of Allred–Rochow's electronegativities $(\chi \propto Z_{eff} / r^2)$ allows us to write [54]

$$\eta^* = k\sqrt{\chi^*} \qquad \text{[A.6]}$$

where k is the constant of fit.

Electronegativity Equalization Principle. The formation of a bond between atoms A and B of different electronegativities induces an electron transfer from B to A if A is more electronegative than B. Atom A acquires a negative charge and its electronegativity decreases, whereas atom B acquires a positive charge and its electronegativity increases. The electronegativities, which are different when the atoms are without interaction, converge when they are bounded toward an average value χ. This is the electronegativity equalization principle postulated by R. T. Sanderson [56].

When electronegativity is defined as a potential, equalization that concerns electron transfers is equivalent to equalization of chemical potentials $(\mu = \partial G / \partial n)$ related to transfers of matter. The direction of electron transfers is governed by the relative values of electronegativities, and the intensity of the transfers is dependent on the hardness of bounded atoms. The principle may be extended to more complex atomic structures in which electron transfers cease when each atom in the combination has acquired the same average electronegativity, χ, which is also that of the structure. Knowledge of this average value allows, in principle, evaluation of the charges on the atoms of the structure.

Calculation of the Average Electronegativity. In a first approximation, the perturbation induced by the nearby atoms of the i atom is neglected compared to the effect of the charge δ_i. The electronegativity of the i atom in any combination can therefore be written as [54]

$$\chi_i = \chi_i^* + \eta_i^* \delta_i \quad \text{with } \eta_i^* = k\sqrt{\chi_i^*} \qquad \text{[A.7]}$$

The constant k is calibrated from the charges in NaF (δ(Na) = $-\delta$(F) = 0.75) as proposed by Sanderson [67]. When we use Allred–Rochow's scale, that gives k = 1.36. At equilibrium in the combination, the equalization principle leads to $\chi_i = \chi_i^* + \eta_i^* \delta_i = \chi$, and the charge of the i atom is expressed as

$$\delta_i = (\chi - \chi_i^*) / \eta_i^* \qquad \text{[A.8]}$$

The conservation of the global charge z of the combination is expressed as

$$\sum_i \delta_i = z \qquad \text{[A.9]}$$

which finally gives

$$\chi = \frac{\Sigma_i \sqrt{\chi_i^*} + 1.36z}{\Sigma_i \left(1/\sqrt{\chi_i^*}\right)} \qquad \text{[A.10]}$$

χ may therefore be easily calculated for any molecular structure from the composition, charge, and electronegativities of each atom (Fig. 2.12), thereby allowing us to obtain the partial charge distribution.

An example of charge distribution in some metal aquocomplexes $[M(OH_2)_6]^{z+}$ calculated in using the model is given in Table A1. The average electronegativity is calculated by using Eq. [A10] and the charges of atoms from Eqs. [A8] and [A7]. The electronegativities of elements are presented in the Figure 2.12.

The increase of positive charge of hydrogen and the decrease of negative charge of oxygen atoms in the coordinated water molecules as the oxidation state of the cation increases highlight the increasing polarization of the ligands and acidity of these complexes. In the "free" water molecule, the calculated charges are −0.40 and +0.20 on oxygen and hydrogen atoms, respectively (see § 2.1.1).

APPROXIMATIONS AND LIMITATIONS OF THE MODEL

- To consider that χ_i varies linearly with the charge δ_i as in relation [A.7] implies that composition only governs the charge distribution within the combination. Structural isomerism is ignored, which is obviously a strong limitation of the model. In fact, as the geometries of studied species in solution are usually not known with any precision, the simplified approach consists of neglecting the term $(\Sigma\, \delta_j / R_{ij})$ compared to χ_i^* in Eq. [A.5].

TABLE A1 } Charge distribution calculated for some metal hexaaquo complexes.

Complex	χ	δ(M)	δ(0)	δ(H)	δ($H_2$0)
$[Mn(OH_2)_6]^{2+}$	2.657	+0.59	−0.33	+0.28	+0.23
$[Cr(OH_2)_6]^{3+}$	2.762	+0.68	−0.29	+0.34	+0.39
$[Ti(OH_2)_6]^{4+}$	2.848	+0.98	−0.25	+0.38	+0.51

H 2.10																	He 3.20
Li 0.97	Be 1.57											B 2.02	C 2.50	N 3.07	O 3.50	F 4.10	Ne 5.10
Na 1.01	Mg 1.29											Al 1.47	Si 1.74	P 2.11	S 2.48	Cl 2.83	Ar 5.10
K 0.91	Ca 1.04	Sc 1.23	Ti 1.32	V 1.56	Cr 1.59	Mn 1.63	Fe 1.72	Co 1.75	Ni 1.80	Cu 1.75	Zu 1.66	Ga 1.82	Ge 2.00	As 2.20	Se 2.50	Br 2.69	Kr 3.10
Rb 0.89	Sr 0.99	Y 1.19	Zr 1.29	Nb 1.45	Mo 1.56	Tc 1.67	Ru 1.78	Rh 1.84	Pd 1.85	Ag 1.68	Cd 1.60	In 1.49	Sn 1.89	Sb 1.98	Te 2.15	I 2.33	Xe 2.60
Cs 0.87	Ba 0.89	*	Hf 1.36	Ta 1.50	W 1.59	Re 1.88	Os 1.99	lr 2.05	Pt 2.00	Au 2.02	Hg 1.80	Tl 1.60	Pb 1.92	Bi 2.03	Po 2.12	At 2.28	Rn 2.30
Fr 0.86	Ra 0.95	**															

*	La 1.18	Ce 1.17	Pr 1.18	Nd 1.19	Pm 1.20	Sm 1.20	Eu 1.13	Gd 1.27	Tb 1.24	Dv 1.26	Ho 1.28	Er 1.30	Tm 1.30	Yb 1.24	Lu 1.36
**	Ac 1.12	Th 1.24	Pa 1.22	U 1.24	Np 1.22	Pu 1.24	Am 1.25	Cm 1.20	Bk 1.20	Gf 1.20	Es 1.20	Fm 1.20	Md 1.20	No 1.20	Lw 1.20

FIGURE 2.12 Allred–Rochow's electronegativities χ^* of the elements compiled from data in [65–69].

- The calculated charges are dependent on hardness η_i^* through the calibrating constant *k*. That means the calculated charges are relative and depend on the reference used for calibration.
- Even if χ^* is an electronegativity taking into account the mean structural and surrounding effects of an atom in a combination, the results of the model have to be carefully considered since structural peculiarities take place, such as π-overlaps in borates (§ 3.3.1) and in oxometalates (§ 3.3.2). Similar difficulties are encountered for atoms of identical nature but with different oxidation states or with different coordination numbers that are not discriminated.

It must be clearly stressed that the model does not allow for any distinction between isomers; therefore, it must not be used to identify any property related to the structure of a combination. Although the model is simple in its approach, it provides an interesting guide and appears to be quite useful for study of reactivity in solution. It is especially valuable in predicting and anticipating hydrolysis and condensation reactions.

References

[1] Chabanel M., Gressier P. *Liaison chimique et spectroscopie*. Paris: Ellipses Marketing, 1991.

[2] Turner D.W., Baker C., Backer A.D., Brundle C.R. *Molecular Photoelectron Spectroscopy*. London: Wiley-Interscience, 1970.

[3] Rahman A., Stillinger F.H. Hydrogen-bond patterns in liquid water. *J. Am. Chem. Soc.* 1973, 95, 7943–7948.

[4] Stillinger F.H. Water revisited. *Science* 1980, 209, 451–457.

[5] Geissler P.L., Dellago C., Chandler D., Hutter J., Parrinello M. Autoionization in liquid water. *Science* 2001, 291, 2121–2124.

[6] Symons M.C.R. Water structure and reactivity. *Acc. Chem. Res.* 1981, 14, 179–187.

[7] Head-Gordon T., Hura G. Water structure from scattering experiments and simulation. *Chem. Rev.* 2002, 102, 2651–2670.

[8] Richards R.E., Smith J.A.S. Nuclear magnetic resonance spectra of some acid hydrates. Trans. *Faraday Soc.* 1951, 47, 1261–1274.

[9] O'Reilly D.E., Peterson E.M., Williams J.M. NMR of the aquated proton: hydronium perchlorate. *J. Chem. Phys.* 1971, 54, 96–98.

[10] Ardon M., Bino A. A new aspect of hydrolysis of metal ions: the hydrogen-oxide bridging ligand $H_3O_2^-$. Structure and Bonding 65, 1–28. Berlin: Springer Verlag, 1987.

[11] Magini M., Licheri G., Paschina G., Pinna G. *X-ray Diffraction of Ions in Aqueous Solutions: Hydration and Complex Formation*. Boca Raton, FL: CRC Press, 1988.

[12] Ohtaki H., Radnai T. Structure and dynamics of hydrated ions. *Chem. Rev.* 1993, 93, 1157–1204.

[13] Ohtomo N., Arakawa K., Takeuchi M., Yamaguchi T., Ohtaki H. Neutron diffraction study of aqueous hydrochloric and hydrobromic acid solutions. *Bull. Chem. Soc. Jpn* 1981, 54, 1314–1319.

[14] Lee H.G., Matsumoto Y., Yamaguchi T., Ohtaki H. Diffraction studies on the structures of hydrated oxonium ion, and the chlorocobalt(II) and tetrachlorocobaltate(II) complexes in aqueous solutions. *Bull. Chem. Soc. Jpn* 1983, 56, 443–448.

[15] Marx D., Chandra A., Tuckerman M.E. Aqueous basic solutions: hydroxide solvation, structural diffusion and comparison to the hydrated proton. *Chem. Rev.* 2010, 110, 2174–2216.

[16] Kropman M.F., Bakker H.J. Dynamics of water molecules in aqueous solvation shells. *Science* 2001, 291, 2118–2120.

[17] Basolo F., Pearson R.G., *Mechanism of Inorganic Reactions*, 2nd ed. New York: J. Wiley and Sons, 1958.

[18] Franck H.S., Wen W.Y. Structural aspects of ion-solvent interaction in aqueous solutions: a suggested picture of water structure. *Discus. Faraday Soc.* 1957, 24, 133–140.

[19] Hesse W., Jansen M. Polymorphie von tetramethylammoniumhydroxid-pentahydrat, $NMe_4OH \cdot 5H_2O$, und kristallstrukturen der raumtemperatur- und tieftemperaturform. *Z. anorg. allg. Chem.* 1991, 595, 115–130.

[20] Krishnan C.V., Friedman H.L. Solvation enthalpies of various ions in water, propylene carbonate, and dimethylsulfoxide. *J. Phys. Chem.* 1969, 73, 3934–3940.

[21] Rabenau A. The role of hydrothermal synthesis in preparative chemistry. *Angew. Chem. Int. Ed. Engl.* 1985, 24, 1026–1040.

[22] Stein A., Keller S.W., Mallouk T.E. Turning down the heat: design and mechanism in solid-state synthesis. *Science* 1993, 259, 1558–1564.

[23] Barrer R.M. *Hydrothermal Chemistry of Zeolites*. London: Academic Press, 1982.

[24] Haar L., Gallagher J.S. NBS/NRC Steam Tables. Washington DC: Hemisphere Publ., 1984.

[25] Akiya N., Savage P.E. Roles of water for chemical reactions in high-temperature water. *Chem. Rev.* 2002, 102, 2725–2750.

[26] Seward T.M. Chemistry and geochemistry of solutions at high temperatures and pressures. In D. Rickard and F. Wickman, Eds., *Physics and Chemistry of the Earth*, vol. 13/14. New York: Pergamon Press, 1981.

[27] Tödheide K. *Water, A Comprehensive Treatise*. Vol. 1, F. Francks, Ed. New York: Plenum, 1972, 463–514.

[28] Holzapfel W., Franck E.U. Leitfähigkeit und lonendissoziation des wassers bis 1000°C und 100 kbar. *Ber. Bunsenges.* 1966, 70, 1105–1102.

[29] Charlot G. *Chimie analytique générale, solutions aqueuses et non aqueuses*. Paris: Masson, 1967.

[30] Ferey G. Hybrid porous solids: past, present, future. *Chem. Soc. Rev.* 2008, 37, 191–214.

[31] Dietrich B., Viout P., *Lehn J.M. Aspects de la chimie des composés macrocycliques*. Paris: InterEditions, 1991.

[32] Baes C.F., Mesmer R.E. *The Hydrolysis of Cations*. New York: Wiley, 1976.

[33] Charlot G. *L'analyse qualitative et les réactions en solution*. 6ème éd. Paris: Masson, 1969.

[34] Sillen L.G. *Stability Constants of Metal-ion Complexes*. The Chemical Society. London: Special Publication n°17, 1964.

[35] Sillen L.G. *Stability Constants of Metal-ion Complexes*. The Chemical Society. London: Special Publication n°25, 1971.

[36] Kragten J. *Atlas of Metal Ligand Equilibrium in Aqueous Solution*. E. Horswood, Ed. Chichester, UK: Wiley, 1978.

[37] Ringbom A. *Les complexes en chimie analytique*. Paris: Dunod, 1967.
[38] Jorgensen C.K. *Inorganic Complexes*. New York: Academic Press, 1963.
[39] Kepert D.L. *The Early Transition Metals*. New York: Academic Press, 1972.
[40] Wells A. F. *Structural Inorganic Chemistry*, 5th ed. Oxford: Oxford University Press, 1991.
[41] Purcell, K.F., Kotz J.C. *Inorganic Chemistry*. Holt Saunders Japan Intern. Ed., Philadelphia, 1985.
[42] Pope, M.T. *Heteropoly and Isopolyoxometallates*. Berlin: Springer-Verlag, 1983.
[43] Ballhausen C.J., Gray H.B. The electronic structure of the vanadyl ion. *Inorg. Chem.* 1962, 1, 111.
[44] Selbin J. The chemistry of oxovanadium (IV). *Chem. Rev.* 1965, 65, 153–175.
[45] Kevan L. Comparison of electron magnetic resonance and X-ray studies of the structure of $VO(H_2O)_5^{2+}$. *J. Phys. Chem.* 1984, 88, 327–328.
[46] Boas L.V., Pessoa J.C. *Comprehensive Coordination Chemistry*. Sir G. Wilkinson, Ed. Oxford: Pergamon Press, 1987.
[47] Cotton F.A., Wilkinson G. *Advanced Inorganic Chemistry*, 4th ed. New York: Wiley, 1980.
[48] Dodgen H.W., Liu G., Hunt J.P., Water exchange with ferric ion and oligomerized iron in acidic aqueous solutions. *Inorg. Chem.* 1981, 20, 1002–1005.
[49] Grant M., Jordan R.B., Kinetics of solvent water exchange on iron(III). *Inorg. Chem.* 1981, 20, 55–60.
[50] Akitt J.W. Multinuclear studies of aluminium compounds. Progress in NMR *Spectroscopy* 1989, 21, 1–149.
[51] Matijevic E. Colloidal chemical aspects of the corrosion of metals. *Pure Appl. Chem.* 1980, 52, 1179–1193.
[52] Matijevic E. Monodisperse colloids: art and science. *Langmuir* 1986, 2, 12–20.
[53] Souchay P., Lefebvre J. *Equilibres et réactivité des complexes en solution*. Paris: Masson, 1969.
[54] Henry M. Thèse. Université Pierre et Marie Curie. Paris, France, 1988.
[55] Henry M., Jolivet, J. P., Livage J. Aqueous chemistry of metal cations: hydrolysis, condensation and complexation. *Struct. Bond.* 1992, 77, 153–206.
[56] Sanderson R.T. An interpretation of bond lengths and a classification of bonds. *Science* 1951, 114, 670–672.
[57] Viste A., Gray H.B. The electronic structure of permanganate ion. *Inorg. Chem.* 1964, 3, 1113–1123.
[58] Pauling L. *The Nature of Chemical Bond*, 1st ed. Ithaca, NY: Cornell University Press, 1939.
[59] Chermette H., Lissilour R. Une vieille notion toujours d'actualité: l'électronégativité, et sa relation avec le concept de dureté et molesse des acides et des bases. *L'Actualité Chimique* 1985, 4, 59–69.
[60] Parr R.G., Donnelly R.A., Levy M., Palke W.E. Electronegativity: the density functional viewpoint. *J. Phys. Chem.* 1978, 68, 3801–3807.
[61] Iczkowski R.P., Margrave J.L. Electronegativity. *J. Am. Chem. Soc.* 1961, 83, 3547–3551.
[62] Pearson R.G. Density functional theory: electronegativity and hardness. *Chemtracts Inorg. Chem.* 1991, 3, 317–333.
[63] Pearson R.G. Hard and soft acids and bases. *J. Am. Chem. Soc.* 1963, 85, 3533–3539.

[64] Allred A.L., Rochow F.G. A scale of electronegativity based on electrostatic force. *J. Inorg. Nucl. Chem.* 1958, 5, 264–268.
[65] Mortier W.J., Ghosh S.K., Shankar S.J. Electronegativity equalization method for the calculation of atomic charges in molecules. *J. Am. Chem. Soc.* 1986, 108, 4315–4320.
[66] Sanderson R.T. *Inorganic Chemistry*. New York: Reinhold, 1976.
[67] Little E.J., Jones M.M. A complete table of electronegativities. *J. Chem. Educ.* 1960, 37, 231–233.
[68] Batsanov S.S. A geometrical system of electronegativities. *Russ. J. Struct. Chem.* 1968, 37, 263–270.
[69] Zhang Y. Electronegativities of elements in valence states and their applications. 1. Electronegativities of elements in valence states. *Inorg. Chem.* 1982, 21, 3886–3889.

3 }

Condensation in Solution

POLYCATIONS AND POLYANIONS

Condensation of metal complexes in solution forms entities in which the cations are linked by hydroxo (HO^-) or oxo (O^{2-}) bridges. The reaction is initiated by the addition of a base to an aquocomplex:

$$2\left[Cr(OH_2)_6\right]^{3+} + 2\,HO^- \rightarrow \left[Cr_2(OH)_2(OH_2)_8\right]^{4+} + 2\,H_2O$$

or by the addition of an acid to an anionic complex:

$$2\left[CrO_4\right]^{2-} + 2H^+ \rightarrow \left[Cr_2O_7\right]^{2-} + H_2O$$

Thus, purely aquo- and purely oxocomplexes are stable in solution, and the condensation of cations is initiated by hydroxylation. With regard to electrically charged hydroxylated complexes, the reaction forms discrete and soluble entities—polycations and polyanions with a molecular complexity which depends on acidity conditions. This chapter presents a detailed study of their formation and structure. With regard to noncharged hydroxylated complexes, the condensation reaction is no longer limited and leads to the formation of a solid (a subject that is examined in the following chapters).

3.1 Hydroxylation and Condensation of Cations

The hydroxylation reaction is the key stage to initiate the condensation of cations in solution. It is thus important to precise the mechanism of the successive steps of the process, in order to understand why the behavior of a cation is closely related to its oxidation state, and why the reaction product may be a discrete molecular species or a solid.

3.1.1 MECHANISMS AND STRUCTURAL CONSIDERATIONS

a) Mechanism

As a cation generally exhibits its maximum coordination number in the initial monomeric complex and in condensed species, the condensation reaction is a substitution that proceeds according to one of three basic mechanisms: dissociation, association, and interchange or direct displacement [1, 2].

Dissociative substitution is a two-step process involving the formation of a reduced-coordination intermediate: In the first step, a labile ligand, the leaving group, breaks its bond in the starting complex before a nucleophilic entering group completes, in the second step, the cation coordination (Fig. 3.1 a). Associative substitution is also a two-step process in which the intermediate temporarily has increased coordination. The bond with the nucleophilic entering group (first step) occurs prior to the release of the leaving group (second step) (Fig. 3.1 b). In an interchange or direct displacement mechanism, the substitution is a one-step process in which the leaving group and the entering nucleophilic group contribute simultaneously to the transition state. Bond formation with the entering group and breaking of the bond with the leaving group are synchronized (Fig. 3.1 c). The associative and interchange mechanisms are often difficult to distinguish, unless an intermediate can be directly identified. In these mechanisms, the transition states are very reactive, and therefore, the first step is rate-controlling. Consequently, the dissociative process is known as a unimolecular nucleophilic substitution (SN_1), whereas the associative and interchange processes requiring two partners to form the transition state are called bimolecular nucleophilic substitution SN_2.

Condensation reactions involving substitution require a charge donor, the nucleophile, which can attack the cation bearing a positive charge and being an electron acceptor. The coordination sphere of the cation must also contain a ligand that can be eliminated (the leaving group). The ligands that are most easily removable bear a positive partial charge, δ^+. In the absence of complexing species in solution, three kinds of ligands may exist in the coordination sphere of the cation: the aquo (H_2O), hydroxo (OH), and oxo (O) ligands. What is their role in nucleophilic substitution?

First, let us consider aquo species $[M(OH_2)_N]^{z+}$. The partial charges on the cation and on the coordinated water molecule are both positive:

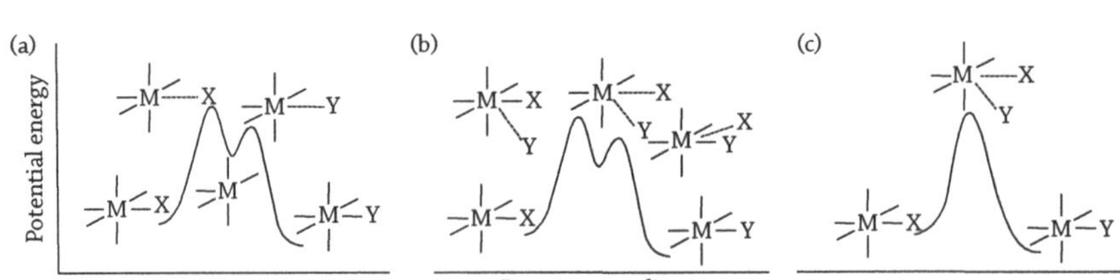

FIGURE 3.1 Reaction pathways for nucleophilic substitutions: dissociative (a), associative (b), and interchange (c).

$$\left[Mn(OH_2)_6\right]^{2+} \quad \chi = 2.657 \quad \delta(Mn) = +0.59 \quad \delta(H_2O) = +0.23$$

$$\left[Cr(OH_2)_6\right]^{3+} \quad \chi = 2.756 \quad \delta(Cr) = +0.68 \quad \delta(H_2O) = +0.39$$

The aquo ligand does not have a nucleophilic character; it may only act as a leaving group. Therefore, condensation of purely aquo species does not occur.

In oxo species $[MO_N]^{(2N-z)-}$, the partial charge on the cation is positive, whereas that on the oxo ligand is strongly negative:

$$\left[MnO_4\right]^{-} \quad \chi = 2.533 \quad \delta(Mn) = +0.52 \quad \delta(O) = -0.38$$

$$\left[CrO_4\right]^{-} \quad \chi = 2.055 \quad \delta(Cr) = +0.27 \quad \delta(O) = -0.57$$

The oxo ligand may be an excellent nucleophile. In fact, the calculated charges are erroneous because of the strong π-character of the M–O bonds. Therefore, the basic properties and the nucleophilic character of the oxygen atoms are strongly decreased. Consequently, condensation of purely oxo species does not occur, mainly because the oxo ligand never acts as a leaving group.

In an aquohydroxo complex, $[M(OH)_h(OH_2)_{z-h}]^{(z-h)+}$, nucleophilic ($OH^-$) and leaving ($H_2O$) groups are simultaneously present. The hydroxo ligand usually bears a negative partial charge; otherwise it would be acid and would lose its proton (the charge of oxygen, always negative, would not be sufficient to shield a cation–proton repulsion). Therefore, the hydroxo ligand is a nucleophile, and the presence of the aquo ligand allows condensation to occur, leading to the formation of hydroxo bridges:

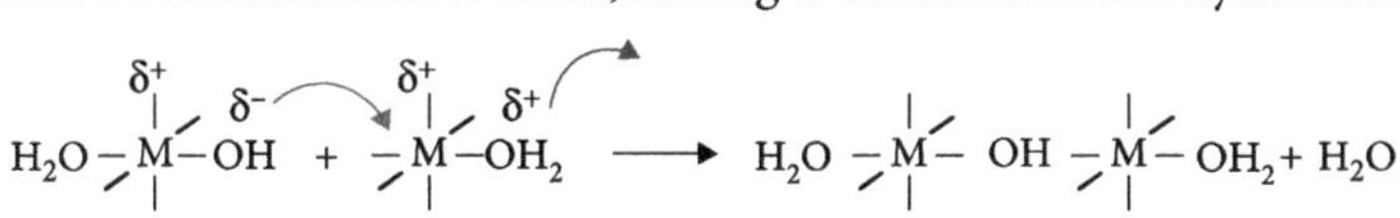

As the aquo ligand is often very labile (§ 3.2), the reaction likely proceeds through a dissociative SN_1 mechanism with a reduced-coordination intermediate in which the breaking of the $M–OH_2$ bond occurs before the formation of an M-OH-M bridge. The bridging OH ligands are called "*ol*" groups in order to distinguish them from terminal hydroxo ligands [3, 4]. The reaction leading to the formation of such a hydroxo bridge is consequently named "olation" (§ 3.2). In the case of transition metal ions strongly stabilized in octahedral symmetry by the crystal field, such as Cr^{3+} and Ni^{2+} ions, the lability of the coordinated water is very low, and condensation proceeds via an associative mechanism with formation of the bridging ligand $[H_3O_2]^-$ (§ 3.2.1).

In oxohydroxo complexes, $[MO_{N-h}(OH)_h]^{(2N-h-z)-}$, there is no preexisting leaving group such as the aquo ligand. An hydroxo ligand may, however, act as a leaving group if, through a proton transfer within the transition state of an SN_2 mechanism, the formation of an aquo ligand is possible. The proton in the *ol* bridge is more acidic than that of a terminal hydroxo ligand, and it may migrate to one of them, forming an aquo ligand that will be eliminated if its partial charge is positive. In this case, condensation proceeds with formation of an *oxo* bridge, which is referred to as oxolation [3]:

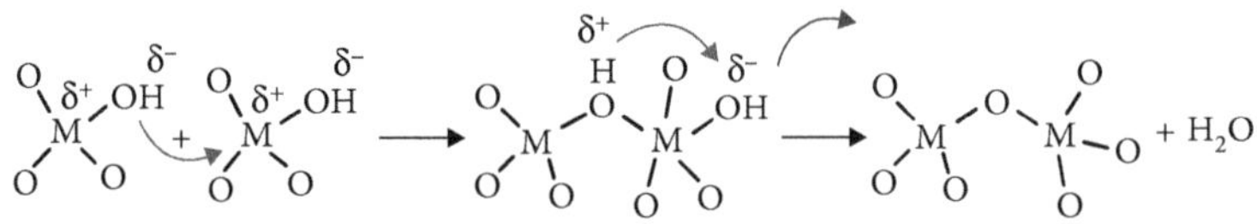

As there is no labile group in the coordination sphere of the reactants, oxolation reaction proceeds following an associative mechanism in which the coordination in the transition state has to increase before the leaving group can be released.

Experimentally, oxolation is often much slower than olation. For instance, condensation by oxolation of vanadate, tungstate, or silicic acid, $Si(OH)_4$, is much slower than condensation by olation of aquohydroxo complexes of iron, aluminum, or titanium. The difference in mechanism likely makes olation intrinsically faster than oxolation. A comparison is difficult, however, because it would require study, everything else being equal, of the kinetics of each reaction with the *same* chemical element. Comparison is also made difficult because oxolation, which strongly depends on the acidity of the medium, is very sensitive to acid–base catalysis. This effect is well illustrated with silicates (§ 3.3).

Thus, the presence of a hydroxo ligand bearing a negative partial charge (nucleophile) in the coordination sphere of a cation is required for condensation to take place. However, this is not a sufficient requirement because the nucleophilic attack by the hydroxo ligand may occur only if the cation can undergo this attack; thus, its electrophilic character has to be high enough. As a rule, many examples show that condensation occurs only if the partial charge on the cation in the precursor complex is equal to or higher than 0.3 (for charges calculated with the partial charge model, see § 2.4). This empirical value defines the threshold for the onset of condensation, whatever the partial charge on the hydroxo ligand may be. Therefore, one may define the criterion for condensation in solution as the double condition

$$\delta(\mathrm{OH}) < 0 \qquad \delta(\mathrm{M}) > +0.3$$

The validity of this requirement may be demonstrated with a few examples. In the species $[Cr(OH)(OH_2)_5]^{2+}$, $\delta(OH) = -0.02$ and $\delta(Cr) = +0.64$. The complex condenses by olation, and various polycations are observed in solution (§ 3.2.2). In the most basic form of silicates in solution, $[SiO_2(OH)_2]^{2-}$, the partial charges are $\delta(OH) = -0.55$ and $\delta(Si) = +0.20$. Although this species contains strongly nucleophilic hydroxo groups, it does not condense because the partial charge on silicon is too low to undergo nucleophilic attack. On the other hand, the species $[SiO(OH)_3]^-$, in which $\delta(OH) = -0.30$ and $\delta(Si) = +0.35$, condenses and through oxolation forms various polysilicates (§ 3.3.1). With phosphates, calculations give the following:

$$\left[\mathrm{PO_3(OH)}\right]^{2-} \quad \chi = 2.18 \quad \delta(\mathrm{P}) = +0.03 \quad \delta(\mathrm{OH}) = -0.48$$

$$\left[\mathrm{PO_2(OH)_2}\right]^{-} \quad \chi = 2.49 \quad \delta(\mathrm{P}) = +0.19 \quad \delta(\mathrm{OH}) = -0.20$$

$$\left[\mathrm{PO(OH)_3}\right] \quad \chi = 2.71 \quad \delta(\mathrm{P}) = +0.35 \quad \delta(\mathrm{OH}) = 0.0$$

Condensation does not occur in solution because the P^V atom in phosphates is too weakly electrophilic. Although the partial charge on phosphorus in phosphoric acid is higher than 0.3, it does not condense further because the hydroxo ligands have lost their nucleophilic character. Indeed, the first acidity of phosphoric acid is high enough, and the hydroxo ligand is unstable. Therefore, condensation of phosphate does not spontaneously occur in aqueous solution. It takes place, however, in the solid state by heating, which forces dehydration [5], but soluble polyphosphates are metastable in water. They are more or less rapidly hydrolyzed in acidic medium to form monomeric species.

b) Structure

The hydroxo ligand is able to form one, two, or three bonds through the weakly 2σ bonding and π nonbonding MOs (Fig. 3.2). Therefore, the most frequent oxo or hydroxo bridges are represented in Scheme 3.1.

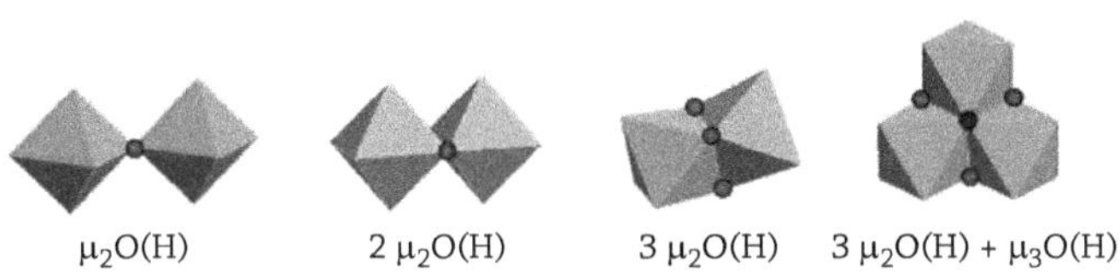

SCHEME 3.1

The single bridge μ_2 – O(H) corresponds to corner-sharing coordination polyhedra of the cations, whereas the double bridge $2\mu_2$ – O(H) corresponds to edge-sharing polyhedra. The connection of coordination polyhedra through a face ($3\mu_2$ – O(H) bridge) is rare. The μ_3 – O(H) bridge that links three polyhedra by a single hydroxo or oxo bridge is often formed. As the number of shared corners grows, the electrostatic repulsions between the cations become increasingly strong.

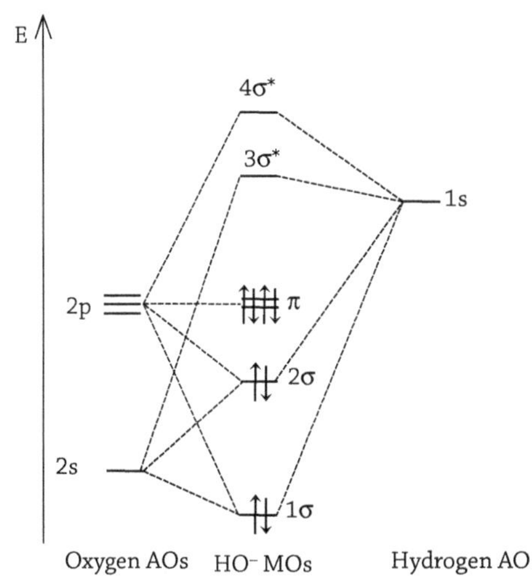

FIGURE 3.2 Schematic molecular orbitals energy diagram for HO^- ion.

Repulsions are very high in edge-sharing tetrahedra. Face-sharing tetrahedra are never observed, either in solution or in the solid state. The repulsions are weaker between octahedra and edge sharing is usual; face sharing sometimes occurs, for instance, in the corundum structural type.

The hydroxylation ratio, which indicates the number of hydroxo ligands in the coordination sphere of the cation, characterizes the functionality of the complex toward condensation. It directly influences the structural complexity of condensed species.

c) Extension of the Reaction

Under given acid–base conditions, the growth of condensed species may stop at various stages. The reaction can be limited to the formation of oligomers, or it may extend to precipitation of a solid. For instance, the condensation of chromate $[CrO_4]^{2-}$ does not exceed the formation of the dimer $[Cr_2O_7]^{2-}$, and Cr^{VI} never forms a solid from solution. On the other hand, Cr^{III} forms a series of soluble polycations as long as the hydroxylation ratio, h, remains smaller than 2.5 (§ 3.2.2), but a solid precipitates for $h = 3$.

Condensation of hydroxylated and electrically charged complexes always ends at a more or less advanced stage, leaving discrete species, polycations or polyanions, depending on whether the monomeric complex precursor is a cation or an anion. Indeed, electrical charges do not indefinitely accumulate on an oligomer. But when does condensation have to stop? The answer is as soon as the conditions allowing the nucleophilic substitution on the growing oligomer are no longer present. The charges within the precursors, which fulfill the conditions defined above as $\delta(OH) < 0$ and $\delta(M) > 0.3$, may evolve during condensation. Indeed, the condensation involving water elimination leads to a change in the average electronegativity of the growing species due to a change in composition, with a charge redistribution and a modification in the reactivity of the functional groups. Therefore, the OH ligands in the oligomer may lose their nucleophilicity ($\delta(OH)$ becomes zero or positive) and/or the cations may lose their electrophilicity ($\delta(M)$ becomes lower than +0.3). The nucleophilic character of hydroxo ligands usually cancels itself in polycations, and the electrophilic character of the cation cancels itself in polyanions. With regard to electrostatically neutral complexes, elimination of water never leads to a change in average electronegativity high enough to cancel the reactivity of the functional groups. Condensation occurs indefinitely until the precipitation of a solid, a hydroxide, oxyhydroxide, or a more or less hydrated oxide, or a basic salt in the presence of complexing ligands.

In summary, the condensation of cations in solution is initiated by the presence of the hydroxo ligand in their coordination sphere. Hydroxylation of the cation may be obtained in an acid–base reaction (e.g., neutralization, thermolysis). It also may result from a redox reaction. Indeed, reduction of an oxo species lowers the charge of the cation and the polarization of the oxo ligands allowing their protonation. Conversely, oxidation of an aquo cation, by increasing the charge of the cation and the polarization of the ligands, leads to their deprotonation with formation of the hydroxo ligand (Fig. 3.3). Condensation can then occur, involving two reactions,

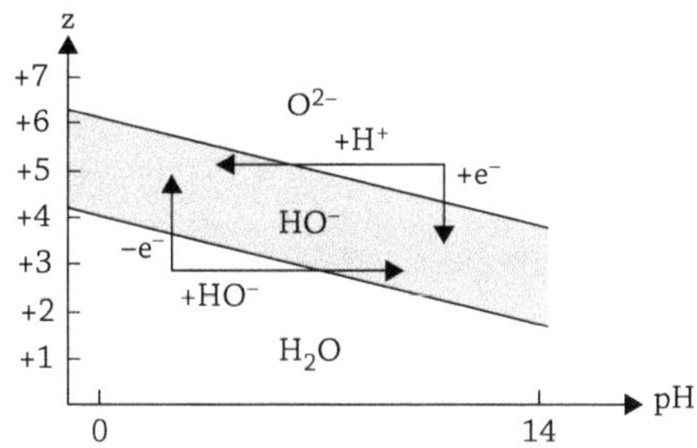

FIGURE 3.3 Schematic processes for initiation of condensation reactions in solution: acidification or reduction of oxocomplexes and alkalinization or oxidation of aquocomplexes.

oxolation and olation, respectively. Condensation of hydroxylated cationic and anionic complexes is always limited, forming polycations and polyanions, respectively. The formation of a solid requires the presence of uncharged complexes.

3.1.2 THE DIFFERENT BEHAVIORS OF CATIONS AGAINST CONDENSATION

At first glance, the behavior of the elements with respect to condensation and precipitation within the periodic table may appear quite disparate. The classification according to the formal charge, *z*, of the elements (Figs. 2.9 and 3.3) has to be improved because other parameters, such as electronegativity, size, and electron configuration, have to be taken into account in any attempt to characterize their acid–base properties and their behavior in solution. For instance, boron and aluminum ($z = +3$) exhibit radically different chemistries in solution, as is the case for silicon and titanium ($z = +4$) and for phosphorus, antimony, and vanadium ($z = +5$). To systematize the behavior of the elements, a charge-electronegativity diagram derived from the partial charge model (Fig. 3.4) is used. A brief presentation is given as follows; detailed calculations required for constructing the diagram are given elsewhere [6].

Let us consider an element of formal charge *z*, under the form of the noncharged molecular species MO_NH_{2N-z}. This form may have various behaviors. It may be a strong base, such as NaOH (low values of *z* and χ_M) or a strong acid, such as ClO_3OH (high values of *z* and χ_M). In these cases, any condensation reaction may occur, and the element may exist only as a monomer, in either a cationic form (Na^+, xH_2O, zone I) or an anionic form ($[ClO_4]^-$, zone V) (Fig. 3.4). In between these extreme behaviors is a negative partial charge on the hydroxo ligands in the MO_NH_{2N-z} form, which avoids their acid dissociation, and condensation takes place. The zone II corresponds to elements of moderated electronegativity with low oxidation states that condense by olation to form stable hydroxides, such as Ag^I, Ca^{II}, Mg^{II}, Mn^{II}, and Al^{III}. In zone III, hydroxides are not stable, and condensation through olation and oxolation leads to oxyhydroxides (Fe^{III}, Cr^{III}) or oxides (Zr^{IV}, Ti^{IV}). Elements from the zone IV condense by oxolation and form polyacids

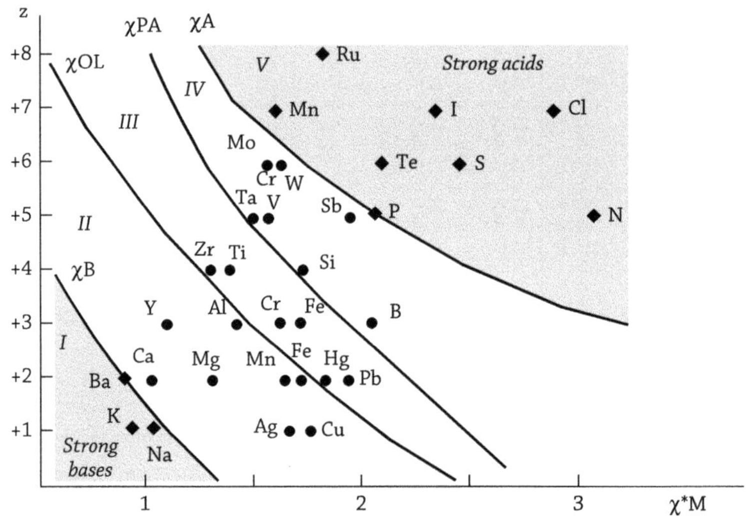

FIGURE 3.4 Charge–electronegativity diagram showing the five classes of behavior for the noncharged form, MO_NH_{2N-z}, of a M^{z+} cation. In zones I and V, the element remains monomeric and soluble. Elements in zone II condense by olation only, and those in zone IV by oxolation. Both condensation processes may be simultaneous for the elements present in zone III. The lines delimiting the different zones have been calculated using the partial charge model [6].

of various condensation degrees. Some transition elements with a high formal charge (V^{V}, Cr^{VI}, Mo^{VI}, W^{VI}) are strongly polarizing and form oxopolyanions. Some of these elements (V^{V}, Mo^{VI}, W^{VI}) may also form solids after aging (hours, days) of solutions, but the polyanions are not implicated in the formation of solids. Such formation seems always to require at least one olation step between hydroxoaquo monomer or oligomer species. These last-named species result from decondensation of the polyanion in acidic conditions (corresponding formally to the acidity range of existence of the noncharged species). Some examples concerning hydrates of vanadium and tungsten oxides are discussed in § 4.1.5. Similarly, antimony Sb^{V} and tin Sn^{IV} form polyacids and oxides by oxolation, but one or several olation steps are involved because the noncharged precursors, $Sb(OH)_5(OH_2)$ and $Sn(OH)_4(OH_2)_2$, contain aquo ligands (see § 4.1.4). Although silicic acid $Si(OH)_4$ condense only by oxolation, it forms a hydrated silica in which terminal hydroxo ligands exhibit a strong acid character. In these different cases, the solid may be described as high-molecular-weight polyanions/polyacids. In the case of boron, B^{III}, although the hydroxo ligands in boric acid $B(OH)_3$ bear a negative partial charge, it does not condense in solution because the partial charge on boron is too low. The monomeric form $B(OH)_3$ is stable in solution, and the solid phase obtained after drying is made of monomeric units connected by hydrogen bonds. The 3-coordination of boron in boric acid is stabilized by a strong π character of the B–OH bonds. In alkaline medium, boron in borates adopts tetrahedral coordination and forms polyanions (§ 3.3.1).

3.2 Olation: Formation of Polycations

Condensation of cations by olation proceeds through the formation of hydroxo bridges (*ol* bridges). The OH ligand of a complex acts as a nucleophilic entering group in the substitution of a water molecule in another complex. In passing from a terminal ligand in a monomer complex to a bridging ligand in condensed species, the OH ligand undergoes only a change in coordination, according to the following scheme:

$$-M-OH + -M-OH_2 \rightarrow -M-OH-M- + H_2O$$

This mechanism, which requires the existence of an aquo leaving group, thus concerns cationic or noncharged aquohydroxo $[M(OH)_h(OH_2)_{(N-h)}]^{(z-h)+}$ complexes ($h \leq z$), formed by elements of the oxidation state (formal charge) lower than or equal to 4. Condensation leads to soluble polycations for $h < z$ and to solids for $h = z$.

3.2.1 MECHANISMS AND STRUCTURAL CONSIDERATIONS

The olation reaction involving elimination of an aquo ligand is kinetically controlled by its lability. The lability of the aquo ligand is generally high. It increases when the formal charge z of the cation decreases and its size increases, owing to low polarization exerted by the cation [7, 8] (Table 2.1). Then, olation proceeds via a dissociative (SN_1) mechanism involving a reduced-coordination intermediate (Fig. 3.5). The activation energy is low, and the olation reaction is usually very fast, being practically diffusion-controlled.

The mechanism may be modified, however, by factors such as the presence of strongly complexing ligands, as well as steric or electronic effects. That is the case for transition elements, for which crystal field stabilization energy (CFSE) in octahedral symmetry is at maximum (Cr^{3+} d^3, Ni^{2+} d^8). Such cations form low-reactive aquocomplexes, inert toward substitution [2]. The large variation in CFSE resulting from symmetry change during substitution contributes to the activation

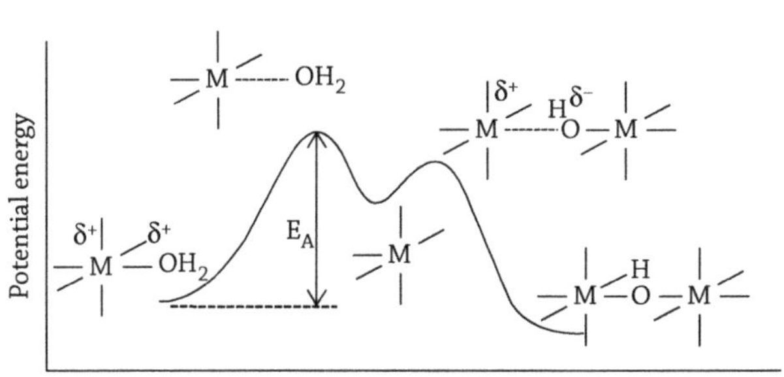

FIGURE 3.5 Scheme of the energy profile in dissociative mechanism of a nucleophilic substitution. E_A is the activation energy of the reaction.

energy of the reaction. So, for the formation of a pentacoordinated transient species of C_{4v} symmetry:

$$\begin{array}{lccc} & \left[M(OH_2)_6\right]^{3+} & \rightarrow \left[M(OH_2)_5\right]^{3+} & + H_2O \\ \text{Symmetry} & O_h & C_{4v} & \end{array}$$

the variation in CFSE is written as $\delta_{CFSE} = CFSE_{C4v} - CFSE_{Oh}$.

For Cr^{3+}, $\delta_{CFSE} = + 0.2\Delta$, ($\Delta$ is the splitting energy of the *d* orbitals of the cation in octahedral symmetry; see Fig. 2.5), and the rate constant of the above reaction is approximately 10^{-6} s^{-1}. By comparison, for Fe^{3+} (d^5 high spin), $\delta_{ESCC} = 0$. Consequently, the Fe–OH_2 bond is very labile, and the reaction rate constant is approximately 10^2 s^{-1}, allowing an SN_1 substitution mechanism. For the elements strongly stabilized in octahedral symmetry and inert to substitution, the condensation mechanism is more complex. Typically, the structure of condensed species of Co^{III} (electron configuration d^6 low spin) and Cr^{III} (d^3) reveals the existence of the bridging ligand $[H_3O_2]^-$ [9] (Fig. 3.6).

This ligand is formed by a hydrogen bond between the hydroxo ligand of one cation and an aquo ligand of the other. The existence of such a $[H_3O_2]^-$ ligand, albeit temporary, shows that condensation for these elements proceeds according to

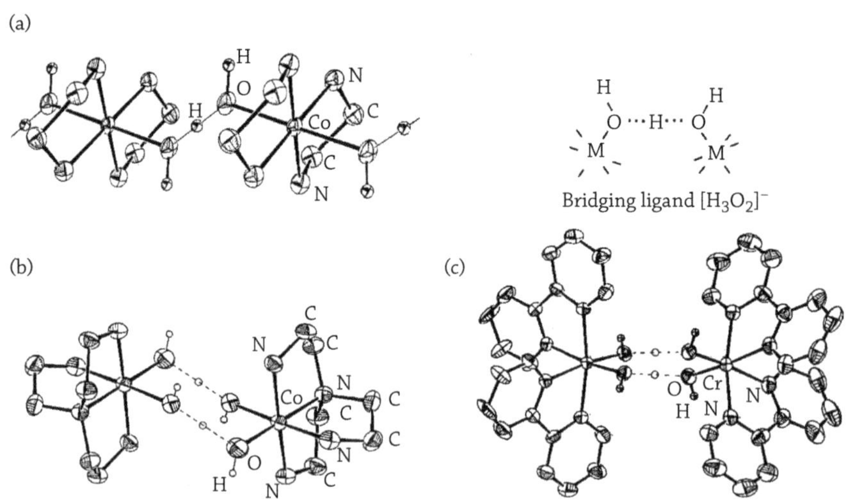

FIGURE 3.6 (a) Chain fragment of trans-$[Co(en)_2(H_3O_2)]_n^{2n+}$,2n ClO_4^- including the $[H_3O_2]^-$ ligand in which the proton appears approximately in a central position along the O-O axis (distances O-O: 2,44 Å, O-H: 1,22 Å, O-$H_{nonbridging}$ about 0,95 Å). Crystal structure of $[(tren)Co(H_3O_2)_2\text{-}Co(tren)]^{4+}$ (b) and Cis$[(bipy)_2Cr(H_3O_2)_2Cr(bipy)_2]^{4+}$ (c) [en: ethylenediamine; bipy: bipyridine; tren: triaminotriethylamine] (reproduced with permission from [Ardon M., Bino A., Role of the H_3O_2 bridging ligand in coordination chemistry. 1. Structure of hydroxoaquametal ions, Inorg. Chem. 1985, 24, 1343–1347] Copyright 1985 American Chemical Society).

an associative mechanism, with formation of an outer-sphere complex or an ion pair in the transition state:

$$M{-}OH + H_2O - M \rightarrow M{-}\overset{\displaystyle H}{\overset{|}{O}} \cdots H{-}\underset{\displaystyle H}{\underset{|}{O}}{-}M$$

The $[H_3O_2]^-$ bridge has a long lifetime because of the chemical inertness of the cation, but it is not very stable. With more reactive cations, the OH bridge is quasi instantaneously formed, and such intermediate species cannot be isolated. Molecular dynamic calculations show that the formation of an $[H_3O_2]$ bridge leads to the formation of a second bridge (edge sharing), according to an autocatalytic process (one bond favors the formation of another) [10]. The subsequent evolution toward the di-μOH complex occurs in involving the elimination of an axial or equatorial water molecule coordinated to one of the cations. That leads to deprotonation and to breaking one of the $[H_3O_2]$ bridges with formation of a transient structure in which the OH ligand completes the coordination of the undercoordinated cation (Fig. 3.7).

Condensation of monohydroxylated species may lead to chains of corner-shared polyhedra, or preferentially to dimers in which polyhedra are linked by a double *ol* bridge (edge sharing) resulting from intramolecular olation after the formation of the first hydroxo bridge. If a bridging OH ligand keeps a nucleophilic character high enough, the same mechanism may occur with a third cation to form a μ_3–OH bridge. However, the presence of bulky complexing ligands, through steric effects, may limit the edge sharing between coordination polyhedra (Fig. 3.6).

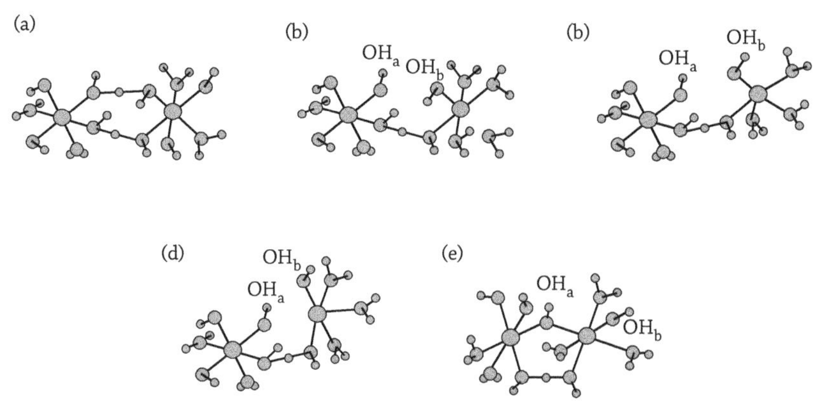

FIGURE 3.7 Scheme of the transformation of a M-H_3O_2-M bridge (a) into a M-OH-M bridge (e); (b) loss of a water molecule and deprotonation of the H_3O_2 bridge; (c) formation of a structure with the OH_a-OH_b contact; (d) rotation of OH_b and nucleophilic attack of OH_a on the undercoordinated cation (reproduced with permission from [10] Copyright 2006 American Chemical Society).

The structure of condensed species is conditioned by several factors:

- *The hydroxylation ratio, h*, corresponds to the number of ligands forming simple, double, or triple *ol* bridges. It represents, in a first approach, the functionality of the cation throughout the condensation process.
- *The geometry of the coordination polyhedra* limits the type of assembling because the sharing of corners, edges, or faces shortens the distance between cations. Edge sharing is the most common for octahedra. Face sharing is rather rare. It exists in some aminohydroxo complexes of Co^{III} ($[Co_2(OH)_3(NH_3)_6]^{3+}$, $[Co_3(OH)_6(NH_3)_6]^{3+}$) (Fig. 3.8). However, these complexes do not form by hydroxylation because of the chemical inertness of Co^{3+} ion (d^6 low spin); rather, they result from oxidation of Co^{II} complexes [2, 11]. Face-sharing octahedra is a characteristic of oxides with the corundum structure (M_2O_3) formed by some trivalent elements.
- *Cyclization* stabilizes the corner-sharing μ_2–OH bridge. This bridge, which usually is weakly stable, is found in a few polycations of bi- and trivalent elements, for instance, Be^{3+} and Au^{3+} [4]. The $[Bi_6(OH)_{12}(OH_2)_x]^{6+}$ polycation of Bi^{3+} (Fig. 3.8) is an interesting species consisting of single hydroxo bridges forming the edges of an octahedra of bismuth atoms. Metal–metal interactions ($d_{Bi\text{-}Bi} = 3.71$ Å) likely contribute to the stabilization of such geometry [12]. It is perhaps the same type of interaction in the Pb^{2+} polycation, $[Pb_6O(OH)_6(OH_2)_4]^{4+}$ (Fig. 3.8), in which one of the distances, 3.44 Å, is very short, whereas the others vary from 3.67 to 4.09 Å [13]. In the absence of such factors, the single corner-sharing bridge is weakly stable, and edge sharing is much more frequent. For entropic and structural reasons, the double hydroxo bridge in discrete species appears to be much more likely than a linear chain condensation leading to high-molecular-weight species.

Condensation of cationic $[M(OH)_h(OH_2)_{N-h}]^{(z-h)+}$ complexes ($h < z$) is always limited and stops spontaneously when a finite condensation degree is reached. The condensed species remain soluble in form of polycations, which usually are of molecular size. As the degree of condensation increases, the accumulation of electrostatic charges on the condensed species, as well as the change in composition due to dehydration, lowers the nucleophilic character of the hydroxo ligands. The average electronegativity of the polycation increases, which raises the charges $\delta(OH)$ and

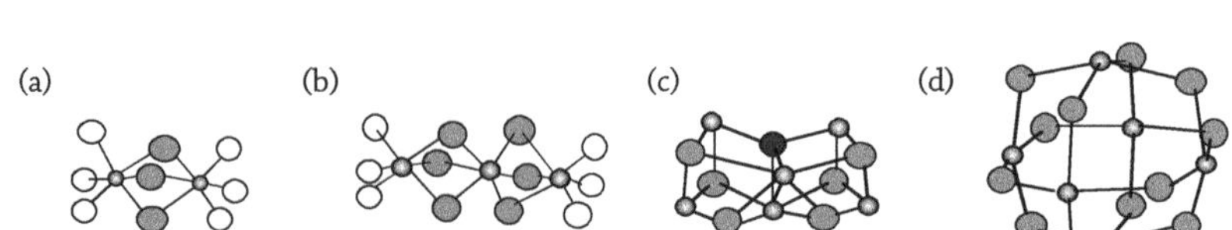

FIGURE 3.8 Structures of some polycations with different connections of coordination polyhedra: (a) $[Co_2(OH)_3(NH_3)_6]^{3+}$; (b) $[Co_3(OH)_6(NH_3)_6]^{3+}$; (c) $[Pb_6O(OH)_6(OH_2)_4]^{4+}$; (d) $[Bi_6(OH)_{12}(OH_2)_x]^{6+}$.

$\delta(M)$. The condensation stops as soon as the condensation criteria are no longer fulfilled. Then the size of polycations increases with the hydroxylation ratio because the hydroxo ligands, which are more numerous in less charged complexes, retain their nucleophilicity until more advanced condensation stages are reached. For an example, see the case of chromium (III) discussed in the following section.

3.2.2 CHROMIUM III POLYCATIONS

Although the study of condensation of Cr^{III} in solution is complicated because of the slowness of reactions, the chemical inertness is beneficial for the separation and analysis of the formed species. Several species have been isolated on ion-exchange resin from solution hydroxylated at $h = HO^-/Cr \approx 1$ at a pH close to 4. These species are oligomers with acidity constants reported in Table 3.1.

The first step of Cr^{III} condensation is the formation of outer-sphere binuclear complexes, involving the bidentate $[H_3O_2]^-$ ligand formed by an intramolecular hydrogen bond (Fig. 3.9). Elimination of a water molecule forms the dimer in which the polyhedra are linked by corner. Elimination of a second water molecule leads to the di-μ-OH complex where the coordination octahedra share an edge [14, 15]. These transformations are associated with a change in color: the precursor ($h = 0$) and the hydrated binuclear complex ($h = 1$) are green, and the di-μ-hydroxo complex is blue [16]. The dimers–monomer equilibria have been thermodynamically characterized [17].

For the monomer $h = 1$, $[Cr(OH)(OH_2)_5]^{2+}$, we have: $\chi = 2.69$, $\delta(Cr) = +0.64$, $\delta(OH) = -0.02$; and for the dimer, $[Cr_2(OH)_2(OH_2)_8]^{4+}$ we have: $\chi = 2.73$, $\delta(Cr) = +0.66$, $\delta(OH) = +0.01$. Thus, the condensation of the $h = 1$ complex stops at the dimer stage because the hydroxo bridging groups have lost their nucleophilicity. The crystal structure of the dimer has been resolved with X-ray diffraction on monocrystals that are obtained by dissolution of hydroxide in paratoluene sulfonic acid or in mesitylene-2 sulfonic acid [18]. The acidity of the dimer $[Cr_2(OH)_2(OH_2)_8]^{4+}$ is stronger than that of monomer (Table 3.1) because of

TABLE 3.1 } Acidity of chromium (III) polycations. (Source: Stünzi H., Spiccia L., Rotzinger F.P., Marty W., Early stages of the hydrolysis of chromium (III) in aqueous solution. 4. Stability constants of the hydrolytic dimer, trimer and tetramer at 25 °C and I = 1.0 M, Inorg. Chem. 1989, 28, 66–71.) 19]).

Complexes	pK
$[Cr(OH_2)_6]^{3+}$	4.3
$[Cr_2(OH)_2(OH_2)_8]^{4+}$	3.7
$[Cr_3(OH)_2(OH_2)_9]^{5+}$	4.3
$[Cr_4(OH)_6(OH_2)_{10}]^{6+}$	3.5
$[Cr_4O(OH)_5(OH_2)_{10}]^{5+}$	0.9

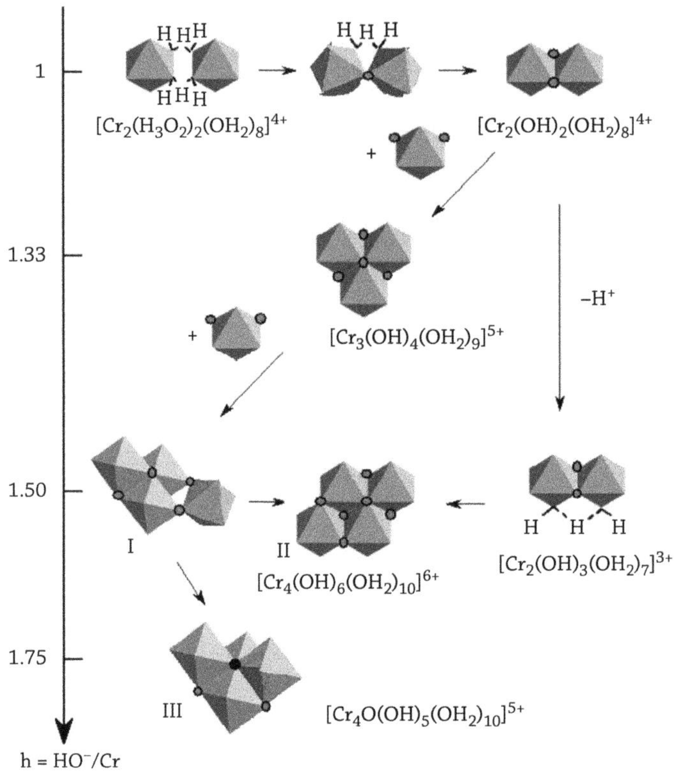

FIGURE 3.9 Scheme of interconversions of Cr^{III} polycations in aqueous solution as a function of the hydroxylation ratio *h*.

an intramolecular hydrogen bond between two water molecules in *cis* configuration with respect to the equatorial plane of the polycation. This bond stabilizes the deprotonated $[Cr_2(OH)_3(OH_2)_7]^{3+}$ form ($h = 1.5$) with the bridging $[H_3O_2]^-$ ligand (Fig. 3.9). The same exists for the *cis*-form $[Cr_2(OH)_2(OH_2)_2(NH_3)_6]^{4+}$, which is more acidic than the *trans*-configuration form for which an intramolecular hydrogen bond cannot be developed [16].

The deprotonated dimer $[Cr_2(OH)_3(OH_2)_7]^{3+}$ may form other polycations because the decrease of its charge reinforces the nucleophilicity of the hydroxo groups ($\chi = 2.68$, $\delta(OH) = -0.03$). It may react in several ways:

- With the $h = 1$ species, it forms the trimer $[Cr_3(OH)_4(OH_2)_9]^{5+}$, including three μ_2–OH bridges and a central μ_3–OH bridge (Fig. 3.9). This species, for which $\chi = 2.72$, $\delta(Cr) = +0.66$, $\delta(OH) = +0.01$, is stable. It is also the most chemically inert of all chromium polycations [19–22].
- It may dimerize to form the planar tetramer $[Cr_4(OH)_6(OH_2)_{10}]^{6+}$ (structure II, Fig. 3.9). The same planar tetramer may also be formed from the trimer $[Cr_3(OH)_4(OH_2)_9]^{5+}$ and a monomer $h = 2$ via an intermediate form $[Cr_4(OH)_6(OH_2)_{11}]^{6+}$ (form I, Fig. 3.9) [21], which

undergoes an internal olation. At higher pH, the deprotonation of the μ_3–OH bridge in form I leads, by internal oxolation, to the tetramer $[Cr_4O(OH)_5(OH_2)_{10}]^{5+}$ (structure III; Fig. 3.9). Kinetics studies have indeed shown that interconversion between the (II) and (III) forms is dependent on the acidity of the medium [23]. This structural analysis is confirmed by the fact that acidification of the tetramer leads to the mixture monomer–trimer, regardless of the way it was formed. The tetramer synthesized by reaction of the ^{51}Cr-tagged monomer with a large excess of nontagged trimer (to prevent the formation of tetramer from the monomer alone) yields, after separation in an ion-exchange resin and degradation in acid medium, a mixture of monomer and trimer, both of which contain the same amount of ^{51}Cr [21]. This result allows us to exclude other configurations that could not permit ^{51}Cr exchange, such as:

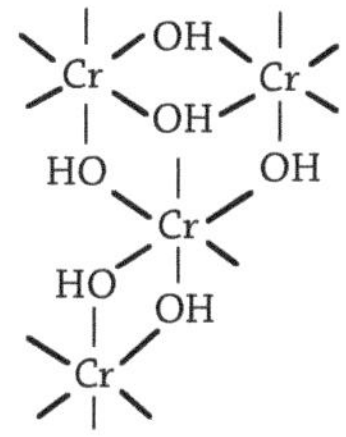

SCHEME 3.2

X-ray absorption spectroscopy of hydroxylated solutions of Cr^{III} has confirmed, via the determination of Cr-Cr distances, the coexistence of tetrameric $[Cr_4(OH)_6(OH_2)_{11}]^{6+}$ (structure I), $[Cr_4(OH)_6(OH_2)_{10}]^{6+}$ (structure II) and $[Cr_4O(OH)_5(OH_2)_{10}]^{5+}$ (structure III) polycations [24].

Similar polycations exist with aluminum (Chap. 6) and very likely with ferric iron as well. The polycations of iron, however, are not as well known as those of chromium and aluminum because of the very high lability of ferric complexes. Only a few polycations (dimer, trimer) have been characterized in acidic solutions ($pH < 1{,}5$) [25]. The existence of aquocomplexes $[(H_2O)_4Fe_2(OH)_2(OH_2)_4]^{4+}$ and $[(H_2O)_5Fe_2O(OH_2)_5]^{4+}$ seems very likely in spite of the lack of structural data. Many polynuclear complexes of Fe^{III} have been characterized, but their stabilization in solution needs coordination with strongly complexing multidentate ligands (Chap. 7).

With regard to the difference seen in electrostatically charged hydroxylated complexes, the condensation of noncharged complexes $M(OH)_z(OH_2)_{N-z}$ ($h = z$) is unlimited and forms a solid, most often as nanosized particles. The increase in the growing object's average electronegativity, which is due solely to dehydration, is never able to sufficiently lower the nucleophilicity of the hydroxo ligands. It results in the precipitation of a solid phase, which in principle is a hydroxide $M(OH)_z$ (Chap. 4). Notably, the polycations are never the germs or nuclei of the solid obtained for higher hydroxylation ratios. In many systems, a structural relationship between the polycations and the solid does not exist. Indeed, the increase in the hydroxylation ratio more often leads to rapid modification and interconversion of labile-condensed species, so that the polycationic structural trace is lost once the precipitation occurs.

3.3 Oxolation: Formation of Polyanions

Oxolation is involved in the formation of oxo bridges between cations when no water molecule is present in their coordination sphere. Therefore, the reaction concerns elements of high formal charge ($z \geq 4$) that exist as oxo or hydroxo anion monomers in diluted alkaline or neutral solutions and that condense by acidification.

The ligands in the anionic oxohydroxo form of an element carry a high negative partial charge. The formation of an oxo bridge between such entities in which the coordination of the cation is at maximum requires a substitution and therefore the formation of a water molecule as a leaving group. Since this water molecule does not exist in the coordination sphere of the partners, oxolation likely involves a two-step associative mechanism SN_2 (Fig. 3.10): (1) nucleophilic addition with formation of an *ol* bridge leading to a transition state with an increased coordination; and (2) proton transfer from the *ol* bridge to a terminal hydroxo ligand OH forming a leaving aquo ligand.

The eliminated water preexists in the coordination sphere of the cations as hydroxo ligands and not as coordination water. This makes the oxolation rate strongly sensitive to the acidity of the medium. The reaction may be catalyzed, and from a kinetics standpoint, one may distinguish basic, acid, and neutral oxolation. Neutral oxolation is usually the slowest of the three, as evidenced by the behavior of silicates: their condensation rate is 10 times faster at pH 1 than at pH 2, the slowest at pH 3, and 100 times faster at pH 6 than at pH 4 (Fig. 3.11).

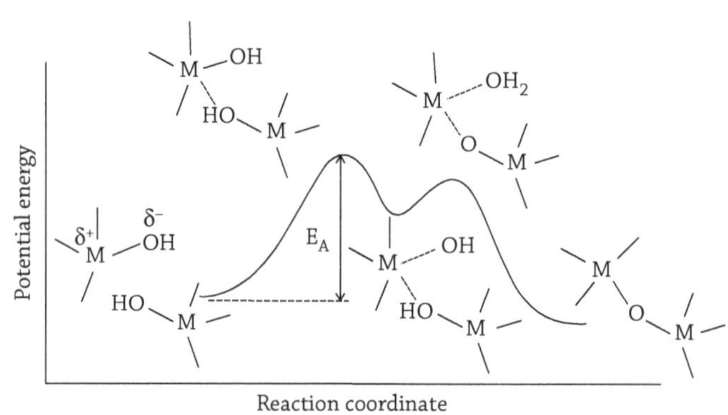

FIGURE 3.10 Scheme of the energy profile of an associative mechanism of nucleophilic substitution. E_A is the activation energy of the reaction.

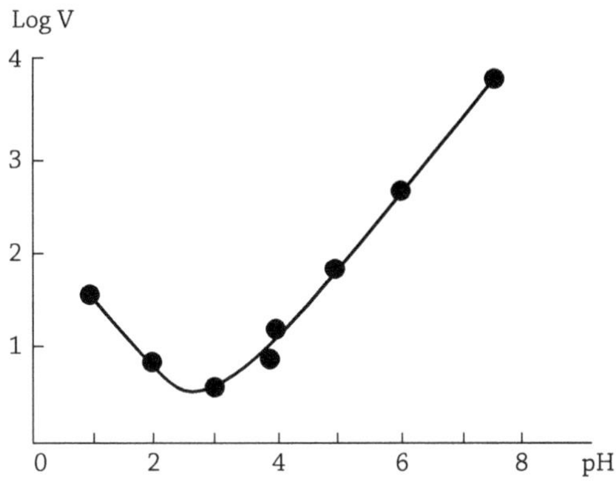

FIGURE 3.11 Rate of disappearance of monomeric silicic acid $Si(OH)_4$ as a function of pH (6 g.l^{-1} SiO_2, 20°C) (data from Coudurier M., Baudru B., Donnet J.B., Etude de la polycondensation de l'acide disilicique, Bull. Soc. Chim. Fr. 1971, 9, 3147–3165).

Basic catalysis is due to the reaction of hydroxyl ions with positively charged centers of reactants such as the cations themselves or the protons of hydroxo ligands [26]. For the oxohydroxo and hydroxo forms of cations that have a saturated coordination, they are the protons of the hydroxo group which are attacked by hydroxyl ions. That increases the negative charge of nucleophilic oxygen, thereby easing the first step of oxolation. This process is illustrated by the following concerted mechanism involving hydroxyl ions and water molecules:

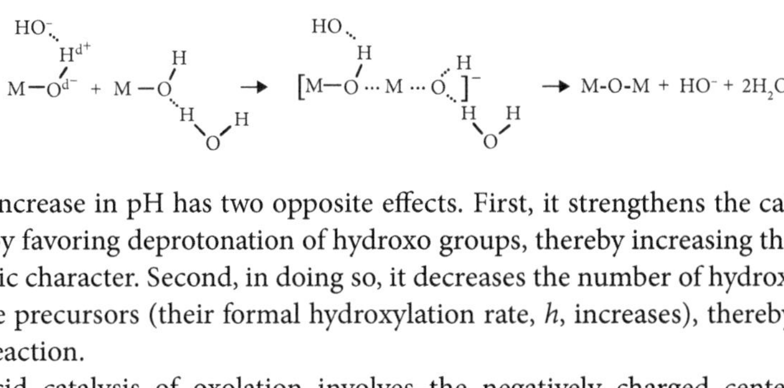

The increase in pH has two opposite effects. First, it strengthens the catalysis effect by favoring deprotonation of hydroxo groups, thereby increasing their nucleophilic character. Second, in doing so, it decreases the number of hydroxo ligands in the precursors (their formal hydroxylation rate, *h*, increases), thereby limiting the reaction.

Acid catalysis of oxolation involves the negatively charged centers of the reactants. Protonation of hydroxo ligands facilitates the formation of an aquo ligand, which strengthens the acidity of the *ol* bridge. The second step of the mechanism is therefore favored:

If the acidity of the medium increases, the formation of positively charged aquohydroxo species is possible. These species are even more prone to nucleophilic attack by hydroxo forms and to proton elimination, allowing a decrease in the charge of the transition state. However, an acid excess inhibits the nucleophilic character of hydroxo ligands and therefore limits the reaction.

Oxolation may occur on a wide range of acidity. The reaction rate will be at a minimum when the reacting species are uncharged (neutral oxolation). Such charge effects do not occur in olation because (in the absence of special electronic effects) the condensation rate is limited by the lability of the coordinated water molecule.

As for olation, condensation by oxolation of negatively charged species is always limited.

- Condensation may stop because all hydroxo ligands have reacted and there is no more leaving group on the growing species. Condensation is limited by the acidification rate of the precursor complexes, as, for instance, for vanadates acidified at $H^+/V = 1$, where the reaction stops at the formation of pyrovanadic dimers:

$$2\left[VO_3(OH)\right]^{2-} \rightarrow \left[V_2O_7\right]^{4-} + H_2O$$

Further condensation of pyrovanadates requires higher acidification (see § 3.3.2).

- Hydroxo ligands of a condensed species may become too acid and therefore dissociate. Condensation stops because of the lack of a leaving group. That is typically the case of chromate $[CrO_4]^{2-}$. It forms bichromate $[Cr_2O_7]^{2-}$ by acidification, but condensation does not progress because the chromic acid $Cr_2O_5(OH)_2$ is a strong acid, always dissociated in solution. A similar situation is observed with other transition elements that have a high oxidation state (see later). Condensation leads to an increase in the average electronegativity of condensed species—hence an increase in the partial charge of hydroxo ligands and the loss of their nucleophilic character.

Some illustrative examples are presented next.

3.3.1 ELEMENTS OF THE *p*-BLOCK

a) Silicates

The chemistry of silicates in solution is highly complicated because of the large number of species and the complexity of the involved equilibria [27, 28]. In most silicon oxygenated compounds, the basic structural unit is the tetrahedra SiO_4, which may share corners but not edges. Sharing edges would give rise to exceedingly high Si · · · Si repulsions. Silicate solutions are obtained through dissolution of metasilicate Na_2SiO_3, which is made of infinite chains of SiO_4 tetrahedra [29]. These chains break in water almost immediately, giving way to fast dissolution. The monomeric silicic forms are summarized in Table 3.2.

TABLE 3.2 } Some characteristics of monomeric silicates in solution.

Form	pK	χ	δ(OH)	δ(Si)
$[SiO_2(OH)_2]^{2-}$		2.10	−0.55	+0.20
	13			
$[SiO(OH)_3]^-$		2.37	−0.30	+0.35
	9.9			
$[Si(OH)_4]^0$		2.58	−0.12	+0.47
	2			
$[Si(OH)_3(OH)_2]^+$		2.74	+0.03	+0.56

The $[SiO_2(OH)_2]^{2-}$ species is a strong base that exists in aqueous solution only in highly alkaline medium. Although the basicity of hydroxo groups is very high, this species does not condense because of the too low partial charge of silicon that occurs following an intense ligand–cation electron transfer.

The decrease in basicity of the medium forms $[SiO(OH)_3]^-$ species in which the partial charge of the cation increases because of the presence of three hydroxo groups. At low concentration, this species does not condense with itself, but it may undergo the nucleophilic attack of coexisting $[SiO_2(OH)_2]^{2-}$ species to form the pyrosilicate made of two corner-sharing tetrahedra:

$$\left[SiO(OH)_3\right]^- + \left[SiO_2(OH)_2\right]^{2-} \rightarrow \left[Si_2O_4(OH)_3\right]^{3-} + H_2O$$

This dimer is stable because of the too low electrophilic character of silicon (χ = 2.183, δ(Si) = +0.25, δ(OH) = −0.22). However, $[SiO(OH)_3]^-$ is able to autocondense:

$$2\left[SiO(OH)_3\right]^- \rightarrow \left[Si_2O_3(OH)_4\right]^{2-} + H_2O$$

to form a dimer for which: χ = 2.346, δ(OH) = −0.33, δ(Si) = +0.34. Therefore, condensation does not stop at this stage. High-resolution ^{29}Si NMR techniques have allowed in situ characterization of polysilicate species in concentrated solutions [30–39]. At around pH ≈ 12, oxolation between $[SiO_2(OH)_2]^{2-}$ and $[SiO(OH)_3]^-$ leads to about 20 anionic oligomers made of corner-sharing tetrahedra, from the dimer to decamers (Fig. 2.22). Polysilicates are mainly cyclic and branched with a connection degree at maximum of 3.

The countercation in solution has a deep influence on the structure of cyclic oligomers. With tetraalkylammonium ions $[NR_4]^+$ (e.g., R = CH_3, C_2H_5, C_3H_7), cages made of double cycles from 3 to 5 SiO_4 tetrahedra have been identified in solution [40, 41]. Tetramethylammonium (TMA) hydroxide promotes formation of the cubic $[Si_8O_{20}H_x]^{(8-x)-}$ octamer (Fig. 3.12), which is indirectly obtained after formation of the prismatic $[Si_6O_{15}H_x]^{(6-x)-}$ hexamer. This last species corresponds to the end of evolution in the presence of tetraethylammonium hydroxide. With tetrabutylammonium

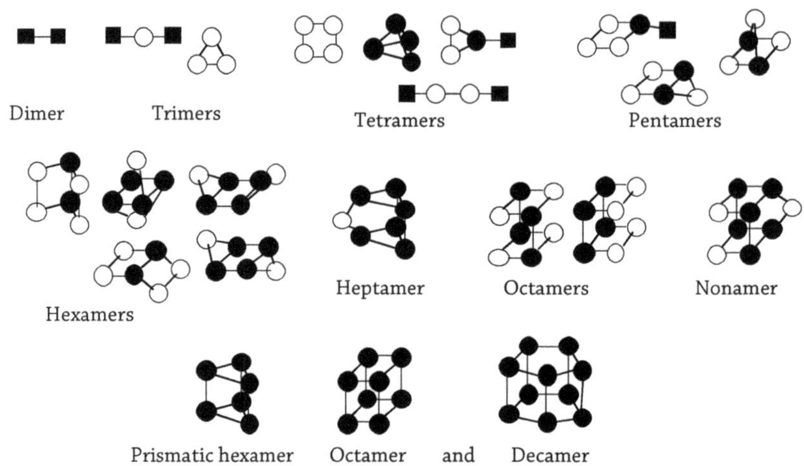

FIGURE 3.12 Main schematized structures of silicates in aqueous solution (silicon concentration 1.5 mol.l^{-1}, KOH/Si = 1, 22 °C). SiO_4 tetrahedra shown using different symbols share 1 (■), 2 (○) or 3 (●) corners. The cages made of double cycles are formed with the presence of tetralkylammonium cations (adapted from [30].)

hydroxide, the decamer $[Si_{10}O_{25}H_x]^{(10-x)-}$ appears to be preferably formed. However, the degree of control decreases as the methyl groups are replaced with larger organic substituents, and it disappears completely when all four alkyl groups are C_4 or bigger [42]. Why does TMA cation favor the Si_8 and Si_6 cagelike species?

It has been observed that the presence of TMA^+ induces anomalous ^{29}Si chemical shifts and relaxation rates in silicic species [42]. In addition, it has been demonstrated that the concentration of TMA^+ explicitly appears in the formation rate equation of the Si_8 anion [43]. That finding proves that the organic cations participate in the formation and subsequent stabilization of these cagelike species. This phenomenon may be understood through considering the effect of TMA^+ on the structure of liquid water in the extended-hydration region of silicate anions. It has been shown that electrolytes disrupt the normal hydrogen-bonded structure of water due to their ability to orient solvent dipoles, a phenomenon known as *hydrophilic hydration* (§ 2.1.3c). On the other hand, ions containing nonpolar alkyl groups induce *hydrophobic hydration*, whereby the hydrogen bond network of water molecules in the immediate hydration region is modified to form short-lived cage systems, which are believed to be comparable to the hydration structures of clathrates (Fig. 2.6b). As the concentration of organocations increases, long-range solute–solute interactions can occur. Hydration cages overlap and mutually enhance one another's stability, leading to regions of microheterogeneity in the solution and, ultimately, to the formation of clathrates. Thus, organocations directly associate with cagelike polyanions to form a protective shell of hydrophobic hydration. Polysilicate species with open or irregular structures are subject to considerable thermal motion, such that any association formed between cations will exist only briefly. On the other hand, the Si_6 and Si_8 anionic cages are fairly rigid and will yield longer-lived TMA^+ assemblies comprising greater numbers of cations. Moreover, the nearly spherical charge distribution on these cages would allow TMA^+ hydration spheres to join all around them, creating continuous (or near-continuous) shells of hydrophobic hydration that isolate the oligomers from their aqueous environment.

b) Borates

Boron is a highly electronegative element ($\chi_B = 2.02$) forming a very small cation B^{3+} (ionic radius 0.27 Å). In aqueous solution, boron exhibits two coordinations: tetrahedral in borate $[B(OH)_4]^-$ and trigonal in boric acid $B(OH)_3$. The pK_a of this acid–base couple is 9.5 [27]. Both forms are stable as monomer—borate at pH > 11, boric acid at pH < 7.

Acidification of borate leads to a reduction in the coordination of boron:

$$\left[B(OH)_4\right]^- + H^+_{solvated} \rightarrow B(OH)_3 + H_2O$$

This explains why boric acid behaves as a Lewis acid acceptor of OH^- ions rather than as a proton donor:

$$B(OH)_3 + OH^- \rightarrow \left[B(OH)_4\right]^-$$

A water molecule in the coordination sphere of boric acid may act as a proton donor:

$$B(OH)_3(OH_2) + H_2O \rightarrow \left[B(OH)_4\right]^- + H_3O^+$$

In fact, liquid ^{11}B-NMR [44, 45] and the solid-state structures [46] show boron with D_{3h} symmetry in boric acid, which leads us to consider the $B(OH)_3$ form. The decrease in coordination of boron during acidification in solution—a phenomenon never observed with any other element—is due to a strong π-donor effect from the three hydroxo ligands, as in the BX_3 halides [2].

The $[B(OH)_4]^-$ ion has the T_d symmetry. Its MO's diagram (Fig. 3.13a) shows that the π-donor character of the ligands is weak and that their π symmetry AOs are essentially nonbonding. Protonation of borate forms a H_2O ligand with extension of the B–O bond and lowering of symmetry ($T_d \rightarrow C_{3v}$). Hence, the water molecule can easily be eliminated from the coordination sphere of the $B(OH)_3(OH_2)$ species. Indeed, the boron ion is too small to retain the 4-coordination ($r_{B3+}/r_{O2-} = 0.19 < 0.225$, the typical value for tetrahedral coordination in the ionic model). In addition, the D_{3h} symmetry allows the three hydroxo ligands to establish $p(\pi) - p(\pi)$ interactions between the empty p_z boron orbital and an AO of π symmetry from each ligand (Fig. 3.13 c). It results in three strong σ bonds (labeled a'_1 and e' Fig. 3.13 b) and in a π bond (labeled a''_2). The delocalized π bonds in boric acid cause a decrease in partial charges on boron and hydroxo ligands compared with a σ system alone. Therefore, condensation of boric acid does not occur in solution as well as in solid state. (This example also explains why the partial charge model cannot be used to explain the stability of this species.)

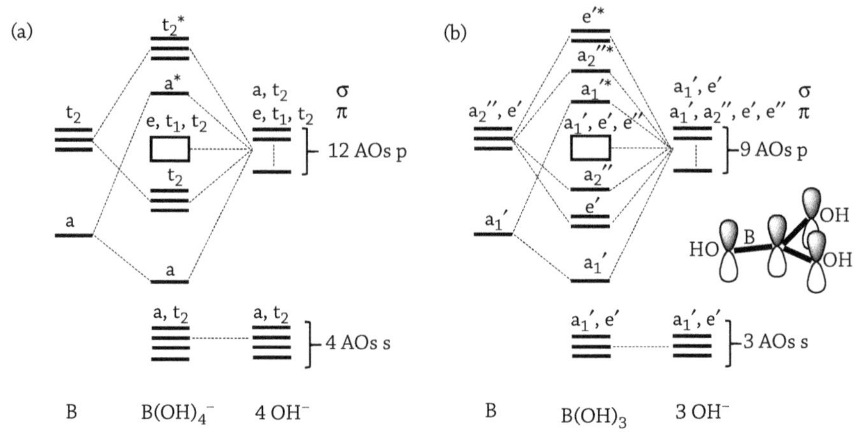

FIGURE 3.13 Schematic MO's energy diagram for (a) $[B(OH)_4]^-$ ion and (b) $B(OH)_3$ molecule. The Lowest Unoccupied Molecular Orbital (LUMO) are the a* and $a_1'^*$ MOs respectively. (c) Scheme of the $B(OH)_3$ molecule showing the $p(\pi)$ orbitals of OH ligand and the p_z boron orbital involved in π overlap.

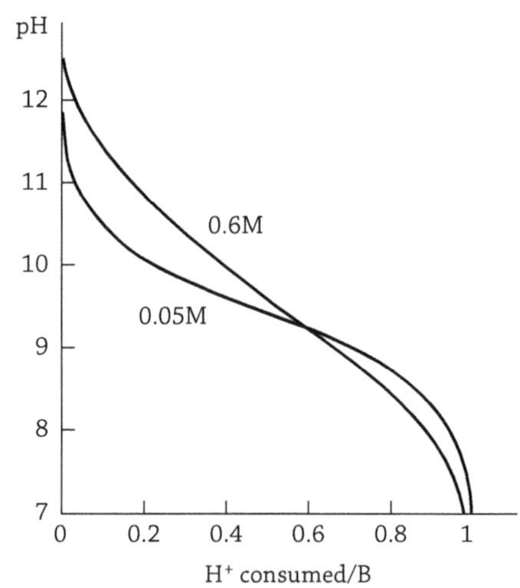

FIGURE 3.14 Protometric titration curves of borate at two concentrations, showing an intersection point corresponding to a maximum in condensation. The curves are adjusted using the formation function, taking into account the amount of protons effectively consumed [47].

Potentiometry measurements [47–49] show that borate and boric acid co-condense. The borate acidification ratio has to be between 0 and 1, so that the two species are coexisting (11 > pH > 7). The intersection of titration curves for two concentrations (Fig. 3.14) shows that condensation experiences a maximum.

Acidification of borate forms a dimer in dilute solution and a tetramer in concentrated solution for $p = H^+_{added} / B = 1/2$, and a trimer for $p = 2/3$.

For $p = 1/2$, the formation of the dimer is written as

$$[B(OH)_4]^- + B(OH)_3 \rightarrow [B_2O(OH)_5]^- + H_2O$$

At high concentration (0.4 mol l^{-1}), the dimer itself dimerizes to form the cyclic tetramer $[B_4O_5(OH)_2]^{2-}$ (Fig. 3.15):

$$2[B_2O(OH)_5]^- \rightarrow [B_4O_5(OH)_2]^{2-} + 3\ H_2O$$

For higher acidification ($p = 2/3$), formation of the trimer occurs according to

$$[B(OH)_4]^- + 2B(OH)_3 \rightarrow [B_3O_3(OH)_4]^- + 2\ H_2O$$

This species is stable in solution owing to cyclization (Fig. 3.15).

For $p > 2/3$, decondensation occurs, and monomeric boric acid is predominant in solution. It seems, however, that at around $p = 4/5$, condensation proceeds as far as the pentamer (Fig. 3.15):

$$[B(OH)_4]^- + 4\ B(OH)_3 \rightarrow [B_5O_6(OH)_4]^- + 6\ H_2O$$

The existence of this species, as deduced from ^{11}B-NMR spectroscopy investigations, remains controversial [48, 49]. The lifetime of this species is likely very short since the hydrated potassium pentaborate $K[B_5O_6(OH)_4]\ 2H_2O$ is rapidly hydrolyzed in solution.

In most condensed boron compounds, the basic structure is a six-membered ring (three boron and three oxygen atoms), which always contains one or two

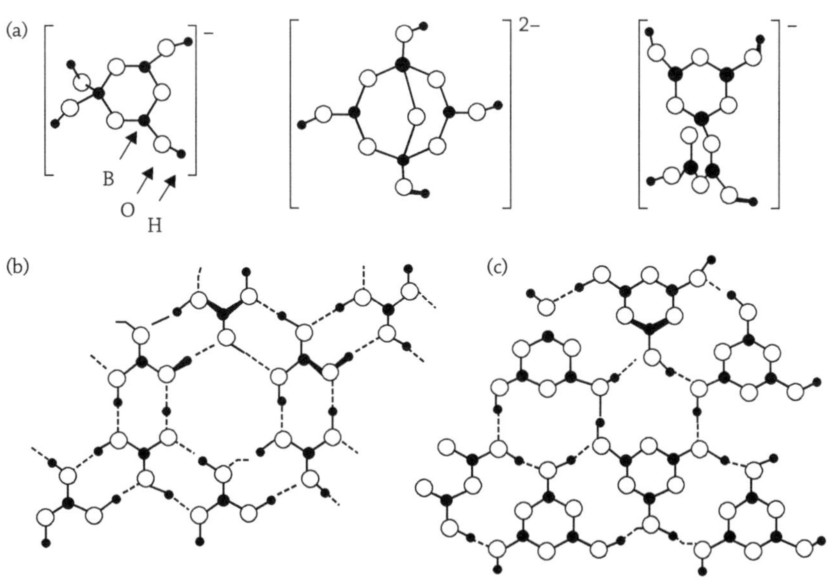

FIGURE 3.15 Schemes of structures of (a) polyborates, (b) boric acid, and (c) metaboric acid.

tetrahedral boron atoms that form connections between rings [29] (Fig. 3.15). Cyclization explains the condensation degrees that occur in solution. In a solid state, boric acid also forms rings of monomers linked by hydrogen bonds. Since boron adopts a trigonal coordination, a layered structure develops (Fig. 3.15 b). Dehydration of trimers beyond 100°C forms metaboric acid, HBO_2, which exists in different crystal structures. The orthorhombic phase obtained by quenching is made of cyclic trimers linked by hydrogen bonds in a layered structure (Fig. 3.15 c). The monoclinic phase is made of chains $[B_3O_4(OH)(OH_2)]_n$ in which 3- and 4-coordinated boron atoms coexist. In the cubic phase, tetrahedral BO_4 groups are linked by hydrogen bonds [29]

Polyborates have important industrial applications: a million tons of borax $Na_2B_4O_5(OH)_4 \cdot 8H_2O$, made of chains of tetramers linked by hydrogen bonds, are produced per year worldwide. This material is widely used in the synthesis of ceramics, glass fibers, enamels, and the like. Boron oxide B_2O_3, made of a partially ordered network of B_2O_3 rings, is used mostly in the fabrication of borosilicate glasses (pyrex). Owing to its small thermal expansion coefficient, pyrex exhibits a good resistance to thermal shocks.

3.3.2 TRANSITION ELEMENTS UNDER HIGH OXIDATION STATES: POLYOXOMETALATES

In their maximum oxidation state, transition elements such as V^V, Cr^{VI}, Mo^{VI}, W^{VI}, and Mn^{VII} of d^0 electron configuration exist in alkaline medium as tetraoxo $[MO_4]^{(8-z)-}$ monomers [27].

These complexes condense by protonation, increasing when the corresponding acid, $[MO_{4-y}(OH)_y]^{(8-z-y)-}$, is weak; that is, the OH ligand is stable in the complex. In the series V^V, Cr^{VI}, Mn^{VII}, the strength of the acid increases as the formal charge of the cation increases and the condensation degree decreases (Table 3.3).
Mn^{VII} exists only as a monomer $[MnO_4]^-$ in permanganate. Polarization of the OH ligand is extremely high, and the permanganic acid, $MnO_3(OH)$, does not exist in molecular form. For the same formal charge, the condensation degree increases in the series Cr^{VI}, Mo^{VI}, W^{VI}. In this column of the periodic classification, the strength of the acid decreases, but the variation in condensation degree cannot be so simply justified inasmuch as the size of the cation increases, allowing the coordination to increase. So, Mo^{VI} and especially W^{VI} do not form condensed species from tetrahedral structural units.

In their tetraoxo form, d^0 transition elements may involve their d and p AOs in π bonds with oxo ligands. There are mainly the e-labeled AOs because the t_2 (d and p) AOs, already involved in σ bonds, may give rise only to weak π overlaps (Fig. 3.16) [2, 50, 51]. The π bonds may be very strong in oxoanions, especially in $[MnO_4]^-$, owing to the high polarizing power of the small-sized and highly charged Mn^{VII} cation. This explains why oxo ligands in $[MnO_4]^-$ are not basic and cannot be protonated even in strongly acidic medium. With larger d^0 cations of less high formal charge, tetrahedral coordination is still favored because the antibonding

TABLE 3.3 } Some characteristics of oxo complexes of transition elements of high oxidation states (d^0 electron configuration).

	$[VO_4]^{3-}$	$[CrO_4]^{2-}$	$[MnO_4]^{-}$	$[MoO_4]^{2-}$	$[WO_4]^{2-}$
pK	13.2	5.6	(−2)	3.9	4
	7.5	0.8	–	3.9	4
r (Å)	0.56	0.52	0.46	0.62	0.67
r_M/r_O [a]	0.40	0.37	0.33	0.44	0.48
N	4–5–6	4	4	4–6	4–6
Condensation [b]	10	2	1	7–8	7–12

[a]The radius of the oxide ion O^{2-} is taken equal to 1.40 Å. The cation radii are those given by R. D. Shannon and C. T. Prewitt [Effective ionic radii in oxides and fluorides, Acta Cryst. B 1969, 25, 925–946].
[b]Highest condensation degree of isopolyanions in aqueous solution.

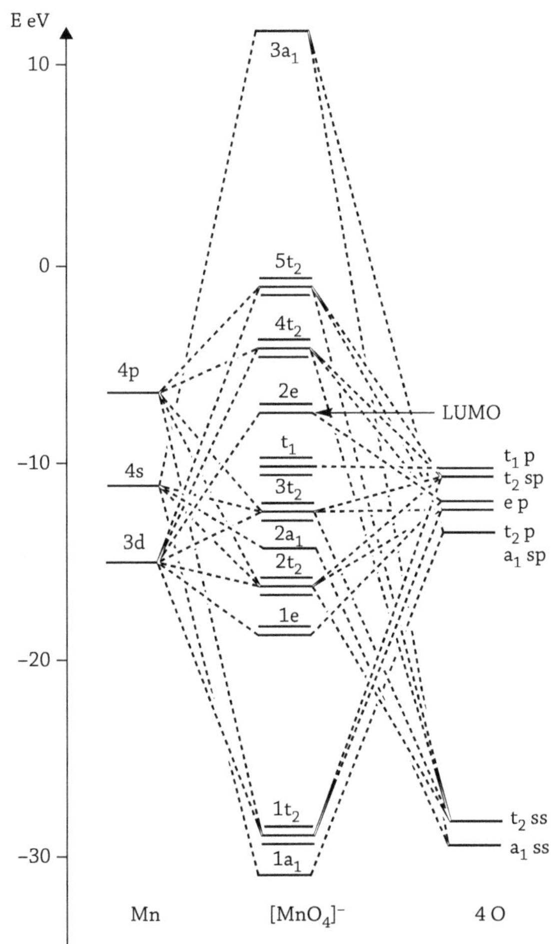

FIGURE 3.16 Molecular orbital energy diagram for $[MnO_4]^{-}$.

MOs are empty. However, the π overlaps are less strong, and the basic character of the oxo ligands is more pronounced. Their protonation becomes possible, and formation of hydroxo ligands further weakens the π transfers. As a result, the electrophilic demand of the largest cations is no longer balanced by strong enough π bonds, giving way to an increase in coordination. This increase may proceed by condensation via tetrahedra addition or by solvation in acidic medium. Then, in the octahedral surroundings of O_h, D_{4h}, or C_{4v} symmetry and even with a single σ bonding system, the cation acquires better ligand field stabilization than occurs in 4-coordination with σ and π bonds. By comparison, Ti^{IV} with a size similar to that of Mo^{VI} and W^{VI}, but with a smaller formal charge, does not involve π bonds. Ti^{IV} remains in 6-coordination in aquohydroxo complexes as in solids in alkaline medium.

The protonation stage for which the coordination change takes place is directly dependent on the size and charge of the cation because the basic character of the oxo ligands decreases when the cation is polarizing. For V^V, the expansion of coordination from 4 to 6 occurs for $H^+/V = 3$ corresponding to $[VO(OH)_3]^0$ species, whereas for Mo^{VI} and W^{VI}, the change takes place earlier, when $1< H^+/M < 2$, corresponding to formation of $[MO_3(OH)]^-$ and $[MO_2(OH)_2]^0$ species. That explains why W^{VI} never forms condensed species containing tetrahedral units. Cr^{VI} and Mn^{VII} are too small to exceed 4-coordination. The change in coordination requires a cation-to-oxygen ionic radii ratio close to 0.414, the value corresponding to the size of octahedral sites in a close compact stacking of oxide ions (Table 3.3). The limitaions of the Chemistry Cr^{VI} and Mn^{VII} ions in solution is likely due to their inability to exceed the 4-coordination.

Expansion of cation coordination may occur through two mechanisms according to the acidity range.

1. The protonated anionic species $[MO_a(OH)_b]^{x-}$ (a + b = 4) may condense through nucleophilic addition (V^V, Mo^{VI}, W^{VI}). Protonation lengthens the M–O bond and lowers its π character. This mechanism leads to edge-sharing polyhedra [52, 53] (Fig. 3.17 a).
2. In uncharged species, the cation may increase its coordination by solvation, that is, by aquo ligand addition (Fig. 3.17 b). With regard to this mechanism, water is a better nucleophilic agent than the OH ligand itself in these species. Such a mechanism explains the formation of vanadium and tungsten oxides, as will be discussed in Chapter 4.

FIGURE 3.17 Scheme of (a) tetrahedra addition during acidification of oxo species, and (b) expansion of coordination by solvation of an uncharged tetrahedral species for transition elements.

a) Isopolyoxometalates

VANADATES V^V

Condensation of vanadates is clearly evidenced in the shift of protometric titration curves at different concentrations [26, 27, 54] (Fig. 3.18). The successive condensation stages of these curves correspond to well-differentiated equivalent points.

From $p = H^+_{consumed}/V = 0$ to 1, the shift of the titration curves is such that $\Delta pH/\Delta \log C > 0$. That corresponds to condensation, and the equivalent point indicates the formation of pyrovanadate:

$$2\left[VO_4\right]^{3-} + 2\,H^+ \rightarrow \left[V_2O_7\right]^{4-} + H_2O$$

or: $$2\left[VO_3(OH)\right]^{2-} \rightarrow \left[V_2O_7\right]^{4-} + H_2O$$

Pyrovanadate is made of two corner-sharing tetrahedra, as for $[P_2O_7]^{4-}$ and $[Cr_2O_7]^{4-}$. At pH ≥ 12, the pyrovanadate ion is stable.

From $p = 1$ to 2 (9 > pH > 6), the $\Delta pH/\Delta \log C$ term is positive, indicating that condensation goes further and that the second equivalent point corresponds to the formation of metavanadates, mainly as tetramer rings $[V_4O_{12}]^{4-}$:

$$4\left[VO_4\right]^{3-} + 8\,H^+ \rightarrow \left[V_4O_{12}\right]^{4-} + 4\,H_2O$$

or: $$4\left[VO_2(OH)_2\right]^{-} \rightarrow \left[V_4O_{12}\right]^{4-} + 4\,H_2O$$

The composition of metavanadate solutions is complex, as indicated by ^{51}V NMR spectroscopy coupled with potentiometric investigations [55]. In alkaline medium,

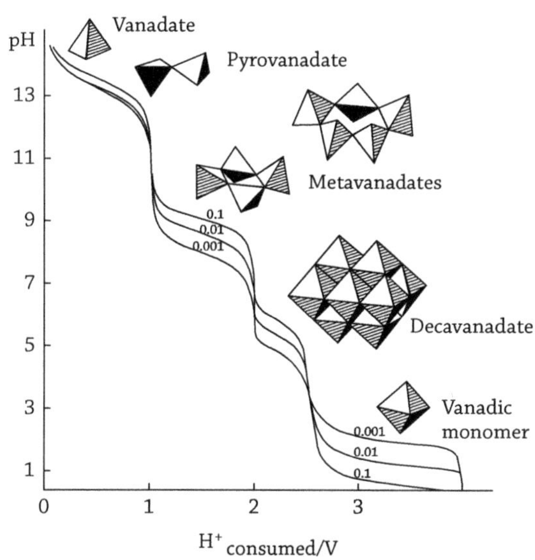

FIGURE 3.18 Titration curves of vanadate at different concentrations (mol.l⁻¹) in solution and structure of the main species in solution.

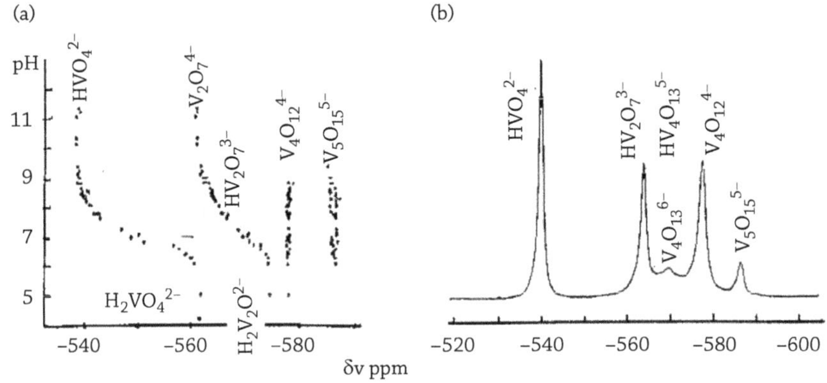

FIGURE 3.19 (a) ^{51}V NMR chemical shifts of the major vanadic species as a function of pH. (b) ^{51}V NMR spectrum of a solution at pH 8.3 ($C_V = 10^{-2}$ mol.l^{-1}) (reproduced from [55] with permission of The Royal Society of Chemistry).

a single peak is assigned to the monomer $[VO_4]^{3-}$ (Fig. 3.19a). During acidification, additional peaks appear (Fig. 3.19b). The displacement of the chemical shift of some of these peaks with the pH indicates the presence of protonated forms of vanadate ($[HVO_4]^{2-}$, $[H_2VO_4]^-$) and of pyrovanadate ($[HV_2O_7]^{3-}$, $[H_2V_2O_7]^{2-}$). These peaks remain single because the protonation equilibria are faster than NMR observation and an average situation (an exchange peak) is observed.

Within the formation range of metavanadate (9 > pH > 6), two peaks are observed, with a chemical shift independent of pH. The corresponding species are not protonable, and the peaks are attributed to the tetramer $[V_4O_{12}]^{4-}$ and to the pentamer $[V_5O_{15}]^{5-}$. The small linewidth suggests high symmetry around structurally identical vanadium atoms forming rings of four and five corner-sharing VO_4 tetrahedra. A small peak at 570 ppm (Fig. 3.19 a), which may be assigned to a linear tetramer $[V_4O_{13}]^{6-}$ and its protonated form $[HV_4O_{13}]^{5-}$, merges at pH < 9, with the underlying broad resonance due to cyclic $[V_4O_{12}]^{4-}$. In the solid state, metavanadates exhibit linear anionic structures: infinite chains of corner-sharing tetrahedra (KVO_3) or distorted trigonal bipyramids (KVO_3, H_2O) in which vanadium is pentacoordinated [56, 57].

By acidification at $p > 2$, from pH 6 to 3, vanadate solutions become orange, and the shift of the titration curves is such that $\Delta pH/\Delta \log C > 0$. Condensation further proceeds, and the equivalent point around pH 3 ($H^+/V = 2.5$) corresponds to the formation of decavanadate:

$$10\left[VO_4\right]^{3-} + 25\,H^+ \rightarrow \left[HV_{10}O_{28}\right]^{5-} + 12\,H_2O$$

$$\text{or:}\quad 5\left[VO_2(OH)_2\right]^{-} + 5\left[VO(OH)_3\right]^{0} \rightarrow \left[HV_{10}O_{28}\right]^{5-} + 12\,H_2O$$

In this acidity range, the equilibria are slow, and the coordination of vanadium increases. The structure of decavanadate consists of 10 edge-sharing VO_6 octahedra (Figs. 3.18 and 3.20). ^{51}V NMR investigations have shown that decavanadates have

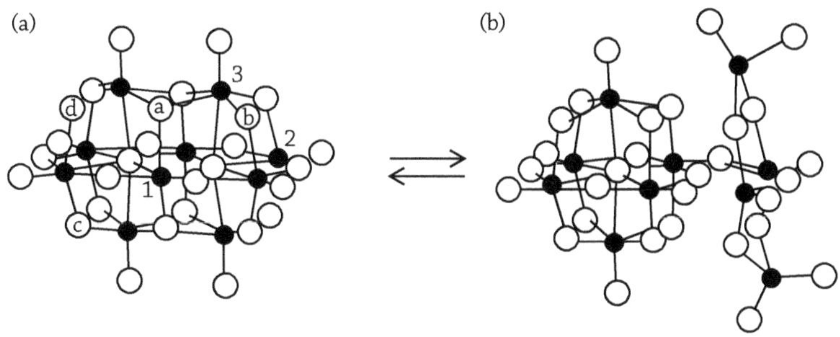

FIGURE 3.20 (a) Structure of $[V_{10}O_{28}]^{6-}$. The protonation site of $[HV_{10}O_{28}]^{5-}$ is labeled *a*; those of $[H_2V_{10}O_{28}]^{4-}$ are labeled *b* and *c*; those of $[H_3V_{10}O_{28}]^{3-}$ in H_2O/CH_3CN medium are labeled *a,b,d*. (b) Scheme of the "half-bonded" intermediate form involved in the oxygen atom exchange.

the same structure in both solution and solid state [58]. Three types of vanadium ions (labeled 1, 2, 3 in Figure 3.20a) exist in 1:2:2 ratios. Their coordination polyhedra share 7, 4, and 5 edges, respectively. Such a compact stacking of octahedra gives rise to important repulsions between vanadium atoms, which are strongly shifted toward the outside of the polyanion. V–O bond lengths vary from 1,60 Å (V – $O_{terminal}$) to 2,32 Å (V – $O_{central}$). The strong distortion of vanadium coordination polyhedra within the polyanion explains the low basicity of the external oxygen atoms. In water, the most protonated form of decavanadate is $[H_2V_{10}O_{28}]^{4-}$ (equivalent point at $H^+/V = 2.6$, Fig. 3.18). In a water/acetonitrile mixture ($\varepsilon_{CH3CN} = 36.2$), which is a less dissociative medium than pure water, the $[H_3V_{10}O_{28}]^{3-}$ form exists. The different protonated sites have been identified by X-ray diffraction and multinuclear NMR (^1H, ^{17}O, ^{51}V) [58]. No oxygen atoms of the very short V = O bonds are protonable. In $[HV_{10}O_{28}]^{5-}$, the oxygen atoms in μ_3 (labeled *a* in Figure 3.20a) appear to be the most basic and one of them is protonated. That causes the stretching of the involved V–O bond (+0,1Å) and an important charge redistribution within the polyanion, so that the sites that are further protonated are different (Fig. 3.20a). $[H_2V_{10}O_{28}]^{4-}$ may easily be deprotonated by the addition of sodium hydroxide in forming $[V_{10}O_{28}]^{6-}$. The slowness of equilibria in this acidity range prevents the formation of metavanadates [54]. Although the decavanadate is a thermodynamically stable species with a compact structure and V–O bonds are strongly distorted, it gives rise to an important fluxionality. All the oxygen atoms of the structure are readily exchangeable with those of water molecules of the solution with similar rates ($6\ h \le t_{1/2} \le 11\ h$ at 298 K) [59]. A ^{17}O NMR study performed with ^{17}O-enriched decavanadate dissolved in natural water (pH 5.5) has shown that the exchange proceeds with formation of a "half-bonded" intermediate species, involving the breaking of some bonds without loss of fragments (in opposition to an "open" mechanism in which the decavanadate would be degraded in two or more fragments) (Fig. 3.20b).

More advanced acidification ($H^+/V > 2.6$) does not permit $[H_2V_{10}O_{28}]^{4-}$ to form the expected protonated species. Within this acidity range, the $\Delta pH/\Delta logC$ ratio is negative (Fig. 3.18), and the system decondenses to release the octahedral vanadic cation $[VO_2(OH_2)_4]^+$ [26, 54] according to the formal equilibrium:

$$\left[H_2V_{10}O_{28}\right]^{4-} + 14\,H_3O^+ + 18\,H_2O \Leftrightarrow 10\left[VO_2\left(OH_2\right)_4\right]^+$$

The titration curve intersection located at $H^+/V = 2.6$ then indicates maximum condensation. The decondensation in this acidity range may be explained by considering the nature of the V^V coordination sphere in the monomer. Because of the high polarizing power of V^V, it appears better stabilized in a dioxoaquo $[VO_2(OH_2)_4]^+$ form involving π bonds than in oxohydroxoaquo $[VO(OH)_2(OH_2)_3]^+$ or hydroxoaquo $[V(OH)_4(OH_2)_2]^+$ forms. Hence, condensation is inhibited because both oxo groups are engaged in very short V–O bonds with neither a nucleophilic nor a basic character.

Decavanadate is the largest isopolyanion formed by V^V in water in the presence of alkaline cations or proton. A change of solvent and/or in the nature of cations may cause the formation of original structures of vanadic polyanions. Refluxing in acetonitrile for 1–2 minutes of decavanadate $H_2V_{10}O_{28}(N^nBu_4)_4$ allows us to crystallize $[CH_3CN{\cdot}V_{12}O_{32}]^{4-}$ anion as a tetraphenylphosphonium salt [60]. The compound is formed of V^VO_5 pyramids and exhibits a surprising structure in the form of an open basket containing in quasi-totality an acetonitrile molecule (Fig. 2.31a). On the other hand, prolonged refluxing of $H_3V_{10}O_{28}(N^nBu_4)_3$ in acetonitrile yields a compact polyanion $[V_{13}O_{34}]^{3-}$ with a structure based on strongly distorted V^VO_6 octahedra [61] (Fig. 3.21b). The structure resembles that of

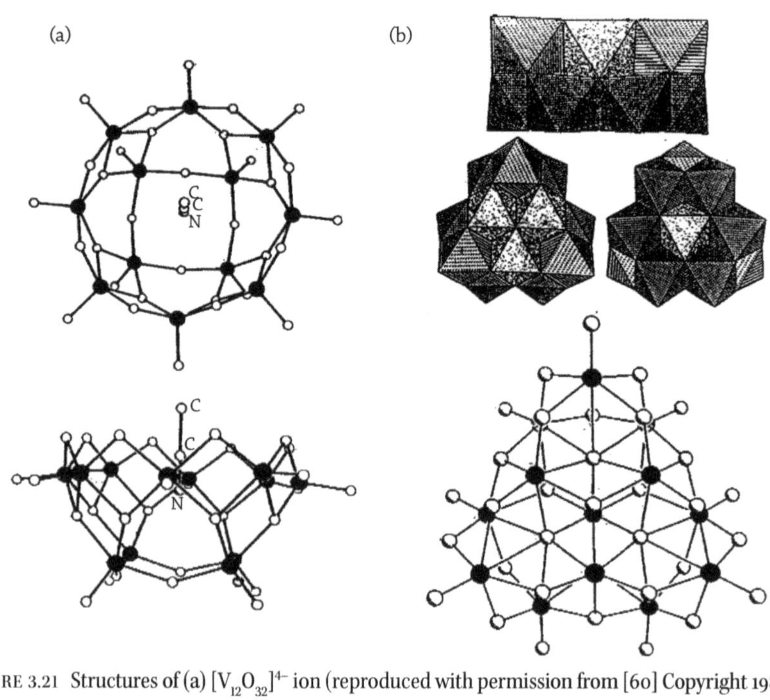

FIGURE 3.21 Structures of (a) $[V_{12}O_{32}]^{4-}$ ion (reproduced with permission from [60] Copyright 1989 American Chemical Society) and (b) $[V_{13}O_{34}]^{3-}$ ion (reproduced with permission from [64] Copyright 1992 American Chemical Society). The vanadium atoms are shown as filled circles.

metatungstate (Fig. 3.24), but the trivanadic groups and the thirteenth vanadium atom form much more compact two-layer stacking.

MOLYBDATES MO^{VI}

At pH > 7, molybdates and tungstates exist in solution as tetraoxo $[MO_4]^{2-}$ ions, forming many polyanions upon acidification. These are complex systems, and so only some of the most stable compounds will be examined here. Molybdates in solution reach their condensation equilibria much faster than tungstates, which may take days or months at RT and form stable polytungstates after the evolution of many intermediate species. Octahedral coordination is always reached in polyanions, at least for polytungstates. A few polymolybdates contain both MoO_4 tetrahedra and MoO_6 octahedra.

The protometric titration curve of molybdates does not exhibit a clearly marked equivalent point, but a quite widespread equivalence zone is evident around H^+/Mo = 1.5 [54] (Fig. 3.22). That indicates the coexistence of many equilibria and species in solution.

The predominant species is a heptamolybdate (paramolybdate) formed according to the formal reaction

$$7\left[MoO_4\right]^{2-} + 8\,H^+ \rightarrow \left[Mo_7O_{24}\right]^{6-} + 4\,H_2O$$

This species has been characterized in many studies [56], but no intermediate between monomer and heptamer has been observed. Between pH 5 and pH 3, this

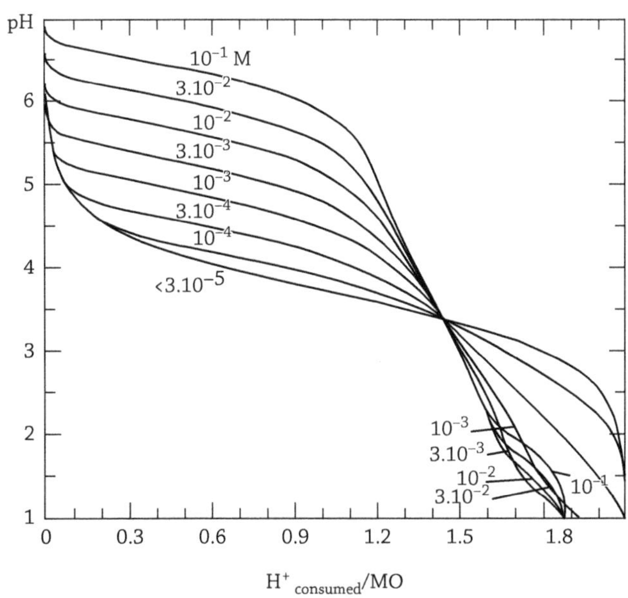

FIGURE 3.22 Protometric titration curve of molybdate $[MoO_4]^{2-}$ at various molybdenum concentrations [29].

species coexists with its protonated forms $[HMo_7O_{24}]^{5-}$ (pK = 4.4), $[H_2Mo_7O_{24}]^{4-}$ (pK = 3.5), $[H_3Mo_7O_{24}]^{3-}$ (pK = 2.5).

The structure of the paramolybdate is made up of edge-sharing octahedra (addition mechanism) (Fig. 3.23). It may be considered as deriving from the structure of the decavanadate $M_{10}O_{28}$, with elimination of three octahedra [62].

The composition of the more acidic solutions (pH 2–3, H^+/Mo = 1.5) remains unclear. It may correspond to a series of protonated heptamolybdates or a mixture of hepta and octamolybdates $[Mo_8O_{26}]^{4-}$ [56]. The latter has been isolated, and its structure (β–$[Mo_8O_{26}]^{4-}$) also derives from the $M_{10}O_{28}$ unit by elimination of two octahedra (Fig. 3.23). An α-isomer of the octamolybdate may be stabilized in aqueous solution in the presence of tetrabutylammonium. This polyanion is formed of a six-member ring of octahedral molybdenum atoms capped on each side with two tetrahedral molybdenum atoms (Fig. 3.23). Addition of K^+ ions leads to α→β isomerization, probably owing to solvent structuring effects (see § 3.3.1a).

In nonaqueous medium (dimethylformamide, dimethyl sulfoxide), acidification of molybdate forms a compact hexacondensed structure $[Mo_6O_{19}]^{2-}$ (Fig. 3.23). This unit is common to many systems (W^{VI}, Nb^V, Ta^V). It contains highly distorted octahedra in which the Mo–O bond lengths vary greatly: Mo–$O_{terminal}$ 1.68 Å, Mo–$O_{bridging}$ 1.93 Å, Mo–$O_{central}$ 2.32 Å [56]. As usual in most polyanions, the cations are strongly pushed toward the outside of the unit.

A large polyanion Mo_{36} has been observed in more acidic medium (H^+/Mo = 1.8) [63] (Fig. 3.23). This polyanion likely represents a minor part of soluble

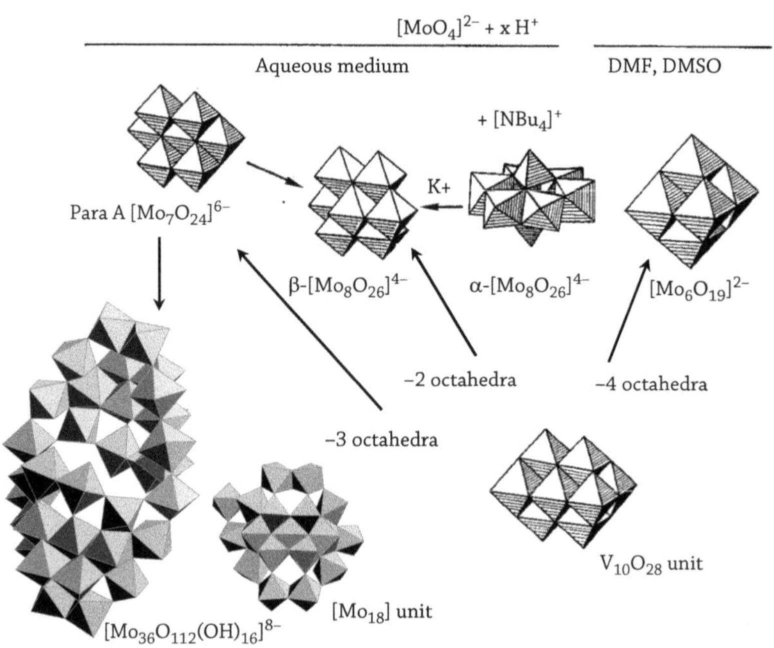

FIGURE 3.23 Structure of some isopolymolybdates.

species. It is made up of two identical Mo_{18} units, each of which is formed of a crown of 11 octahedra surrounding an Mo_7 group in which two molybdenum atoms exhibit a 7-coordination. The two Mo_{18} units are connected by four shared oxygen atoms around an inversion center.

At a higher acidification degree ($H^+/Mo > 2$), soluble polycationic species are formed without the development of solid phases, even after aging. Molybdic hydrates obtained in strongly acidic medium are described in Chapter 4. Upon base addition, all these compounds rapidly decompose to yield the monomer $[MoO_4]^{2-}$.

TUNGSTATES W^{VI}

Acidification of tungstate in aqueous medium forms many polytungstates. Only two are thermodynamically stable. They are

1. The paratungstate B ($n = 12$) formed at around pH 6:

$$12\left[WO_4\right]^{2-} + 14\,H^+ \rightarrow \left[H_2W_{12}O_{42}\right]^{10-} + 6\,H_2O$$

2. The α-metatungstate (also $n = 12$) formed at around pH 4:

$$12\left[WO_4\right]^{2-} + 18\,H^+ \rightarrow \left[H_2W_{12}O_{40}\right]^{6-} + 8\,H_2O$$

These two species represent the final stage in the evolution of acidified solution at $H^+/W = 1.17$ and 1.5, respectively. Paratungstate B crystallizes after 1 to 2 days at RT. α-metatungstate is completely formed in 15 days at 50°C. The structure of these species (Fig. 3.24) contains linear W_3O_{14} and cyclic W_3O_{13} tritungstic groups made up of edge-sharing octahedra. These groups are connected by corner sharing: two linear and two cyclic groups form paratungstate B, and four cyclic groups form α-metatungstate (Keggin's structure) [52, 56]. In both cases, tritungstic groups form a central cavity containing two protons, as indicated by neutron diffraction and ^1H NMR [56]. These protons are bounded to the common oxygen atom in the cyclic groups (μ_3–O bridge). In α-meta, these internal protons are not titratable by a base, nor, because of their nonaccessibility, are they exchangeable with the solvent [64]. On the other hand, those protons of para B are more accessible and are quickly exchanged with water [65]. In the α-meta structure, the four μ_3 oxygen atoms form a tetrahedral cage that may be occupied by various elements such as P^V, Si^{IV}, and Ge^{IV}, to form heteropolyanions [54, 56].

The two polyanions (para B and α-meta) are produced from a series of complex transformations. Only some intermediates have been structurally characterized in solution in using ^{183}W and ^{17}O NMR [66, 67]. The large difference in kinetics (ranging from a few seconds to several weeks) of numerous equilibria makes the system highly complex [54, 56].

Acidification of $[WO_4]^{2-}$ at around pH 6 rapidly forms a heptatungstate (paratungstate A, $[W_7O_{24}]^{6-}$):

$$7\left[WO_4\right]^{2-} + 8\,H^+ \rightarrow \left[W_7O_{24}\right]^{6-} + 4\,H_2O$$

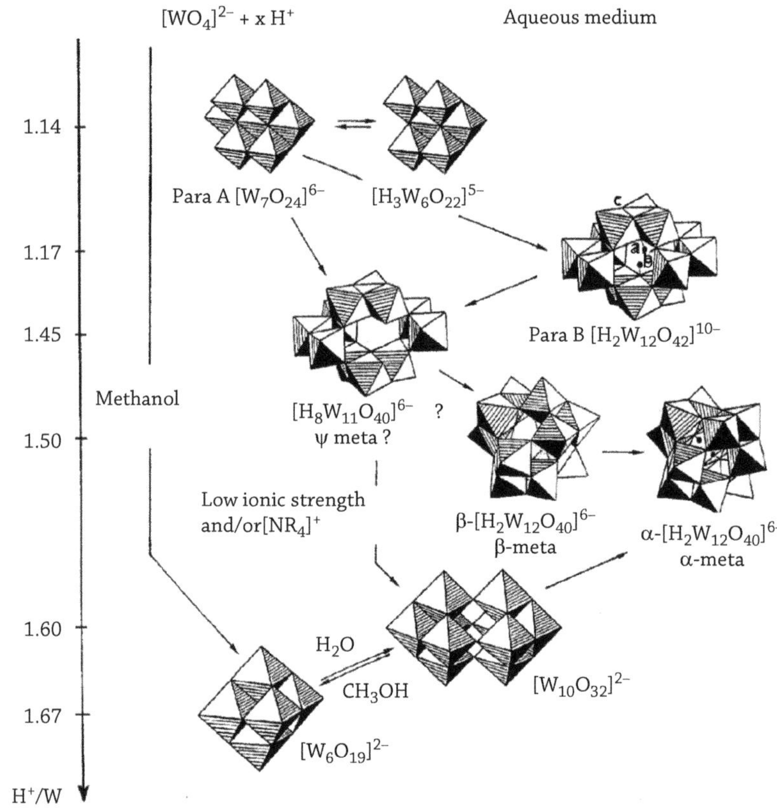

FIGURE 3.24 Structure of some isopolytungstates.

The crystal structure of the polyanion obtained from the piperidinium salt $(C_5H_{10}NH_2)_6[W_7O_{24}]$ [56] is similar to that of the paramolybdate. It has long been assumed that the paratungstate A was a hexatungstate [54]. After a long debate, it was established that, for the ratio 1.15H$^+$/W (8H$^+$/7W), the most likely polyanion was the heptatungstate [68]. Although a hexatungstate $[H_3W_6O_{22}]^{5-}$ has been identified in solution by infrared (IR) and Raman spectroscopies and has been structurally characterized by X-ray diffraction as sodium salt [69] (Fig. 3.24), it seems to be a very minor component in solution. Para A does not protonate at least until pH 5.5 (there is no shift of the NMR signal [67]), and it slowly transforms into para B. This gives us an idea about the number of possible species within such a narrow acidity range and about the complexity of their interconversion equilibria.

The formation of α-metatungstate in more acidic medium is not direct either. It seems to take place from the rapid formation of para A. It transforms itself into an unknown species, the ψ-metatungstate, which would in turn form the β-metatungstate isomer that slowly transforms into the stable α-metatungstate. The α and β isomers differ by a 120° rotation of one of the tritungstic groups around the C_3 axis (Fig 3.24).

^{183}W and ^{17}O NMR investigations show that acidification of concentrated solutions of para B [67] forms $[H_3W_{12}O_{42}]^{9-}$ ion in fixing an extra proton ($pK_a = 4.59$). Previously, we saw that two protons are present on the μ_3–O bridges inside the central cavity of the polyanion (*a* sites). The shift of the NMR signals with acidity shows that the additional protonated sites are mainly μ_2–O bridges of *b* type inside the polyanion and external μ_2–O bridges of *c* type (Fig. 3.24). Several protonated $[H_3W_{12}O_{42}]^{9-}$ isomers are assumed to exist. Some carry three internal protons, with two of them on *b* sites, causing the displacement of one of the protons from the μ_3–O bridges. The same phenomenon is observed with decavanate (Fig. 3.20), in which the second and third protonation stages modify all the protonation sites. Other isomers of $[H_3W_{12}O_{42}]^{9-}$ would carry the third proton in rapid exchange on external μ_2–O sites of *c* type. Interconversion of isomeric forms seems to be limited by the proton exchange rate between the internal and external sites of the polyanion. These metastable protonated species evolve toward an intermediate having any symmetry element that seems to be a lacunary derivative of para B, $[H_8W_{11}O_{40}]^{6-}$, obtained by removal of a WO_6 octahedron from one of the linear tritungstic groups [67] (Fig. 3.24). Such a species could be one of the intermediates allowing the transformation by acidification into the α-metatungstate or into the yellow decatungstate $[W_{10}O_{32}]^{2-}$, the unique colored isopolytungstate.

Decatungstate $[W_{10}O_{32}]^{2-}$ is formed in water in more acidic medium ($H^+/W > 1.5$, $pH < 2$), at low ionic strength and/or in the presence of quaternary ammonium cations [54] (Fig. 3.24). It is metastable and transforms slowly in α-meta. It can be stabilized in nonaqueous solvent (methanol, DMSO). Acidification of tungstate in methanol leads to a compact hexatungstate $[W_6O_{19}]^{2-}$, a species also formed by Mo, Nb, and Ta (Fig. 3.24). Addition of water causes its rapid transformation into decatungstate $[W_{10}O_{32}]^{2-}$ (with a half-reaction time of 8 minutes at 25°C). The reverse transformation of decatungstate in solution in methanol is much slower ($t_{1/2}$ = 4 days at 35°C). In mixed solvent (water addition), both ions are in equilibrium.

Similarly to polyvanadates and polymolybdates, coordination polyhedra in polytungstates are strongly distorted: in WO_6 octahedra belonging to tritungstic cycles, the W–$O_{terminal}$ bond is especially short (1.70 Å), whereas the W–$O_{central}$ bond in *trans* position is very long (2.26 Å). The cation is moved toward the outside of the polyanion because of the π character of the W–O_t bond. This short bond creates an external layer of oxygen atoms that is strongly polarized toward the inside of the polyanion by $d\pi$–$p\pi$ interactions, as in $V_{10}O_{28}$. Consequently, the terminal oxygen atoms are weakly basic with a limited protonation, thereby preventing subsequent condensation and water elimination. If condensation were to continue, it would be more difficult for cations to move in order to reduce their mutual repulsions and so the edge sharing would stop. Therefore, the smaller the cation is, the more likely the relaxation of interactions will occur and hence the larger the number of edge-shared octahedra per polyanionic unit:

$$R_{V(V)} \quad > \quad R_{Mo(VI)} \quad \approx \quad R_{W(VI)}$$

$$[V_{10}O_{28}]^{6-} \quad [Mo_8O_{26}]^{4-} \quad [W_7O_{24}]^{6-}$$

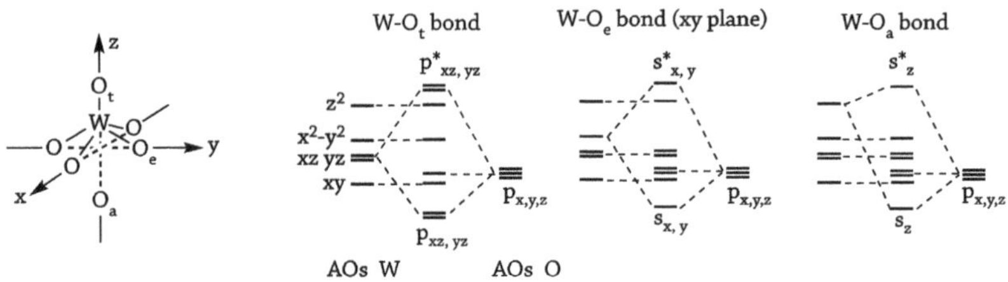

FIGURE 3.25 Simplified MO diagrams for different W–O bonds in a C_{4v} WO_6 complex.

Larger polyanions of tungsten and molybdenum may be formed ($[H_2W_{12}O_{40}]^{6-}$, $[Mo_{36}O_{112}(H_2O)_{16}]^{8-}$), but they involve the corner sharing of octahedra in order to limit the electrostatic repulsions between metal cations.

It is interesting to highlight the difference in behavior between Mo^{VI} and W^{VI}. There are few isostructural polymolybdates and polytungstates: the para A $[M_7O_{24}]^{6-}$ and the compact hexa $[M_6O_{19}]^{2-}$. Most polymolybdates are of *cis*-dioxo type (two *cis* terminal oxygen atoms), whereas most polytungstates are of mono-oxo type (only one terminal oxygen atom per octahedron). The only *cis*-dioxo tungstate (para A) is unstable compared to molybdate. It is difficult to understand this difference in behavior. Since the single M–O bonds in tungstates and molybdates have the same length (1.92 Å), it is likely that a better orbital overlap occurs with tungsten since 5*d* orbitals are larger than 4*d*. Molybdates could therefore compensate for the weakness of single bonds by an increase in the number of π bonds [56].

The difference in the electronic configuration of these two elements also has interesting consequences for the electrochemical reduction of the cation. Distortion of WO_6 octahedron implies C_{4v} symmetry because of the shortening of the W–O_t bond. This bond, too short for a σ character, may be considered a double π bond. A simplified molecular orbital diagram (Fig. 3.25) shows that, for this symmetry, the d_{xy} orbital always remains nonbonding, whereas in the case of molybdenum, this d_{xy} orbital is used to form the second π bond corresponding to the *cis*-dioxo configuration. It is therefore easy to understand that two-electron reduction is reversible in tungstate: electron population in the empty nonbonding d_{xy} orbital has little influence on crystal structure. In the *cis*-dioxo configuration, the electron must occupy an available antibonding orbital, leading to an important structural change and irreversibility of the reduction [70]. The reversible nature of the reduction in tungstates gives them important catalytic potential: the polyanion allows electron transfer during the redox reactions in solution, and the initial form is regenerated by reoxidation in air [56]. Their high Brönsted acidity strengthens the interest in polyoxometalates in this field. The biological activity of some polyoxometalates also makes them potential candidates for medical applications, especially heteropolyanions [71], which form large, various families of compounds.

b) Heteropolyoxometalates

Polyanions larger than those previously studied may be formed by condensation of vanadic, molybdic, or tungstic anions in the presence of various kinds of heteroatoms:

- Other elements, for instance, P^{V}, As^{V}, Si^{IV}, and B^{III} in the form of phosphate, arsenate, silicate, and borate, respectively, able to complex the transition cation and to be embedded in the structure during condensation and formation of the polyanion.
- Metal cations at low oxidation degree (transition elements, lanthanides), able to assume the assembling of more or less lacunary fragments of polyoxometalates.
- Reduced forms of the elements themselves such as V^{IV}, Mo^{V}, and W^{V} which take part in the polycondensation process.
- The sulfide ligand, S^{2-}, able to form in association with the oxo ligand and various metals, polyanionic species called oxothiometalates.

The large structural diversity of heteropolymetalates is briefly illustrated next. These entities indeed exhibit noteworthy properties in various fields, including catalysis, magnetism, and luminescence [72].

POLYVANADATES

V^{IV} oxopolyvanadates and mixed-valence V^{IV}–V^{V} compounds synthesized in aqueous solution exhibit hollow-shaped architectures containing various ions, molecules, or units as templates.

The $[V_{18}O_{42}]^{12-}$ polyvanadate (IV) is formed by treatment of sulfate $VOSO_4$ in alkaline medium (pH 14). The polyanion forms a hollow sphere (4.5 Å in diameter) resulting from corner and edge sharing $V^{IV}O_5$ pyramids (Fig. 3.26a). A water molecule is positioned at the center of the sphere [73]. In this species, called Schlemper's structure, the V^{IV} ions (d^1), at average distance of 2.95 Å, are antiferromagnetically coupled.

Reduction of KVO_3 with hydrazine hydroxide and addition of HCl, HBr, or HI yield the compound $K_9[H_4V^{IV}_{18}O_{42}(X)]\cdot16H_2O$ (X = Cl, Br or I) [74]. The halogenide is located at the center of the hollow sphere with a structure similar to that of Schlemper (Fig. 3.26b). A similar cage containing a nitrate anion [75] also exists in $[HV^{IV}_{12}V^{V}_{6}O_{42}(NO_3)]^{10-}$ (Fig. 3.26c). The treatment of V_2O_5 by $LiCO_3$ and then by hydrazine sulfate forms $Li_7[V^{IV}_8V^{V}_7O_{36}(CO_3)]\cdot39H_2O$ in which the $V_{15}O_{36}{}^{5-}$ shell contains the carbonate anion [74] (Fig. 3.26d).

Partial reduction of NH_4VO_3 by hydrazine sulfate in aqueous solution forms $[V_{19}O_{41}(OH)_9]\cdot8NH_4\cdot11H_2O$. Twelve $V^{IV}O_6$ octahedra and six $V^{V}O_4$ tetrahedra form an ellipsoid containing a $V^{V}O_4$ tetrahedra, sharing its corners with four octahedra of the shell [76] (Fig. 3.26e). This compound heated at 75°C in aqueous solution in the presence of NEt_4BF_4 and NaN_3 forms $[H_2V^{IV}_8V^{V}_{10}O_{44}(N_3)](NEt_4)_5$ [77] (Fig. 3.26f). The hollow cage contains an N_3^- anion. Treated with NEt_4ClO_4 under the same conditions, the V_{19} compound forms $[HV^{IV}_8V^{V}_{14}O_{54}(ClO_4)](NEt_4)_6$ [77] (Fig. 3.26g) in which the perchlorate ion is located inside the hollow cage. Finally, the reduction of KVO_3 at 90°C in aqueous solution with hydrazine hydroxide and the treatment with acetic acid forms $K_{10}[V_{34}O_{82}]\cdot20\ H_2O$. Thirty pyramidal pentacoordinated vanadium atoms (12 V^{IV}, 18 V^{V}) form a cage containing a $[V^{IV}_4O_4]O_4$ core [78] (Fig. 3.26h).

In these structures, the encapsulated anions (ClO_4^-, N_3^-, NO_3^-) are not part of the framework. Their position within the cavity of the polyanion derives from

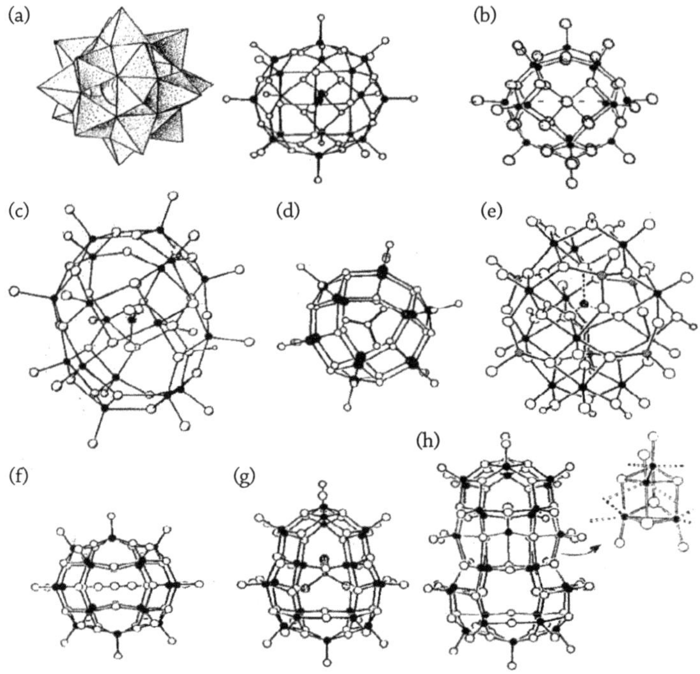

FIGURE 3.26 Structures of polyoxovanadic cages: (a) Schlemper's structure $[V_{18}O_{42}(H_2O)]^{12-}$ (reproduced from [56] with permission of Springer Nature); (b) $[H_4V_{18}O_{42}(X)]^{9-}$ (X = Cl, Br or I); (c) $[HV_{18}O_{42}(NO_3)]^{10-}$; (d) $[V_{15}O_{36}(CO_3)]^{7-}$; (e) $[V_{19}O_{41}(OH)_9]^{8-}$; (f) $[H_2V_{18}O_{44}(N_3)]^{5-}$; (g) $[HV_{22}O_{54}(ClO_4)]^{6-}$; (h) $[V_{34}O_{82}]^{10-}$ (reproduced with permission from [74–77] Copyright Wiley-VCH Verlag GmbH & Co. KGaA).

weak interactions with vanadium and oxygen atoms of the cage with no crystallographic disorder [77]. In $[H_2V_{18}O_{44}(N_3)]^{5-}$, the shortest distances N···O are 3.05 Å. In $[HV_{22}O_{54}(ClO_4)]^{6-}$, the shortest distances V···OClO$_3$ are 2.96 Å. The $[V^{IV}_{12}V^{V}_{2}O_{2}$ $_2(OH)_2(OH_2)_2(C_6H_5PO_3)_2]^{6-}$ anion [79] exhibits an even more interesting internal structure. The cage, with a framework containing in this instance $C_6H_5PO_3$ ligands (Fig. 3.27), includes two Cl^- ions, two NH_4^+ cations, and two water molecules. Inside the cage, these ions form a planar ring and the distances between them are similar to that existing in the NH_4Cl lattice. The hydrogen bonds within the cavity likely make an important contribution to interactions allowing such an organization.

From a strictly topological standpoint, this family of polyoxovanadates illustrates a peculiar aspect of its molecular organization in inorganic chemistry. The encapsulation of anions or molecules inside cages, themselves anionic in nature, appear to control the molecular architecture. That illustrates the concept of template effect or the guest–host supercomplex, which resembles the chemistry of zeolites and supramolecular chemistry [80] and themselves are based on an association of strong and weak interactions. Formation of this kind of hollow shell made of pyramids seems to be imposed by the anion (halogenide, nitrate,

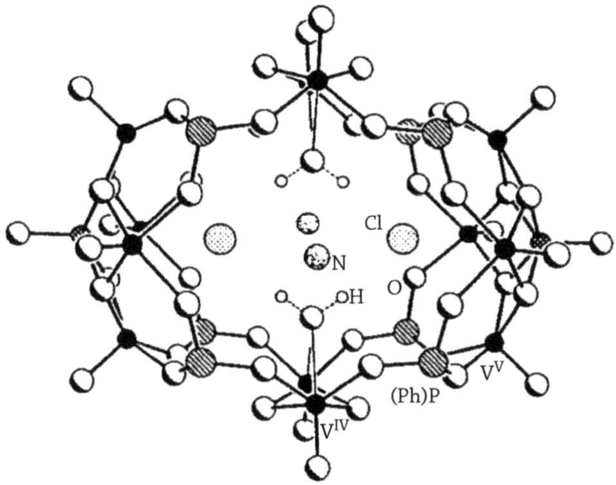

FIGURE 3.27 Structure of the $[V_{14}O_{22}(OH)_2(OH_2)_2(C_6H_5PO_3)_2]^{6-}$ anion (reproduced with permission from [79] Copyright Wiley-VCH Verlag GmbH & Co. KGaA).

perchlorate, or V_4O_4 group), which is bulky and is able to combine weak repulsive interactions with oxygen atoms coordinated with vanadium and attractive interactions with the cation itself. The latter are made possible through the pyramidal coordination of vanadium and the trans effect of the V = O terminal bond.

PHOSPHATOTUNGSTATES

Many anionic species such as silicate, germanate, arsenate, borate, and especially phosphate can be incorporated in the structure of polyanions in forming heteropolyoxometalates. The phosphate is the heteroanion that forms the largest variety of compounds with Mo^{VI}, W^{VI}, and V^{V}. The phosphatotungstates are interesting to consider for their structural diversity.

Most phosphatotungstates belong to three structural types. The first is Keggin's polyanion α-$[PW_{12}O_{40}]^{3-}$ [81] (Figs. 3.28 and 3.29). Its structure is analogous to that of α-metatungstate $[H_2W_{12}O_{40}]^{6-}$ (Fig. 3.24), with a phosphate tetrahedron occupying the central cage linked by its corners to the four μ_3 oxygen atoms in the cage. The polyanion is formed by the addition of an excess of phosphoric acid to a hot aqueous solution of sodium tungstate, Na_2WO_4, after the addition of hydrochloric acid. It crystallizes upon cooling of the solution [54].

Keggin's polyanion has five potential isomers (Fig. 3.28). They correspond to a π/3 rotation of one, two, three, or four tritungstic groups in forming the β, γ, δ, and ε-isomers, respectively, and leading to a change in connection with trimeric groups, from corner sharing to edge sharing. The α-form seems to be the most stable because the cation–cation repulsions are minimum and *dπ–pπ* overlaps are favored [56]. The β-isomer is unknown, and the γ-isomer is rare. The δ-form exists in δ-Al_{13} and Al_{30} polycations. The well-known Al_{13} polycation is the ε-isomer in which the four Al_3O_{13} groups have turned and are linked by edge sharing. The broadened ^{27}Al NMR-MAS (Magic Angle Spinning) signal indicates the strongly

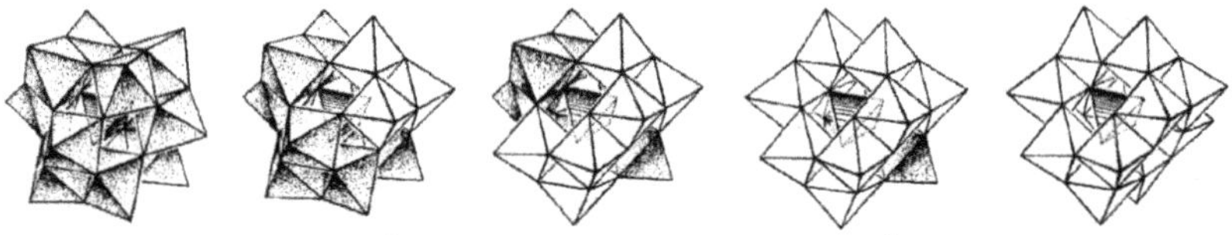

FIGURE 3.28 Isomers of Keggin's structure. The trimers having rotated π/3 from the α-form are drawn in white (reproduced from [56] with permission of Springer Nature).

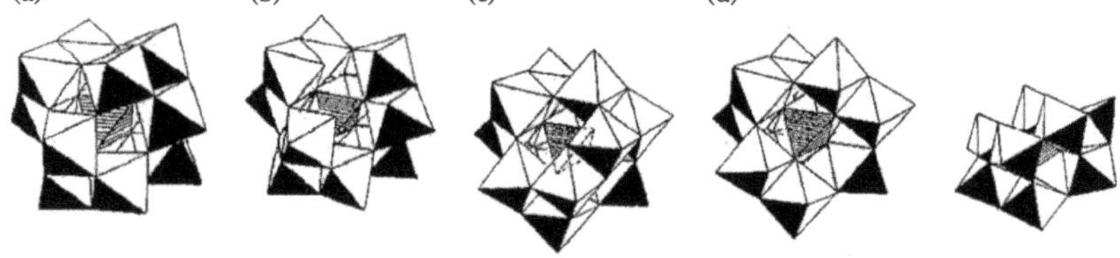

FIGURE 3.29 Structure of (a) α-$[PW_{12}O_{40}]^{3-}$ (Keggin's structure) and lacunary derivatives; (b) α-$[PW_{11}O_{39}]^{7-}$; (c) Aα-$[PW_9O_{34}]^{9-}$; (d) structure of $[PW_{10}O_{36}]^{7-}$.

distorted octahedral environment of the Al^{3+} ions in this isomer. The α-isomer also exists among the polycationic forms of aluminum, in zunyite, an aluminosilicate (Chap. 6).

α-$[PW_{12}O_{40}]^{3-}$ is highly soluble in water, but its stability range is limited (pH < 1). With loss of tungsten atoms, it transforms into lacunary derivatives. It is thus at the origin of a family of heteropolyanions, having a structure related to Keggin's. Through alkalinization, it leads rapidly and reversibly to the following species [82]:

$$\alpha\text{-}\left[PW_{12}O_{40}\right]^{3-} \Leftrightarrow \alpha\text{-}\left[PW_{11}O_{39}\right]^{7-} \Leftrightarrow A\alpha\text{-}\left[PW_9O_{34}\right]^{9-} \Leftrightarrow \left[HPO_4\right]^{2-} + \left[WO_4\right]^{2-}$$

Most of their structures (Fig. 3.29) have been deduced from investigations in solution [82] as well from ^{31}P NMR [83]. All these species can be directly synthesized by acidification of mixtures of tungstate and phosphate [82].

The reaction of tungstic acid with cesium hydroxide, followed by neutralization of the solution up to pH 7 with phosphoric acid, forms $Cs_6P_2W_5O_{23}$. The $[P_2W_5O_{23}]^{6-}$ anion is made of a ring of five edge-sharing WO_6 octahedra, except for two of them which share only one corner (Fig. 3.30). The phosphate tetrahedra are tricoordinated [84]. Heated at 100°C in water, this compound forms rapidly $[PW_{10}O_{36}]^{7-}$ (Fig. 3.29), which derives from Keggin's structure by a π/3 rotation of two tritungstic groups and the removal of two octahedra. The structural modifications suggest the great complexity of the equilibria in solution and stress the importance of the countercation. In the presence of sodium or potassium instead of cesium, the synthesis leads to $[PW_9O_{34}]^{9-}$.

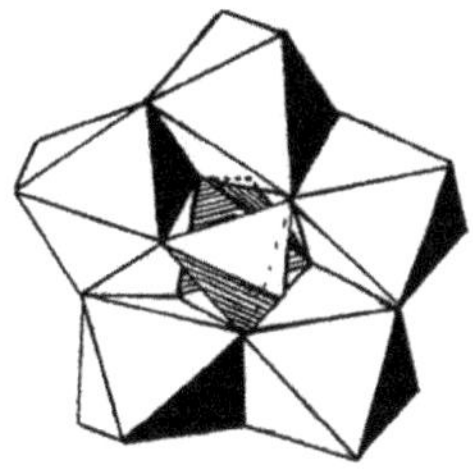

FIGURE 3.30 Structure of the $[P_2W_5O_{23}]^{6-}$ anion.

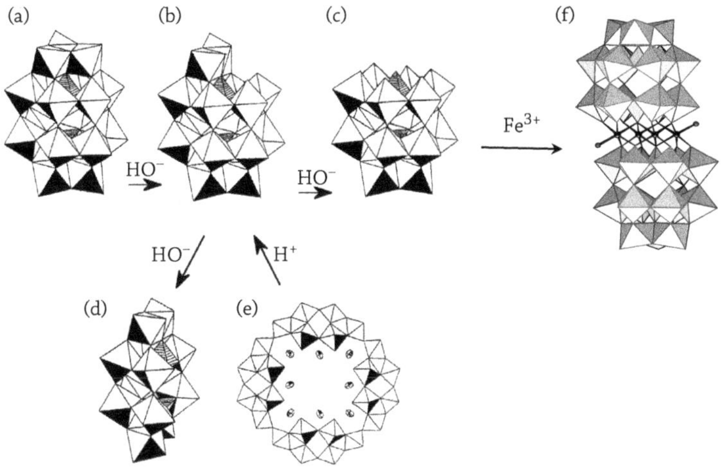

FIGURE 3.31 Structures of (a) α-$[P_2W_{18}O_{62}]^{6-}$ (Dawson's structure) and lacunary derivatives; (b) α-$[P_2W_{17}O_{61}]^{10-}$; (c) α-$[P_2W_{15}O_{56}]^{12-}$; (d) $[H_2P_2W_{12}O_{48}]^{12-}$; (e) $[P_8W_{48}O_{184}]^{40-}$ (the potassium ions inside the crown are drawn as ellipsoids); (f) $[Fe^{III}{}_4(H_2O)_2(P_2W_{15}O_{56})_2]^{12-}$.

The second structural type of phosphatotungstates is that of the α-$[P_2W_{18}O_{62}]^{6-}$ polyanion (Dawson's structure), formed of two α-PW_9 units connected by corners of octahedra (Fig. 3.31a) [85, 86]. It is obtained through refluxing sodium tungstate in the presence of an excess of phosphoric acid. It exists as a β-isomer in which one of the two polar W_3O_{13} groups has rotated by 60° [87, 88]. Both isomers can be separated by fractioned crystallization of the ammonium salts. The alkalinization of α-P_2W_{18} forms a series of lacunary compounds through removal of polar octahedra (Fig. 3.31) [89]:

$$\alpha\text{-}\left[P_2W_{18}O_{62}\right]^{6-} \Leftrightarrow \alpha\text{-}\left[P_2W_{17}O_{61}\right]^{10-} \Leftrightarrow \alpha\text{-}\left[P_2W_{15}O_{56}\right]^{12-} \Leftrightarrow \left[HPO_4\right]^{2-} + \left[WO_4\right]^{2-}$$

After a few days, addition of α-$[P_2W_{15}O_{56}]^{12-}$ into a solution of ferric chloride in the presence of concentrated NaCl and heating of the mixture at 80°C for 5 minutes form crystals of $Na_{12}[Fe^{III}{}_4(H_2O)_2(P_2W_{15}O_{56})_2]\cdot 58H_2O$ [90]. Two lacunary α-P_2W_{15} polyanions are connected with a ferric tetramer in which the $Fe^{III}O_6$ octahedra are edge-sharing (Fig. 3.31f). In fact, the cations that connect the two phosphatotungstic units are coordinated to the 12 corners of tungstic octahedra and to 2 corners of phosphoric tetrahedra. Two water molecules fulfill the coordination of cations. Similar structures are obtained with divalent cations (Co, Cu, Zn) [91].

Through loss of polar and equatorial octahedra, α-$[P_2W_{17}O_{61}]^{10-}$ forms $[H_2P_2W_{12}O_{48}]^{12-}$ (Fig. 3.31d). In a lithium acetate buffer and in the presence of potassium exclusively, $[H_2P_2W_{12}O_{48}]^{12-}$ crystallizes as $K_{28}Li_5H_7P_8W_{48}O_{184}$ [88]. In fact, $[P_2W_{17}O_{61}]^{10-}$ and $[H_7P_8W_{48}O_{184}]^{33-}$ are concurrently formed during α-P_2W_{12} acidification according to complex equilibria involving several species and isomerization processes. The cyclic structure of $[H_7P_8W_{48}O_{184}]^{33-}$ (Fig. 3.31e) is formed by the end-to-end connection of four P_2W_{12} units. Eight K^+ ions are located inside the wheel at variable distances (2.7 to 3.1 Å) from internal oxygen atoms. The remaining charge of the polyanion is balanced by that of K^+ and Li^+ cations located outside the crown. The polyanion forms only in the presence of potassium, but all alkaline cations (K^+, Li^+) are exchangeable by protons. The potassium ions likely play a key role in the synthesis. The phenomenon seems to be similar to that observed with polyvanadic cages, as seen above, and it is possible that its formation involves a template effect.

Preyssler's anion $[NaP_5W_{30}O_{110}]^{14-}$ is another example of phosphatotungstic wheel [92, 93] (Fig. 3.32). It encloses a Na^+ cation, required for its formation, which is statistically distributed on two sites located on the C_5 axis of the wheel. The sodium cation may be exchanged by calcium (not by proton) after treatment at 120°C for a few hours. This anion is an apparent by-product of the synthesis of P_2W_{18} [92].

The third structural type of phosphatotungstates corresponds to $[H_6P_2W_{21}O_{74}]^{6-}$. Its structure formally derives from that of P_2W_{18} by insertion of three 6-coordinated tungsten atoms between the two PW_9 halves [94] (Fig. 3.33a). One of these tungsten atoms in the equatorial plane is pushed out toward the outside of the polyanion. It is coordinated to the oxygen atom of an internal water molecule, which itself is H-bonded to the two other oxygen atoms in the same plane. Each of the two other tungsten atoms in the equatorial plane carries a coordinated water molecule

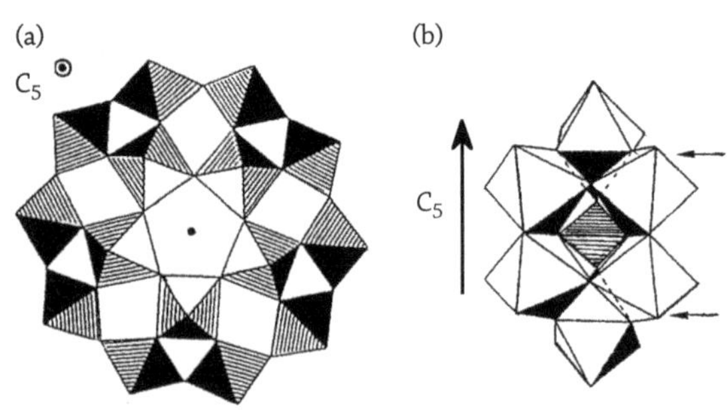

FIGURE 3.32 (a) Structure of $[NaP_5W_{30}O_{110}]^{14-}$ (reproduced with permission from [93] Copyright 1985 American Chemical Society); (b) PW_6O_{32} fragment shown along the C_5 axis. The arrows perpendicular to the C_5 axis indicate the planes where the sodium ion is located inside the wheel.

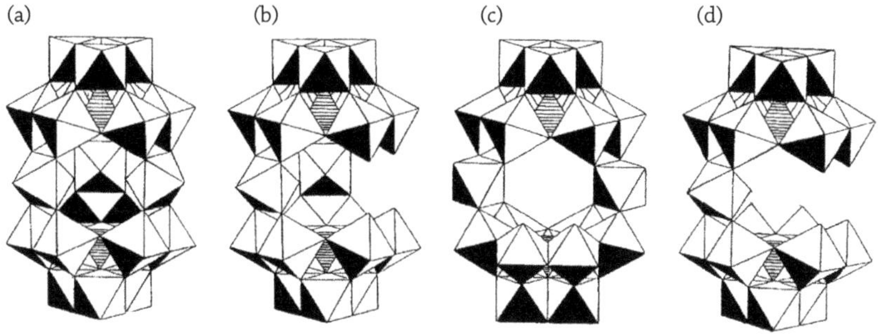

FIGURE 3.33 (a) Idealized structure of $[P_2W_{21}O_{71}(OH_2)_3]^{6-}$; structures of lacunary derivatives; (b) $[P_2W_{20}O_{70}(OH_2)_2]^{10-}$ and (c) $[P_2W_{20}O_{72}]^{14-}$; (d) structural scheme proposed for $[P_2W_{19}O_{69}(OH_2)]^{14-}$.

at the outside of the polyanion, so that the formula of the compound is $[P_2W_{21}O_{71}(OH_2)_3]^{6-}$. The polyanion is formed by acidification of α-$K_9PW_9O_{34}$ or $K_7PW_{11}O_{39}$ with the stoichiometric amount of tungstate [82, 94], but the presence of potassium or rubidium is essential. The polyanion, stable in acidic medium (pH < 2), is destroyed at higher pH in forming $[P_2W_{20}O_{70}(OH_2)_2]^{10-}$ and $[P_2W_{19}O_{69}(OH_2)]^{14-}$ [82]. Their structure derives from that of P_2W_{21} by elimination of one and two tungsten atoms, respectively, joining the two PW_9 halves (Fig. 3.33b, d).

P_2W_{20} exists as two "isomers" that distinguish themselves by rotation of 60° of one of the PW_9 groups (Fig. 3.33b, c) and by the presence of coordinated water molecules in *trans*-dioxo configuration on the joining tungsten atoms [82, 95]. One of these ions (Fig. 3.33b) behaves as a tetradentate ligand for a transition element such as Co^{II}, the cobalt ion occupying the tungsten vacancy [82]. $[P_2W_{19}O_{69}(OH_2)]^{14-}$ results from alkalinization of P_2W_{20} around pH 7 [82] or from acidification of a boiling stoichiometric mixture of tungstate and phosphate in the presence of potassium at pH slightly lower than 7. It is not very stable (it decomposes yielding PW_{11}), but it nonetheless forms complexes with Co^{II}. Like the other polyanions in this series, it requires potassium to form. The presence of cations inside the polyanion allows the structure to be stable by keeping both jaws of the polyanion open.

Finally, there are a few phosphate-rich phosphatotungstates such as $[P_5W_{18}O_{75}(OH)_3]^{20-}$ [96], $[P_4W_{14}O_{58}]^{12-}$ [95, 97], $[P_6W_{18}O_{79}]^{20-}$ [96], and $[P_4W_8O_{40}]^{12-}$ [98] (Fig. 3.34). These polyanions are formed in concentrated mixtures of Na_2WO_4 and Na_2HPO_4 in the ratio of 3:1, acidified with acetic or perchloric acid in various solvents. Their structure is based on edge- and corner-sharing distorted octahedra and does not seem to be related to any of the previously mentioned families. Although it is always possible to recognize more or less large fragments of Keggin's structure in these assemblings, they contain different tri- and tetracoordinated phosphate groups.

These phosphatotungstic polyanions exemplify the structural diversity produced by the phosphate anion acting as a tri- or tetrabridging ligand. The presence of various cations in solution (Fe^{3+}, Co^{2+}, Cu^{2+}, Zn^{2+}) increases the structural variety due to the assembling possibilities of lacunary entities. As a

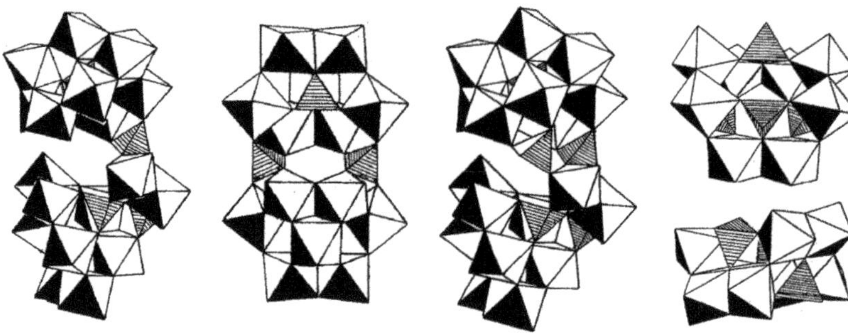

FIGURE 3.34 Structure of (a) $[P_5W_{18}O_{75}(OH)_3]^{20-}$; (b) $[P_4W_{14}O_{58}]^{12-}$; (c) $[P_6W_{18}O_{79}]^{20-}$; (d) $[P_4W_8O_{40}]^{12-}$.

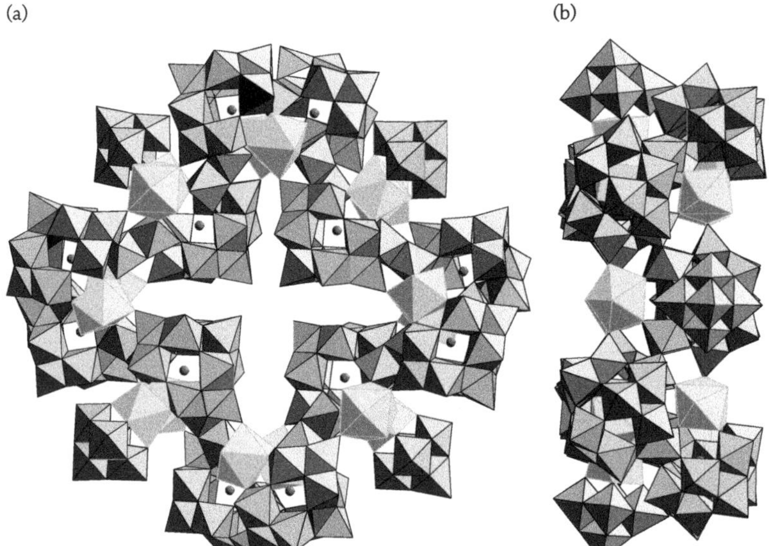

FIGURE 3.35 Structure of $[As^{III}_{12}Ce^{III}_{16}(H_2O)_{36}W_{148}O_{524}]^{76-}$ of face (a) and profile (b) (WO_6 octahedra in dark gray, CeO_x polyhedra in light gray, As atoms are represented by dots).

last example of the role of cations in the formation of giant polytungstates, there is the effect of arsenic (III) and cerium (III) on the condensation of tungstate. The heteroanion $[As^{III}_{12}Ce^{III}_{16}(H_2O)_{36}W_{148}O_{524}]^{76-}$ is obtained from the mixture of tungstate, As_2O_3, and cerium nitrate at pH 7 [99] (Fig. 3.35). It is the largest known polytungstate.

POLYMOLYBDATES

Mixed-valent polymolybdates are spectacular examples of giant polyanions [100, 101]. They constitute the "molybdenum blues," which are complex systems, known and studied for over two centuries, especially by Danish chemists Carl Scheele (1742–1786) and Jacob Berzelius (1779–1848), but their structural characterization is very recent. Achim Müller and his coworkers at the University of

Bielefeld in Germany have made an exceptional contribution to the chemistry of giant polyanions. Only a brief overview is given here.

The presence of reduced Mo^{V} ions expands and diversifies the solution chemistry of Mo^{VI} in a surprising manner, favoring the building of giant spherical and circular structures. The reasons for their formation have not yet been clearly established, but their growth and their structures are known to be sensitive to numerous factors, including acidity, reduction rate, the nature of reductant and present anions, and temperature. The reduction of Mo^{VI} leads to an increasing charge of the polyanion and thus its solubility. Its nucleophilicity is also increased, allowing additional electrophilic groups to be fixed and consequently favoring its growth and size. This phenomenon is illustrated as follows.

Acidification of molybdate $[MoO_4]^{2-}$ in the presence of phosphate produces, as with tungstate, the α-form of Keggin's polyanion $[PMo^{VI}_{12}O_{40}]^{3-}$. On the other hand, after part of the molybdate is reduced, Keggin's ion is still obtained but in its ε-form. The stability of the polyanionic core, however, requires the protonation and coordination of metal cations, as in the compound $[PMo^{V}_{8}Mo^{VI}_{14}O_{36}(OH)_4(La(OH_2)_4)_4]^{5-}$ [102]. Complete reduction of molybdate in an acidic medium also leads to the ε-Keggin $[H_xMo^{V}_{12}O_{40}]^{(20-x)-}$ polyanion stabilized by four metal centers, for instance, $[Mo^{V}_{12}O_{30}(OH)_{10}H_2(Ni(OH_2)_3)_4]$ [103]. In the presence of molybdate, the ε-Keggin core may fix four molybdic $Mo^{VI}O_3$ groups to form the (Mo_{16}) entity, $[H_xMo^{V}_{12}O_{40}(Mo^{VI}O_3)_4]^{(20-x)-}$, which crystallizes as $(CH_3NH_2)_6[H_2Mo_{16}(OH)_{12}O_{40}]$ [107] (Fig. 3.36). The four Mo^{VI} centers of these last groups can be in turn reduced, which allows them to serve as anchor points of additional (Mo_{10}) molybdic groups, $[H_3Mo^{V}_{4}Mo^{VI}_{6}O_{29}]^{+}$, and (Mo_{11}), $[H_5Mo^{V}_{6}Mo^{VI}_{5}O_{31}]^{3+}$, to form the (Mo_{37}) polyanion [104] (Fig. 3.36).

The existence of two different fragments (Mo_{10}) and (Mo_{11}) may be explained by the asymmetric protonation of three of the four μ_3O bridges that form the central cavity of the polyanion, giving them an asymmetrical charge distribution. This example clearly shows the role of the reduced species. However, these different complexes are not successively obtained as the above model would suggest. (Mo_{16}) is formed by reduction with hydrazine at 85°C of heptamolybdate dissolved in water in the presence of phenylphosphonic acid and crystallized by the subsequent addition of dimethylammonium chloride dissolved in dimethylformamide

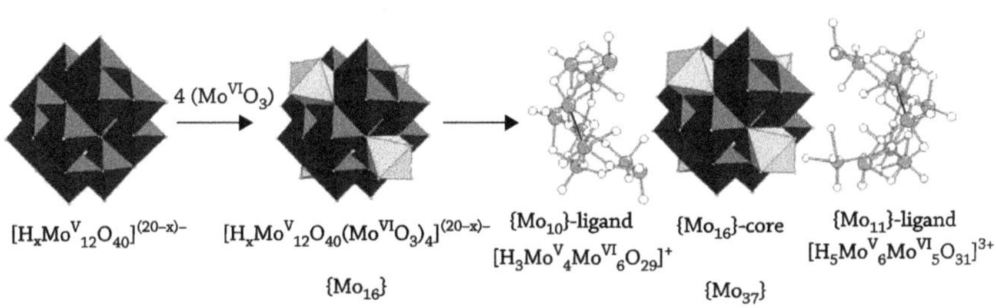

FIGURE 3.36 Formation scheme of the molybdic core ε-Keggin $[H_xMo^{V}_{12}O_{40}]^{(20-x)-}$ by fixation of 4 $Mo^{VI}O_3$ groups and of the (Mo_{39}) complex by fixation of (Mo_{10}) and (Mo_{11}) ligands (reproduced from [101] with permission of Elsevier).

FIGURE 3.37 Scheme of the structure of $[Mo_{132}O_{372}(CH_3COO)_{30}(H_2O)_{72}]^{42-}$, including 12 (Mo_6) units (dark gray) linked by (Mo_2) bridges (light gray).

[105]. (Mo_{37}) is obtained by reduction with hydrazine at 100°C of heptamolybdate solubilized in acetic acid [104].

Special conditions of molybdate reduction in solution form the (Mo_6) unit made of a MoO_7 pentagonal bipyramid sharing coplanar edges with five MoO_6 octahedra (Fig. 3.37). This unit appears as a building block in the structure of special polyanions exhibiting different architectures of hollow sphere or wheel. Some examples follow.

Reduction of molybdate with hydrazine at room temperature and in concentrated acetic acid forms a polyanionic hollow sphere, including 132 molybdenum atoms $[Mo^{VI}_{72}Mo^{V}_{60}O_{372}(CH_3COO)_{30}(H_2O)_{72}]^{42-}$. Its structure (Fig. 3.37) is built with (Mo_6) units located on the 12 corners of an icosahedron and linked by $(Mo^{V}_2O_4(CH_3COO))^+$ bridges. The red–brown coloration of the polyanion indicates that the two Mo^V ($4d^1$) ions in the bridges form diamagnetic pairs, but no $Mo^V \rightarrow Mo^{VI}$ charge transfer occurs within the structure. This polyanion belongs to the family of keplerates, so called because of their similarity to the first model for the universe created by Johannes Kepler (1571–1630)[104]. Formation of such a polyanion likely requires a set of reactions that allow the pentagonal units to be formed and their linkage on the right positions of the icosahedrical symmetry. The acidity, reduction, and complexation conditions of molybdates offer numerous possibilities allowing such an auto-organization of the system.

The circular architecture of (Mo_{154}) $[Mo^{VI}_{126}Mo^{V}_{28}(NO)_{14}O_{448}H_{14}(H_2O)_{70}]^{28-}$ [106] provides another example of linkage of the same pentagonal (Mo_6) units in addition to those existing in the (Mo_{132}) keplerate (Fig. 3.38). Fourteen of these units are linked in the sharing corners of MoO_6 octahedra. The compound is dark blue, indicating an electron mobility by $Mo^V \rightarrow Mo^{VI}$ charge transfer. It is formed, among other polyanionic species, by reduction of molybdate with hydroxylamine at 70°C in chlorhydric acid.

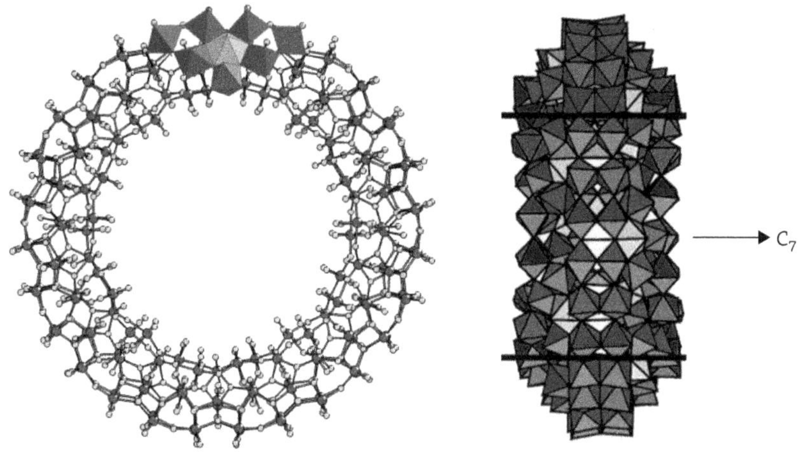

FIGURE 3.38 Scheme of the structure of the (Mo_{154}) wheel. Left, view perpendicular to the C_7 axis. Polyhedral representation of a (Mo_6) unit with the linkage MoO_6 octahedra. Right, polyhedral representation along the C_7 axis (reproduced from [Müller A., Kögerler P., Kuhlmann C., A variety of combinatorially linkable units as disposition: from a giant icosahedral keplerate to multi-functional metal–oxide based network structures, Chem. Commun. 1999, 1347–1358] with permission of The Royal Society of Chemistry).

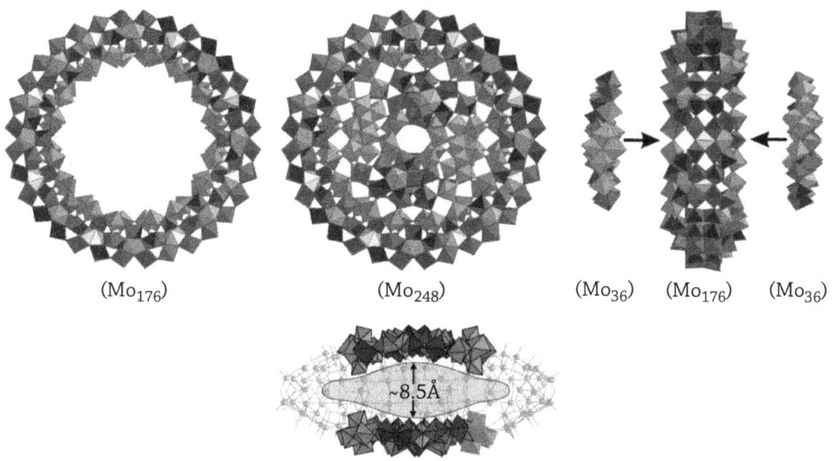

FIGURE 3.39 Schematic view of (Mo_{176}) and (Mo_{248}) (the MoO_7 polyhedra are in Turkish blue). View of the cavity in (Mo_{248}) and sketch of the association of the two (Mo_{36}) fragments with (Mo_{172}) (reproduced from [107] with permission of the Nature Publishing Group).

The same (Mo_6) units linked by corner sharing with MoO_6 octahedra find themselves in the (Mo_{176}) and (Mo_{248}) wheels [107] (Fig. 3.39). (Mo_{176}) is quasi quantitatively formed in a solution of molybdate acidified by HCl and with ascorbic acid as a reductant. The aging of the solution in the presence of an excess of molybdate forms (Mo_{248}). It may be considered to be built by the addition to (Mo_{176}) of two (Mo_{36}) entities constituting the flasks of the wheel (Fig. 3.39). These compounds are also dark blue.

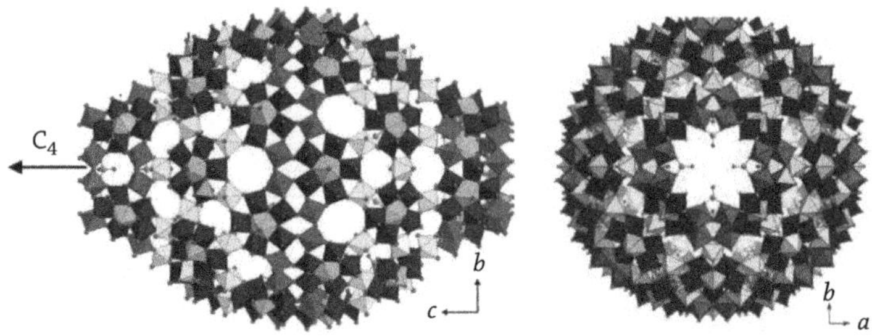

FIGURE 3.40 Scheme of the structure of (Mo_{368}) along th C_4 axis. The MoO_7 polyhedra are in Turkish blue (reproduced from [101] with permission of Elsevier).

Reduction of molybdate with tetrathionate $Na_2S_2O_4$ at RT in sulfuric acid forms (Mo_{368}) $[H_xMo^{V}{}_{112}Mo^{VI}{}_{256}O_{1032}(H_2O)_{240}(SO_4)_{48}]^{48-}$, still including the ($Mo_6$) pentagonal unit [108] (Fig. 3.40). It is the biggest polyanion known to date.

These heteropolymolybdates containing Mo^V and Mo^{VI} centers exemplify the wealth of their structural chemistry. Many others species exist, especially the heteropolymolybdates, which contain not only Mo^V and Mo^{VI} centers but also various other metallic centers [100].

OXOTHIOMOLYBDATES

Metal sulfides (MoS_2, WS_2) exhibit interesting electronic properties which are used, for instance, in hydrotreating catalysis in the petroleum industry. The activity of many metalloenzymes is also due to the Mo–S or W–S bond. Introduction of the S^{2-} sulfido (as oxo or aquo) ligand as substitution for an oxo ligand into polyoxometalates could result in significant modifications of their structural and electronic characteristics, leading to specific properties. Francis Sécheresse and his team at the University of Versailles-Saint-Quentin, France, have made a major contribution to the chemistry of such compounds.

The use of $[MS_4]^{2-}$ sulfided precursors (M = Mo, W) has the drawback of involving strong redox transfers between sulfur and the metal, which leads to reduced M^V or M^{IV} centers within weakly condensed species. The direct sulfuration of polyoxometalates also causes the reduction of metal centers. An elegant and versatile approach to this subject is based on the use of the $[Mo_2S_2O_2(OH_2)_6]^{2+}$ cation [109]. This species can autocondense to form a large variety of cyclic architectures and also serves as ligand linking various preformed lacunary polyoxometalates.

The $[Mo_2S_2O_2(OH_2)_6]^{2+}$ dication contains two octahedral Mo^V centers, and the two sulfido ligands form an edge bond between the two octahedra (Fig. 3.41).

The very short Mo–Mo distance, 2.8Å, in comparison to the Mo^{VI}–Mo^{VI} distance, 3.5Å, reveals metal–metal interactions that stabilize the dication in solution, even in highly acidic medium. It is synthesized by oxidation with iodine of terminal disulfur ligands of the entity $[(S_2)MoOS_2MoO(S_2)]^{2-}$. The $[Mo_2S_2O_2(OH_2)_6]^{2+}$ dication may autocondense by olation during alkalinization, which generates equatorial hydroxo ligands. By recrystallization in pure water, a noncharged [Mo

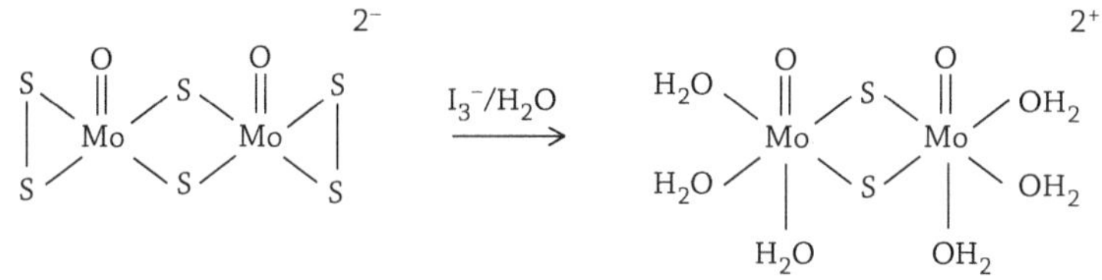

FIGURE 3.41 Structural schemes of $[(S_2)MoOS_2MoO(S_2)]^{2-}$ and $[Mo_2S_2O_2(OH_2)_6]^{2+}$.

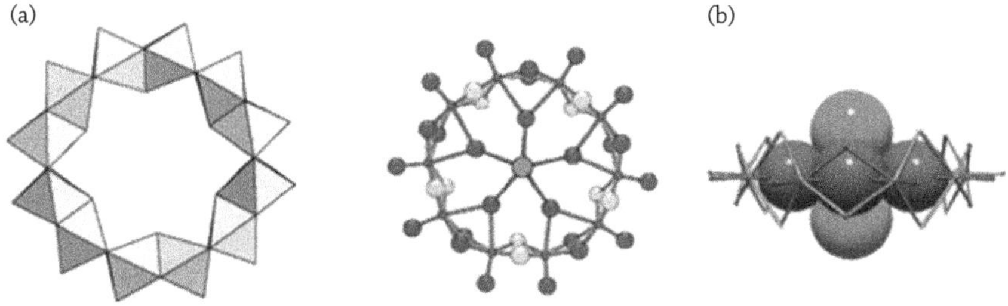

FIGURE 3.42 Structural schemes of (a) $[Mo_{12}S_{12}O_{12}(OH)_{12}(OH_2)_6]$ and (b) $X_2[Mo_{10}S_{10}O_{10}(OH)_{10}(OH_2)_5]^{2-}$ (X = Cl, I). The space-filling representation highlights the formation of a compact cluster $(X_2(OH_2)_5)^{2-}$ inside the ring (courtesy of F. Sécheresse, Univ. Versailles-St-Quentin, France).

$_{12}S_{12}O_{12}(OH)_{12}(OH_2)_6]$ ring made of six (Mo_2) units enchained by double hydroxo bridges is obtained (Fig. 3.42a). The dication is reversibly re-obtained through acidification. The cyclization process may be efficiently controlled using various species acting as templates. In the presence of halides, the interaction between water molecules inside the ring with the Mo^V centers and with two halide ions allows recrystallization of a ring including five units (10 Mo^V). The two halide ions are on either side of the ring (Fig. 3.42b).

Various dicarboxylates allow modification of the size and shape of the molybdic rings [109]. Oxalate ($C_2O_4^{2-}$), glutarate ($H_6C_5O_4^{2-}$), and pimelate ($H_{10}C_7O_4^{2-}$) induce the formation of rings with 8, 10, and 12 Mo atoms, respectively, in which carboxylate is coordinated with two Mo^V ion pairs that are diametrically opposed (Fig. 3.43). In Mo_{10}-glu and Mo_{12}-pim compounds, there is no more water, and the Mo^V centers noncoordinated to carboxylate exhibit the 5-pyramidal coordination. On the other hand, with tricarboxylate 1,3,5 methylcarboxylatobenzene (trimesate) of ternary symmetry, the ring including 12 octahedral Mo atoms is perfectly circular.

The ability of molybdenum to adopt 6- and 5-coordination probably gives a large structural flexibility to these species. It is noteworthy that this flexibility is associated with a strong fluxionality. The dynamics has been characterized by the thermal evolution from 320 to 226 K of ^1H NMR spectra in the range of OH bridges and of CH_2 groups of alkyl chain resonance, with compounds dissolved in CD_3CN [110]. At room temperature, glutarate randomly wheels inside the ring owing to the concerted hopping of the two terminal carboxylate groups over the 10 molybdenum atoms. There is a dynamical synergy in

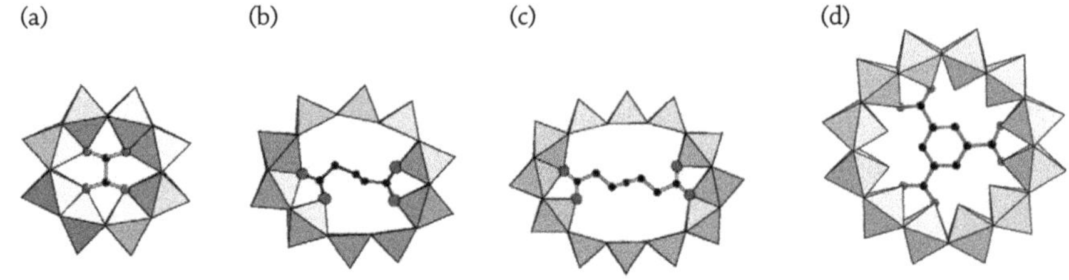

FIGURE 3.43 Structural schemes of the $[Mo_{2n}S_{2n}O_{2n}(OH)_{2n}(OH_2)_n]$ rings in the presence of different ligands: oxalate (a), glutarate (b), pimelate (c), and trimesate (d) (courtesy of F. Sécheresse, Univ. Versailles-St-Quentin, France).

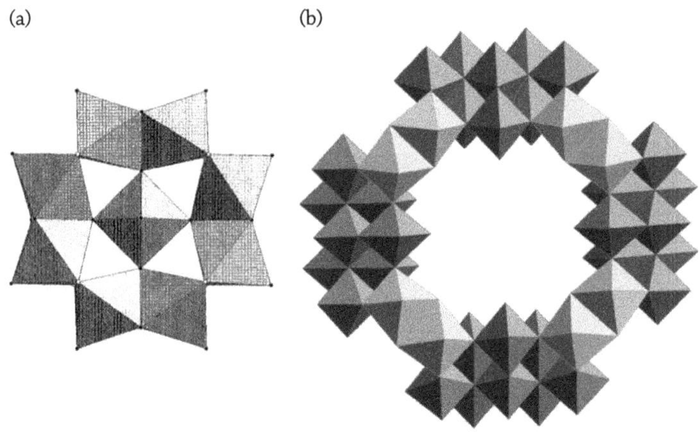

FIGURE 3.44 Structural scheme of (a) $[Mo_9S_8O_{12}(OH)_8(H_2O)_2]^{2-}$ in which one $Mo^{VI}O_6$ octahedron is included inside a $Mo^V{}_8$ ring (reproduced from [111] with permission of The Royal Society of Chemistry) and (b) $[(Mo^{VI}{}_8O_{28})_4(Mo^V{}_2O_2S_2)_4]^{24-}$.

the motion of glutarate in the ring: the hopping of a carboxylate group from an Mo atom to the closest one causes deformations in the wheel, and, reciprocally, deformations in the wheel induce the internal rotation of the carboxylate fragment. At low temperature (226 K), the dynamics is very slow on the NMR time scale, which gives a signal of a frost compound. The behavior of pimelate is analogous, whereas the (Mo_8–ox) compound exhibits no temperature dependence. Oxalate linked to all molybdenum atoms ensures a strong rigidity to the ring in which all Mo in the octahedral environment prevents any concerted internal rotation of oxalate.

At pH > 11, deprotonation of the dication forms a stable dianion $[Mo_2S_2O_2(OH)_4(OH_2)_2]^{2-}$. In the presence of molybdate $MoO_4{}^{2-}$, at around pH 5, condensation of the $Mo_2S_2O_2$ unit still occurs, forming a ring that includes a Mo^{VI} octahedra in the compound $[Mo_9S_8O_{12}(OH)_8(H_2O)_2]^{2-}$ (Fig. 3.44a) [111]. This example once again shows the strong tendency of the $Mo_2S_2O_2$ dimer to condense because of its bifunctionality due to deprotonable aquo ligands in equatorial positions. Under different experimental conditions (such as mixing of the acidic

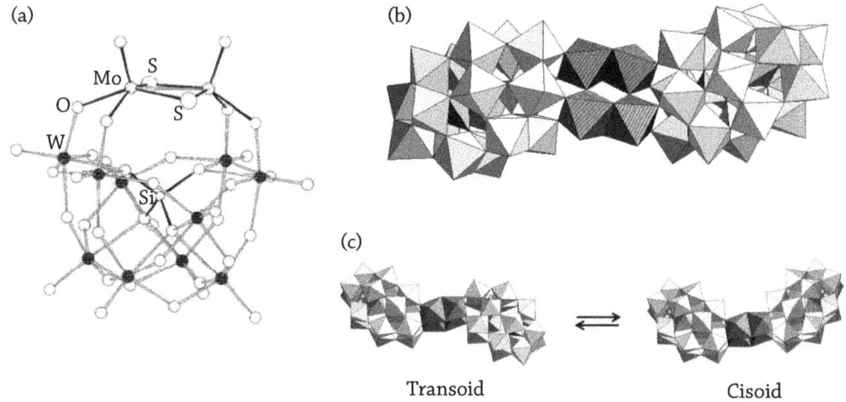

FIGURE 3.45 Structure of (a) γ-$[SiW_{10}Mo_2S_2O_{38}]^{6-}$ (reproduced with permission from [115] Copyright 1996 American Chemical Society). (b) $[(P_2W_{17}O_{61})_2(Mo_4S_4O_4(OH_2)_2)]^{16-}$ showing the Mo_4 central group (adapted from [113]); (c) perpendicular view showing the transoïd and cisoïd conformations of the P_2W_{17} moieties. In solution at RT, this compound gives rise to transoïd (30%)-cisoïd (70%) isomerization (courtesy of E. Cadot, Univ. of Versailles-St-Quentin, France).

solution of $Mo_2S_2O_2$ with a concentrated molybdate solution in ammonium acetate buffer, pH 4), the macrocycle $[(Mo^{VI}{}_8O_{28})_4(Mo^V{}_2O_2S_2)_4]^{24-}$ is obtained. It is composed of four rare isomers of octamolybdate $[Mo_8O_{28}]^{8-}$ linked by four $Mo_2S_2O_2$ fragments (Fig. 3.44b) [112]. In this case, the geometry of the $Mo^{VI}{}_8$ clusters obviously does not allow cyclization of the thiomolybdic fragments.

Combination of the $[Mo_2S_2O_2(OH_2)_6]^{2+}$ dication with lacunary polyoxometalates allows formation of a large variety of polyoxothiometalates in which the oxothio fragment may behave as a chelatant or bridging ligand [113].

With divacant tungstic polyanions $[XW_{10}O_{32}]^{8-}$ (X = Si, P) derived from the Keggin structure (cf. Fig. 3.29), the dication exactly adjusts to the vacancy in forming four W–O–Mo bonds, retaining the initial C_{2v} symmetry [114, 115] (Fig. 3.45). Such thioderivatives are interesting because the sulfur/oxygen substitution allows modulation in energy of the frontier orbitals [115]. With monolacunary derivatives, such as $[PW_{11}O_{39}]^{7-}$ (Fig. 2.39) and $[P_2W_{17}O_{61}]^{10-}$ (Fig. 3.31), the dication behaves as a bridging ligand. Two lacunary units are so linked with a $(Mo_2S_2O_2)_2$ double bridge, constraining the two $Mo_2S_2O_2$ fragments to interact and eliminate two water molecules (Fig. 3.45).

Association of trivacant polyoxometalates with the sulfured dication leads to more complex structures. For instance, in $[PW_9O_{34}]^{9-}$ (Fig. 3.29c), the vacancy forms a crown of six terminal oxygen atoms (21Å in perimeter), allowing coordination of three $Mo_2S_2O_2$ fragments arranged in a pillar to link two polyanions (Fig. 3.46) [116]. In $[P_2W_{15}O_{56}]^{12-}$ (Fig. 3.31), the crown is smaller (17Å in perimeter), and the oxygen atom of the P–O bond prevents dimerization. That leads to a more complex tetrameric plane arrangement. Four P_2W_{15} polyanions are linked to two $(Mo_8S_8(OH)_8(OH)_2(H_2O)_5)^{6+})$ bands containing $(Mo_2O_2S_2)$ fragments

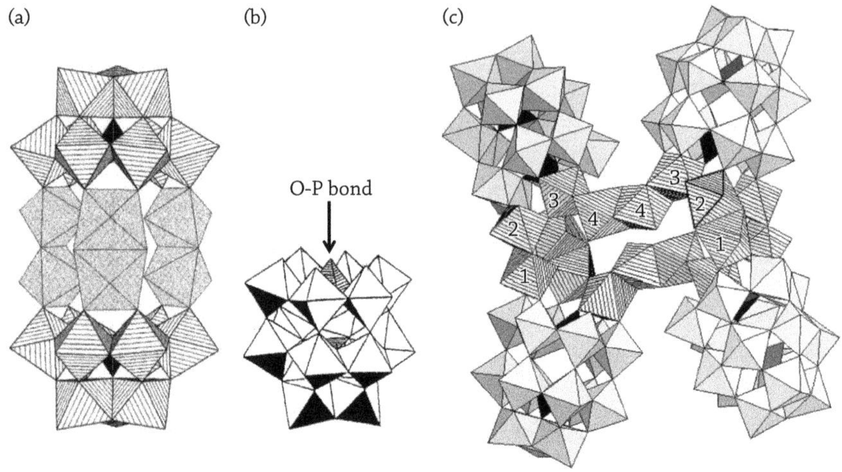

FIGURE 3.46 Schemes of the structure of (a) $[(PW_9O_{34})_2(Mo_2S_2O_2(OH_2)_3)$ (reproduced with permission from [116] Copyright 1999 American Chemical Society), (b) $[P_2W_{15}O_{56}]^{12-}$ and (c) $[(\alpha\text{-}H_2P_2W_{15}O_{56})_4\{Mo_2O_2S_2(H_2O)_2\}_4\text{-}\{Mo_4S_4O_4(OH)_2(H_2O)\}_2]^{28-}$ (the MoO_4S_2 octahedra are connected by S-S common edges (1–2 and 3–4 junctions), by O corners (2–3 junctions) and by faces (4-4 junctions)) (courtesy of E. Cadot, Univ. of Versailles-St-Quentin, France).

bounded by double hydroxo bridges, as in the rings previously examined (Fig. 3.46) [116].

These examples highlight the potentialities of polyoxothiometalates synthesized from the $[Mo_2S_2O_2(OH_2)_6]^{2+}$ fragment. There are others, for instance, $[Mo_3S_4(OH_2)_9]^{4+}$ [113], that permit expansion of the structural diversity and that demonstrate the wealth and variety of the chemistry of heteropolyoxometalates.

References

[1] Basolo F., Pearson R.G. *Mechanism of Inorganic Reactions*, 2nd ed. New York: Wiley, 1958.
[2] Purcell K.F., Kotz J.C. *Inorganic Chemistry*. Philadelphia: Holt Saunders Japan International, 1985.
[3] Rollinson C.L. Chemistry of the coordination compounds. Am. Chem. Soc. Master Ser., Bailar Ed., New York, 1956, pp. 131, 448.
[4] Baran V. Hydroxyl ion as a ligand. *Coord. Chem.* Rev. 1971, 6, 65.
[5] Van Wazer J.R. *Phosphorus and Its Compounds*. Vol. 1. New York: Wiley-Interscience, 1958.
[6] Henry M., Jolivet J.P., Livage J. Aqueous chemistry of metal cations: hydrolysis, condensation and complexation. *Struct. Bond.* 1992, 77, 153–206.
[7] Eigen M. Fast elementary steps in chemical reaction mechanisms. *Pure Appl. Chem.* 1963, 6, 97–115.
[8] Kruger H. Techniques for the kinetic study of fast reactions in solution. *Chem. Soc. Rev.* 1982, 11, 227–255.

[9] Ardon M., Bino A. A new aspect of hydrolysis of metal ions: the hydrogen-oxide bridging ligand $H_3O_2^-$. *Struct. Bond.* 1987, 65, 1–28.

[10] Rustad J.R., Casey W.H. A molecular dynamics investigation of hydrolytic polymerization in a metal-hydroxide gel. *J. Phys. Chem. B* 2006, 110, 7107–7112.

[11] Jones M.N. *Elementary Coordination Chemistry*. Englewood Cliffs, NJ: Prentice-Hall, 1964.

[12] Maroni V.A., Spiro, T. G. Normal coordinate analysis for $Bi_6(OH)_{12}^{6+}$. Evidence for bismuth-bismuth bonding. *Inorg. Chem.* 1968, 7, 183–188.

[13] Spiro T.G., Templeton D.H., Zalkin A. The crystal structure of a hexanuclear basic lead(I1) perchlorate hydrate: $Pb_6O(OH)_6(ClO_4)_4{\cdot}H_2O$. *Inorg. Chem.* 1969, 8, 856–861.

[14] Thompson M., Connick R.E. Hydrolytic polymerization of chromium (III). 1. Two dimeric species. *Inorg. Chem.* 1981, 20, 2279–2285.

[15] Rotzinger F.P., Stünzi H., Marty W. Early stages of the hydrolysis of chromium(III) in aqueous solution. 3. Kinetics of dimerization of the deprotonated aqua ion. *Inorg. Chem.* 1986, 25, 489–495.

[16] Mønsted L. Mønsted O., Springborg J. Evidence for "classical" hydroxo-bridged polymers in hydrolyzed hexaaquachromium(III) solutions. *Inorg. Chem.* 1985, 24, 3496–3498.

[17] Merakis T., Spiccia L. A new route to crystalline salts of the hydrolytic dimer of chromium(III). *Aust. J. Chem.* 1989, 42, 1579–1589.

[18] Spiccia L., Stoeckli-Evans H., Marty W., Giovanoli R. A new "active" chromium (III) hydroxide: $Cr_2(\mu\ OH)_2\ (OH)_4\ (OH_2)_4.2H_2O$. Characterization and use in the preparation of salts of the $(H_2O)_4Cr(\mu\text{-}OH)_2Cr(OH_2)_4^{4+}$ ion. Crystal structure of $[(H_2O)_4Cr(\mu\ OH)_2Cr(OH_2)_4]\ [(H_3C)_3C_6H_2SO_3]_4.4H_2O$. *Inorg. Chem.* 1987, 26, 474–482.

[19] Stünzi H., Spiccia L., Rotzinger F.P., Marty W. Early stages of the hydrolysis of chromium (III) in aqueous solution. 4. Stability constants of the hydrolytic dimer, trimer and tetramer at 25 °C and $I = 1.0$ M. *Inorg. Chem.* 1989, 28, 66–71.

[20] Spiccia L., Marty W., Giovanoli R. Hydrolytic trimer of chromium (III). Synthesis through chromite cleavage and use in the preparation of the "active" trimer hydroxide. *Inorg. Chem.* 1988, 27, 2660–2666.

[21] Stünzi H., Marty W. Early stages of the hydrolysis of chromium (III) in aqueous solution. 1. Characterization of a tetrameric species. *Inorg. Chem.* 1983, 22, 2145–2150.

[22] Finholt J.E., Thompson M.E., Connick R.E. Hydrolytic polymerization of chromium (III). 2. A trimeric species. *Inorg. Chem.* 1981, 20, 4151–4155.

[23] Stünzi H., Rotzinger F.P., Marty W. Early stages of the hydrolysis of chromium (III) in aqueous solution. 2. Kinetics and mechanism of the interconversion between two tetrameric species. *Inorg. Chem.* 1984, 23, 2160–2164.

[24] Roussel H., Briois V., Elkaim E., de Roy A., Besse J.P., Jolivet J.P. Study of the formation of the layered double hydroxide [Zn-Cr-Cl]. *Chem. Mater.* 2001, 13, 329–337.

[25] Van Der Woude J.H.A., De Bruyn P.L. Formation of colloidal dispersions from supersaturated iron (III) nitrate solutions. 1 Precipitation of amorphous iron hydroxide. *Colloids Surf.* 1983, 8, 55–78.

[26] Souchay P., *Lefebvre J. Equilibres et réactivité des complexes en solution*. Paris: Masson, 1969.

[27] Baes C.F., Mesmer R.E. *The Hydrolysis of Cations*. New York: Wiley, 1976.

[28] Iler R.J. *The Chemistry of Silica*. New York: Wiley, 1979.
[29] Wells A. F. *Structural Inorganic Chemistry*, 5th ed. Oxford: Oxford University Press, 1991.
[30] Knight C.T.C. A two-dimensional silicon-29 NMR spectroscopic study of the structure of the silicate anions present in an aqueous potassium silicate solution. *J. Chem. Soc. Dalton Trans*. 1988, 1457–1460.
[31] Knight C.T.C., Kirkpatrick R.J., Oldfield E. Silicon-29 multiple quantum filtered NMR spectroscopic evidence for the presence of only six single site silicate anions in a concentrated potassium silicate solution. *J. Chem. Soc. Chem. Commun*. 1989, 919–921.
[32] Harris R.K., Newman, R.H. ^{29}Si N.M.R. studies of aqueous silicate solutions. *J. Chem. Soc. Faraday Trans*. II 1977, 73, 1204–1215.
[33] Harris R.K., Knight C.T.C., Hull W.E. Nature of species present in an aqueous solution of potassium silicate. *J. Am. Chem. Soc*. 1981, 103, 1577–1578.
[34] Harris R.K., Knight C.T.C., Hull W.E. NMR studies of the chemical structure of silicates in solution. ACS Symposium Series 1982, Vol. 194, 79–93.
[35] Harris R.K., Knight C.T.C. Silicon-29 NMR studies of aqueous silicate solutions Part 5. First-order patterns in potassium silicate solutions enriched with silicon-29. *J. Chem. Soc. Faraday Trans*. II 1983, 79, 1525–1538.
[36] Harris R.K., Knight C.T.C. Silicon-29 NMR studies of aqueous silicate solutions Part 6. Second-order patterns in potassium silicate solutions enriched with silicon-29. *J. Chem. Soc. Faraday Trans*. II 1983, 79, 1539–1561.
[37] Harris R.K., O'Connor M.J., Curzon E.H., Howarth O.W. Two-dimensional silicon-29 NMR studies of aqueous silicate solutions. *J. Magn. Reson*. 1984, 57, 115–122.
[38] Moravetski V., Hill J.-R., Eicher U., Cheetham A.K., Sauer J. ^{29}Si NMR chemical shifts of silicate species: ab initio study of environment and structure effects. *J. Am. Chem. Soc*. 1996, 118, 13015–13020.
[39] Cho H., Felmy A.R., Craciun R., Keenum J.P., Shah N., Dixon D.A. Solution state structure determination of silicate oligomers by ^{29}Si NMR spectroscopy and molecular modeling. *J. Amer. Chem. Soc*. 2006, 128, 2324–2335.
[40] Harris R.K., Knight C.T.C. Silicon-29 NMR studies of aqueous silicate solutions: Part IV Tetraalkylammonium hydroxide solutions. *J. Mol. Struct*. 1982, 78, 273–278.
[41] Knight C.T.C., Kirkpatrick R.J., Oldfield E. The unexpectedly slow approach to thermodynamic equilibrium of the silicate anions in aqueous tetramethylammonium silicate solutions. *J. Chem. Soc. Chem. Commun*. 1986, 66–67.
[42] Kinrade S.D., Knight C. T. G., Pole D.L., Syvitski R.T. Silicon-29 NMR studies of tetraalkylammonium silicate solutions. 1. Equilibria, ^{29}Si chemical shifts, and ^{29}Si relaxation. *Inorg. Chem*. 1998, 37, 4272–4277.
[43] Kinrade S.D., Knight C. T. G., Pole D.L., Syvitski R.T. Silicon-29 NMR studies of tetraalkylammonium silicate solutions. 2. Polymerization kinetics. *Inorg. Chem*. 1998, 37, 4278–4283.
[44] Momii R.K. Nachtrieb N.H. NMR study of borate-polyborate. Equilibria in aqueous solution. *Inorg. Chem*. 1967, 6, 1189–1192.
[45] Smith Jr H.D., Wieserma R.J. Boron-11 NMR study of polyborate ions in solution. *Inorg. Chem*. 1972, 11, 1152–1154.
[46] Farmer J.B. Metalborates. *Adv. Inorg. Chem. Radiochem*. 1982, 25,187–225.

[47] Lefebvre J. Méthode de la surface potentiométrique. II Application à l'étude de systèmes à deux composants: borates, vanadates, oxalate de magnesium. *J. Chim. Phys.* 1957, 54, 567–580.
[48] Ingri N. Equilibrium studies of polyanions. *Acta Chem. Scand.* 1962, 16, 439–448.
[49] Messmer R.E., Baes C.F., Sweeton, F.H. Acidity measurements at elevated temperatures. VI. Boric acid equilibria. *Inorg. Chem.* 1972, 11, 537–543.
[50] Ballhausen C.J., Gray H.B. *Molecular Orbital Theory*. New York, W.A. Benjamin, 1964.
[51] Figgis B.N. *Introduction to Ligand Fields*. New York: Wiley-Interscience, 1966.
[52] Kepert D.L. Isopolytungstates. *Prog. Inorg. Chem.* 1962, 4, 199–274.
[53] Tytko K.H., Glemser O. Isopolymolybdates and isopolytungstates. *Adv. Inorg. Chem. Radiochem.* 1976, 19, 239–315.
[54] Souchay P. *Ions minéraux condensés*. Paris: Masson, 1969.
[55] Heath E., Howard O.W. Vanadium-51 and oxygen-17 nuclear magnetic resonance study of vanadate (V) equilibria and kinetics. *J. Chem. Soc. Dalton Trans.* 1981, 1105–1110.
[56] Pope M.T. *Heteropoly and Isopolyoxometallates*. Berlin: Springer-Verlag, 1983.
[57] Kepert D.L. *The Early Transition Metals*. New York: Academic Press, 1972.
[58] Day V.W., Klemperer W.G., Maltbie D.J. Where are the protons in $H_3V_{10}O_{28}{}^{3-}$? *J. Am. Chem. Soc.* 1987, 109, 2991–3002.
[59] Comba P., Helm L. The solution structure and reactivity of decavanadate. *Helv. Chim. Acta* 1988, 71, 1406–1420.
[60] Day V.W., Klemperer W.G., Yaghi O.M. Synthesis and characterization of a soluble oxide inclusion complex, $[CH_3CN \subset (V_{12}O_{32}{}^{4-})]$. *J. Am. Chem. Soc.* 1989, 111, 5959–5961.
[61] Hou D., Hagen K.S., Hill C.L. Tridecavanadate, $[V_{13}O_{34}]^{3-}$, a new high-potential isopolyvanadate. *J. Am. Chem. Soc.* 1992, 114, 5864–5866.
[62] Kepert D.L. The structures of polyanions. *Inorg. Chem.* 1969, 8, 1556–1558.
[63] Krebs B., Paulat-Böschen I. The structure of the potassium isopolymolybdate $K_8[Mo_{36}O_{12}(H_2O)_{16}].nH_2O$ (n = 36-40). *Acta Cryst. B* 1982, 38, 1710–1718.
[64] Pope M.T., Varga Jr, G.M. Proton magnetic resonance of aqueous metatungstate ion: evidence for two central hydrogen atoms. *Chem. Commun* (London) 1966, 653–654.
[65] Launay J.P., Boyer M., Chauveau F. High resolution NMR of several isopolytunsgtates and related compounds. *J. Inorg. Nucl. Chem.* 1976, 38, 243–247.
[66] Maksimovskaya R.I., Burtseva K.G. ^{17}O and ^{183}W NMR studies of the paratungstate anions in aqueous solutions. *Polyedron* 1985, 4, 1559–1562.
[67] Hasting J.J., Howarth O.W. A ^{183}W, ^{1}H and ^{17}O nuclear magnetic resonance study of aqueous isopolytungstates. *J. Chem. Soc. Dalton Trans.* 1992, 209–215.
[68] Fuchs J., Flindt E.P. Dartellung und strukturuntersuchung von polywolframaten. Ein betrag zur aufklärung des parawolframations A. *Z. Naturforsch.* 1979, 34 b, 412–422.
[69] Hartl H., Palm R., Fuchs J. A new type of paratungstate. *Angew. Chem. Int. Ed. Eng.* 1993, 32, 1492–1494.
[70] Launay J.P. Reduction of the metatungstate ion. High levels of reduction of $[H_2W_{12}O_{40}]^{6-}$, derivatives of the ion $[HW_{12}O_{40}]^{7-}$ and general discussion. *J. Inorg. Nucl. Chem.* 1976, 38, 807–816.
[71] Pope M.T., Müller A. Polyoxometalate chemistry: an old field with new dimensions in several disciplines. *Angew. Chem. Int. Ed. Engl.* 1991, 30, 34–48.

[72] Sécheresse F. *Polyoxometallate Chemistry: Some Recent Trends*. River Edge, NJ: World Scientific, 2013.

[73] Johnson G.K., Schlemper E. O. Existence and structure of the molecular ion 18-vanadate(IV). *J. Am. Chem. Soc.* 1978, 100, 3645–3646.

[74] Müller A., Penk M., Rohlfing R., Krickemeyer E., Döring J. Topologically interesting cages for negative ions with extremely high "coordination number": an unusual property of V-O clusters. *Angew. Chem. Int. Ed. Eng.* 1990, 29, 926–927.

[75] Müller A. Induced molecule self organization. *Nature* 1991, 352, 115.

[76] Müller A., Penk M., Krickemeyer E., Bögge H., Walberg. $[V_{19}O_{41}(OH)_9]^{8-}$ An ellipsoid-shaped cluster anion belonging to the unusual family of V^{IV}/V^{V} oxygen cluster. *Angew. Chem. Int. Ed. Eng.* 1988, 27, 1719–1721.

[77] Müller A., Krickemeyer E., Penk M., Rohlfing R., Armatage A., Bögge H. Template-controlled formation of cluster shells or a type of molecular recognition: synthesis of $[HV_{22}O_{54}(ClO_4)]^{6-}$ and $[H_2V_{18}O_{44}(N_3)]^{5-}$. *Angew. Chem. Int. Ed. Eng.* 1991, 30, 1674–1677.

[78] Müller A., Rohlfing R., Döring J., Penk M. Formation of a cluster sheath around a central cluster by a "self-organization process": the mixed valence polyoxovanadate $[V_{34}O_{82}]^{10-}$. *Angew. Chem. Int. Ed. Eng.* 1991, 30, 588–590.

[79] Müller A., Hovemeier K., Rohlfing, R. A novel host/guest system with a nanometer large cavity for anions and cations: $[2NH_4^+Cl^-]\subset V_{14}O_{22}(OH)_4(H_2O)_2\text{-}(C_6H_5PO_3)_8]^{6-}$. *Angew. Chem. Int. Ed. Eng.* 1992, 31, 1192–1195.

[80] Dietrich B., Viout P., *Lehn J.M. Aspects de la chimie des composés macrocycliques*, InterEditions. Paris, 1991.

[81] Keggin J.F. The structure and formula of 12-phosphotungstic acid. *Proc. Roy. Soc. A* 1934, 144, 75–100.

[82] Contant R. Relations entre les tungstophosphates apparentés à l'anion $PW_{12}O_{40}^{3-}$. Synthèse et propriétés d'un nouveau polyoxotungstophosphate lacunaire $K_{10}P_2W_{20}O_{70}.24H_2O$. *Can. J. Chem.* 1987, 65, 568–573.

[83] Massart R., Contant R., Fruchart J.M., Ciabrini J.P., Fournier M. ^{31}P NMR studies on molybdic and tungstic heteropolyanions. Correlation between structure and chemical shift. *Inorg. Chem.* 1977, 16, 2916–2921.

[84] Knoth W.H., Harlow R.L. New tungstophosphates: $Cs_6W_5P_2O_{23}$, $Cs_7W_{10}PO_{36}$ and $Cs_7Na_2W_{10}O_{37}$. *J. Am. Chem. Soc.* 1981, 103, 1865–1867.

[85] Dawson B. The structure of the 9(18)-heteropoly anion in potassium 9(18)-tungstophosphate, $K_6(P_2W_{18}O_{62}).14H_2O$. Acta Cryst. 1953, 6, 113–126.

[86] D'Amour H. Vergleich der heteropolyanionen $[PMo_9O_{31}(H_2O)_3]^{3-}$, $[P_2Mo_{18}O_{62}]^{6-}$ und $[P_2W_{18}O_{62}]^{6-}$. *Acta Cryst. B* 1976, 32, 729–740.

[87] Acerete R., Harmalker S.P., Hammer C.F., Pope M.T., Baker L.C.W. Concerning isomerisms and interconversions of 2:18 and 2:17 heteropoly complexes and their derivatives. *J. Chem. Soc. Chem. Commun.* 1979, 777–779.

[88] Contant R., Tézé A. A new crown heteropolyanion $K_{28}Li_5H_7P_8W_{48}O_{184}.92H_2O$: Synthes is, structure and properties. *Inorg. Chem.* 1985, 24, 4610–4614.

[89] Contant R., Ciabrini J.P. Preparation and solution properties of some "defect" heteropolyanions related to 18-tungsto 2 phosphates (α and β isomers). *J. Chem. Res.* Miniprint 1977, 2601–2617.

[90] Zhang X., Chen Q., Duncan D.C., Campana C.F., Hill C.L. Multiiron polyoxoanions. Syntheses, characterization, X-ray crystal structures and catalysis of

H_2O_2-based hydrocarbon oxidations by $[Fe^{III}_4(H_2O)_2(P_2W_{15}O_{56})_2]^{12-}$. *Inorg. Chem.* 1997, 36, 4208–4215.

[91] Finke R.G., Droege M. W. Trivacant heteropolytungstate derivatives. 2. Synthesis, characterization and ^{183}W NMR of $P_4W_{30}M_4(H_2O)_2O_{112}{}^{16-}$ (M = Co, Cu, Zn). *Inorg. Chem.* 1983, 22, 1006–1008.

[92] Preyssler C. Etude de l'existence de l'anion phospho-3 tungstique 18. *Bull. Soc. Chim. Fr.* 1970, 30–36.

[93] Alizadeh M.H., Harmalker S.P., Jeannin Y., Martin-Frère J., Pope M.T. A heteropolyanion with fivefold molecular symmetry that contains a nonlabile encapsulated sodium ion. The structure and chemistry of $[NaP_5W_{30}O_{110}]^{14-}$. *J. Am. Chem. Soc.* 1985, 107, 2662–2669.

[94] Tourné C.M., Tourné G.F., Weakley T.J.R. Triaquahenicosatungstodiphosphate (6–) heteropolyanion, $[P_2W_{21}O_{71}\ (OH_2)_3]^{6-}$: X-ray crystallographic and ^{183}W NMR structural studies. *J. Chem. Soc. Dalton Trans.* 1986, 2237–2242.

[95] Fuchs J., Palm R. Struktur und schwingungsspektrum der heteropolyverbindung $K_{13}(KP_2W_{20}O_{72})[H_2O]x$. Z. Naturforsch. 1984, 39 b, 757–762.

[96] Fuchs J., Palm R. Strukturen und schwingungsspektren neuer wolframatophosphate. *Z. Naturforsch.* 1988, 43b, 1529–1537.

[97] Thouvenot R., Tézé A., Contant R., Hervé G. Synthesis, structure and ^{31}P and ^{183}W NMR spectra of $P_4W_{14}O_{58}{}^{12-}$. *Inorg. Chem.* 1988, 27, 524–529.

[98] Gatehouse B.M., Jozsa A.J. Preparation and structure of sodium octatungstotetraphosphate (12-)icosahydrate, $Na_{12}P_4W_8O_{40}.20H_2O$, a 1:2(X:M) heteropolyanion. *Acta Cryst. C* 1983, 39, 658–661.

[99] Wassermann K., Dickman M.H., Pope M.T. Self-assembly of supramolecular polyoxometalates: the compact, water-soluble heteropolytungstate anion $[As^{III}_{12}Ce^{III}_{16}(H_2O)_{36}W_{148}O_{524}]^{76-}$. *Angew. Chem. Int. Ed. Engl.* 1997, 36, 1445–1448.

[100] Müller A., Peters F., Pope M.T., Gatteschi D. Polyoxometalates: very large clusters-nanoscale. *Magnets, Chem. Rev.* 1998, 98, 239–271.

[101] Müller A., Roy S. En route from the mystery of molybdenum blue via related manipulatable building blocks to aspects of materials science. *Coord. Chem. Rev.* 2003, 245, 153–166.

[102] Mialane P., Dolbecq A., Lisnard L., Mallard A., Marrot J., Sécheresse F. $[\varepsilon\text{-}PMo_{12}O_{36}(OH)_4\{La(H_2O)_4\}_4]^{5-}$: The first $\varepsilon\text{-}PMo_{12}O_{40}$ Keggin ion and its association with the two-electron-reduced $\alpha\text{-}PMo_{12}O_{40}$ isomer. *Angew. Chem. Int. Ed. Engl.* 2002, 41, 2398–2401.

[103] Müller A., Beugholt C., Kögerler P., Bögge H., Bud'ko S., Luban M. $[Mo^V_{12}O_{30}(\mu_2\text{-}OH)_{10}H_2\{Ni^{II}(H_2O)_3\}_4]$, a highly symmetrical ε-Keggin unit capped with four Ni^{II} centers: synthesis and magnetism. *Inorg. Chem.* 2000, 39, 5176–5177.

[104] Müller A., Meyer J., Krickemeyer E., Beugholt C., Bögge H., Peters F., Schmidtmann M., Kögerler P., Koop M.J. Unusual stepwise assembly and molecular growth: $[H_{14}Mo_{37}O_{112}]_{14}$ and $[H_3Mo_{57}V_6(NO)_6O_{189}(H_2O)_{12}(MoO)_6]_{21}$. *Chem. Eur. J.* 1998, 4, 1000–1006.

[105] Khan M.I., Müller A., Dillinger S., Bogge H., Chen Q., Zubieta J. Cation inclusion within the mixed-valence polyanion cluster $[(Mo^{VI}O_3)_4Mo^V_{12}O_{28}(OH)_{12}]^{8-}$: syntheses and structures of $(NH_4)_7[NaMo_{16}(OH)_{12}O_{40}].4H_2O$ and $(Me_2NH_2)_6[H_2Mo_{16}(OH)_{12}O_{40}]$. *Angew. Chem. Int. Ed. Engl.* 1993, 32, 1780–1782.

[106] Müller A., Krickemeyer E., Meyer J., Bögge H., Peters F., Plass W., Diemann E., Dillinger S., Nonnenbruch F., Randerath M., Menke C. $[Mo_{154}NO_{14}O_{420}(OH)_{28}(H_2O)_{76}]^{(25\pm5)-}$: a water-soluble big wheel with more than 700 atoms and a relative molecular mass of about 24 000. *Angew. Chem. Int. Ed. Engl.* 1995, 34, 2122–2124.
[107] Müller A., Shah S.N.Q., Bögge H., Schmidtmann M., Molecular growth from a Mo_{176} to a Mo_{248} cluster. *Nature* 1999, 397, 48–50.
[108] Müller A., Beckmann E., Bögge H., Schmidtmann M., Dress A. Inorganic chemistry goes protein size: a Mo_{368} nano-hedgehog initiating nanochemistry by symmetry breaking. *Angew. Chem.* 2002, 114, 1210–1215.
[109] Sécheresse F., Dolbecq A., Mialane P., Cadot E. $\{Mo_2O_2X_2\}^{2+}$ (X = O, S), a magic building block for the design of wheel shaped metalates. *C. R. Chimie* 2005, 8, 1927–1938.
[110] Cadot E., Sécheresse F. Cyclic molecular materials based on $[M_2O_2S_2]^{2+}$ cores (M = Mo or W). *Chem. Commun.* 2002, 2189–2197.
[111] Dolbecq A., Cadot E., Sécheresse F. $[Mo_9S_8O_{12}(OH)_8(H_2O)_2]^{2-}$: a novel polyoxothiomolybdate with a Mo^{VI} octahedron encapsulated in a reduced Mo^{V} cyclic octanuclear core. *Chem. Commun.* 1998, 2293–2294.
[112] Korenev V.S., Boulay A.G., Dolbecq A., Sokolov M.N., Hijazi A., Floquet S., Fedin V.P., Cadot E. A new oxomolybdate component extracted from the "Virtual Dynamic Library" yielding the macrocyclic anion $[(Mo^{VI}{}_8O_{28})_4(Mo^{V}{}_2O_2S_2)_4]^{24-}$. *Inorg. Chem.* 2010, 49, 9740–9742.
[113] Cadot E., Sokolov M.N., Fedin V.P., Simonnet-Jégat C., Floquet S., Sécheresse F. A building block strategy to access sulfur-functionalized polyoxometalate based systems using $\{Mo_2S_2O_2\}$ and $\{Mo_3S_4\}$ as constitutional units, linkers or templates. *Chem. Soc. Rev.* 2012, 41, 7335–7353.
[114] Cadot E., Béreau V., Sécheresse F. Sulfur in Keggin heteropolyanions. Syntheses and multinuclear NMR characterizations (^{31}P, ^{183}W and ^{95}Mo) of γ-$[PW_{10}M_2S_2O_{38}]^{5-}$ (M = Mo, W). *Inorg. Chim. Acta* 1996, 252, 101–106.
[115] Cadot E., Béreau V., Marg B., Halut S., Sécheresse F. Syntheses and characterization of γ-$[SiW_{10}M_2S_2O_{38}]^{6-}$ (M = Mo^V, W^V). Two Keggin oxothio heteropolyanions with a metal-metal bond. *Inorg. Chem.* 1996, 35, 3099–3106.
[116] Béreau V., Cadot E., Bögge H., Müller A., Sécheresse F. Addition of $\{M_2S_2O_2\}^{2+}$, M = Mo, W, to A-α-$[PW_9O_{34}]^{9-}$. Synthesis and structural characterizations in the solid state and in solution. *Inorg. Chem.* 1999, 38, 5803–5808.

4 }

Precipitation

STRUCTURES AND MECHANISMS

The formation of a solid, and especially an oxide, from soluble metal complexes is usually called "precipitation." This term is a generic name that includes a set of complex and intricate phenomena. The process is governed by thermodynamic, structural, and kinetic contingencies, which should be examined in detail to understand the role of synthesis conditions and their influence on the solid obtained.

The chemistry of the process involves a condensation reaction, olation or oxolation, between uncharged hydroxylated complexes. It forms particles of widely variable size over the nano- to micrometric range. These particles are portions of a solid identifiable in using techniques such as X-ray diffraction, absorption, and diffusion, electron microscopy, light-scattering, and various spectroscopies. Of course, these particles have the properties typical of the corresponding bulky solid, but they may be modulated because of the size effect, especially in the nanometric range (Chap. 1). Because of their small size, these objects have a large surface area highlighting their surface physicochemistry, such as ability to disperse in aqueous or nonaqueous medium, to aggregate, and to fix various species from solution, that allows the surface energy to be controlled to adjust the shape and size of these objects (Chap. 5). The crystal structure of polymorphic solids can also be controlled by the choice of the pathway of their formation. Thus, knowledge of the processes involved allows us to exploit the large versatility of the nanostructures synthesized in solution.

This chapter has two main objectives. The first is to show that the crystalline structure of the solid may in many cases be anticipated from the characteristics of the precursor in solution, such as functionality, geometry, reactivity, and electron configuration. This point concerns the structural aspect of the formation of the solid. The second objective is to understand why precipitation forms small particles, generally of nano- or micrometric size, and how the crystallization mechanism influences their morphology. These questions concern the kinetics and dynamics of the precipitation phenomenon.

4.1 The Formation of Solids: Thermodynamics and Crystal Structure

Hydroxylation of metal complexes is the requisite condition for their condensation, but the condensation of electrically charged complexes is always limited. The reaction forms molecular-sized species, polycations, or polyanions, depending on the kind of precursors, aquohydroxo or oxohydroxo complexes, respectively (Chap. 3). Condensation is stopped at a low stage of growth because of the increased electrical charge in the condensed entities, which causes the hydroxo ligands' loss of nucleophilicity in polycations or the metal centers' loss of electrophilicity in polyanions. In contrast, the condensation of noncharged complexes occurs with no limitation and leads to a solid. The precipitation may occur over a relatively large range of acidity, sometimes far from the predominance zone of the noncharged species of a given metal. Of course, different forms of a cation generally coexist at a given pH, but if the concentration of the noncharged species is high enough to slightly exceed the solubility of the solid, precipitation can occur in changing the relative proportions of species in equilibria. In a general way, the precipitation of aquohydroxo $[M(OH)_z(OH_2)_{N-z}]^0$ complexes eliminates all the coordinated water and forms a hydroxide $M(OH)_z$ ($1 \leq z \leq 4$), while the oxohydroxo $[MO_a(OH)_b]^0$ complexes that condense by oxolation eliminate their hydroxo ligands in forming an oxide $MO_{z/2}$ ($4 \leq z \leq 6$).

A hydroxide $M(OH)_z$ may be unstable and spontaneously dehydrates more or less quickly to form an oxyhydroxide MOOH or hydrated oxide. The reaction occurs in solid state and generally leads to crystallization or to an important structural change. This reaction is favored all the more as the hydroxo ligands are heavily polarized by the cation in hydroxide; that is, the formal charge of the cation (its oxidation state) is high. The hydroxides of divalent elements, $M(OH)_2$ (M = Mn, Fe, Co, Ni, etc.), are more often stable (in the absence of oxidation), and they can be characterized by X-ray diffraction among other techniques. Trivalent elements exhibit contrasting behaviors. They may form stable hydroxides and oxyhydroxides ($Al(OH)_3$, AlOOH), oxyhydroxides and oxides (FeOOH, Fe_2O_3), hydroxides, oxydroxides and oxides ($Cr(OH)_3(OH_2)_3$, CrOOH, Cr_2O_3). With tetravalent elements (Ti, Zr), the polarization of the hydroxo ligands is so strong that only oxides such as TiO_2 and ZrO_2 are stable.

The elements that condense by oxolation form oxides which often are hydrated and appear to actually be macropolyacids. In the solid state, the residual hydroxo ligands become strong acids, releasing protons located in specific sites within the structure or on the surface of particles. Some examples of oxides of elements with high oxidation states, such as silicon, antimony, vanadium, and tungsten, are presented in the following section.

4.1.1 DIVALENT ELEMENTS

Neutralization of divalent elements around $h = 2$ generally forms a hydroxide $M(OH)_2$ rapidly. For many elements such as Mg, Mn, Fe, Co, and Ni, the hydroxide has the layered brucite-$Mg(OH)_2$ structure of CdI_2-type [1], and the particles are

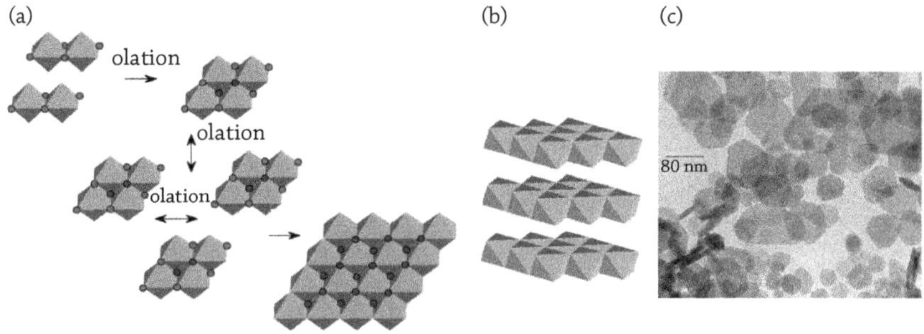

FIGURE 4.1 (a) Schematic formation pathway of brucite-type layers and (b) stacking of the layers in the brucite structure. (c) TEM image of $Mg(OH)_2$ particles.

hexagon-shaped platelets. The sheets of the structure are composed of $M(OH)_6$ octahedra sharing six edges through μ_3–OH bridges (Fig. 4.1).

The noncharged precursor $[M(OH)_2(OH_2)_4]^0$ helps us understand the formation of this structural type. The first step in the condensation of this bifunctional complex is the formation by olation of a dimer through a double μ_2–OH bridge (edge octahedra sharing). Because of the low charge of the cation, the μ_2–OH ligand retains a high nucleophilicity in the dimer, allowing an increase of its coordination in forming μ_3–OH bridges in plane tetramers. They can condense in turn by olation to form the characteristic sheets of the brucite-type structure, which are stacked and linked by van der Waals bonds in forming hexagonal platelets. All coordinated water is eliminated in the process. This global scheme allows us to account for the formation of the brucite-type structure. However, it is interesting to observe the influence of acidity during the formation of hydroxides such as Mg and Ni. It is also useful to understand why some divalent elements such as Cu, Pd, and Zn form hydroxides or oxides with structural types other than brucite.

a) Magnesium

The acidity of the precipitation medium of brucite, $Mg(OH)_2$, notably influences the texture of the particles. They are formed in introducing progressively the solution of magnesium nitrate in a solution of sodium hydroxide, the pH being maintained constant by using a potentiometric device [2]. The solids formed at pH 10, 11, and 12, aged in suspension and dried at 60°C, exhibit the unit-cell parameters of the brucite structure. The diameter of platelets does not significantly vary, but their thickness increases when the pH is increased, as indicated by the width of the *hkl* lines in the X-ray diffraction patterns. The thickness of the platelets increases to around 10 nm for the sample formed at pH 12.

The solid formed at pH 10 and aged at RT resembles a gel made of crumpled layers (Fig. 4.2). Its X-ray diffraction pattern exhibits a split 001 line corresponding to a *c* unit-cell parameter of 7.6 Å instead of 4.77 Å for brucite (Fig. 4.2).

This is compatible with the existence of a "hydrated brucite," a phase containing water molecules in the interlayer space, but without turbostratic disorder in the layer stacking, which should lead to extinction of the *hkl* lines. As the solubility of

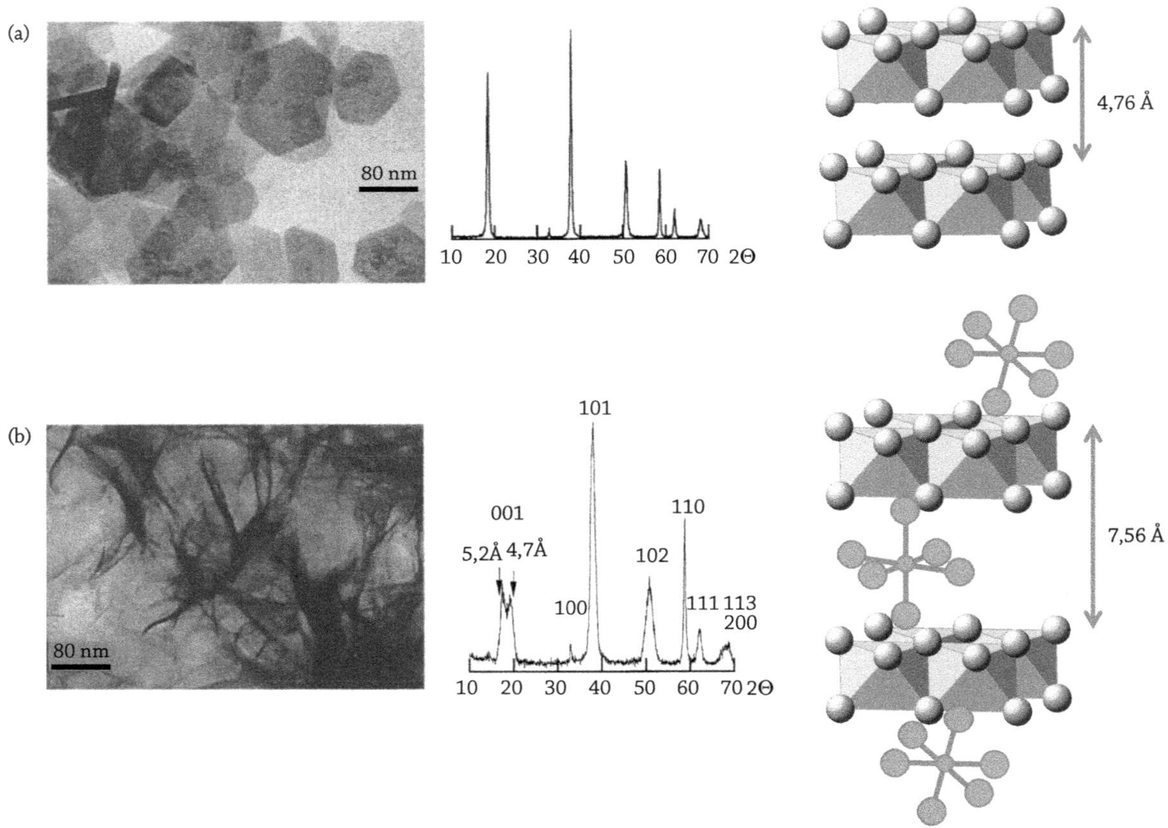

FIGURE 4.2 TEM images, X-ray diffraction patterns and schemes of the structure of magnesium hydroxide particles formed and aged at (a) pH 12, 60°C, (b) pH 10, RT.

the hydroxide suddenly increases at pH < 12 (10^{-3} mol.l^{-1} at pH 10), it seems more likely that partly hydroxylated complexes, such as $[Mg(OH)(OH_2)_5]^+$, are adsorbed on sheets of global $Mg(OH)_2$ composition involving hydrogen bond interactions or substitution of labile water molecules (Fig. 4.2). Adsorption of such electrostatically charged complexes should allow stabilization of the solid as a gel because of the accumulation of positive charges on the sheets, leading to mutual repelling and an increase in their spacing of approximately the size of one water molecule. According to this hypothesis, when the solid formed at pH 10 is then aged at pH 12, the splitting of the 001 line in the diffraction pattern disappears, and well-defined hexagonal platelets are obtained. The increase in pH probably permits eliminating the adsorbed charged complexes and forcing their integration in the crystalline solid.

b) Nickel

Progressive hydroxylation of Ni^{2+} ions with sodium hydroxide has been studied by extended X-ray absorption fine structure (EXAFS), small-angle X-ray scattering (SAXS), and transmission electron microscopy (TEM) [3]. From the beginning of hydroxylation, two phases are found to coexist: at OH/Ni = 0.6, rods 1 to 2 nm in width and around 200 nm in length, aggregated in fibers of around 4 to 15 nm in diameter and 1 μm in length, are in the presence of soluble monomer complexes, as deduced from the EXAFS study. At OH/Ni = 1.6, the solid matter is formed of fibers and thin hexagonal platelets of $Ni(OH)_2$ hydroxide of brucite structure. At OH/Ni = 2, the fibers totally disappear, and only well-defined hexagonal platelets of $Ni(OH)_2$ are present (Fig. 4.3).

The main species existing in solution at the global hydroxylation rate of 0.6 is $h = 1$ among $[Ni(OH)_h(OH_2)_{6-h}]^{(2-h)+}$ ($0 \leq h \leq 2$) complexes. Its condensation should form polycations, but instead of forming compact entities, the process produces polycationic chains. Their diameter, evaluated by SAXS, and their topology, analyzed by EXAFS, correspond well to the association of plane tetramers (Fig. 4.4).

As seen earlier, the partial charge of the μ_2–OH ligand in the $[Ni_2(OH)_2(OH_2)_8]^{2+}$ dimer is negative, and the water molecules bear a positive charge. The hydroxo ligands can then increase their coordination in forming μ_3–OH ligands in plane tetramers with elimination of water molecules. The process continues, but as the

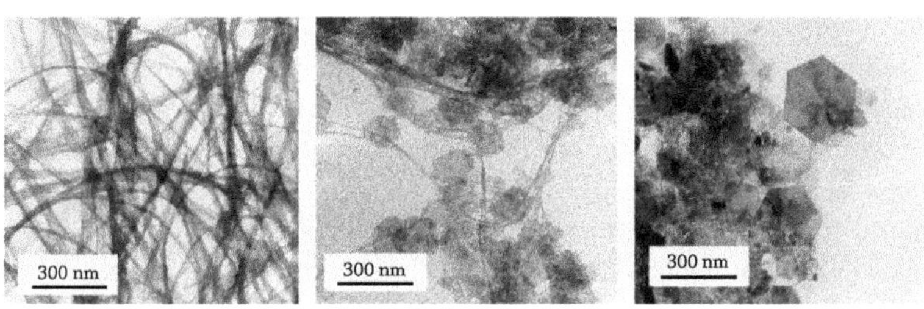

FIGURE 4.3 TEM images of suspensions of nickel hydroxide aged during 1 month (reproduced with permission from [3] Copyright 2003 American Chemical Society).

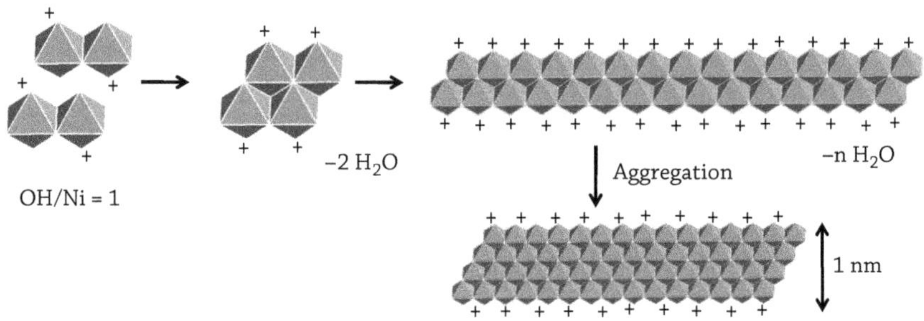

FIGURE 4.4 Scheme of possible formation of polycationic fibers from the $[Ni(OH)(OH_2)_5]^+$ complex.

tetramers are positively charged, the growth is directed by the minimization of electrostatic interactions to form polycationic chains, which in turn, aggregate in fibers, as shown by electron microscopy (Fig. 4.3). However, such structures are fragile because they are destroyed by the addition of base and, for OH/Ni = 2 corresponding to the noncharged $h = 2$ complex existence, condensation occurs in a two-dimensional way, leading to hydroxide in the form of hexagonal platelets. This singularity in the behavior of nickel is likely due to the inertness of Ni^{2+} ions toward nucleophilic substitution reactions. The cation with the d^8 electron configuration is indeed strongly stabilized in octahedral symmetry, so that the changes in geometry of the coordination sphere of Ni^{2+} ions imposed by the substitution reaction involve high-activation energies. Thus, the reconstruction of polycationic units during hydroxylation and the decrease of their charge are likely relatively difficult, leading to the formation of polycationic fibers before their restructuration in platelets of $Ni(OH)_2$.

The inertness of Ni^{2+} ions towards substitution may also explain the formation of a turbostratic compound when $Ni(OH)_2$ precipitation is made by the addition of ammonia into a solution of nickel nitrate. The hydroxide layers are parallel but disoriented in their stacking (the X-ray diffraction pattern exhibits only *hk* bands and 00*l* lines) [4]. An important amount of water associated with the layers by hydrogen bonds and nitrate ligands, strongly bounded and located between the hydroxide layers, hinders the structuration of the solid.

c) Copper

$Cu(OH)_2$ crystallizes not with the brucite-type structure, but with that of boehmite, γ-AlOOH, or lepidocrocite, γ-FeOOH. The structure is made of corrugated layers of edge-sharing octahedra (Fig. 4.5) [1]. This peculiarity of copper likely derives from the distortion of the coordination polyhedra of Cu^{2+} ions by the Jahn–Teller effect due to the d^9 electron configuration. The hexaaquo $[Cu(OH_2)_6]^{2+}$ complex exhibits four short (1.94 Å) and two long (2.40 Å) Cu–O bond lengths. That causes distinct polarization and consequently a strongly differentiated acidity of the coordinated water molecules in axial or equatorial positions within the coordination polyhedron. As the most acidic water molecules are those with the shortest bond length, hydroxylation of the noncharged complex is then confined in the equatorial plane, so that their condensation by olation is oriented to form chains carrying water molecules only in the axial position. Subsequent interchain condensation leads to the elimination

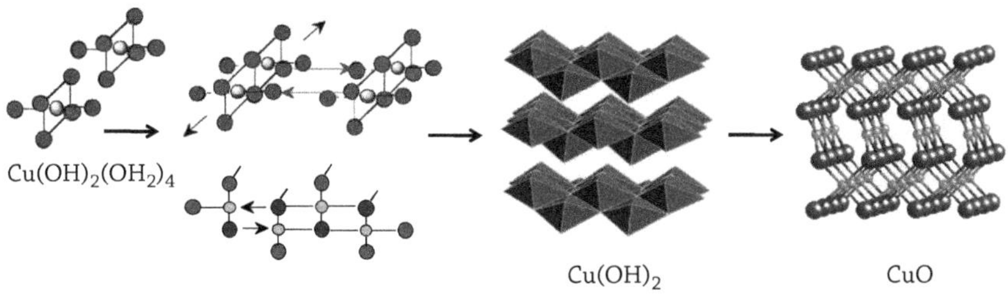

FIGURE 4.5 Scheme of possible formation pathway of copper hydroxide. Structures of $Cu(OH)_2$ and CuO.

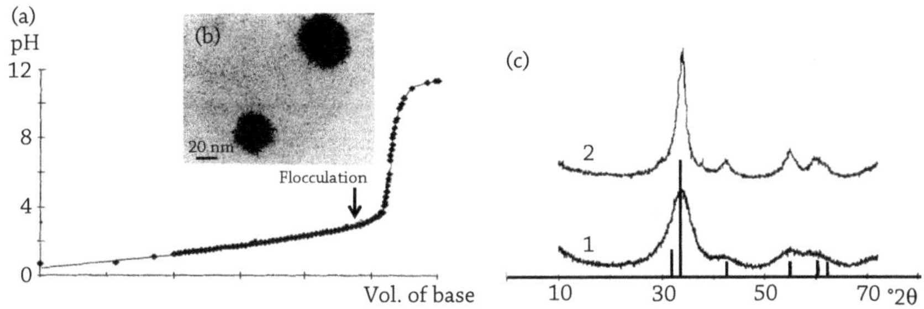

FIGURE 4.6 (a) Titration curve of an acidic solution of palladium nitrate ($C_{Pd} = 5.10^{-2}$ mol.l^{-1}) with sodium hydroxide; (b) TEM image of isolated particles and aggregates of PdO precipitated at pH 2.8 (the nonaggregated particles are 1.8 nm in mean diameter); (c) X-ray diffraction pattern of the solid obtained from a solution of palladium nitrate (1) by sodium hydroxide addition, (2) by thermolysis at 100°C for 5 hours (the vertical bars correspond to the diffraction pattern of PdO) (adapted from [5]).

of these water molecules in forming the corrugated hydroxide planes. Notably, the OH ligands form μ_4–OH bridges (Fig. 4.5) with 5-coordinated oxygen atoms. That explains the instability of copper hydroxide, which easily transforms into tenorite, CuO, at high pH in suspension or by mild heating in solid state. In the oxide, the Cu^{2+} ions are hexa (4 + 2) coordinated: they exhibit the 4-coordination in chains of square-planar complexes (Cu-O distance of 1.955 Å) and two interactions at longer distance with oxygen atoms of the nearby chains (Cu-O distance of 2.776 Å).

d) Palladium and Platinum

Pd^{2+} and Pt^{2+} ions, with the same d^8 electron configuration as Ni^{2+}, do not form stable hydroxides. They precipitate in forming directly an oxide MO (M = Pd, Pt), in which the cation is square-planar coordinated [1]. The same coordination of Pd^{2+} and Pt^{2+} cations in their soluble complexes (§ 2.2) explains the strong acidity of aquo species (pK_a close to 2). Accordingly, hydroxylation and precipitation of palladium in a $Pd(NO_3)_2$ solution are quasi-complete at pH 3 [5] (Fig. 4.6). The X-ray diffraction pattern of the early solid corresponds to that of PdO.

The condensation of noncharged $[M(OH)_2(OH_2)_2]$ complexes proceeds by olation, likely leading to small chains of $M(OH)_2$ composition (Fig. 4.7). In turn, these small chains condense by oxolation with loss of water and formation of

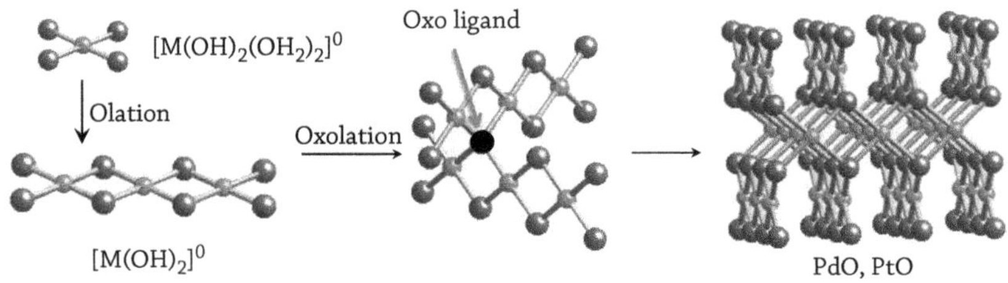

FIGURE 4.7 Scheme of the possible formation pathway of palladium or platinum oxide and the crystal structure of these oxides.

tetrahedrally coordinated oxo ligands, which are characteristic of the structure of PdO and PtO oxides. This structure resembles that of CuO, though it has a higher symmetry.

The particles are spheroidal with a mean diameter close to 2 nm, as evaluated by electron microscopy and diffraction line widening. The particle size is not sensitive to the acidity of the precipitation medium, the nature of the base, or the temperature up to 80°C. It is very difficult to obtain larger particles, which probably indicates that the oxolation reaction leading to oxide is very fast since hydroxide is never detected. However, the thermolysis at 100°C during 5 hours of acidic solutions leads to PdO nanoparticles 4 nm in mean diameter. Under such conditions, the nucleation of the solid is restricted because of the strong acidity. The growth is therefore favored without dissolution–crystallization equilibria take place (§ 4.2).

e) Zinc

The previous examples show that it is possible to understand the behavior of divalent ions and the crystalline structure of their hydroxides and oxides, taking into account electronic effects that influence the reactivity or stereochemistry of the cation. For other metals, especially zinc, the situation is more complicated because the Zn^{2+} cation is too small to form an hydroxide with the brucite-type structure, and precipitation generally leads to the ZnO oxide with the würtzite structure in which the Zn^{2+} ions are tetrahedrally coordinated (Fig. 4.8c).

Zinc exists in acidic medium as hexaaquo $[Zn(OH_2)_6]^{2+}$ complexes that have a weak tendency toward hydroxylation. Zinc has a high solubility in alkaline medium as zincates $[Zn(OH)_3OH_2]^-$ and $[Zn(OH)_4]^{2-}$ because of its propensity to adopt 4-tetrahedral coordination, similar to aluminum ions (§ 2.2). Therefore, zinc's range of precipitation is narrow—approximately between pH 8 and pH 11 [6].

Alkalinization with ammonia in the range of pH 10 to 11 of zinc nitrate at room temperature forms an hydroxide, ε-$Zn(OH)_2$ wülfingite, in which the Zn^{2+} ions are tetrahedrally coordinated by hydroxyl ion within a corner-sharing tetrahedra framework (Fig. 4.8b). It is the main solid phase obtained from ambient to 60°C. At 70°C, the ZnO oxide is exclusively formed, as indicated by X-ray diffraction and infrared spectroscopy (ZnO is transparent to infrared radiation, while the

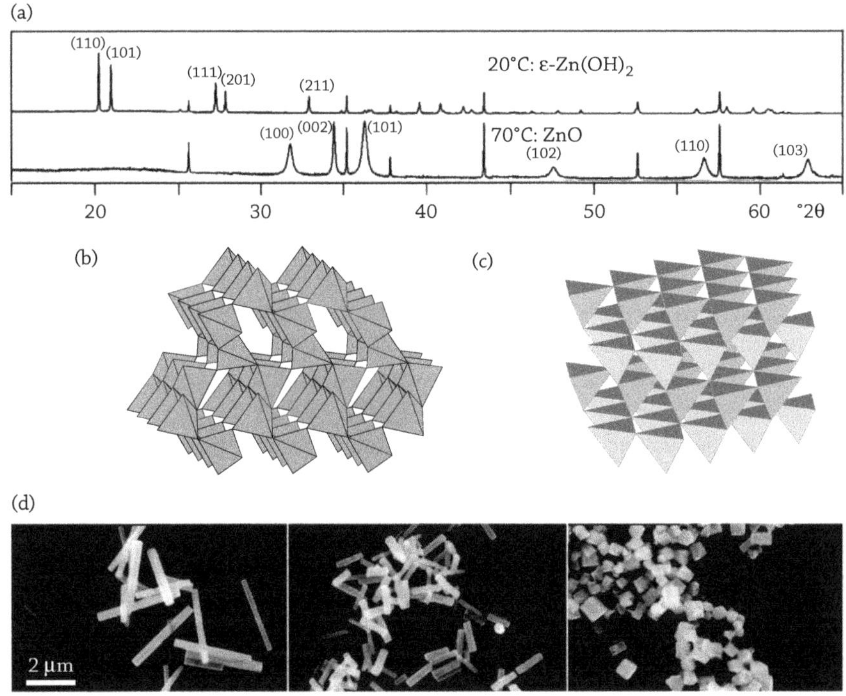

FIGURE 4.8 (a) X-ray synchrotron diffraction patterns of solids precipitated with ammonia in a zinc nitrate solution at various temperatures (reproduced from [7] with permission of The Royal Society of Chemistry). Structures of (b) wülfingite $Zn(OH)_2$ and (c) würtzite ZnO. (d) SEM images of ZnO nanoparticles formed at 90°C by hexamethylene tetramine decomposition in the presence of (from left to right) glycolate, malate, and citrate (mono-, bi-, and tridentate ligands, respectively). The particles formed in the presence of glycolate are indiscernible of the check sample (reproduced with permission from [Meagley K.L., Garcia S.P., Chemical control of crystal growth with multidentate carboxylate ligands: Effect of ligand denticity on zinc oxide crystal shape, Cryst. Growth Des. 2012, 12, 707–713] Copyright 2012 American Chemical Society).

O–H bond strongly absorbs) [7, 8] (Fig. 4.8a). Isotopic exchange and tracking of ^{18}O have shown that ZnO is formed mainly in suspension by in situ transformation of the hydroxide, involving a dehydration reaction in solid state rather than a dissolution–crystallization process. The surface of particles, however, retains its hydroxylation [7].

A morphological design of the ZnO particles can be made by controlling the reaction conditions. The growth of hydroxide can be limited before its transformation into oxide—for instance, in precipitating zinc in a hot solution or in heating the suspension after aging at room temperature [9]. The use of various multidentate ligands, especially citrate, may also be an interesting way to control the particle size and shape (Fig. 4.8d) [8, 10, 11]. An extensive literature has been produced on the synthesis of ZnO nanoparticles, focusing on their various applications, for instance in piezoelectric devices and, since ZnO is a large-gap semiconductor, in light-emitting diodes.

4.1.2 LAYERED DOUBLE HYDROXIDES

Layered double hydroxides (LDHs) form a large family of di- and trivalent mixed hydroxides. Their crystal structure is made of a stacking of brucite-type layers $[M^{II}_{1-x}M'^{III}_{x}(OH)_2]^{x+}$ in which part of the divalent cations is replaced by trivalent cations (Fig. 4.9). The layers bear a positive charge, which is balanced by that of anions present in the interlayer space [12] and are exchangeable [13]. The structural prototype is the mineral hydrotalcite, $[Mg_6Al_2(OH)_{16}]CO_3{\bullet}4H_2O$, so called because of its likeness to talc. Another isostructural mineral is pyroaurite, a compound containing ferric ions, $[Mg_6Fe_2(OH)_{16}]CO_3{\bullet}4H_2O$.

The LDH family includes many compounds with a large variety of divalent (Mg, Zn, Ni, Co, Mn, etc.) and trivalent (Al, Cr, Fe, V, Co, Ga, etc.) cations and also anions (CO_3^{2-}, SO_4^{2-}, Cl^-, NO_3^-, etc.). Water molecules are also present inside the interlayer spaces. In fact, many cations may be present in the cationic sites within the brucite-type layers, providing they are the right size. Be^{2+} ions are too small and Ba^{2+} ions are too big to adopt the octahedral coordination and occupy the cationic sites of the framework. Numerous anions, including complex ions, heteropolymetalates, or organic anions, can be introduced into the structure in varying the interlayer distance [12–14].

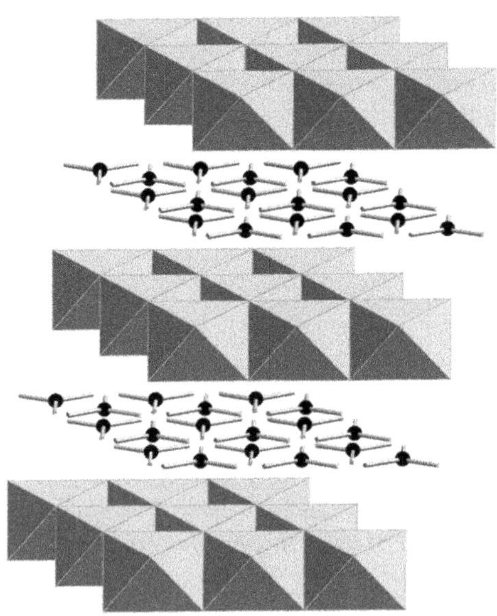

FIGURE 4.9 Crystal structure of hydrotalcite $[Mg_6Al_2(OH)_{16}]CO_3{\cdot}4H_2O$ (carbonate ions are schematically drawn between the layers).

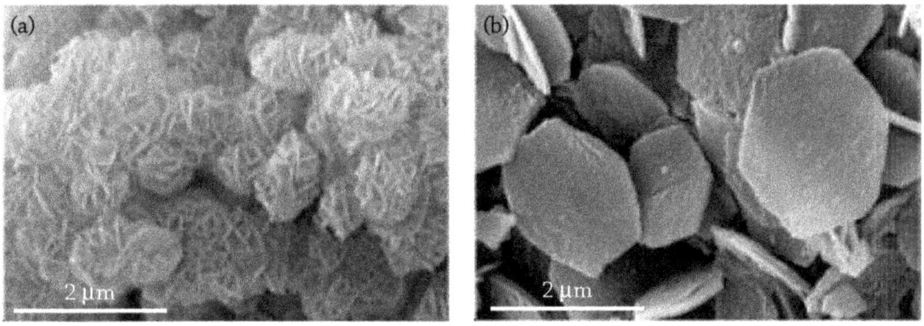

FIGURE 4.10 SEM images of particles of hydrotalcite $[Mg_6Al_2(OH)_{16}]CO_3{\cdot}4H_2O$ synthesized by coprecipitation (a) at 20°C (pH 10, sodium carbonate buffer) and aging for 25 hours, inducing platelet aggregates in the shape of sand roses; (b) at 90°C with urea and aging for 48 hours (reproduced from [17] with permission of The Royal Society of Chemistry).

LDHs are easily formed by coprecipitation on the composition range of $0.2 \leq M'^{III}/M^{II} \leq 0.33$. Alkalinization has been carried out with sodium hydroxide, sodium carbonate [12, 15], or urea [14, 16, 17]. The aging of suspensions under heat (90–100°C) or in hydrothermal conditions allows us to obtain well-defined and well-crystallized particles (Fig. 4.10).

The titration curve of the $(2M^{2+} + Al^{3+})$ mixture [18] shows that alkalinization begins by precipitation of pure $Al(OH)_3$, followed by LDH formation during subsequent hydroxylation of the divalent cation, as shown by the two successive steps in the titration curve (Fig. 4.11) and confirmed by the structural study of the solid.

The mechanism involved in LDH formation requires the transformation of aluminum hydroxide, which is formed early during adsorption of partly hydroxylated divalent ions. The pH at which LDH is formed is indeed far below that corresponding to the precipitation of the divalent cation alone as hydroxide $M(OH)_2$. Therefore, the LDH phase results from a dissolution–crystallization mechanism, as evidenced by EXAFS, SAXS, and TEM studies of the formation of the Ni_2Ga phase containing cations chosen for their strong absorption of X-rays [19, 20]. A similar mechanism is involved in the crystallization of green rusts, which are isostructural ferric–ferrous compounds (§ 7.2.3). Cr^{III}-based LDH involves a different formation mechanism, as indicated by the titrating curve of $(M^{2+} + Cr^{3+})$ mixture exhibiting only one step at pH 3–4 and limited by a single equivalent point indicating the end of LDH formation (Fig. 4.11) [21]).

The formation of $Zn_2Cr(OH)_6Cl{\cdot}2H_2O$ at pH 5 has been studied by EXAFS at Cr K-edge and Zn K-edge [22]. It has been clearly shown that the liquid phase contains Zn^{2+} hexaaquo complexes and various Cr^{III} complexes and polycations. EXAFS spectra of the solid phase contains only signals characteristic of LDH. As there was no precipitation of chromium trihydroxide (at the difference with aluminum), it was concluded that Zn_2Cr LDH is formed in only one step during the addition of the cations in solution. Because zinc hexaaquo complexes are unable

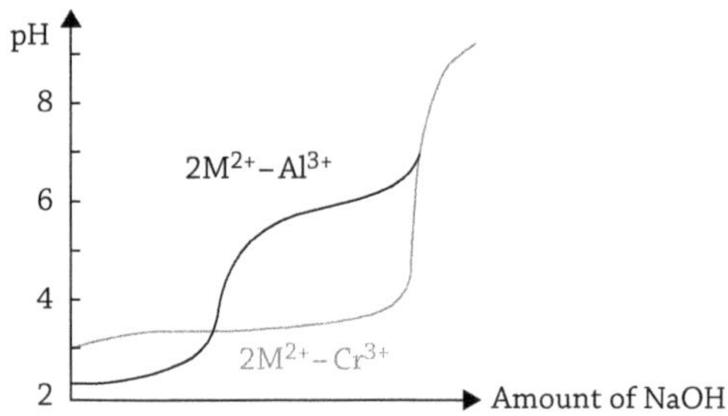

FIGURE 4.11 Titration curve of $2M^{2+}+Al^{3+}$ mixture: the two successive equivalent points correspond to additions of base in the 3:4 ratio. They indicate the complete precipitation of Al^{3+} ions at pH4, followed by that of the M^{2+} ions at pH 8. The titration curve of the $2Zn^{2+}+Cr^{3+}$ mixture with sodium hydroxide exhibits only one equivalent point at pH 8, corresponding to the same total amount of base as that in the previous case. This indicates simultaneous hydroxylation of the two cations (adapted from [18, 21]).

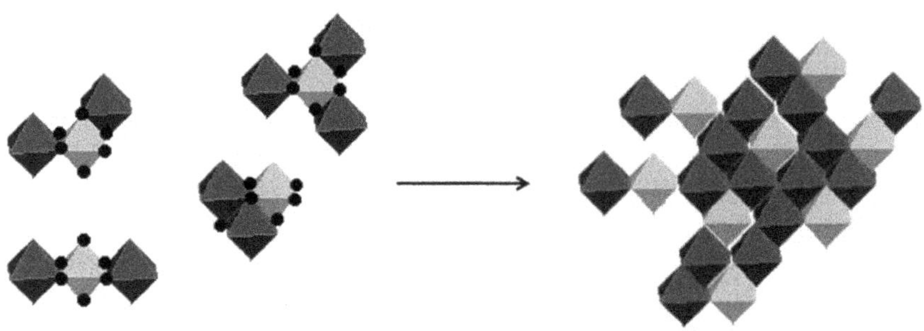

FIGURE 4.12 Association of $[Zn_2Cr(OH)_6(OH_2)_8]^+$, $[Zn_2Cr(OH)_6(OH_2)_7]^+$, and $[Zn_3Cr(OH)_6(OH_2)_{12}]^{3+}$ complexes in the formation of $[Zn_2Cr(OH)_6]^+$ LDH layers. Dark and light octahedra represent the coordination polyhedra of zinc and chromium, respectively (reproduced with permission from [22] Copyright 2001 American Chemical Society).

to condense and because hydroxylated Cr^{3+} complexes, inert toward condensation (§ 3.2 and § 4.1.3), are strongly nucleophilic species, the formation of mixed complexes with the same stoichiometry as the initial mixture (Zn/Cr = 2) should occur according to the following scheme:

$$\left[Cr(OH)_x(OH_2)_{6-x}\right]^{(3-x)+} + 2\left[Zn(OH_2)_6\right]^{2+} + (6-x)\,OH^- \rightarrow$$
$$[Zn_2(\mu_2-OH)_4Cr(OH)_x(OH_2)_8]^+ + (10-x)\,H_2O$$

Subsequent condensation of these mixed complexes has to occur by elimination of all water coordinated to zinc in forming μ_3–OH bridges, which will lead to positively charged $[Zn_2Cr(OH)_6]^+$ layers (Fig. 4.12). Such a pathway helps us understand how cation ordering takes place within the layer, in which a chromium ion is always surrounded by six zinc ions, and to recognize that the Zn/Cr = 2 stoichiometry is the unique composition observed for this LDH type.

4.1.3 TRIVALENT ELEMENTS

Aluminum, chromium, and iron are the most interesting examples of trivalent elements because of their rich crystal chemistry, the various properties of their oxides, and their multiple uses for extremely varied technological applications. Aluminum and iron are also important elements in the environment, where they play a major role because of their abundance in the Earth's crust and their role in many biogeochemical processes. Their detailed study is presented in Chapter 6 (Aluminum) and Chapter 7 (Iron). In this section, we present the specific details important in understanding the differences in behavior of aluminum, chromium, and iron, which also exhibit strong likenesses.

The solution chemistry of trivalent elements is strongly diversified (§. 3.2). Chromium forms small polycations that are kinetically stable, whereas iron, being very labile, leads in similar conditions to nanoparticles. Aluminum is the only one of these three elements that forms the Al_{13} polycation by direct hydroxylation in water. The crystal chemistry of these elements is also extremely differentiated, even if they form isostructural MOOH oxyhydroxides and M_2O_3 oxides, but more often under most different conditions. These specific behaviors may be understood on the basis of size and the electron configuration of these cations.

The acidity of the cations in solution increases from Cr, Al to Fe, as indicated by the pH at the beginning of precipitation (Fig. 4.13). Some characteristics of these ions are given in Table 4.1.

Hydroxylation of aquo cations:

$$\left[M(OH_2)_6\right]^{3+} + H_2O \Leftrightarrow \left[M(OH)(OH_2)_5\right]^{2+} + H_3O^+{}_{\text{solvated}}$$

is fast for the three ions ($k \approx 10^5\ s^{-1}$) [6]. Hydrolysis is indeed a simple acid–base reaction that concerns the exchange of protons between coordinated water molecules

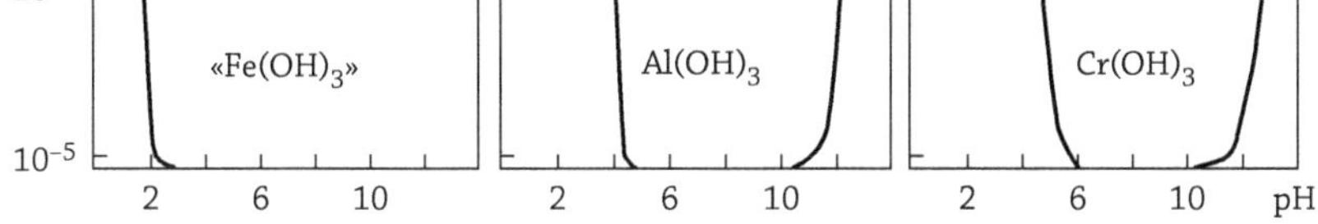

FIGURE 4.13 Solubility diagrams of Fe^{III}, Al^{III}, Cr^{III} [from Charlot G., L'analyse qualitative et les réactions en solution, 6ème éd., Masson, Paris, 1969].

TABLE 4.1 } Some characteristics of trivalent elements.

	Al^{3+}	Cr^{3+}	Fe^{3+}
Electron configuration	$2p^6$	$3d^3$	$3d^5$
Ionic radius (Å)	0.53	0.65	0.64
$\chi_M{}^*$	1.47	1.59	1.72

and the solvent, involving a low activation energy (§ 2.3). On the other hand, the lability of water molecules coordinated to these cations is variable, causing olation kinetics to be very different. The exchange rate constants of coordinated water molecules are around 10^{-6}, 1 and 10^{2} s^{-1} for Cr^{3+}, Al^{3+}, and Fe^{3+} respectively [23]. Chromium (III), with the d^3 electron configuration strongly stabilized by the crystal field in octahedral symmetry, exhibits a strong chemical inertness toward substitution reactions (§ 3.2). For iron (III) in high spin d^5 electron configuration typical of oxygenated surroundings, there is no crystal field effect, whatever the symmetry of the complex. Consequently, the coordinated water molecules are highly labile. For aluminum, the lability is less than that for iron because of the smaller size of the Al^{3+} cation and a higher polarization of the coordinated water molecules.

Precipitation of the three cations occurs at the hydroxylation rate $h \approx 2.5$ from which the concentration of noncharged species is high enough that nucleation of solids takes place. Hydroxylation of the cation may be performed according to several different procedures: addition of a base, thermohydrolysis, or hydrothermal treatment. These procedures can often lead to solid phases with different crystal structures and various particle morphologies. The precipitation pathways may also be modified because kinetic and/or thermodynamic factors induced by these different procedures can change the behavior of complexes in solution as well as the properties of the solvent itself.

Fast alkalinization of an aluminum nitrate solution at pH 6 to 7 with soda or ammonia immediately provokes the formation of a translucent gel, which is amorphous in X-ray diffraction. ^{27}Al NMR indicates only hexacoordinated aluminum. The transparency of the gel is due to the absence of particles or fluctuations in density within the bulk which are light scatters. However, the gel results from the rapid aggregation of various sized species, which did not have time to be ordered at large distance because of the sudden increase in viscosity blocking their motion within the bulk. A gel is a highly metastable material because of its disorder and heavy hydration. The solvent is trapped in a loose, solid network, and the solvent prevents the network from collapsing. The driving force to evolve is the crystallization of a solid, allowing the lattice energy to be released in lowering the free energy of the system. The nature of the solid phases obtained is greatly dependent on the acidity: at pH 6–7, oxyhydroxide γ-AlOOH (boehmite) forms, while at pH ≤ 5 or pH ≥ 8, hydroxide $Al(OH)_3$ as gibbsite or bayerite, respectively, is obtained (Chap. 6). The pathway followed in the transformation of the gel is dependent on the solubility of aluminum: dehydratation and structuration in the solid state at pH 6–7, dissolution and crystallization on both sides of this acidity range. At pH < 6 and pH > 8, the gel transforms involving the transport of matter within the liquid phase (§ 4.2.3). Soluble species slowly generated at low concentration and likely as monomers feed the growth of hydroxide $Al(OH)_3$, which is thermodynamically more stable than boehmite at room temperature ($\Delta G^{\circ}_{f\,298K}$ (kJ.mol.l^{-1}): $Al(OH)_3 = -546$, γ-AlOOH = − 436).

The behavior of ferric ions is somewhat similar to that of Al^{3+} ions. Addition of a base in a solution of ferric nitrate forms from acidic media (pH 2 – 2.5) a gelatinous precipitate because condensation of ferric ions does not easily stop owing to their high lability and reactivity. The ferric gel is made of an hydrated and loosely organized oxyhydroxide (ferrihydrite), with a $Fe_4(O, OH, H_2O)_{12}$ approximative composition (Chap. 7). Aging of the gel at room temperature or under mild heating leads, at $5 \leq pH \leq 10$, mainly to hematite, α-Fe_2O_3, and, outside of this acidity range, to goethite, α-FeOOH. At the difference of $Al(OH)_3$, ferric hydroxide $Fe(OH)_3$ does not exist.

The behaviors of aluminum and iron are similar in forming quasi-immediately nonstructured gels that evolve according to analogous mechanisms: transformation of the early solid in situ or by dissolution–crystallization. The two cations are highly reactive toward condensation and precipitation, but because iron has a higher polarizing power than aluminum toward oxygenated ligands, hydroxides and oxyhydroxides form with Al^{3+} instead of oxyhydroxides and oxide forming with Fe^{3+}.

Chromium (III) differs from its two congeners by an exceptional chemical inertness. A fast addition of base into a solution of Cr^{3+} ions also forms a gel which results from the charge cancelling of the soluble complexes and polycations in solution. There is no ligand substitution between species of zero charge at least in the early stages. Various hydrated hydroxides composed of aggregates of the molecular species preexisting in solution are obtained in this way.

The $[Cr(OH_2)_6]^{3+}$ monomeric complex precipitates by base addition the hydrated hydroxide $Cr(OH)_3(OH_2)_3$, blue-gray in color, in which the $Cr(OH)_3(OH_2)_3$ octahedra are linked by $[H_3O_2]^-$ double bridges resulting from hydrogen-bonded hydroxo and aquo ligands (Scheme 4.1) [24, 25]. Chromium hydroxide can also be obtained by acidification of alkaline solutions of chromite $[Cr(OH)_4(OH_2)_2]^-$ [26]. Its structure consists of planes of hexagons similar to those existing in aluminum hydroxides $Al(OH)_3$ (Fig. 4.14). In this phase, the $[H_3O_2]^-$ groups form chelating ligands coordinated with the monomer precursor $[M(OH)_3(OH_2)_3]^0$ or $[M(O_2H_3)_3]^0$ (Scheme 4.1a). The coordination mode of the $[H_3O_2]^-$ ligands can easily become bridging through a simple change in the orientation of hydrogen bonds between the OH and OH_2 groups (Scheme 4.1b). The formation of chains and rings is possible in this way. Their size is dependent on sterical constraints imposed by the bridging ligands. The rings can easily aggregate and form embryos of the solid phase.

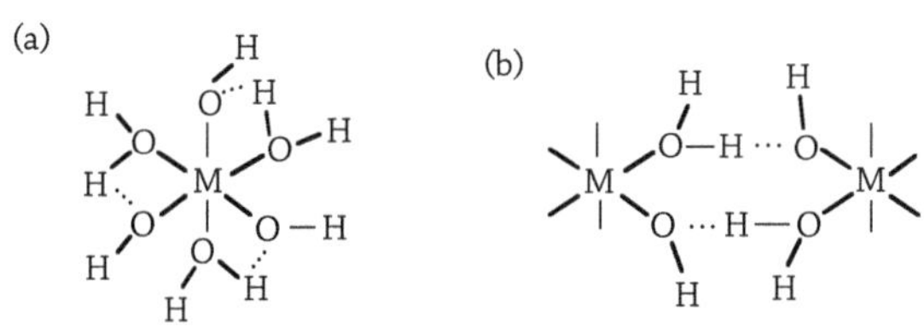

SCHEME 4.1

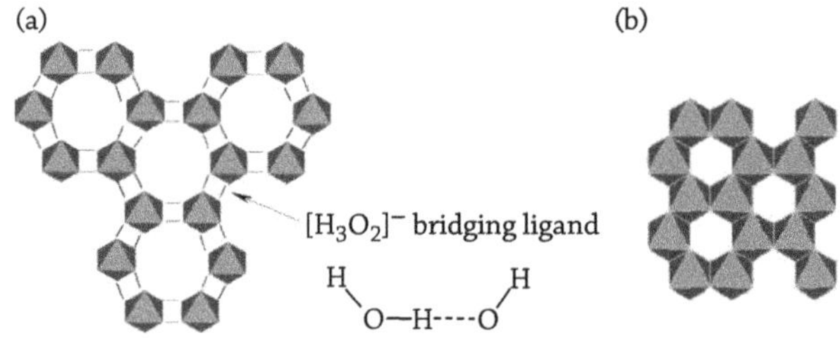

FIGURE 4.14 (a) Scheme of layers in chromium hydroxide $Cr(OH)_3(OH_2)_3$ in which the $[Cr(OH)_3(OH_2)_3]$ octahedra are linked by the $[H_3O_2]^-$ bridging ligand. (b) Structure of layers in aluminum hydroxide $Al(OH)_3$ formed of $Al(OH)_6$ octahedra sharing three edges.

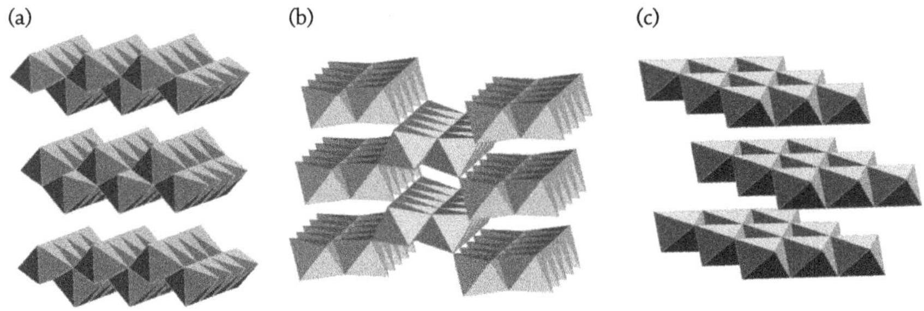

FIGURE 4.15 Structures of boehmite γ-AlOOH (a), goethite α-FeOOH (b), and α-CrOOH (c).

Because of the chemical inertness of chromium (III), the structure of dimer and trimer polycations (Fig. 3.9) is retained during their neutralization in a pyridine buffer (pH ≈ 6.9) that forms the blue–green $Cr_2(\mu\text{–}OH)_2(OH)_4(OH_2)_4{\cdot}2\,H_2O$ [27] and the light green $Cr_3(\mu\text{–}OH)_4(OH)_5(OH_2)_4{\cdot}4\,H_2O$ [26] hydroxide gels. In their structures, the dimer and trimer entities, respectively, are linked by $[H_3O_2]^-$ ligands. These hydroxides and also $Cr(OH)_3$ are not stable phases because they transform from 60°C or by aging at room temperature into a green-colored amorphous phase characteristic of chromium gels [28] and then into oxyhydroxide CrOOH and more or less hydrated oxide Cr_2O_3 [24, 26, 29–31]. The formation of hydrated chromium(III) hydroxides results from the inertness of the cation toward ligand substitution, but finally, chromium gets closer to iron than to aluminum in forming oxyhydroxides and oxide in solution.

Thermolysis at 80 to 100°C of acidic solutions of the hexaaquo complexes of the trivalent elements leads to their hydroxylation and precipitation. Aluminum forms boehmite γ-AlOOH on a large range of acidity (Fig. 4.15, Chap. 6) [32]. Iron in acidic solution (1 < pH < 3) forms hematite α-Fe_2O_3 and goethite α-FeOOH at higher pH. Thermolysis of chromium in acidic solutions forms oxyhydroxides with poorly defined composition and structure. Heating for long durations (some months) leads to α-CrOOH [2] and to an hydrated oxide Cr_2O_3 [31, 33]. Hydrothermal treatment of solutions (T > 100°C) allows us to obtain

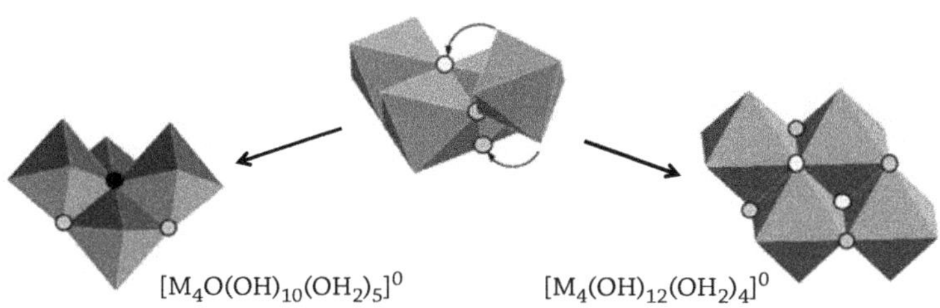

FIGURE 4.16 Possible conformations of the tetramers formed by M^{III} (Al, Cr, Fe) elements.

other oxyhydroxides of aluminum such as diaspore α-AlOOH isostructural with goethite, and those of chromium such as α-CrOOH having the brucite-type structure [34, 35], β-CrOOH isostructural with InOOH [36], and γ-CrOOH [37, 38].

In the thermolysis process, deprotonation and progressive hydroxylation are slow, so that noncharged entities structurally analogous to polycations may be generated and then form crystalline solid phases involving aggregation and condensation. In this way, it is possible to imagine how these solids can be built from $[M_4O(OH)_{10}(OH_2)_5]^0$ and $[M_4(OH)_{12}(OH_2)_4]^0$ tetramers (Fig. 4.16). These species could indeed be right structural models of oxyhydroxide embryos. As seen in the study of the first stages of the condensation of trivalent elements (§ 3.2.2), these tetramers result from the rotation of an octahedron linked to the compact trimer (Fig. 4.16).

The tetramer containing a μ_4–O bridge seems to be favored with aluminum because of the strong acidity of the μ_3–OH bridge, which forms a μ_3–O⁻ bridge having a nucleophilicity higher than that of μ_2–OH bridges. Condensation of the $[Al_4O(OH)_{10}(OH_2)_5]^0$ unit by olation and oxolation could then directly lead to the AlO(OH) corrugated layers, including μ_4–O and μ_2–OH bridges. These layers characteristic of the boehmite γ-AlOOH phase are linked by hydrogen bonds between hydroxo groups in adjacent layers (Fig. 4.17a).

In the case of iron and chromium, the planar configuration of the $[M_4(OH)_{12}(OH_2)_4]^0$ tetramer is favored, as shown by X-ray absorption spectroscopy of hydroxylated solutions of Cr^{III} [22]. This unit results from a downward rotation of the fourth octahedron in the early species (Fig. 4.16). It is then possible to justify the structural differences between oxyhydroxides, taking into account the kinetics in condensation of these units. The high lability of ferric ions favors the formation by olation of double chains of octahedra and then, their linking by oxolation, leading to corner sharing to create the μ_3–O bridges of goethite (Fig. 4.17b). With chromium (III), the olation and oxolation reactions become kinetically competitive owing to its high chemical inertness. Instead, to form the double chains by fast olation, both reactions occur simultaneously so that the layers containing the μ_3–OH and μ_3–O bridges characteristic of the α-CrOOH phase (Fig. 4.17c) can be formed.

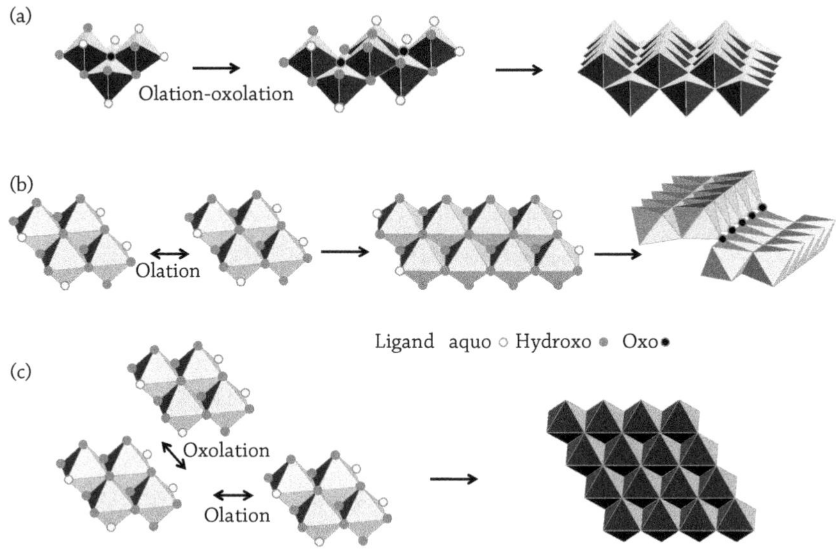

FIGURE 4.17 Possible pathways for the formation by thermolysis of boehmite γ-AlOOH (a), goethite α-FeOOH (b), and α-CrOOH (c).

TABLE 4.2 } Some characteristics of tetravalent elements.

	Ti	Zr	Hf	Ce	Th	V	Mn	Si	Ge	Sn	Pb
χ^*_M	1.32	1.29	1.36	1.17	1.24	1.56	1.63	1.74	2	1.89	1.92
Electron config.	....	....	d^0	....	....	d^1	d^3	....	p^6	....	....
Ionic radius (Å)	0.60	0.72	0.71	1.03	1.0	0.59	0.53	0.26	0.54	0.69	0.77
Coordination	6					4-5-6	6	4	4-6	6	6
Solubility	....	....	pH < 0.5 – 1	....	....			.....	pH > 10–12	....	....

4.1.4 TETRA- AND PENTAVALENT ELEMENTS

The tetravalent elements belong mainly to two groups of the classification: the *d* block (family of titanium) of low electronegativity ($\chi \approx 1.25$) and the *p* block (family of silicium) of much higher electronegativity ($\chi \approx 1.90$). These elements exhibit distinct behaviors in solution where they exist only in narrow acidity ranges (Table 4.2): those belonging in the *d* block are soluble only in strongly acidic medium as cationic hydroxoaquo species and condense by olation, while elements of the *p* block, soluble only in strongly alkaline medium, form, by oxolation, oxo or oxohydroxo polyanions [6, 39].

The position of these elements in the middle of the charge–pH diagram ($z = 4$) (Fig. 2.9) explains their hydroxylation on the whole acidity range and consequently their heavy tendency to precipitate. The charge–electronegativity diagram (Fig. 3.4) provides for the formation of oxides (Ti, Zr, Hf, Ce, Th, V, Mn) and polyanions (Si, Ge, Sn, Pb). As previously noted, $M(OH)_4$ hydroxides are of low stability because of the strong polarization of the hydroxo ligands by highly charged cations. They spontaneously dehydrate to form more or less hydrated $MO_2 \cdot nH_2O$ oxides, which are often polymorphic.

Titanium, manganese, and zirconium, with their rich crystal chemistry, are studied in Chapter 8. The case of vanadium (IV) is examined here with the transition elements under high oxidation states (§ 4.1.5). The behavior of silicium in silicates, which condense by oxolation, is described in the following section with, for purposes of comparison, a discussion of antimony, a pentavalent element from the *p* block.

a) Si^{IV} Silicate

Silicate has considerable importance in mineralogy because of its abundance in the Earth's crust and because it combines with a number of metallic elements in forming vast classes of compounds with different 1-, 2-, and 3-D architectures [1]. Here, only "silica" arising from precipitation in solutions of soluble anionic silicates (Chap. 3) is considered. Some aluminosilicates are examined in Chapter 6.

The addition of an acid to an aqueous sodium silicate up to pH ≤ 9 forms silicic acid, $Si(OH)_4$, the monomeric noncharged species that predominates in dilute solution (Fig. 2.8) [6]. Acidification can be achieved with an ion-exchange resin to avoid the presence of sodium salt in the solution. From silicic acid, the oxolation reaction leads to amorphous hydrated silica gels, $SiO_2 \cdot xH_2O$ [40]. An oxide does not really form, but a heavily hydrated gel containing many silanol terminal groups, Si-OH, which are strongly acid (pK around 2), is obtained. The gel may be considered to contain large-sized polyanions, aggregated through secondary condensation between silanol groups belonging to different units. SAXS and light-scattering studies of the early stages of condensation of silicates in neutral medium (pH ≈ 7) show that nuclei, around 2 nm in size, grow to around 8 nm after some hours in forming amorphous "silica" nanoparticles. They exhibit an open polymeric structure of low density, characterized by a fractal dimension of 2.2 [41]. The solid contained into this type of silica gel is never crystallized. The chains of corner-shared SiO_4 tetrahedra do not spontaneously organize to form compact and crystalline solids, so that the material is highly hydrated, with a stoichiometry roughly corresponding to $SiO(OH)_2$. Crystallization of such a silica gel into quartz or cristobalite requires heating above 1000°C.

At around pH 3, in the range of predominance of $Si(OH)_4$ in solution, the condensation is slow. It is strongly accelerated when the acidity deviates from pH 3. The variations in the consumption rate of monomers account for acid or basic catalysis of the reaction in the respective acidity ranges (§ 3.3). At around pH 8, gelation is instantaneous thanks to a catalysis due to the $[SiO(OH)_3]^-$ species favoring the nucleophilic attack. At pH < 2, the presence of the $[Si(OH)_3(OH_2)]^+$ species favors water elimination, which increases the reaction rate.

The silica particles arising from alkaline or acidic media exhibit quite different morphologies [40, 42]. At pH > 3, the molecular weight of the soluble species increases quite rapidly following the onset of the reaction. Spherical particles with a diameter of some tens of nanometers and made of highly branched chains forming a tridimensional framework are obtained as aqueous opaque particular

TABLE 4.3 } Partial charges calculated (§ 2.4) for different groups in a silic polymer such as sketched in Scheme 4.2.

Site	χ	δ(Si)	δ(OH)
A	2.64	+0.50	–0.06
B	2.78	+0.58	+0.06
C	2.71	+0.54	0

gels (within the gel, the nanoparticles are domains of higher density inducing light scattering). At pH < 3, the molecular weight remains relatively low throughout the process. Condensation especially forms oligomers (condensation degree of 2 to 6), which slowly aggregate and form a gel containing diffuse piles of minute particles of around 5 nm in size leading to transparent polymeric gels (the silica species do not create high contrast in density within the gel, so that there is no scattering center able to diffuse light). The role of acidity in particle morphology may be understood in considering the mechanisms involved in the catalysis of oxolation (§ 3.3). It is accordingly possible to interpret the architecture of the final polymer in considering, at an early stage of condensation, a small chain more or less branched, including different types of silicic groups (Scheme 4.2).

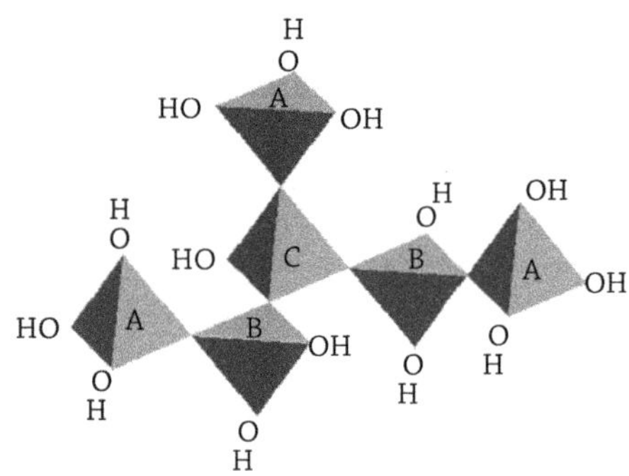

SCHEME 4.2

Scheme 4.2 contains $Si(OH)_3(OX)$ [**A**], $Si(OH)_2(OX)_2$ [**B**], $Si(OH)(OX)_3$ [**C**] groups in which X stands for the rest of the chain having the average composition of the raw material $SiO(OH)_2$. The X group is a ligand with the same average electronegativity as that of the raw material: $\chi(X) = \chi(SiO(OH)_2) = 2.63$. It is then possible to calculate the partial charges on the Si atoms and OH ligands in the different groups (Table 4.3).

In alkaline medium, catalysis concerns the first step of the condensation mechanism, which is the nucleophilic attack due to anionic species (or HO^-). It can be observed that the positive charge on the silicon atom in all groups of the growing chain is higher than in the monomer $Si(OH)_4$ ($\delta(Si) = 0.47$). Therefore, the nucleophilic attack by silicic species must proceed on the chain, preferentially on atoms carrying the highest positive charge located inside the chains (B and C groups). The condensation should produce many branching points, in agreement with the experience.

In acidic medium, catalysis concerns the second step of the oxolation pathway in which water is eliminated. The departure of the proton from the *ol* bridge in the transition state is eased by protonation of an OH ligand, which favors the formation of the leaving aquo ligand. The concerned OH ligands are those located at the ends of chains because they carry the highest negative partial charge, or those of the $Si(OH)_4$ monomer itself. It results in the growth of chains loosely crosslinked and of low molecular weight forming transparent gels. The morphology of particles is therefore heavily dependent on the acidity of the synthesis medium.

Silica particles of highly uniform size are obtained from hydroxylation of silicon alkoxides $Si(OR)_4$ solubilized in the parent alcohol ROH (R stands for an alkyl group). Such metal-organic precursors [43, 44] react with water to form hydroxylated derivatives that condense by oxolation [45]. Hydroxylation and condensation are schematically represented as

$$M(OR)_z + z\,H_2O \rightarrow M(OH)_z + z\,ROH \qquad M(OH)_z \rightarrow MO_{z/2} + z/2\,H_2O$$

With metal complexes in aqueous solution, hydroxylation is an acid–base reaction, and condensation is a nucleophilic substitution. For alkoxides in an organic solvent, water is a reactive, so that both hydroxylation and condensation proceed according to an associative bimolecular SN_2 mechanism. It involves intermediates in which the coordination of the metal center is increased according to the following scheme:

Hydroxylation

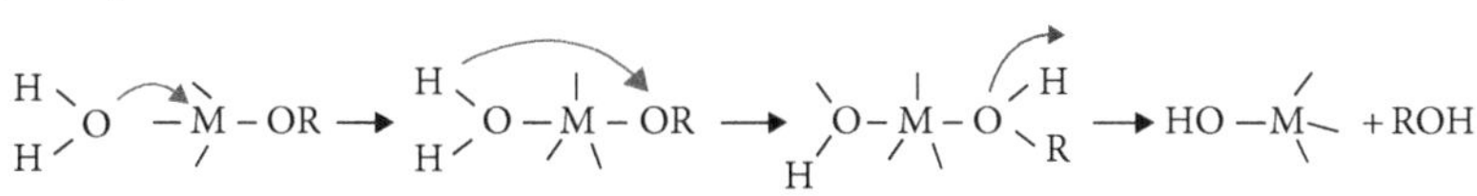

Condensation

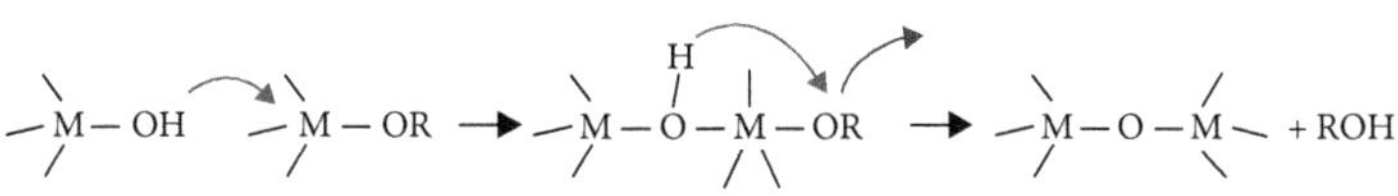

Consequently, the reactivity of metal alkoxides toward both hydrolysis and condensation is governed by three main factors: the electrophilic character of the metal (its polarizing power); the sterical effect of the alkoxy ligands; and the molecular structure of the alkoxide (monomer or oligomer). This last-named factor strongly depends on the difference between the oxidation degree of the metal, z, and its preferred coordination, N. Silicon alkoxides ($\chi_{Si} = 1.74$, $z = N = 4$) are always monomers and are weakly reactive in water, whereas titanium alkoxides ($\chi_{Ti} = 1.32$, $z = 4$, $N = 6$) are often oligomeric and extremely sensitive to moisture because of the frustration in coordination (4 or 5) of the titanium in alkoxides, which is preferentially 6.

The reactivity of silicon alkoxides strongly decreases when the size of the OR groups increases because of their sterical effect and hydrophobicity. In the presence of an excess of water at pH 7, the gelation time at 20°C is 44 hours with methoxide $Si(O\text{-}CH_3)_4$, 240 hours with ethoxide $Si(O\text{-}CH_2CH_3)_4$, and 550 hours with butoxide $Si(O\text{-}(CH_2)_2CH_3)_4$. The acidity strongly influences the rate of hydrolysis and condensation and, as in aqueous medium, induces deep changes in the morphology of the products. In an excess of water and in acidic conditions ($pH \leq 4$), silicon alkoxides form transparent polymeric gels while in basic conditions ($pH \geq 8$), condensation is faster than in neutral medium and forms hydrated silica particles that are perfectly spherical and uniformly sized (Fig. 4.18a). This last procedure, called Stöber's method, is used to produce silica 5 to 2000-nm-sized nanoparticles from the reaction of alkoxides in solution in alcohol with an excess of aqueous ammonia [46]. The particle-size monodispersity results from slow hydroxylation and fast condensation reactions. ^{29}Si and ^{13}C NMR allow us to follow the variation in concentration of the alkoxide and of the produced hydroxylated monomer (Fig. 4.18b). These techniques show that there is no accumulation of hydroxylated species in solution, whereas primary particles are quickly detected by SAXS [47]. Thus, the working conditions allowing the decoupling of nucleation and growth steps in the formation of particles lead to an high uniformity in particle size (§ 4.2).

Similar results are obtained by hydroxylation of silicon ethoxide $Si(OEt)_4$ in contact with a solution of arginine ($pH \approx 10$, 80°C) without mixing the organic and aqueous phases [48, 49]. From SAXS measurements, it has been observed that the amorphous particles (Fig. 4.36) undergo densification during growth. This phenomenon could be due to the completion of reticulation resulting from slow elimination of the last coordinated alkoxo groups.

The silica particles and silicic polymers can remain dispersed in aqueous suspension in forming homogeneous sols or, on the contrary, aggregate to form gels more or less quickly. The dispersion state of oxide nanoparticles depends on the

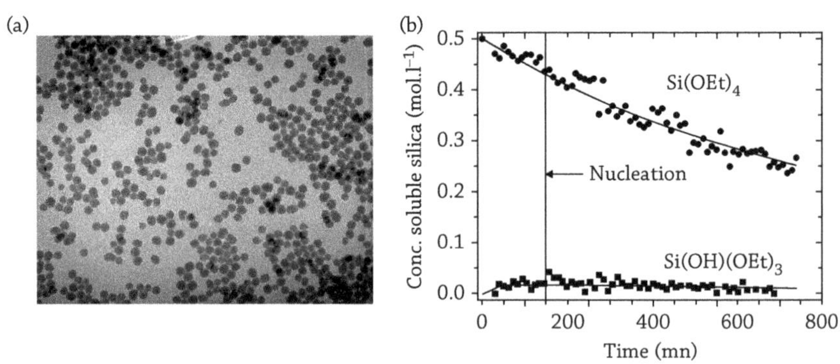

FIGURE 4.18 (a) SEM image of silica nanoparticles (mean diameter 23 nm) synthesized according to Stöber's method by hydrolysis of $Si(O\text{-}CH_2CH_3)_4$ in water/methanol/NH_3 (courtesy of C. Chanéac, Univ. P. & M. Curie, Paris, France). (b) Variation of the concentration of tetraethoxyorthosilicate solubilized in ethanol (initial concentration 0.5 $mol.l^{-1}$) in the presence of water (2.2 $mol.l^{-1}$) and ammonia (0.1 $mol.l^{-1}$) (reproduced from [47] with permission of Elsevier). The line labeled "nucleation" indicates the beginning of detection of primary particles by SAXS.

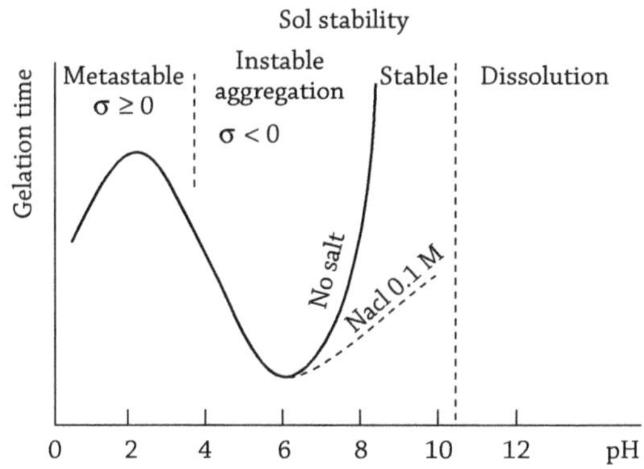

FIGURE 4.19 Variation in gelation time and in stability of silica sols as a function of pH. σ is the surface charge density (adapted from [40]).

surface electrostatic charge density, σ (§ 5.1). Aggregation and flocculation occur rapidly when the surface charge density is very low or zero because the attractive van der Waals forces overcome the electrostatic repulsions between particle surfaces. When the particle charge density is high, positive, or negative, depending on the acidity of the medium, the particles are repelling and the sol retains its stability (§ 5.3). The behavior of suspensions of silica is different because at the zero point charge (pH $\approx$ 2), the stability of the sol is at maximum and gelation is extremely slow (Fig. 4.19). This behavior can be explained by the presence of dangling silanol groups, able to condense and form chemical bridges between chains or particles. On either side of the zero point charge (pH $\approx$ 2), the surface charge density is too small to provide efficient repulsions between particles, so that the acid or base catalysis of oxolation between surfaces in contact takes place. The reaction rate is maximum at pH 6 and the gelation time minimum. At pH > 6, the negative surface charge density becomes high enough to form a stable sol [40].

The silica gels are often used as dispersion medium to form silica-based composite materials. For such an application, it is better to polymerize silica in acidic medium in order to form after gelation and drying a transparent homogeneous polymeric matrix. On the other hand, polymerization in alkaline medium produces well-defined particles that can be used as support of ions, molecules, or particles and usable in catalysis, for instance.

Opals are minerals made of hydrated amorphous silica, $SiO_2 \cdot nH_2O$, formed by deposit of silicates resulting from aqueous weathering of siliceous rocks. The material is made up of spherical particles of a uniform size around 200 nm in diameter. Depending on their place of geological formation, these particles may be primary particles or monodisperse spherical aggregates of particles around 25 to 50 nm in diameter, themselves monodispersed. The monodisperse spheres or aggregates are regularly stacked in forming small crystalline domains (crystals of crystals) able to diffract the visible light and giving rise to play-of-color (a variation of only 5% in sphere diameter is sufficient to preclude diffraction) [50] (Fig. 4.20). In the geological conditions of formation, the transport of matter

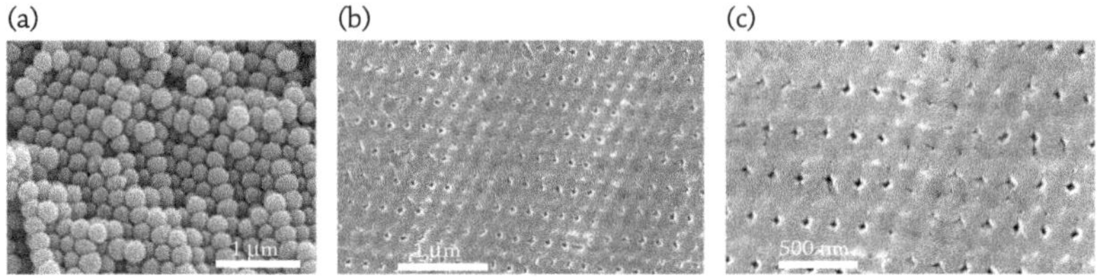

FIGURE 4.20 SEM images of amorphous sedimentary opals from Australia. Depending on the samples, the spherical silica particles can be perfectly individualized (a) or more or less cemented by a silica deposit (b, c). On the (c) image, the spherical particles near the surface of the material are clearly distinguishable. These cemented materials are true gem opals, unlike sample (a), which exhibits a chalky aspect with no play-of-color (courtesy of E. Fritsch, IMN, Univ. of Nantes, France).

resulting from weathering and dissolution of the bedrock likely occurred at very low concentration, so that precipitation at neutral or slightly alkaline pH was carried out in distinct stages of nucleation and growth, allowing uniform-sized particles to be formed (§ 4.2).

b) Sb^V Antimonate

Antimony (V) exists in alkaline medium as $[Sb(OH)_6]^-$ hexacoordinated species. It is stable as a monomer owing to the too weak electrophilic character of the cation. A calculation gives $\chi = 2.478$, $\delta(OH) = -0.21$, $\delta(Sb) = +0.26 < 0.3$ (§ 3.1.1). Its condensation occurs by acidification and forms polyanions and gels. The polyanions of Sb(V) are poorly known. Potentiometry seems to indicate for the $H/Sb = 0.6$ ratio the formation of a species 12-time condensed, but its structure is unknown [51].

Acidification of potassium antimonate $KSb(OH)_6$ in solution using an ion-exchange resin K^+/H^+ to avoid an increase of ionic strength and flocculation immediately forms an opalescent suspension that becomes clear after about 30 minutes at room temperature. After 24 hours of aging, antimonic acid, $HSbO_3 \cdot 1.5H_2O$ crystallizes with the pyrochlore structure [52–54]. The formation of this solid may be understood in considering the noncharged hydroxoaquo species $Sb(OH)_5(OH_2)$, which is able to condense by olation and oxolation ($\chi = 2.60$, $\delta(Sb) = +0.32$, $\delta(OH) = -0.10$, $\delta(H_2O) = +0.16$). As in many systems, such a reactive species, suddenly produced in high concentration, immediately condenses in forming an amorphous chaotic solid, giving rise to opalescent suspension. Opalescence disappears during aging, and crystallization occurs [53]. It involves likely dissolution–crystallization equilibria allowing the formation of the noncharged $[Sb(OH)_5(OH_2)]^0$ complex. Because it contains only one water molecule, it preferentially condenses by olation to form small rings of corner-sharing octahedra, $Sb_3(OH)_{15}$, and is likely more stable and statistically probable than longer chains (Fig. 4.21). Olation and then oxolation allow these rings to be rapidly capped with a fourth octahedron to form $Sb_4O_2(OH)_{16}$ tetramers, which in turn condense by oxolation to form the $SbO_2(OH) \cdot H_2O$ pyrochlore network. It is interesting to observe that, in this process, hydroxo ligands remaining within

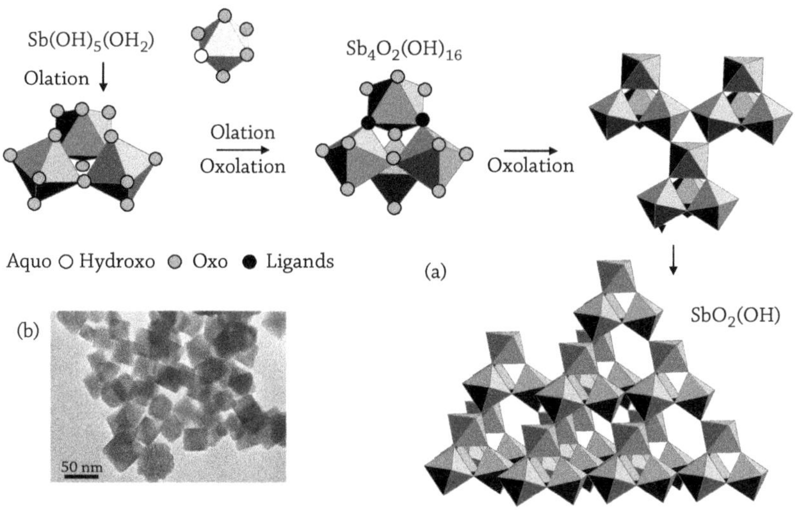

FIGURE 4.21 (a) Possible formation pathway in solution of antimonic acid. In the pyrochlore structure, the SbO_6 octahedra are corner-shared through oxobridges. The solvated protons are located in hexagonal channels within the structure. (b) TEM image of antimonic acid nanoparticles obtained by potassium hexahydroxoantimonate acidification at pH 3 and heating the suspension at 95°C for 7 days (courtesy of T. Garry and C. Chanéac, Univ. P. & M. Curie, Paris, France).

the solid are strongly acid ($\chi = 2.819$, $\delta(OH) = +0.10$). These ligands are therefore dissociated by the hydration water into the solid. Thus, antimonic acid is more accurately represented by the formula $[H_3O^+ \cdot SbO_3^-]$ 0.5 H_2O, in accordance with the ^1H NMR study [55]. The hydrated oxide is in fact a strong polyacid with solvated protons delocalized within the channels of the structure, giving a high ionic conductivity [55, 56] and easily exchanged by alkaline cations and silver ion [57, 58]. Although the proposed mechanism is speculative, it logically accounts for the observed phenomena.

4.1.5 TRANSITION ELEMENTS UNDER HIGH OXIDATION STATES

Transition elements in their highest oxidation state, such as V^V, Mo^{VI}, and W^{VI}, give rise to a rich aqueous chemistry with the formation of polyoxometalates and polyacids (Chap. 3). They are able to form solids, such as vanadic, tungstic, and molybdic hydrates, with a predictable structure in considering the characteristics of the noncharged molecular precursors. Moreover, phases such as vanadium dioxide (V^{IV}) and hexagonal tungsten bronzes (W^{IV}, W^V), exhibiting interesting (opto)electronic properties, can be obtained by incomplete reduction in solution of the molecular precursors.

a) Vanadate V^V

During the titration of sodium or ammonium vanadate at high ionic strength, no solid phase is formed around $H^+/V = 3$ corresponding to the existence of

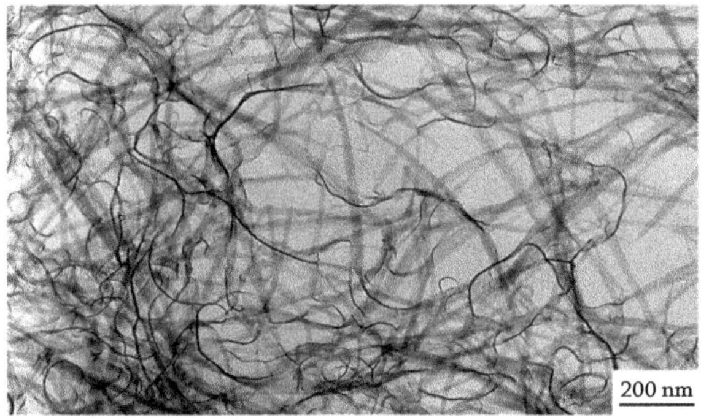

FIGURE 4.22 TEM image of vanadium oxide gel $V_2O_5{\cdot}nH_2O$ obtained from acidification of sodium metavanadate with an ion-exchange resin (courtesy of J. Livage, Univ. P. & M. Curie, Paris, France).

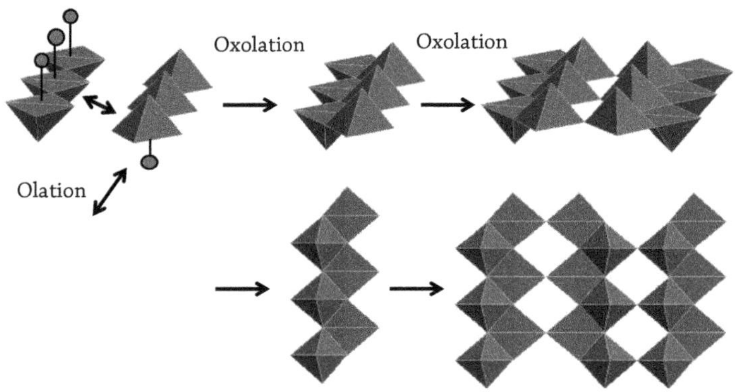

FIGURE 4.23 Possible pathway for $V_2O_5{\cdot}nH_2O$ gel formation from the neutral precursor $VO(OH)_3(OH_2)_2$.

the noncharged species (§ 3.3.2a). Decavanadate is in equilibrium with the noncondensed $[VO_2(OH_2)_4]^+$ cation, and a direct transformation from one form to the other occurs because of kinetic constraints [6, 59]. However, if solution ($C_V > 0.1$ mol.l^{-1}) is acidified at $H^+/V = 3$ at low ionic strength in using an ion-exchange resin, a gel is formed, becoming dark red in a few hours [60, 61]. The vanadium oxide gel $V_2O_5{\cdot}nH_2O$ is made of entangled ribbons several hundred nanometers long, about 10 nm wide, and 1 nm thick [62] (Fig. 4.22).

The crystal structure of the ribbons (Fig. 4.23) is entirely different from that of the decavanadic polyanion immediately formed upon acidification. This polyanion is also too acid to form a zero-charged species, and certainly, it cannot serve as the building unit for the solid. Rather, as in many cases, the noncharged and noncondensed species resulting from slow decondensation equilibria are formed and in turn removed from solution because of their unlimited condensation leading to the solid. It is for that reason that acidification needs to take place at low ionic strength to favor formation of the weakly charged species and avoid their flocculation [63].

The formation of V_2O_5 oxide may be understood by considering the zero-charged monomers $[VO(OH)_3(OH_2)_2]^0$ and $[VO_2(OH)(OH_2)_3]^0$ in equilibrium with the decavanadate ($H^+/V = 3$). As soon as these species appear in solution, they condense by olation since both hydroxo and aquo ligands exist. The formation of linear entities results from the structure of the precursors due to the short V = O bonds preventing condensation along their axis. Different linear structures are possible, depending on the precursor (Scheme 4.3).

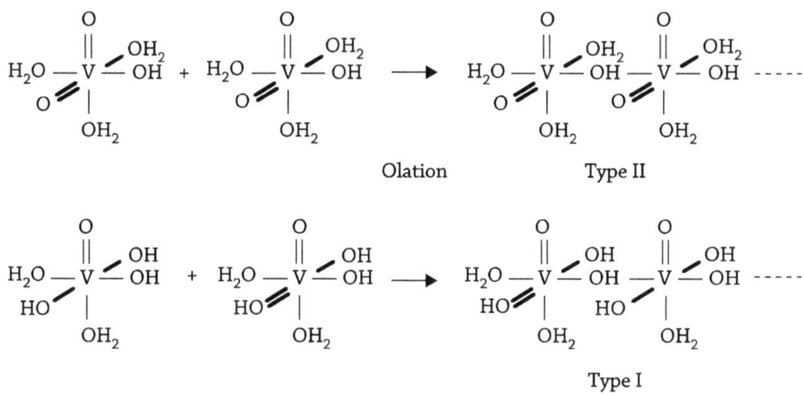

SCHEME 4.3

Type II chains (Scheme 4.3) based on the dioxo precursor cannot condense between them because they have no terminal hydroxo ligands. On the other hand, subsequent condensation by oxolation of the chains of type I (Scheme 4.3) also stabilizes the early μ_2–OH bridges into μ_3–O bridges in forming ribbons [64] (Fig. 4.23). The stacking of the ribbons involves hydrogen bonds, and upon thermal dehydration, V_2O_5 crystallizes.

This mechanism is oversimplified because gelation of vanadic acid requires the presence of traces of V^{IV}, which spontaneously appear during acidification with ion-exchange resin or by addition of a small amount of alcohol [62, 65]. The role of V^{IV} as a catalyst is not yet completely clear [66]. It may act as an initiator of the initial condensation by olation of vanadic complexes or in oxolation between chains. Indeed, the nucleophilicity of the hydroxo ligands in the complex of V^{IV}, $[VO(OH)_2(OH_2)_3]^0$, is higher than that of the hydroxo ligands present in V^V species because of the smaller formal charge of the cation. Interestingly, the mild hydrothermal treatment (150–200°C) of V_2O_5 in the presence of various organic compounds allows partial reduction of vanadium V in leading to a collection of oxides with various crystal structures, mostly lamellar, exhibiting interesting electrochemical properties [67].

The conductivity of vanadic gels has both ionic and electronic components. The ionic conductivity due to the mobility of protons in the water-swollen space between ribbons occurs in wet gels. The electronic counterpart due to intervalent V^{IV}–V^V transfers in the lattice occurs solely in dry gels [68]. It results in interesting applications, such as antistatic layers on photographic films [69].

Although qualitative and schematic, the proposed mechanism stresses that the formation of vanadium oxide does not result solely from an oxolation reaction. That is also true for some other elements (Sn^{IV}, Sb^{V}, W^{VI}, etc.). It is important to remember that the formation of metal oxides by precipitation always needs at least one olation stage because oxolation *alone* leads only to the formation of polyoxoanions (Chap. 3).

b) Vanadyle V^{IV}

Vanadium (IV) exists in acidic medium (pH $\leq$ 2) as the light-blue $[VO(OH_2)_5]^{2+}$ complex of C_{4v} symmetry [70] (§ 2.2). Hydroxylation and condensation begin at pH 2.5, inducing a green color. Precipitation occurs at around pH 3 in forming a brown solid with a composition and structure not clearly established because of spontaneous oxidation. Thermolysis at 95°C for a few days of vanadyle chloride solutions forms different solids, depending on the acidity initially adjusted by addition of sodium hydroxide [71]:

- At pH around 2.5, a green solid made of flat rods about 100 nm thick and 0.5 µm long is obtained (Fig. 4.24a). The X-ray diffraction pattern of this phase contains a series of lines of 00*l* indices that could correspond to a lamellar phase with an interplanar spacing of 10.8 Å. The diffraction diagram is close to that of fernandinite (Ca, Na, K)$_x$$V_2O_5$, H_2O, a natural bronze in which the average oxidation state of vanadium is 4.8 and with a structure made of V_2O_5-type layers [72] (Fig. 4.24b);
- At 3<pH<6, a pink solid, so-called Gain's hydrate with the composition $VO_2\cdot(1 + x)\ H_2O$ ($0.05 < x < 0.17$) is obtained [73]. The solid is formed of ribbons around 50 nm in section and 1 µm in length (Fig. 4.24c). Its crystal structure is still unknown;
- At pH around 8, small particles more or less spheroidal and around 5 nm in diameter are formed. The X-ray diffraction pattern, which is rather poorly defined, is, however, similar to that of the bronze $Na_{1.8}V_2O_5$ [74] in which the average oxidation state of vanadium is 4.1.

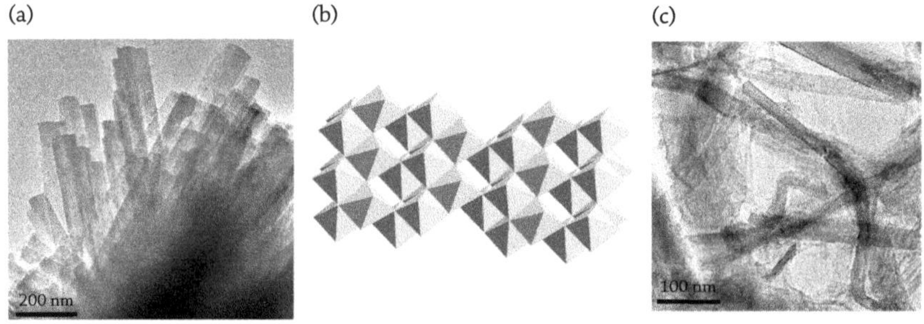

FIGURE 4.24 (a) TEM image of nanorods resulting from thermolysis at pH 2.6 of vanadyle chloride; (b) structure of sheets in fernandinite; (c) TEM image of ribbons in Gain's hydrate (adapted from [71]).

The aqueous chemistry of V^{IV} is rather complex, and precipitation of the expected VO_2 oxide with rutile-type structure does not occur in these conditions. For that purpose, an attempt was made to chemically reduce vanadate precursors. Through this procedure, two phases were clearly identified: häggite $V_2O_3(OH)_2$ and duttonite $VO(OH)_2$, among many other known oxides and oxyhydroxides of V^{IV} [73, 75]. These two phases exhibit lamellar structures made of single or double chains of VO_6 corner-sharing distorted octahedra (Fig. 4.25). Häggite exhibits an insulator–semiconductor transition at 66 K involving a large thermal variation in resistivity. This material could be an interesting component of various nanodevices such as Mott field-effect transistors, temperature sensors, and energy-efficient smart windows [76]. Both phases are obtained in close conditions: hydrazine is added to a solution of sodium metavanadate, $NaVO_3$ (0.15 mol.l^{-1}) in the molar ratio $N_2H_4/V = 2.5$. The acidity is then controlled by addition of concentrated chlorhydric acid, and the mixture is heated at 95°C for 4 to 5 days. At pH 3.6, häggite is obtained in the form of rods 10 nm in section and 200 nm to 1 μm in length, while duttonite is formed at pH 4 as sticks 5 nm in section and 5 to 25 nm in length [75, 77] (Fig. 4.25).

In both cases, the particle growth direction is parallel to the axis of edge-shared octahedra chains. The brutal effect of pH on the structural change seems to result from its influence on the solubility of vanadium (IV): at pH 4, the solubility is very low and the solution in the presence of duttonite is colorless, whereas at pH 3.6 during the formation of häggite, the solution is blue, indicating the presence of soluble V^{IV}. The formation of solids does not proceed in a straightforward way. Heating of the reactional mixture during 1 hour at pH 3.6 forms a solid gray in color and is amorphous in X-ray diffraction. After 5 hours, the Gain's hydrate (having the same composition as duttonite) is present, and then häggite appears

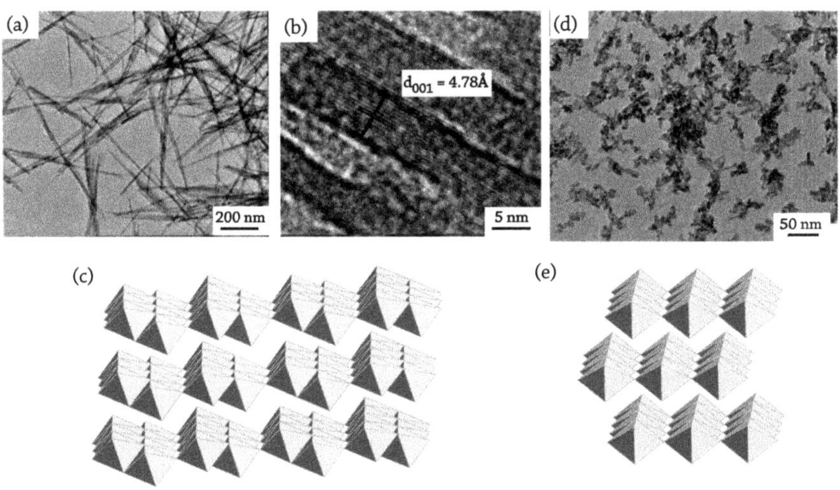

FIGURE 4.25 (a) and (b) TEM images of häggite particles; (c) structure of the sheets in häggite; (d) TEM image of particles of duttonite; (e) structure of the sheets in duttonite (adapted from [77]).

FIGURE 4.26 TEM images of (a) (b) asterisk-shaped VO_2M microparticles; (c) a VO_2M microrod in the presence of VO_2B nanorods; (d) a branch of VO_2M asterisk-shaped microcrystal and its electron diffraction pattern; (e) VO_2M nanoparticles (reproduced with permission from [79] Copyright 2010 American Chemical Society).

after heating for 1 day. Its proportion increases with time, up to the entire disappearance of Gain's hydrate after 4 days. The oxidation state of 4 for vanadium in häggite has been confirmed by V K-edge EXAFS [75, 77] as recently suggested [76], at the difference of the 3.5 value proposed in the first structural studies [78]. The same technique has also shown that vanadium is reduced in a two-step process before heating: around half of the vanadium before the hydrazine addition (pH 7.2) and the rest during acidification at pH 3.6 in also leading to precipitation. Because of the solubility of vanadium IV during the synthesis, the formation of häggite at the expense of Gain's hydrate seems to result from a dissolution–crystallization process. The formation of duttonite could result from a transformation in situ of the early amorphous solid.

The monoclinic VO_2 phase (VO_2M), which gives rise to an insulator–metal transition, can be obtained as micro- and nanoparticles at higher temperatures in hydrothermal conditions [79]. The process involves reduction of commercial V_2O_5 with hydrazine in sulfuric acid solution at 60°C (or vanadyle sulfate, $VOSO_4$, in acidic solution in the presence of hydrazine), and then adjustment of the pH between 4 and 10 by addition of sodium hydroxide, which induces the precipitation of a gray–brown solid. The noncrystalline precipitate, likely containing hydrazine, is washed and dispersed in water and then heated at 220°C for 48 hours. If the pH of the suspension before the hydrothermal treatment is adjusted from 4 to 5.5, asterisk- or rod-shaped VO_2M microcrystals are obtained with variable amounts of VO_2B nanorods, whereas at higher pH, hexagonal platelets of VO_2M nanocrystals are formed (Fig. 4.26). (VO_2B is a monoclinic phase, differing from the usual rutile phase and isotype of the Na_xTiO_2 phase (§ 8.2.2) [80]). Because of

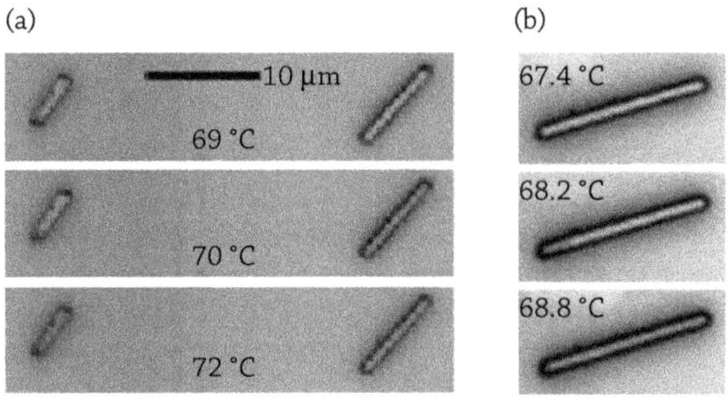

FIGURE 4.27 Bright-field optical microscope images of VO_2M microrods deposited on a silica substrate at a series of warming temperatures. The dark color of rods corresponds to their conducting state. On the right of image (a), the conducting state appears between 69 and 70°C, while on the left, it appears between 70 and 72°C. On the image (b), the rod 20 μm in length becomes conducting between 68.2 and 68.8°C (reproduced with permission from [79] Copyright 2010 American Chemical Society).

the difference in particle size of the VO_2M and VO_2B phases, their separation by decantation or dissolution in dilute acidic medium is relatively easy.

The asterisk-shaped particles are made of six platelets with thickness varying from 50 to 200 nm and growth occurring from their center. They are 1 to 5 μm in width and 1 to 20 μm in length. The morphology of particles depends on many factors, especially vanadium concentration, pH, and temperature. Whatever the precursor, vanadic oxide or vanadyle sulfate, the presence of hydrazine prior to hydrothermal treatment is required for the formation of VO_2M; otherwise only VO_2B is obtained [79]. It seems, then, that in addition to being a reducing agent, hydrazine likely plays an important role as ligand in directing the crystal growth. Investigations of the reaction mechanism and the role of hydrazine, however, have to be thoroughly examined to allow any conclusion.

The insulator–metal transition exhibited by VO_2M phase microrods occurs at around 68°C, according to the particle size. The phenomenon is clearly visible in optical microscopy, with rods deposited on a warmed substrate (Fig. 4.27). The conducting phase appears darker because it absorbs light more strongly than the insulating phase.

The high-temperature phase involved in the insulator–metal transition adopts the rutile structure, VO_2R, in which the V^{4+} ions are regularly spaced in the chains of edge-sharing octahedra, while the insulating low-temperature phase undergoes a structural distortion leading to the alternating long and short distances between the V^{4+} in chains. Such an effect, known as Peierls's distortion [81], provokes a change in the electronic structure in opening a forbidden band precisely at the Fermi level, the solid becoming an insulator (Fig. 4.28).

c) Molybdate Mo^{VI}, Tungstate W^{VI}, and Tungsten Bronzes

The difference in behavior between Mo^{VI} and W^{VI} ions in polyanions (§ 3.3.2) is also observed in the formation of the hydrated oxide phases. The greater versatility

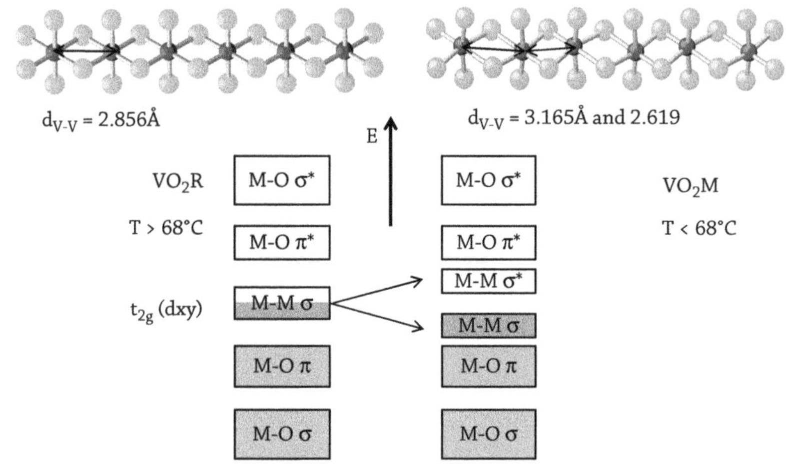

FIGURE 4.28 Scheme of edge-sharing VO_6 octahedra chains in the rutile VO_2R phase (T > 68°C), where the V^{4+} ions are regularly spaced (left), and in the VO_2M monoclinic (T < 68°C), where V^{4+} ions are alternatively spaced with short and long distances (right). The scheme of the electronic structure shows that interaction between t_{2g} orbitals of V^{4+} ions in the chains of the rutile phase form a M-Mσ band partly filled by *d* electrons that ensure conductivity. Because of the structural change, this band is split in a totally full M-Mσ band (insulating) and an empty M-Mσ^* band, separated by a forbidden band gap (Peierls's distortion).

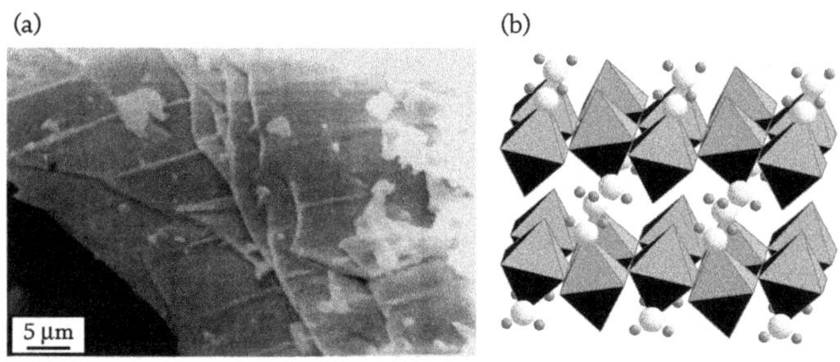

FIGURE 4.29 (a) SEM image (courtesy of J. Livage, Univ. P. & M. Curie, Paris, France) and (b) structure of sheets in the tungstic gel $WO_3{\cdot}H_2O$ obtained from acidification of sodium tungstate with an H-form exchanger resin.

of tungsten compared with molybdenum is reflected in the larger variety of oxides formed from solutions, as clearly shown by Michel Figlarz, University of Picardie, France [82, 83].

As in the case of vanadate, aging or heating of tungstic acid solutions of low ionic strength obtained from acidification of sodium tungstate with a protonated ion-exchange resin leads to a turbid suspension and a precipitate after a few hours. At low concentration ($C_W < 0.5$ mol.l^{-1}), the pale yellow hydrate $WO_3{\cdot}2H_2O$ is obtained, while at $C_W > 0.7$ mol.l^{-1}, the dark yellow $WO_3{\cdot}H_2O$ is formed [84]. Their structures are lamellar, and the particles form ordered stackings of platelets (tactoids) [85, 86] (Fig. 4.29a). The sheets in the structure of these hydrates are composed of corner-sharing $WO_5(OH_2)$ octahedra, the water molecule being alternatively above and below the plane (Fig. 4.29b). In the dihydrate, the second water molecule lies between the sheets. These hydrates are also obtained by acidification of tungstate $NaWO_4$ with concentrated HCl (HCl ≈ 3 mol.l^{-1}) [87]. $WO_3{\cdot}2H_2O$ precipitates at room temperature, and $WO_3{\cdot}H_2O$ is formed at 100 °C.

The crystal structure of the tungstic hydrates is easily understood in considering the species generated from acidification of tungstate. It immediately forms decatungstate in equilibrium with the octahedral $[WO(OH)_4(OH_2)]^0$ monomer. At the difference of vanadate previously examined, only one water molecule exists in the coordination sphere: in the *trans* position of the oxo ligand. Condensation along this axis is therefore blocked, and the only possible reaction is oxolation, allowing two-dimensional planar growth through formation of μ_2–O bridges between corner-sharing octahedra. The sheets so formed are piled on top of each other in a more or less ordered manner and are linked by hydrogen bonds between water molecules and oxo ligands on each side of the tungstic sheets.

With molybdenum, $MoO_3{\cdot}2H_2O$ hydrate precipitates slowly after about 3 weeks at room temperature from a solution of sodium molybdate in concentrated nitric acid (≈ 4 mol.l^{-1}). Dissolution of this hydrate in 12 mol.l^{-1} hydrochloric acid followed by dilution to 3 mol.l^{-1} of the solution leads to crystallization of $MoO_3{\cdot}H_2O$. Both hydrates, made of sheets of corner-sharing octahedra, are isomorphous with their tungstic homologues [82]. Their formation is more difficult, however, since drastic synthesis conditions are required for corner sharing between molybdic octahedra. This behavior is similar to that observed with tungstic and molybdic polyanions.

Hydrothermal treatment at 120°C of a tungstic gel or of a suspension of $WO_3{\cdot}2H_2O$ forms another hydrate, $WO_3{\cdot}1/3H_2O$, with needle or platelet morphologies, depending on the precursor used [82, 83] (Fig. 4.30a, b). Hexagonal nanoplatelets of this hydrate (Fig 4.30c) can also be obtained by hydrothermal treatment from 100 to 200°C of the peroxocomplex formed from dissolution of tungstic acid in hydrogen peroxide [88]. The orthorhombic structure of $WO_3{\cdot}1/3H_2O$ is made of planar sheets containing two kinds of octahedra, WO_6 and $WO_4(OH_2)_2$, bounded by 6 and 4 corners, respectively, and forming hexa- and trigonal rings (Fig. 4.30d). The planes, offset from one another in their stacking, are connected by corner sharing of WO_6 octahedra (Fig. 4.30e).

$WO_3{\cdot}1/3H_2O$ is stable up to 250°C. It dehydrates at higher temperatures when forming hexagonal WO_3 oxide [89]. Its structure is similar to that of the hydrate, but the planes are eclipsed from one another in their stacking (Fig. 4.31b), so that the structure contains hexagonal channels in which hydrogen or alkaline cations (Li^+, Na^+) can be (electro)chemically inserted to form hexagonal M_xWO_3 bronzes [83]. Chemical insertion of *n*-butyl-Li or naphthalene-Li(Na) is a redox reaction, reversible up to $x = 1$. Beyond this value, the constraints due to insertion are too strong, and the oxide becomes amorphous.

The transformation of the hydrate $WO_3{\cdot}1/3\,H_2O$ into hexagonal WO_3 retains the morphology of particles. Although a direct structural relationship could be assumed between these two phases, the transformation actually occurs with reconstruction of the structure [83]. Elimination of water induces the rearrangement of W–O

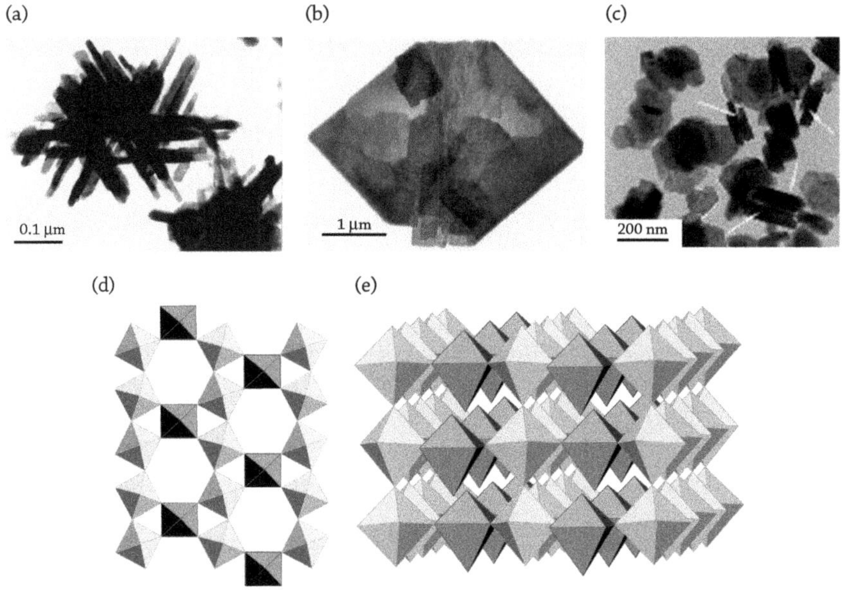

FIGURE 4.30 TEM images of (a) $WO_3{\cdot}1/3H_2O$ needle-shaped particles obtained by hydrothermal treatment of tungstic gel and (b) platelets obtained from $WO_3{\cdot}H_2O$ (reproduced from [83] with permission of Elsevier). (c) Stacking of sheets forming nanodiscs obtained by hydrothermal treatment of the tungstic peroxocomplex (reproduced with permission from [88] Copyright 2008 American Chemical Society). (d) Structure of sheets including WO_6 (light) and $WO_4(OH_2)_2$ (darker) octahedra and (e) connection of the sheets in $WO_3{\cdot}1/3H_2O$ hydrate (see also Fig. 4.31a).

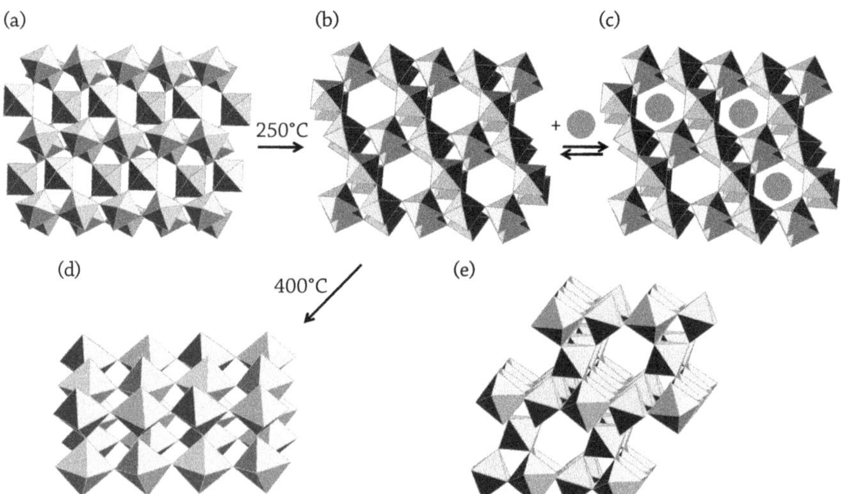

FIGURE 4.31 Structures of tungstic hydrate and oxides: (a) $WO_3{\cdot}1/3\,H_2O$, (001) planes with offset; (b) hexagonal h-WO_3, (001) planes eclipsed, forming hexagonal channels running along the *c* axis; (c) hexagonal bronze M_xWO_3; (d) monoclinic WO_3; (e) pyrochlore WO_3.

bonds, but the arrangement of bonds is similar in the (001) planes of both phases. Therefore, the nucleation energy of the metastable hexagonal oxide is minimum. and nuclei of oxide appear first in the (001) planes of the hydrate. The transformation proceeds with a progressive and oriented displacement of the oxide–hydrate interface. At around 400°C, hexagonal WO_3 transforms into monoclinic WO_3 (Fig. 4.31). The transformation is nonpseudomorphous, but it also proceeds by oriented nucleation of the thermodynamically stable phase.

Anhydrous hexagonal h-WO_3 oxide can be synthesized in soft conditions by heating at 95°C a solution of sodium tungstate Na_2WO_4 acidified with HCl at pH 1.3 [77]. The particles are rods 10–15 nm in diameter and 50–100 nm in length. The hexagonal channels are oriented along the rods (Fig. 4.32a, b). The same synthesis, but in the presence of hydrazine, allows formation of the hexagonal bronze $(NH_4)_xWO_3$ [77]. The acidified initial mixture forms a suspension of milky-white aspect indicating that no reduction of W^{VI} has occurred, but complexation of tungstate takes place, as shown by absorption bands in infrared spectroscopy characteristic of W–O–N bonds. Heating induces the reduction, as indicated by the appearance of the blue color typical of W^{V} ions and the insertion of ammonium ions inside the channels of the structure. Rods 10 nm in diameter and 50 nm in length, similar to those obtained under nonreducing conditions, are formed. The amount of hydrazine does not modify the particle morphology, but it plays a role in the reduction rate, which can reach the W^{V}/W^{VI} ratio of 0.07 and obviously on the blue color. If heating the early acidified milky-white mixture containing hydrazine at 95°C begins only after a maturation time at room temperature of around 12 hours, a new phase of hexagonal bronze, h′-WO_3, is obtained (Fig. 4.32c, d). It differs from the h-WO_3 phase by the arrangement of hexagons of octahedra in the

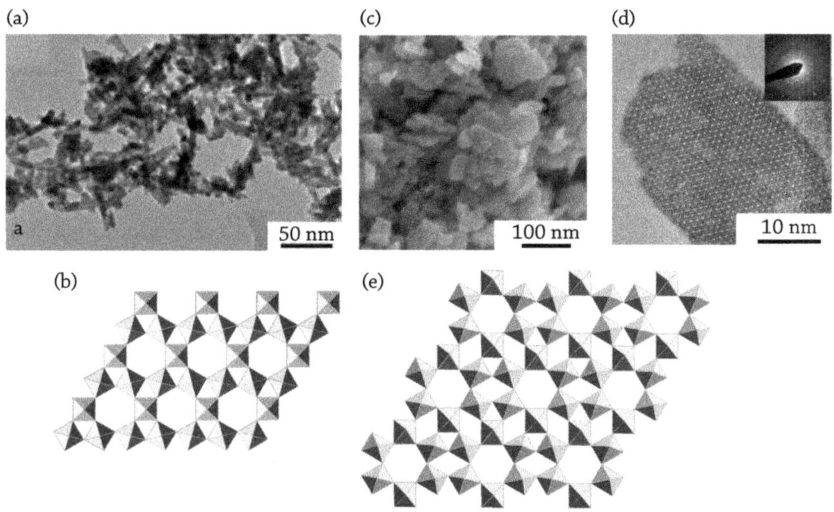

FIGURE 4.32 (a) TEM image of particles and (b) structure of (001) plane of the h-WO_3 phase. (c) FESEM image, (d) METHR image and (e) structure of (001) plane of the h′-WO_3 phase (adapted from [77]).

(001) planes, in forming rings of six, four, and three octahedra (Fig. 4.32e). The particles are platelets 50 to 100 nm in diameter and 5 to 10 nm in thickness, with hexagonal channels perpendicular to the basal faces of platelets [77]. In this new phase, the reduction of tungsten is apparently compensated by proton insertion instead of ammonium or sodium ions. Structural defects are sometimes due to a rotation of 30° of 6-octahedra rings in the (001) planes.

The structural changes in solid phases often induced by cations also occur with tungsten, which forms a pyrochlore $(M_2O)_xW_2O_6$ phase in the presence of Cs^+ or Rb^+, with $0.3 < x < 0.5$ [83] (Fig. 4.31e). The synthesis may be carried out by heating at 90°C a solution of sodium tungstate and cesium or rubidium carbonate at pH 3. Another route is to boil the carbonate and tungstate solution in ethylene glycol acidified with acetic acid [83]. The reaction is completed after 20 hours of heating. Heating B-paratungstate $(NH_4)_2W_{12}O_{41}\cdot5H_2O$ in similar conditions forms the pyrochlore $(NH_4O)_{0.5}W_2O_6$ phase. The three-dimensional channels of the pyrochlore lattice allow the easy and complete exchange of ammonium ions by protons in acidic medium. Cs^+ and Rb^+ are not as easily exchanged because of their size. Water contained in the protoned phase and hydrated at the contact of atmospheric water, $WO_3\cdot0.5H_2O$, is weakly bound to the nonelectrically charged lattice and is very mobile [83]. It behaves as "zeolitic water," contrary to that in antimonic acid of similar pyrochlore structure (§ 4.1.4).

Hydrothermal treatment at 110°C of a solution of molybdic acid, obtained from acidification of molybdate with an exchanger resin or by dissolution of the hydrate $MoO_3\cdot2H_2O$, forms $MoO_3\cdot1/3H_2O$. Its morphology differs from that of its tungstic homologue (Fig. 4.33), but both are isostructural [87].

Dehydration at 300°C of $MoO_3\cdot1/3H_2O$ leads to the monoclinic phase of MoO_3 oxide. The transformation proceeds, as for the corresponding tungstic compounds, according to an oriented nucleation and growth mechanism. However, in the molybdic phases, the arrangement of the Mo–O bonds is of ReO_3 type, which shows that the monoclinic MoO_3 phase has a lower nucleation energy. This phase is metastable, however, and transforms at 400°C in an orthorhombic variety (Fig. 4.34) in which molybdenum exhibits the *cis*-dioxo configuration characteristic

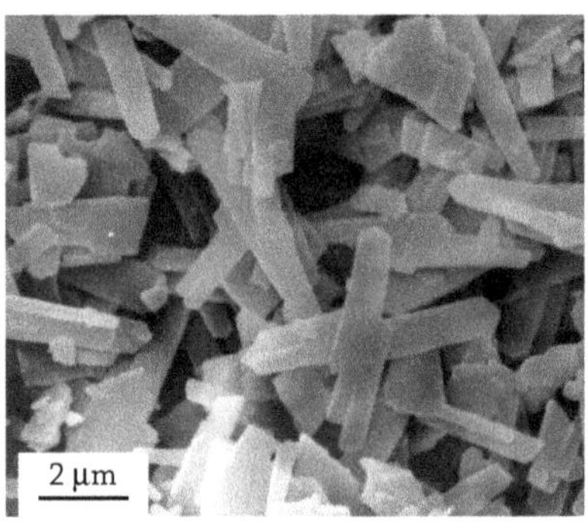

FIGURE 4.33 SEM image of $MoO_3\cdot1/3H_2O$ hydrate obtained in hydrothermal conditions (reproduced from [83] with permission of Elsevier).

FIGURE 4.34 Structure of orthorhombic MoO_3.

of stable polyanions of Mo^{VI}. The difference in behavior between tungstic and molybdic hydrates and oxides likely stems from a difference in the π interactions in W–O and Mo–O bonds, as in polyanions [83, 90].

4.1.6 SUMMARY

This study shows that the increase in the formal charge (oxidation state) of a cation strongly increases the polarization of its oxygenated ligands and reinforces the acidity of the zero-charged complex. The stability of hydroxides gradually decreases, favoring the formation of oxyhydroxides and oxides. The cations of highest formal charge form discrete species, polyanions rather than infinite networks. Two parameters are essential: the size and the ability of the cation to be a good π acceptor. This last characteristic results from the ability of the cation of transition elements to involve low-energy empty *d* orbitals. The M–O π bonds in which oxygen loses its base properties almost entirely are consequently terminal bonds that strongly hinder the condensation and lead to oligomers (polyoxoanions). The size of the metal ion can also be a limiting factor in restricting condensation, as in the case of Cr^{VI} where the too small size avoids extension of coordination and limits the condensation to a dimer.

Elements of the *p* block of the classification, such as Si^{IV}, Sn^{IV}, Ge^{IV}, Sb^{V}, and Te^{VI}, have sizes comparable to those of Mo^{VI} and W^{VI}. However, they cannot form π bonds with oxygen (no terminal bond) because they don't have accessible *d* orbitals. The polyoxoanions formed under specific acidities may, by subsequent acidification, grow and transform, providing an extended framework based on chains, layers, or networks of tetrahedra (Si), tetrahedra and octahedra (Ge), or octahedra (Sn, Sb, Te). There is no short M–O bond in the coordination polyhedron that hinders protonation and condensation. A typical example is P^{V}, for which the double P = O bond decreases the electrophilicity of the cation and prevents the condensation of phosphate in solution.

4.1.7 POLYMETALLIC OXIDES

a) Generalities

The association of several cations that are different in the same crystalline framework during coprecipitation involves different kinds of constraints. From a thermodynamic standpoint, the mixed phase must at least be as stable as the single phases formed by each cation alone. In addition, their pHs of precipitation have to be close to each other in order for their hydroxylation to occur within the same acidity range. From a kinetic standpoint, the reactivities of the cations, especially their condensation rates, have to be comparable for a simultaneous incorporation into the solid phase to take place. Also, the mixed phase must form at least as rapidly as the phases formed by each cation. Lastly, from a structural standpoint, the cations have to be compatible in size, charge, and coordination to allow their coexistence in the same crystalline phase formed at low temperature. These criteria are probably more stringent than what a high-temperature dry process would impose. All these conditions are rarely met by cations with different oxidation states, so that the precipitation of their mixtures more often leads to their segregation between either the solid phase and the solution or between two or more solid phases. The synthesis of polymetallic oxide nanoparticles is possible in soft conditions, however, and various strategies may be involved.

- It is possible to take advantage of the high reactivity of nanoparticles of segregated phases, such as amorphous or crystallized hydroxides or oxyhydroxides, and to force them to react by thermolysis, hydrothermal treatment, or microwave heating of the aqueous suspension. Because of the nanometric size of particles, the mixed phase can be formed by the diffusion of ions from one phase to the other or during a dissolution–crystallization process. Such a mechanism is involved in the formation of LDH (§ 4.1.2) and of green rusts (§ 7.2.3). Another example, (§ 4.1.7 b), is that of mixed oxygenated compounds of bismuth.
- In some cases, it is also possible to precipitate the cations with the same oxidation state and then to oxidize or reduce the solid to form the expected phase.
- The mixed phase may also be formed in assembling the cations with the right stoichiometry inside a molecular complex formed with organic multidentate ligands, for instance, of polycarboxylate type: oxalate, citrate, malonate, and so on. The mixed complex is then thermally treated in order to degrade the organic moieties and to form the expected oxide [91]. A similar technique, known as Pechini's technique, consists of adding a polyol in the solution of cations and containing the organic acid. That allows not only chelation of the cations but also induction of esterification of the acid and formation of a polyester resin, avoiding the segregation of the cations [92]. The subsequent thermal treatment above around 600°C allows degrading all the organic matter and forming an oxide at much lower temperatures than those required in the classical ceramic route involving mixing, grinding and heating oxide powders. The main disadvantage of Pechini's technique is that sometimes charcoals are formed, which are difficult to eliminate during the thermal treatment.

These various procedures allowing the formation of many polymetallic oxides are specifically illustrated in other chapters of this book with the synthesis of three types of ferrites: spinel MFe_2O_4, hexaferrite $BaFe_{12}O_{19}$, iron and yttrium garnet $Y_3Fe_5O_{12}$ (Chap. 7); barium titanate $BaTiO_3$; and stabilized cubic zirconias (Chap. 8). The following section focuses on synthesis of two mixed oxides of bismuth.

b) Bismuth Mixed Oxides

Among the mixed oxygenated compounds of bismuth (III), those involving transition elements with a high oxidation degree (V^V, Nb^V, W^{VI}, Mo^{VI}) are interesting because they are semiconductors with a relatively small band gap (from 2.4 to 2.9 eV instead 3.2 eV for TiO_2). This property results from interaction of 6s Bi and 2p O orbitals at the valence band edge [93]. These compounds therefore appear to be good candidates for photocatalysis, especially in the photodegradation of pollutants by visible light [94, 95]. Here, the point of interest is the synthesis of vanadate $BiVO_4$ and tungstate Bi_2WO_6 from aqueous solutions, which exemplify two different crystallization pathways of mixed compounds. Study of their photocatalytic properties is beyond the scope of this book.

Bismuth is an electronegative heavy element, forming strongly acidic complexes in water [6]. The nonhydroxylated $[Bi(OH_2)_x]^{3+}$ aquo species exist only in strongly acid medium (pH < 0). Over the $0 \leq pH \leq 2$ range, $[Bi(OH)(OH_2)_x]^{2+}$ and $[Bi(OH)_2(OH_2)_x]^{+}$ species have been characterized in very dilute solutions, as well as the $[Bi_6(OH)_{12}(OH_2)_x]^{6+}$ polycation (Fig. 3.8). Some $[Bi_9(OH)_{20+y}(O_2)_x]^{(7-y)+}$ ($0 \leq y \leq 2$) polycations, for which a structure is not yet established, are involved in the potentiometric analysis of further hydroxylation [6]. At the concentration of 0.1 mol.l^{-1}, a hydroxide precipitates from pH 0. Unstable at room temperature, it transforms into oxyhydroxide BiOOH. Bi_2O_3 is obtained by the addition of concentrated soda up to pH 11.5 and aging under stirring at room temperature for 3 days [94]

Bismuth Vanadate $BiVO_4$

By controlling the acidity of the medium and the reaction time, the three existing crystal phases of $BiVO_4$ can be selectively formed [96, 97]. Typically, the addition of sodium metavanadate, dissolved in pure water to a solution of bismuth nitrate in HNO_3 1.5 mol.l^{-1} (Bi/V = 1, final bismuth concentration of 0.1 mol.l^{-1}), immediately forms a solid amorphous to X-ray diffraction. After stirring for 30 minutes at room temperature, the pure tetragonal zircon-type phase of $BiVO_4$ is formed, as spheroidal particles of around 5 μm in mean diameter (Fig. 4.35a,d). A longer aging of the suspension transforms the solid into a monoclinic scheelite-type phase, the thermodynamically most stable phase of $BiVO_4$.

If the sodium metavanadate is dissolved in 1.5 mol.l^{-1} nitric acid and then added to the acidic bismuth solution, precipitation is notably slowed down because of the stronger acidity of the medium. After stirring 2 hours at RT, the pure tetragonal scheelite-type phase of $BiVO_4$ is obtained as polycrystalline spheroidal particles resulting from ordered aggregation of small rods around 50 nm in thickness (Fig.

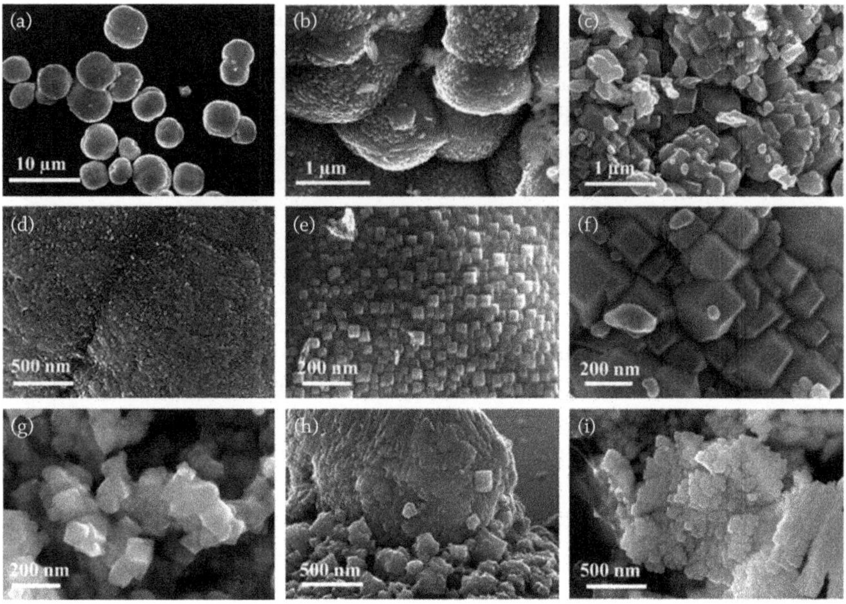

FIGURE 4.35 SEM images of $BiVO_4$ particles: (a, d) tetragonal zircon-type phase; (b, e) tetragonal scheelite-type phase; (c, f) monoclinic scheelite-type phase; monoclinic phase obtained in the presence of xylitol (g), sodium dodecyl sulfate (h) or potassium carbonate at pH 4 (i) (reproduced with permission from [97] Copyright 2015 American Chemical Society).

4.35b,e). The monoclinic scheelite-type phase appears beyond 2 hours' aging of the suspension. This last phase may also be obtained by reaction at room temperature of bismuth in 0.75 mol.l^{-1} nitric acid with solid vanadium oxide V_2O_5 after stirring for 72 hours, in yielding well-faceted prismatic particles around 50 to 200 nm in size (Fig. 4.35c,f).

In the monoclinic scheelite-type phase of $BiVO_4$, bismuth atoms are 8-coordinated in BiO_8 polyhedra (Fig. 4.36c). Each BiO_8 polyhedron shares four edges with four others and is bounded to 8 VO_4 tetrahedra in forming μ_3–O bridges. The VO_4 tetrahedra are corner-linked to BiO_8 polyhedra and are never bounded between them. In the tetragonal zircon-type phase (Fig. 4.36a), the BiO_8 polyhedra are also bonded by edge sharing between them, and the VO_4 tetrahedra are bounded to BiO_8 polyhedra in sharing two edges. In the tetragonal scheelite-type phase (Fig. 4.36b), the BiO_8 polyhedra share between them one edge and six corners, and they share one edge with a VO_4 tetrahedron and six corners with six other VO_4 tetrahedra. In the three polymorphs, the bonding between BiO_8 polyhedra forms 3D frameworks in which are distributed the VO_4 tetrahedra.

All these observations suggest a possible process involved in the formation of $BiVO_4$ from solution. After the mixing of precursors, the acidity of the medium is 0.75 mol.l^{-1}, corresponding to a pH value close to 0. In such conditions, bismuth at the concentration of 0.1 mol.l^{-1} precipitates [6], while vanadate remains stable in solution under the form of the hexacoordinated $[VO_2(OH_2)_4]^+$ monomer

FIGURE 4.36 Scheme of the crystal structures of (a) tetragonal zircon-type, (b) tetragonal scheelite-type, and (c) monoclinic scheelite-type phases of $BiVO_4$.

(§ 3.3.2a). It can be assumed that noncharged $h = 3$ complexes of bismuth condense by olation in sharing edges of coordination polyhedra. The μ_2–OH bridges so formed, bounded to bulky and relatively weakly polarizing Bi^{3+} cations, retain a nucleophilic activity, allowing them to attack the vanadic species and to form μ_3–O bridges by oxolation. The geometric constraint due to 3D crystallization likely forces the decrease in the vanadium coordination and deprotonation. So, the mixed phase appears to result from the nucleophilic character of the bismuth complexes, which, under the right acidity conditions, are able not only to condense but also to associate vanadate in the precipitation of the $BiVO_4$ stoichiometric phase. Such a mechanism, similar to that involved in the formation of Zn_2Cr LDH (§ 4.1.2), allows us to understand why the vanadic tetrahedra remain apart from each other within the structure. When precipitation occurs immediately after the mixing of bismuth and vanadate, the amorphous early solid likely crystallizes via a dissolution–crystallization process to form first the metastable zircon-type and then the tetragonal scheelite-type phases, finally leading to the most stable monoclinic scheelite-type phase as particles of increasing size. In more acidic conditions (HNO_3 1.5 $mol.l^{-1}$), the amount of $h = 3$ complexes of bismuth is obviously much smaller, so that precipitation is slowed considerably. It leads directly to the most stable solid phase of bismuth vanadate in which the BiO_8 polyhedra are edge sharing, whereas the vanadic tetrahedra are corner linked only by Bi_2-O-V bridges. Edge sharing by vanadic tetrahedra occurs only in the less stable zircon and tetragonal phases.

As in many cases discussed in this book, precipitation in the presence of various additives able to be adsorbed onto particles leads to change in their morphology (§ 5.4.4). Thus, precipitation of $BiVO_4$ in the presence of xylitol (§ 6.3.3b) forms octahedral-shaped particles of monoclinic phase with a size of around 100 nm (Fig. 4.35g). In the presence of sodium dodecyl sulfate, micrometric particles seem to be the result of the stacking of platelets (Fig. 4.35h), while particles obtained when the pH of the mixture are initially raised at 4 by the addition of potassium carbonate and stirred for 3 weeks at room temperature appear as micrometric platelets made of the ordered aggregation of 50 nm cubic subunits (Fig. 4.35i). In hydrothermal conditions with various additives, many other morphologies of bismuth vanadate can be obtained [8, 99]. Some examples

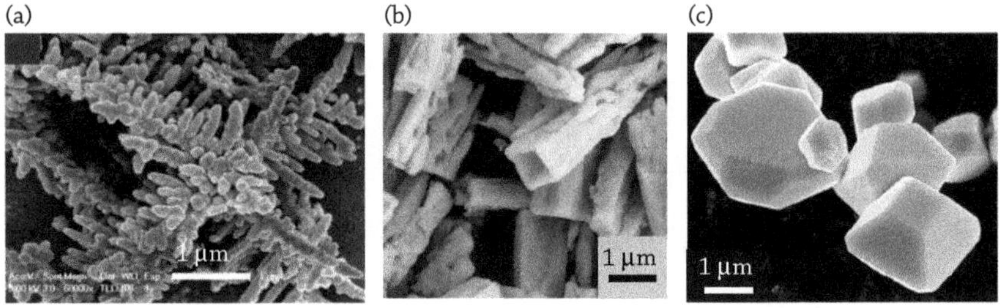

FIGURE 4.37 FESEM and HRSEM images of some morphologies of monoclinic $BiVO_4$ obtained by hydrothermal treatment of bismuth nitrate and vanadate aqueous solutions: (a) hyperbranched structures, 200°C, 6 hours [Zhao Y., Xie Y., Zhu X., Yan S., Wang S., Surfactant-free synthesis of hyperbranched monoclinic bismuth vanadate and its applications in photocatalysis, gas sensing, and lithium-ion batteries, Chem. Eur. J. 2008, 14, 1601–1606] (courtesy of Y. Xie, Hefei Univ., China); (b) nanotubes, 180°C, 6 hours (in the presence of triblock copolymer Pluronic P123); (c) polyhedra, 180°C, 6 hours (reproduced from [99] with permission of Elsevier).

are presented in Figure 4.37. These changes in shape are interesting because they help determine which are the most active crystal faces in the photocatalytic behavior of the material [97].

Bismuth Tungstate Bi_2WO_6

Bismuth tungstate is synthesized by hydrothermal treatment of aqueous mixtures of bismuth nitrate and sodium tungstate, Na_2WO_4, or tungstic acid, H_2WO_4, (Bi/W = 2) in HNO_3 1.5 mol.l^{-1}. The mixture is heated at 200°C during 24 hours in autoclave or during 4 hours in a microwave oven [100]. Platelet-shaped of variable size, Bi_2WO_6 particles are the ones more often obtained (Fig. 4.38), in accordance with the lamellar crystal structure of the solid (Fig. 4.39a). It is indeed made by stacking layers of corner-sharing WO_6 octahedra, similar to those in the hydrate tungstic $WO_3{\cdot}H_2O$ (Fig. 4.29), and sheets of edge-sharing BiO_6 polyhedra that are heavily distorted because of the $6s^2$ inert pair effect of Bi^{3+} ions. The structure of the layers containing bismuth looks somewhat like that in Bi_2O_3 (Fig. 4.39b). Thus, unlike bismuth vanadate, the two elements W and Bi are segregated in adjacent layers. That seems to indicate that the mechanism of Bi_2WO_6 formation is different from that involved in $BiVO_4$.

As we have seen (§4.1.5c), condensation of the tungstic complex $[WO(OH)_4(OH_2)]^0$ existing in acidic medium leads to the $WO_3{\cdot}H_2O$ hydrate. Bismuth also condenses under the same acidity conditions. Condensation of the two elements would consequently occur simultaneously, but separately, before the μ_3–O bridges resulting from oxolation allow the bonding of the two kinds of sheets. Supporting this pathway are the apparently strong similarities between the coordination of bismuth and the structural arrangement of BiO polyhedra in Bi_2O_3 (bismutite) and in Bi_2WO_6 phases. While precipitation of tungstic hydrate occurs at 100°C and that of Bi_2O_3 at room temperature in strongly alkaline medium, the formation of the Bi_2WO_6 mixed phase requires hydrothermal conditions, likely forcing the interaction between the tungstic- and bismuth-containing sheets.

FIGURE 4.38 TEM images of Bi_2WO_6 particles obtained from hydrothermal synthesis (a,b) and from microwave-assisted hydrothermal treatment (c,d).

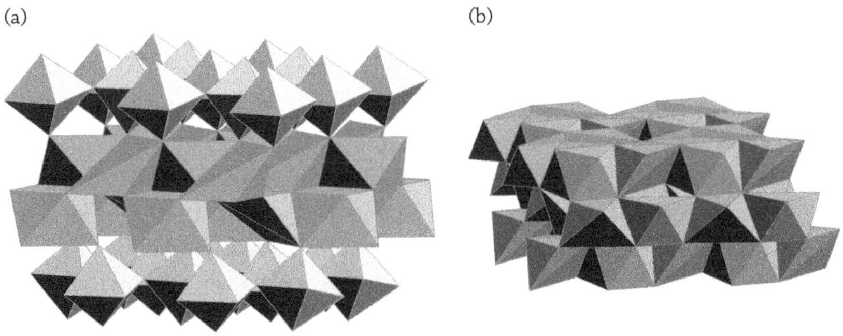

FIGURE 4.39 Scheme of the crystal structure of orthorhombic Bi_2WO_6 (a) and of Bi_2O_3 (b).

4.2 Kinetics of Precipitation and Mechanisms of Crystallization

The solid resulting from precipitation contains particles with a mean size varying from a few nanometers to several tens or hundreds of micrometers and a more or less broad size distribution. The morphology of particles of the same oxide may also be very different, depending on the conditions of its synthesis. Aging of suspensions after precipitation often leads to deep changes in size, morphology, and structure. Understanding these phenomena requires examining the successive kinetic steps, their dynamics, and the various mechanisms involved in the formation of particles.

It is useful to remember that the formation of particles implies, within all kinetic steps, the same chemistry as that previously studied. It is always a matter of olation and/or oxolation condensation reaction occurring between monomeric complexes (nucleation), between monomeric complexes and variable-sized objects (growth), and between nanometric objects (aggregation). The major factor is the concentration of the precursor (supersaturation) that influences the occurrence and rate of the different steps of the process. Other parameters such as acidity, ionic strength, and presence of various ligands also play an important role in determining the reactivity of both soluble species and the surface of growing particles such as sign and density of the electrostatic surface charge, adsorption, and aggregation.

4.2.1 THE STEPS IN THE FORMATION OF A SOLID

The precipitation from an aqueous solution of an oxy(hydroxi)de involves four main steps [101, 102].

a) The Generation of the Zero-Charged Precursor $[M(OH)_z(OH_2)_{N-z}]^0$ Able to Condense until Forming a Solid Phase

Hydroxylation of the cation is a fast acid–base reaction, but the rate of the formation in solution of the noncharged complex may be variable, depending on whether the reaction with a cationic metal complexes takes place through brutal addition of a base, thermolysis in acidic medium, or thermal decomposition of a base such as urea.

b) Creation of Nuclei Through Condensation (Olation or Oxolation) of Zero-Charged Precursor

The condensation rate is highly dependent on its concentration. At the onset of cation hydroxylation, the condensation rate is almost zero (zone I, Fig. 4.40b). With time, accumulation of the precursor in solution occurs and beyond a critical concentration, C_{min}, the condensation rate increases abruptly, producing polynuclear entities, the nuclei, in an “explosive” manner throughout the solution (zone II, Fig. 4.40). Their size is around 1 nanometer or less. Nucleation is a brutal kinetic phenomenon because, in having a high apparent kinetic order (§ 4.2.2a), it is extremely fast or nonexistent within a narrow concentration range. If the rate of generating the precursor is significantly lower than the condensation rate, nucleation sharply reduces the precursor concentration and the condensation rate decreases with equal rapidity. When the precursor concentration is again close to C_{min}, the formation of new nuclei is no longer possible.

c) The Growth of Nuclei Until Reaching the Stage of Primary Particles

At a concentration smaller than C_{min}, nucleation (that is, the autocondensation of precursors) no longer occurs. However, the precursors appearing in solution (hydroxylation is continuing to produce precursors) are captured by the existing

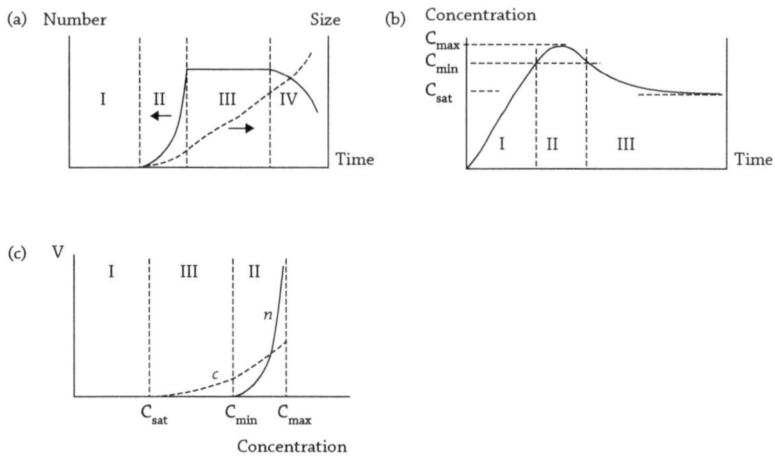

FIGURE 4.40 Scheme of variation (a) of the number and size of particles formed in solution and (b) of the concentration C of the soluble precursor of the solid phase (LaMer's diagram) during precipitation. The condensation rate is zero for $C < C_{min}$ and becomes infinite for $C \geq C_{max}$. C_{sat} is the solubility of the solid phase. (c) Nucleation (n) and growth (c) rates as a function of the precursor concentration in solution. The zones labeled I, II, and III correspond to the three schemes.

nuclei involving the same condensation process, olation, and/or oxolation as during nucleation, but of course, the kinetics of these two steps are governed by different phenomena: diffusion of complexes in solution for nucleation, diffusion and surface reaction for growth. The growth of nuclei consumes the precursors until their concentration reaches solution saturation or, in other words, the solubility of the solid phase (zone III, Fig. 4.40). Growth is a more progressive process than nucleation. To obtain uniformly sized particles, the nucleation and growth steps need to be clearly separated to ensure that only one nucleation phase takes place and that the growth of nuclei involves all the remaining matter. This implies a nucleation rate that is much higher than that of precursor generation (hydroxylation) and a precursor concentration that is low enough. In such conditions, nucleation is brief and clearly decoupled from the growth phase. This corresponds to the classical LaMer's model, which allows accounting for the synthesis of monodispersed suspensions of sulfur through disproportionation of thiosulfate in acidic medium [103]. Fine experimental validations of this model have recently been obtained through the formation of amorphous hydrated silica nanoparticles [104] (Fig. 4.41).

If the nucleation rate is not as high as the rate of the precursor generation and/or if its concentration is too high, the precursor concentration remains permanently higher than the critical concentration C_{min}; therefore, nucleation and growth occur simultaneously. In such a situation, depending on the mechanism of nuclei growth, the dynamics of the process may be very complex. Various mean particle sizes with different size distribution widths may be obtained. Different possibilities are examined (§ 4.2.2 b).

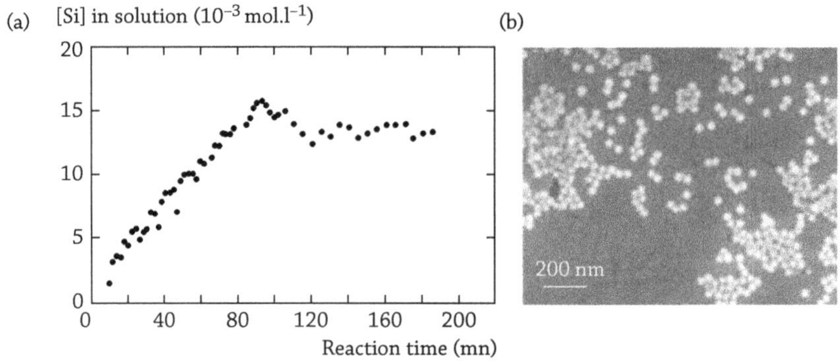

FIGURE 4.41 (a) Evolution of silicon concentration in solution during reaction of $Si(O\text{-}CH_2CH_3)_4$ in contact of an aqueous solution of 40 mM (pH 10, 80°C) (see §4.1.4, Si^{IV}); (b) SEM image of the silica particles at the end of reaction (reproduced with permission from [104] Copyright 2011 American Chemical Society).

d) Aging of Primary Particles That Often Lead to Various Modifications of the Primary Particles in Suspension

At this stage, the primary particles are generally nanosized objects, amorphous or crystallized, and have distinct morphologies. These particles are not in their highest thermodynamic stability state, which should correspond to a minimal surface-to-volume ratio (all the matter in the reactor in only one crystal!) of the crystalline variety with the highest density of the material. Aging of suspensions allows the system solid–solution to evolve toward a thermodynamically more stable state involving varied mechanisms: dissolution–crystallization, or Ostwald's ripening, aggregation, ordered aggregation, and in situ solid-state transformation. That leads to a decrease in surface area (zone IV, Fig. 4.40), structuration of an amorphous early phase, or changes in crystal structure and morphology. Aging is thus one of the most important phenomena occurring during the synthesis because it determines the final characteristics of particles resulting from precipitation.

This basic and global approach of precipitation will allow thorough study of the diverse mechanisms and processes involved in this phenomenon.

4.2.2 NUCLEATION AND GROWTH: ENERGETICS AND DYNAMICS

The precipitation phenomenon is difficult to study experimentally because it generally occurs too fast and abruptly. Many models have been elaborated to account for the formation of small particles in solution. Using the classical theory of nucleation, we adopt a recent approach to characterize the dynamics of precipitation.

a) Nucleation

Let us consider the formation of nuclei, P_n, of size n from n precursors P (zero-charge complexes), according to an autocondensation process in solution (homogeneous nucleation):

$$n\,P \Leftrightarrow P_n$$

The free enthalpy change, $\Delta G°$, of such a reaction includes a term relative to the volume of the P_n nuclei (the solid), depending on the difference in chemical potential (μ_S – μ_L) of the P entity in the solution and in the solid, respectively, as well as a term representing the energy required for creating the surface, A, of the nuclei [105]:

$$\Delta G^{\circ} = \Delta G^{\circ}_{vol} + \Delta G^{\circ}_{surf} = n(\mu_S - \mu_L) + A\gamma \qquad [4.1]$$

γ is the interfacial tension or surface energy of nuclei (§ 5.4.4). $\gamma = \partial G/\partial A$, in assuming isotropic nuclei for simplicity.

In assimilating activity and concentration, the chemical potential difference is written as $(\mu_S - \mu_L) = kTLn(c_S / c_L) = -kTLnS$. c_L is the concentration of the precursor in solution, c_S is the solubility of the solid phase, and $S = c_L/c_S$. S is the supersaturation ratio of the solution. The nucleus, assumed to be spherical, formed from n precursors of molecular volume v, has a radius $r = (3nv/4\pi)^{1/3}$ and a surface area $A = n^{2/3}(36\pi v^2)^{1/3}$. The free enthalpy of nucleation is thus:

$$\Delta G^{\circ} = -nkTLnS + n^{2/3}(36\pi v^2)^{1/3}\,\gamma \qquad [4.2]$$

Considering that the solution is supersaturated ($S > 1$) and the surface tension γ is positive, we see that the first term in Eq. [4.2] is negative and the second one is positive. Therefore, the free enthalpy variation for the formation of a nucleus exhibits a maximum $\Delta G°*$ (Fig. 4.42). It represents the activation barrier for the process, equivalent to the formation of the activated complex in the transition state of any chemical reaction. The kinetics of nucleation appears to be controlled by the relative weight of surface and volume energies.

The size of the critical nucleus formed at the maximum in $\Delta G°$ and able to subsequently grow is given for $\partial(\Delta G°)/\partial n = 0$ by

$$n^{*} = 32\pi\gamma^3 v^2 / 3(kTLnS)^3 \qquad [4.3]$$

The corresponding change in free enthalpy is

$$\Delta G^{\circ *} = (n^{*}/2)kTLnS = 16\pi\gamma^3 v^2 / 3(kTLnS)^2 = (4\pi/3)\gamma r^{*2} \qquad [4.4]$$

and the radius of the critical nucleus, calculated from Eq. [4.3], is given by the Gibbs–Kelvin equation:

$$r^{*} = (3n^{*}v / 4\pi)^{1/3} = 2\gamma v / kTnS \qquad [4.5]$$

The size of the critical nucleus decreases for increasing supersaturation and decreasing surface tension.

Because of an activation barrier in the nucleation process, the nuclei form and resorb statistically, but as the precipitation is spontaneous ($S > 1$), the probability of growth is higher than that of dissolution, which gives a stationary state of nucleation. The nucleation rate, J, represents the number of nuclei formed per unit of time and of volume of solution. It is expressed as [105]

$$J = J_0 \exp(-\Delta G_N / kT) \qquad [4.6]$$

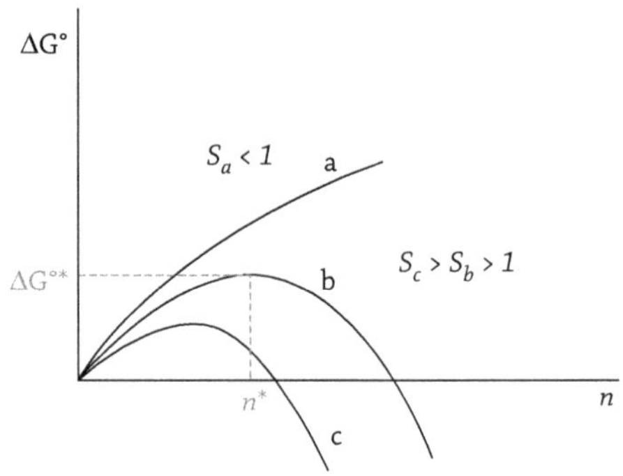

FIGURE 4.42 Variation of the free enthalpy of formation of a nucleus as a function of its size defined as the number n of included precursors. The curve (a) corresponds to a solution nonsupersaturated ($S < 1$); the curves (b) and (c) are for supersaturated solutions with $S_c > S_b > 1$.

J_0 is related to the frequency of collisions between precursors. Its value is between 10^{10} and 10^{35} $s^{-1}.cm^{-3}$. ΔG_N is the nucleation activation energy, which includes two major contributions. The first is the height of the energy barrier ΔG^{o*} required for the formation of a nucleus large enough to be stable (Eq. [4.4]). The second component, ΔG^R, is relative to the chemical reaction (olation or oxolation) involved in the process. It is usually about 35 kJ.mol^{-1}, but acidic or basic catalysis of oxolation (§ 3.3) decreases this activation energy to a few kJ.mol^{-1} [106]. Hence, the nucleation rate may be written as

$$J = J_0 \exp[(-\Delta G^* + \Delta G^R)/kT)] = J_0' \exp[-16\pi\gamma^3 v^2 / 3(kT)^3 (LnS)^2] \qquad [4.7]$$

with $J_0' = J_0 \exp(-\Delta G^R / kT)$. Nucleation is all the easier, that is, rapid and nuclei are all the smaller as the solid–solution interfacial tension is low and supersaturation is high.

It is difficult to investigate nucleation experimentally because the phenomenon depends strictly on concentration. A 300 K, with usual parameters for oxides ($\gamma \approx$ 30 to 100 mJ.m^{-2}, $v \approx$ a few tens of Å^3, $J_0' \approx 10^{30}$ cm^{-3}.s^{-1}), 10^5 nuclei are formed per cm^3 and per second for $S = 100$. That means the reaction is instantaneous, whereas for $S = 10$, nucleation is impossible because only one nucleus is formed every 10^{70} s. Nuclei are also difficult to detect owing to their small size (n^* is typically a few tens of units). Usually, an "induction time," t_{ind}, is determined by light or X-ray scattering when the scattered intensity increases sharply [104, 107, 108] (see Fig. 4.18). From the final particle size and amount of solid obtained, the number of particles, N, can be calculated and a pseudonucleation rate $J = N/t_{ind}$ is determined. Measurements for various supersaturations, S, allow estimate of the preexponential term, J_0', and the surface tension, γ, of the solid [105, 109, 110].

Nucleation does not have a definite kinetic order, in a strict sense, although experimental $\log J = f(\log S)$ curves do fit approximately on a straight line over a few decades of J [105]. Within a limited concentration range, one may write $J = k_n C^n$, where n is roughly equal to n^*, the number of monomers included in the critical nucleus. This is an approximation because n^* is itself a function of S (Eq. [4.3]) and because the probability of a simultaneous efficient collision of n^* precursors is obviously close to zero. The formation of the critical nucleus more probably results from consecutive bimolecular steps, but it may depend on an n^*th power of the concentration, considering that since all steps are very fast, they all contribute to the global reaction [111, 112].

b) Growth

The critical nuclei n^* are in unstable equilibrium in solution because they correspond to a maximum in the function $\Delta G° = f(n)$. Depending on the instantaneous value of supersaturation, they may resorb if they become subcritical ($r < r^*$), or they may grow if they become supercritical ($r > r^*$), because $\partial(\Delta G°)/\partial n < 0$ in both cases. The growth law that correctly describes the phenomenon has to be size dependent, and one may assume that it will be proportional to supersaturation ($S - S^*$), where S^* is the supersaturation corresponding to the critical nucleus r^* (Eq. [4.5]) [113]. The law also depends on the rate-limiting process governing the growth, that is, the capture and incorporation of complexes in solution by nuclei. That may be a surface reaction or the diffusion of complexes in solution. The growth rate of the nucleus radius is written as

$$dr / dt = \kappa(S - S^*) = \kappa[S - \exp(2\gamma v / kTr)] \qquad [4.8]$$

for a surface reaction limited process. The proportionality constant κ would be replaced by κ/r for diffusion-limited growth.

For both nucleation and growth, the key parameter is supersaturation, which, inside a closed reactor, varies (globally decreases) throughout the process. Therefore, except for systems in which the concentration is low enough that a single brief and abrupt nucleation phase occurs, followed by growth-producing uniform-sized particles (LaMer's conditions, Fig. 4.40), nucleation and growth take place simultaneously over a more or less long period. It follows that the decrease in supersaturation produces nuclei with an increasing size because the critical condition evolves. That simultaneously provokes the resorption of the particles formed at the beginning of the process, having insufficient growth to become supercritical at any moment. Thus, nucleation and growth processes involve a feedback effect on supersaturation. This effect is proportional to the amount of units of growth (the metal complexes) which have been transferred between solution and solid. These considerations are the basis of Noguera et al.'s model showing quantitatively the intricate processes involved in precipitation dynamics [113, 114].

Characterizing the evolution of a system requires knowledge of the time dependence of supersaturation and that of particle population. At time t, the population consists of particles that have been formed at all times $t_1 < t$ and have grown in the time interval $\{t_1 - t\}$. It is accounted for through two-variable functions, such as $n(t_1, t)$ and $r(t_1, t)$, representing, respectively, the size and the radius at time t of particles formed at t_1, and calculating their evolution. From Eqs. [4.5] and [4.8], the size $n(t_1, t)$ at time t, of the particles generated at all time t_1 is given by [113]:

$$n(t_1,t)=\frac{2u}{(\ln S(t_1))^3}+\int_{t_1}^{t} w(n(t_1,t'))^{2/3}\left[S(t')-\exp\left(\frac{2u}{n(t_1,t')}\right)^{1/3}\right]dt' \qquad [4.9]$$

with $u=16\pi\gamma^3v^2/3(kT)^3$ and $w=3\kappa(4\pi/3v)^{2/3}$. The first term in Eq. [4.9] represents the size of critical nuclei (assumed spherical) generated at time t_1, and the second term is their subsequent algebraic growth between t_1 and t.

In a closed system (with no advective feeding), the mass conservation is written as

$$c(t)=c_{(t=0)}-Q(t)/N_A$$

Where $c(t)$ represents the concentration of precursor in solution (neglecting the cation speciation), and $Q(t)$ is the amount of cations having changed phase, N_A is the Avogadro number. $Q(t)$ is given by

$$Q(t)=\int_0^t J(t_1)(n(t_1,t)-1)dt_1 \qquad [4.10]$$

$J(t_1)$ is the instantaneous nucleation rate (Eq. [4.7]). The feedback effect of nucleation growth on supersaturation, using the conservation law, is written as

$$S(t)=S_{(t=0)}-\frac{J_0'}{c_S N_A}\int_0^t \exp\left(\frac{u}{(\ln S(t_1))^2}\right)(n(t_1,t)-1)dt_1 \qquad [4.11]$$

Numerical solvation of the set of coupled equations [4.9] and [4.11] was carried out with the NANOKIN code [114–116] using a set of the parameters u, w, J_0' typical of oxides. u varies from 1 to 1000 as γ varies over the range 50–1500 mJ.m^{-2}, v is a few tens of Å^3, and the temperature range is 300–1500 K. w, characterizing the growth rate, varies from 10^{-4} to 10^3 with κ, around 4.10^{-7} m.s^{-1} for the hydrothermal growth of quartz, varying about seven orders of magnitude between 0 and 300°C. The solubilities vary in the range 10^{-3}–10^{-8} mol.l^{-1}. The supersaturation S_c corresponds to a nucleation rate $F(S_c)$ of one nucleus per second and per-volume unit ($S_c = C_{min}/C_{sat}$). In the range $1 < S < S_c$, nucleation is thermodynamically possible but experimentally nonobservable on a reasonable time scale.

In the case of a rapid growth rate and a supersaturation higher than 1, the simulation exhibits three successive stages (Fig. 4.43a):

1. An abrupt dropdown of S. Over this brief time interval, nucleation is efficient ($S_0 >> S_c$), and there is practically no growth (Fig. 4.44, t_2).
2. A quasi-plateau in the variation of S. The two processes occur simultaneously: formation of nuclei with an increasing critical size and an algebraic increase in size of the nuclei previously formed. Figure 4.44 shows the evolution in size of the particle population: the $n(t_1, t)$ curve is drawn as a t_1 function (time of formation of a nucleus) for four successive times: $t_2 < t_3 < t_4 < t_5$. The dashed curves

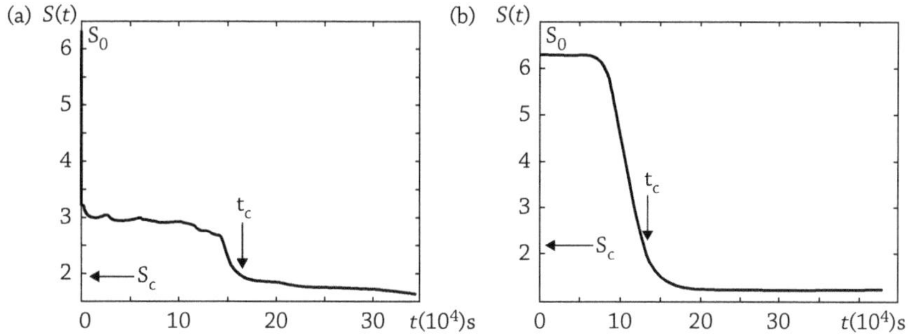

FIGURE 4.43 Time dependence of supersaturation $S(t)$ (Eq. [3.10]) calculated with (a) $u = 24$, $w = 2.10^{-4}$, $J = 1.26.10^{3}$ and $S_0 = 6.3$, and (b) $u = 24$, $w = 2.10^{-4}$, $J = 1.26.10^{-4}$ and $S_0 = 6.3$ (reproduced from [116] with permission of Elsevier).

represent the size of the critical nuclei $n(t_1, t_1)$ which, as expected, is practically constant as long as $S(t)$ remains on the plateau zone. One may observe that almost the early particles have been resorbed and that only some classes of particles formed during the plateau stage (labeled #1 and #2 in Figure 4.44) are able to grow. In fact, the opposite behavior among certain classes of particles formed at different stages of the process results from the size dependence of the growth law: only the particles of supercritical size are able to grow. Thus, at t_2 and t_3 times, all the particles are subcritical, except those labeled #1, which have nucleates around $t \approx 15.10^3$ s and have a slightly larger size than at nucleation. These particles have been supercritical during a certain stage, but at t_3, they are progressively resorbing. They are totally resorbed at t_4 and t_5, while the particles of a second class (labeled #2) nucleated at $t \approx 40.10^3$ s are supercritical and their size will continue to grow to the end of the aging process.

The plateau in the evolution of supersaturation does not mean nothing or little is happening. It actually corresponds to a complex dynamics where all antagonist processes, nucleation growth and nuclei resorption, are simultaneously active and quasi-equilibrated: the nucleation and growth of particles tend to decrease S, while resorption of all other particles tends to increase it. However, small deviations between these two processes (in some time intervals, their equilibrium is not perfect) give rise to oscillations of S (Fig. 4.43a). These oscillations are not artifacts but appear to be a required condition for the growth of particles: all those that nucleated outside the time intervals where $dS/dt > 0$ resorb and will be missing during the aging period. At the end of the plateau stage, the particle size is important ($n(t_1, t) >> n(t, t)$), and their growth rate strongly increases. Resorption no longer equilibrates growth that leads to an abrupt drop in $S(t)$. Nucleation becomes negligible ($S(t) < S_c$, $F < 1$).

3. When $t > t_c$ (Fig. 4.43a), nucleation practically does not occur any more, and so the system reaches the aging period. Precipitation is

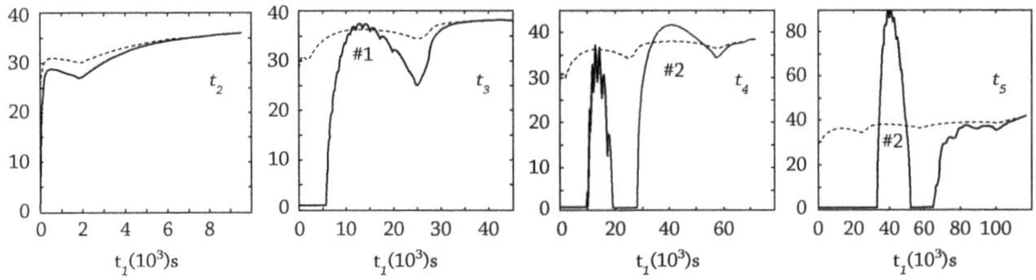

FIGURE 4.44 Time dependence of particle size $n(t_1,t)$ (plain lines) and of nucleus size $n(t_1,t_1)$ (dashed lines) at four successive times $t_2 < t_3 < t_4 < t_5$. ($u = 24$, $w = 2.10^{-4}$, $J = 1{,}26.10^3$ and $S_o = 6{,}3$) (reproduced from [115] with permission of Elsevier).

quasi-complete, the concentration in solution corresponding to the solubility of the solid. In this stage, the evolution is governed by the reduction in surface area of particles, leading to the disappearance of the smallest particles and the increase of the biggest. This process is known as Ostwald's ripening (coarsening) (§ 4.2.3). Theoretically, it should lead to only a single particle including all the matter of the system, but the process is kinetically limited. It leads only to the spreading of the particle size distribution. It is expressed as

$$P(n,t) = \frac{\int_0^t \exp(-u/(\ln S(t_1))^2)\delta(n - n(t_1,t))dt_1}{\int_0^t \exp(-u/(\ln S(t_1))^2)dt_1} \quad [4.12]$$

(δ stands for the Kronecker symbol), and it is represented in Figure 4.45.

It is interesting to compare these results to those resulting from alternative models with other growth laws. The first model involves only nucleation, and the second one includes nucleation and growth, with growth independent of the size. Equation [4.9] giving the particle size as a function of time becomes the first case:

$$n(t_1,t) = \frac{2u}{\ln^3 S(t_1)} \quad [4.13]$$

and the second one is

$$n(t_1,t) = \frac{2u}{(\ln S(t_1))^3} + \int_{t_1}^t w(n(t_1,t'))^{2/3}(S(t') - 1)dt' \quad [4.14]$$

Equation [4.10] describing the feedback effect on supersaturation remains the same.

As regards nucleation alone, supersaturation sharply decreases and rapidly converges toward an asymptotic value higher than S_c, displaying no plateau zone or aging stage (Fig. 4.46a). The S_c value cannot be reached because the nucleation frequency is too low. As the particle size is that of nuclei, the size distribution (Fig. 4.46b) shows that the most numerous particles are those that were

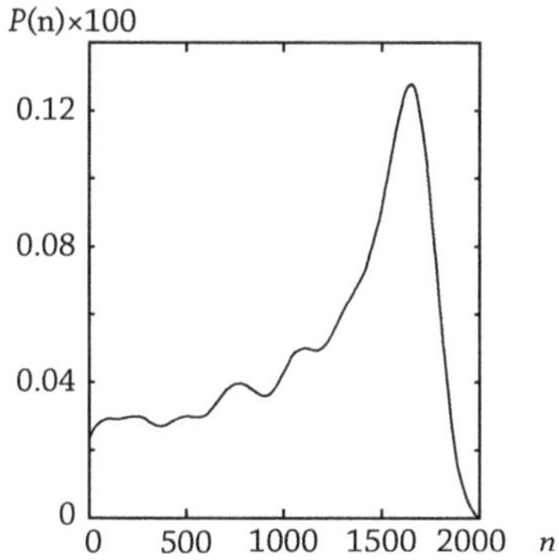

FIGURE 4.45 Normalized particle size distribution function $P(n, t)$ calculated with Eq. [4.12] for $t = 860$s. ($u = 24$, $w = 2.10^{-4}$, $J = 1{,}26.10^3$ and $S_o = 6{,}3$) (reproduced from [115] with permission of Elsevier).

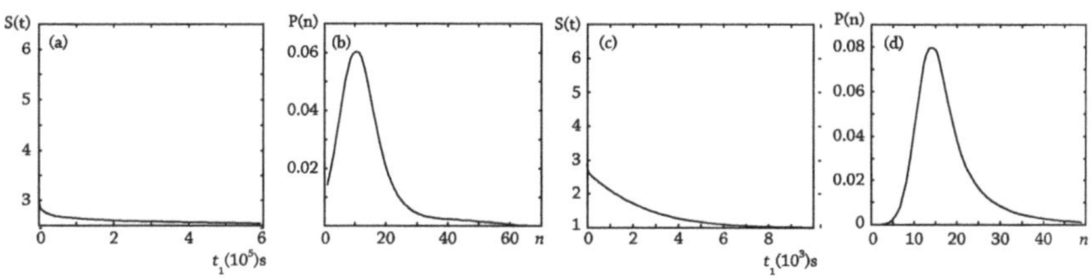

FIGURE 4.46 (a) Time variation of supersaturation $S(t)$ (Eq. [4.13]), and (b) particle size distribution at time $t = 6.10^5$s. (c) Time variation of supersaturation $S(t)$ (Eq. [4.14]) and (d) particle size distribution at time $t = 10^4$s. For the two cases: $u = 24$, $w = 2.10^{-4}$, $J = 1.26.10^3$, and $S_o = 6.3$ (reproduced from [115] with permission of Elsevier).

first formed at the beginning of precipitation, when $S(t)$ was the largest with the highest nucleation rate and the smallest size.

With regard to growth independent of the size, supersaturation also displays an abrupt initial decrease due to nucleation and a second decrease corresponding to the growth process, with no plateau stage. The concentration reaches a value of S much smaller than S_c (Fig. 4.46c). The time scale between these two models is obviously different, and the duration of the second step is shorter when w increases, as growth occurs. The critical time t_c in the decrease of S is about two orders of magnitude shorter than is seen with a size-dependent growth law because resorption that slows down the decrease in S does not take place. All formed particles are present at the end of precipitation, with the size distribution shown in Figure 4.46d.

In the case of a very slow condensation reaction and/or a very low supersaturation, the time dependence of supersaturation displays a plateau at the beginning of evolution and then decreases abruptly (Fig. 4.43b). Such a plateau is sometimes called induction time. As previously seen, it is not a period when nothing occurs; rather, it is a period when nucleation, growth, and resorption are competitive, so that the concentration does not seem to evolve. This phenomenon is observed, for instance, during amorphous silica precipitation from diluted silicic acid solutions [117, 118].

The model presented by Noguera et al. provides original information on the dynamics of precipitation. The ability to track the fate of any particles nucleated in the system highlights the role of resorption of subcritical objects on supersaturation and the impact of the growth law on particle size distribution at the end of the process. The shape of this distribution provides information on the growth mechanism involved in a given system. It is useful to remember that the critical nucleus is small (the precipitation is easy, and the number of nuclei is high for a given amount of matter) all the more because supersaturation is high, the surface tension is low, and the solid phase is metastable, that is, soluble. For many solids, there is a correlation between the lowering of the surface tension and the increase in solubility [110].

The surface tension is strongly influenced by the composition of the solid surface (Gibbs's equation, § 5.4.4). This parameter may be adjusted by the physicochemical conditions of synthesis (for instance, the acidity governs not only the speciation but also the protonation of the oxide surface, the surface charge density, the ligand adsorption equilibria) in order to control the particle size and aging (Chap. 5). For a polymorphic solid such as iron, aluminum, and titanium oxides, the surface tension is low all the more as the phase is less stable and of less density. An empirical relationship is given by $\gamma = (2E/d)(a/\pi)^2$ where E is the Young modulus, d the interplanar spacing between atomic planes normal to the exposed face, and a is the ionic radius of the cation. As d is inversely proportional to the density, γ increases as density increases. The metastable phases are therefore kinetically favored (the free enthalpy of formation of the critical nucleus, $\Delta G°*$, is proportional to $\gamma^3/(LnS)^2$, Eq. [4.4]). Often, the condensation rate of noncharged complexes is so rapid that the solid cannot crystallize and the first stages of precipitation yield an amorphous solid (aluminum and ferric ions, for instance). It is during subsequent aging (beyond t_c in Noguera's model) that evolution toward a thermodynamic equilibrium takes place; the solid may crystallize and undergo a phase transformation as well as a decrease in its specific surface area to lower the free enthalpy.

The initial supersaturation is the master parameter. It may vary considerably for the synthesis of nanoparticles from thermolysis in acidic medium (the precursor is generated during heating of the aqueous solution, the solid being formed under low supersaturation) to a sudden addition of a concentrated solution of a cation into an excess of base. In the first case, the process takes place during some hours or some days, whereas in the second case, it is quasi-instantaneous. Other things being equal, especially the temperature, in the two cases the size of nuclei is most distinct (Fig. 4.42, Eq. [4.3]), as well as dynamic, and possibly the growth process. In the second case, the physicochemical mechanisms, especially particle resorption, do not have time to take place because of the too strong kinetic constraint caused by the huge initial supersaturation. Such a system will likely undergo important transformations during aging.

4.2.3 MECHANISMS OF CRYSTALLIZATION: STRUCTURAL AND MORPHOLOGICAL EVOLUTION OF OXIDE NANOPARTICLES IN SUSPENSION

Nucleation and growth form particles under kinetic control, according to the pathway that involves minimum activation energy. It follows that the products are generally not in their thermodynamic stability state. The aging of suspensions, which may occur over large time scales (hours, days, or months), allows the system to tend toward—or to reach—a more stable state. The driving force in the transformation is the lowering of free enthalpy resulting from the decrease in surface area due to the growth of particles, their change in morphology, and the release of the lattice energy due to crystallization or crystal phase transformation. Two main mechanisms are at work in these conditions: dissolution–crystallization and/or ordered aggregation (oriented attachment).

a) Dissolution–Crystallization, Ostwald's Ripening (Coarsening)

PARTICLE SIZE INCREASE

At the end of the nucleation and growth stages, the solid is in the form of particles with a more or less spread size distribution. In fact, Gibbs–Kelvin's equation (Eq. [4.5]) indicates that for a given supersaturation S^*, only the particles with the unique size r^* are in equilibrium with the solution. For particles such as $r < r^*$, the solution appears to be unsaturated, and they must dissolve. On the other hand, particles such as $r > r^*$ experience a saturated solution and they must grow. These conditions bring about a process involving a transfer of matter from the smallest toward the largest particles via the solution through dissolution–crystallization equilibria. The same process is involved in resorption of subcritical nuclei and growth of the supercritical nuclei during nucleation (§ 4.4.2 b).

A typical example is given in the evolution of the size distribution of magnetite Fe_3O_4 nanoparticles obtained from coprecipitation of ferric and ferrous ions (Fig. 4.47a) (see Chap. 7). The real thermodynamic stability state would be reached if all the precipitated matter was gathered in a single particle, which gives rise to such a spontaneous evolution (in anaerobic conditions). The surface area would then be at a minimum (Eq. [4.1]). Practically, the evolution is rather slow and stops before all the particles have been transformed because it occurs at very low concentration of soluble species ($S \approx 1$). In addition, the increase in particle size slows down their dissolution rate. The process that requires the hydroxo or oxo bonds to break needs activation, so that a pseudo-equilibrium state is reached after some days. This evolution appears as an offset in the distribution due to the disappearance of the smallest particles and the formation of larger particles. That is typically Ostwald's ripening.

Quantitative treatment of the kinetics of the phenomenon [105, 119, 120] shows that this secondary growth process leads to a modification in shape of the absolute particle size distribution, but the normalized distribution is stationary (Fig. 4.47b). Such a test confirms that the secondary particle growth is governed by Ostwald's law, without involving other phenomena such as aggregation or coalescence.

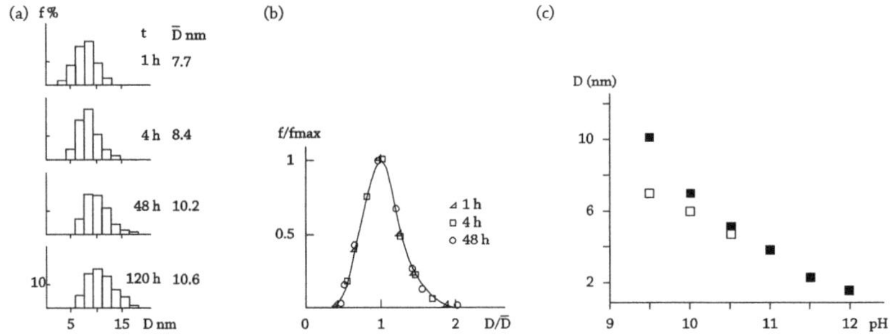

FIGURE 4.47 **(a)** Evolution of the particle size distribution of magnetite Fe_3O_4 during aging in suspension (pH 9). **(b)** Normalized size distribution at different times. (c) Mean particle size of magnetite formed at various pH, immediately after precipitation (□) and after 8 days aging (■).

Ostwald's ripening may be avoided or strongly limited if the interfacial tension, which is the driving force in the process (Eq. [4.1]), is lowered sufficiently. Gibbs's law, $d\gamma = -\Gamma_i d\mu_i$, indicates that interfacial tension is dependent on the surface composition via the adsorption of various species. For that process, precipitation may be carried out at a pH distant from the point of zero charge of the surface (§5.1.1) (Fig. 4.47c). In this case, the proton adsorption–desorption equilibria account for the electrostatic surface charge during precipitation, that leads to a decrease in surface tension (§ 5.4.4). Precipitation may also be carried out in the presence of a specifically adsorbable species. The particle size can then be controlled, and possibly so can the morphology of the nanoparticles if adsorption occurs specifically on special crystal faces (§ 6.3.1b).

Aggregation of particles, which allows the solid–solution interface area to decrease, may also contribute to stabilizing the system. It is favored by physicochemical conditions (pH, ionic strength), which minimize the surface electrostatic charge of particles (§ 5.1). In the absence of an electrical double layer, the particles attract between them, and aggregation, faster than crystal growth, allows a decrease in surface area. It is noteworthy that aggregation does not necessarily occur at a special stage of the evolution. It may take place from nucleation or at any time during growth, and it may lead to the formation of dendrites, mosaic crystals, or fractal structures [121]. Some examples are given in the study of iron (Chap.7), titanium, and manganese oxides (Chap. 8).

Aggregation of primary particles may induce dissolution–crystallization equilibria because the solubility of the zones near the contact points between particles with a negative curvature radius (concave surfaces' so-called necks) is lower than the solubility of convex surfaces with a positive curvature radius (Scheme 4.4).

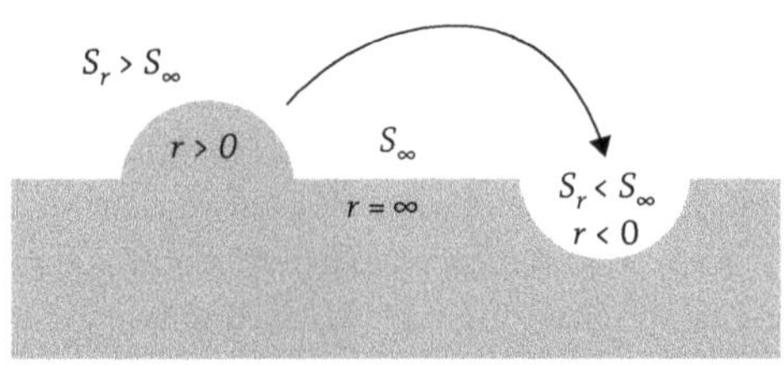

SCHEME 4.4

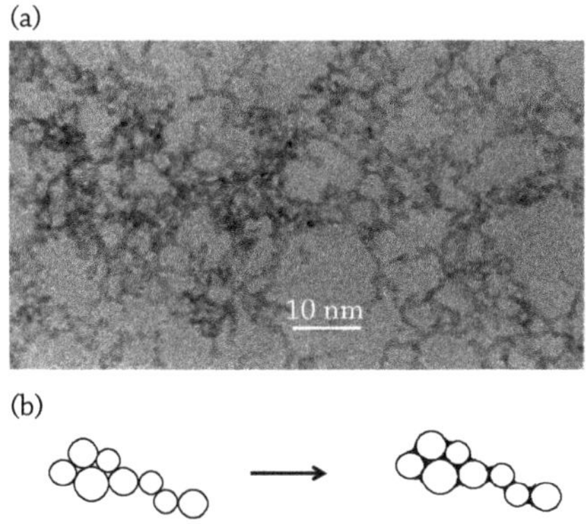

FIGURE 4.48 (a) TEM image of a diluted gel of stannic oxide, SnO_2 (reproduced from [122] with permission of Elsevier). The gel is composed of chains of 2-nm-sized particles. The diluted aqueous suspension gives a scattered light distribution function $P(\theta)$ corresponding to stiff units, such as rods with a mean gyration radius of 130 nm, although the TEM image indicates more or less flexible chains that could behave as coils in light scattering [122]. (b) The stiffness of the scattering units results from sintering of the primary spherical particles within the chains induced by dissolution–crystallization, forming a deposit in the zones of contact between the particles.

That is expressed by the Gibbs–Kelvin equation (Eq. [3.5]) under the following form:

$$S_r / S_\infty = \exp(2\gamma v / kTr) \quad [4.15]$$

S_r and S_∞ are solubilities of particles of radius r and of a flat surface (infinite radius), respectively. Therefore, a transfer of matter occurs from roughnesses to fill the necks between particles to reduce the differences in solubilities within the different zones inside the aggregates. The phenomenon is all the more pronounced as the size of the primary particles is small. It may involve coalescence or sintering of primary particles, stiffening of chains of spherical particles, and even their transformation into fibers or rods, as observed in aggregates of particles of silica [40] and tin oxide (Fig. 4.48) [122].

CRYSTAL STRUCTURE CHANGE

When a solid exists under allotropic varieties, that of least stability and of highest solubility precipitates first. This metastable phase may then transform during aging of the suspension to form another thermodynamically more stable phase. The phenomenon is of kinetic origin. Equations [4.5] and [4.7] show, respectively, that, for a fixed supersaturation S, the size of the critical nucleus is small and the nucleation rate is all the higher, as the interfacial tension γ is low. Consequently, the precipitation of the less stable phase of lowest interfacial tension and of highest solubility is kinetically favored (§ 4.4.2 b). Also, this phase is then sensitive to aging, which induces its transformation into more stable varieties. The phenomenon is typical of many hydrated amorphous phases and of polymorphic solids such as titanium oxide (anatase, brookite, and rutile; Chap. 8), aluminum hydroxides and oxyhydroxides (gibbsite, bayerite, boehmite; Chap. 6),

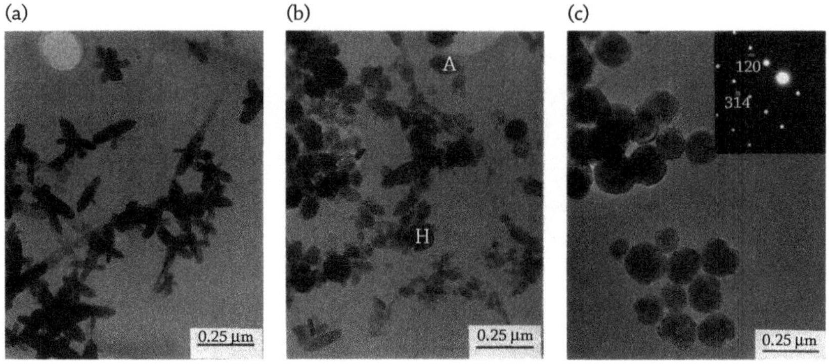

FIGURE 4.49 TEM images of (a) akaganeite particles obtained by heating at 100°C of a $FeCl_3$ solution 0.018 mol.l^{-1}, HCl 0.01 mol.l^{-1} for 12 hours; (b) the same for 24 hours, in which akaganeite (A) coexists with hematite (H) particles; (c) hematite pseudospherical monocrystals formed after a week at 100°C (insert, electron diffraction diagram) (reproduced from [123] with permission of Elsevier).

and iron oxyhydroxides and oxides (ferrihydrite, lepidocrocite, goethite, hematite; Chap. 7).

The process that is more often involved in the transformation is dissolution–crystallization. It leads to an important enlargement of particles because the secondary nucleation step involved in forming the secondary phase occurs at very low supersaturation. A classical example is the transformation of akaganeite, γ-FeOOH, into hematite, α-Fe_2O_3, during the thermolysis of acidic solutions of ferric chloride (Fig. 4.49 and Chap. 7) [123].

The preferential precipitation of the least stable phases may be used to control and regulate the nucleation and growth processes in order to form particles with a uniform size. For instance, the precipitation of a basic salt containing ligands other than oxo or hydroxo constitutes a reservoir able to release slowly a metal cation at a low concentration in solution. That allows nucleation of the stable phase of oxide at low supersaturation with a perfectly decorrelated growth stage, as in La Mer's model (Fig. 4.40), and the formation of particles with a narrow size distribution. A good example, as follows, is the formation of chromium oxide Cr_2O_3 particles from a chromium basic sulfate [124].

Heating around 75°C of diluted solutions of chromium alum, $KCr(SO_4)_2{\cdot}16H_2O$, forms a solid veil with the appearance of crumpled paper (Fig. 4.50). Chemical analysis shows that it contains sulfate. Under similar conditions, chromium nitrate or perchlorate yields only soluble hydrolysis products. During heating, sulfate-free oxide spherical particles form at the expense of the amorphous veil. These phenomena can be understood in relation to noncharged sulfated complexes such as $[Cr(OH)(SO_4)(OH_2)_3]^0$ and $[Cr(OH)_2(HSO_4)(OH_2)_2]^0$ which may be formed in the presence of the potential precursor of the oxide $[Cr(OH)_3(OH_2)_3]^0$. Due to their abundance, the sulfated complexes are stable enough to condense by olation

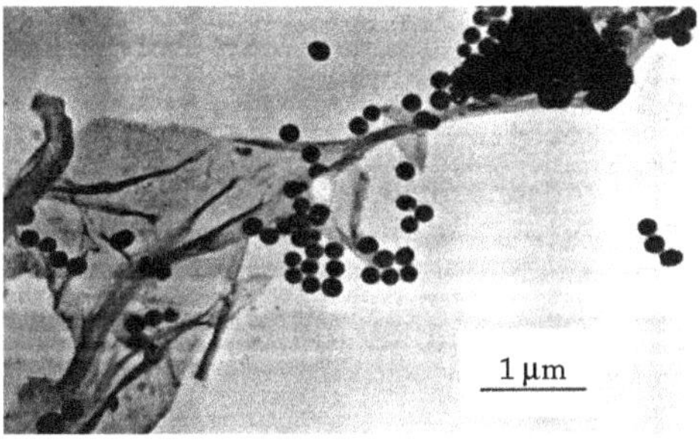

FIGURE 4.50 TEM image of thermohydrolysis products at 75 °C of $KCr(SO_4)_2$ (4.10^{-4} mol.l^{-1}, initial pH = 3.7, final pH = 3.3) [127].

with a functionality of 1 or 2 at the maximum and can lead to entangled linear polymers forming the amorphous veil. The concentration of soluble chromium, especially as $[Cr(OH)_3(OH_2)_3]^0$, is too low to form the oxide by homogeneous nucleation, but the reaction of this complex with the sulfated polymers allows the progressive elimination of sulfate, and so crystallization of sulfate-free oxide nuclei may occur. Thus, the early precipitation of sulfated polymers allows regulation of oxide nucleation in limiting the amount of soluble species. The growth occurs regularly and isotropically, leading to uniform-sized spherical particles. Of course, the techniques yield particles with a relatively important size because nucleation occurs at low supersaturation.

MORPHOLOGICAL CHANGE

The morphology of particles resulting from precipitation is largely dependent on the synthesis conditions, especially on the presence of ligands that control the crystal growth. Many examples demonstrate that the same material may exist under various morphologies, such as oxides of iron, aluminum, titanium, and manganese. It is difficult, however, to predict the geometry of particles of a given solid, and in the absence of specific agents, the morphology of primary particles is determined by the crystal symmetry and the relative growth rates of the faces having the lowest surface energies, as modeled, for instance, by Wulff's construction.

From a thermodynamical standpoint, the sphere is the most favored geometric object because it minimizes the surface-to-volume ratio, and therefore the nucleation-free enthalpy (Eq. [4.1]). In accordance, the very small particles of many amorphous or crystalline solids are often spheroidal. The anisotropy frequently appears during growth because of structural and/or chemical contingencies. The growth may indeed be favored along privileged axes or planes because the reactivity of the surface groups on which the precursors may be bound depends on their structure and electrostatic charge. These two factors are tightly related to the pH of solution and to the presence of complexing species. Various morphologies of hematite, α-Fe_2O_3, are obtained from recrystallization of akaganeite, β-FeOOH [123, 125], or goethite, α-FeOOH [126] (see Chap. 7).

HETEROGENEOUS NUCLEATION: A MECHANISM SIMILAR TO OSTWALD'S RIPENING

The precipitation of a solid may be accelerated when crystals of any solid are preexisting in the solution. That is a catalytic effect proceeding through the lowering of activation energy, which is dependent on the solid–solution interfacial tension γ (Eq. [4.7]). If the surface of preexisting crystals is structurally in accordance with that of the precipitating solid because of a structural adequacy between adsorption sites and the adsorbable precursor, the interfacial tension, γ', between the two solids is less than that of the solid–solution interface, γ, [105]. Nucleation therefore occurs more easily on the surface of the foreign solid than in solution. This heterogeneous nucleation phenomenon allows the growth of a solid onto a preexisting support. Heterogeneous nucleation is favored when homogeneous nucleation is restricted, for instance, in thermohydrolysis in more or less acidic medium when the noncharged precursor is formed slowly with a rather low concentration. Globally, Ostwald's ripening proceeds according to a mechanism similar to that of precipitation initiated by heterogeneous nucleation. The precipitation of gibbsite, $Al(OH)_3$, according to Bayer's process, is a classical example of process induced by heterogeneous nucleation. It is initiated when crystals of gibbsite are introduced in aluminate solutions, $[Al(OH)_4]^-$, which leads to crystallization of gibbsite instead of other aluminic crystal phases (Chap. 6). This is the so-called "seeding" technique in which crystals of the solid to precipitate are introduced in the solution of the precursor [127, 128].

A similar technique may be used to form core-shell systems in covering nanoparticles of a given oxide with another oxide or oxyhydroxide. In a suspension of nanoparticles which will be the cores of nanocomposites, a solution of the precursor of the covering oxide is introduced at low concentration in order to avoid the homogeneous nucleation of this second material. The precipitation of the covering oxide onto the surface of nanoparticles is like this triggered. It is thereby possible to covered nanoparticles of chromium, iron, titanium, silicium oxides with various thicknesses of aluminum, chromium, and yttrium oxides [129–131]. Nanoparticles of many oxides may also be embedded by a more or less thick layer of variously functionalized silica (Fig. 4.51) allowing their dispersion in aqueous or nonaqueous media. Such nanocomposites are used, for instance, as chemical sensors.

Particular morphologies can be formed by heterogeneous nucleation after structural transformation of metastable phases in suspension. For instance, aging of akaganeite, β-FeOOH, in more acidic conditions than in the case shown in Figure 4.49, forms ellipsoid- and double-ellipsoid-shaped hematite particles (Fig. 4.52a,b) [123]. As indicated by electron diffraction at different places on particles, crystallization of hematite onto akaganeite results from heterogeneous nucleation, unlike the case of the Figure 4.49 where akaganeite totally dissolves and disappears during the formation of hematite. An analogous phenomenon occurs during transformation of goethite, α-FeOOH, into hematite by the aging of acidic suspensions (Fig. 4.52c) [126].

FIGURE 4.51 TEM image of hematite α-Fe_2O_3 particles covered with a silica film 1 to 2 nm in thickness (courtesy of N. Chemin, Saint-Gobain, Aubervilliers, France). An aqueous stable sol of hematite particles (pH 2) is introduced in a solution of a silicic precursor $Si(OH)_xR_yF$. (R stands for an organic radical and F for any functionality.) Condensation of the precursor on the surface of the hematite particles forms a layer with controllable thickness (by the amount of the precursor) and properties (by the nature of the R and F functions).

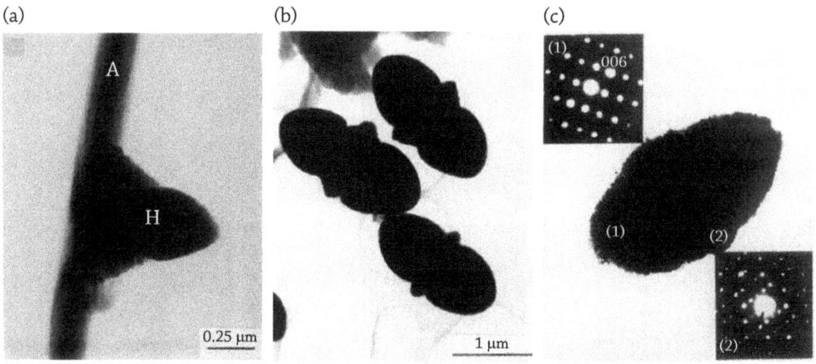

FIGURE 4.52 TEM image of hematite (H) formed onto a rod of akaganeite (A) during aging in suspension ($FeCl_3$ 0.018 mol.l^{-1}, HCl 0.05 mol.l^{-1}, 100°C) (a) for 8 hours; (b) for 1 week: the initial rod of akaganeite is incorporated in the middle of the particle, which is in fact a bicrystal (reproduced from [123] with permission of Elsevier). (c) Hematite formed onto a rod of goethite ($Fe(NO_3)_3$ 0.02 mol.l^{-1}, HNO_3 0.05 mol.l^{-1}, 100°C, 32 hours). The electron diffraction pattern in zone 1 is characteristic of hematite, while the additional spots characteristic of goethite are present in the pattern in zone 2, close to the center of the particle. Also in this case, the initial rod of goethite is incorporated inside the crystal of hematite (reproduced from [126] with permission of Elsevier).

Heterogeneous nucleation is also involved in heteroepitaxy, defined as crystallization of a material onto another with a crystal continuity. A good example may be found among the manganese oxides, MnO_2.

Permanganate reacts with manganous ions at pH 1 (60°C, 7 days) to form nanowires of cryptomelane, α-MnO_2 (Fig. 4.53a), according to the reactions (see Chap. 8):

$$2\left[MnO_4\right]^- + 3\left[Mn(OH_2)_6\right]^{2+} + 4\,H_2O \rightarrow 5\left[Mn(OH)_4(OH_2)_2\right]^0 + 4\,H^+$$

$$\left[Mn(OH)_4(OH_2)_2\right]^0 \rightarrow \alpha\text{-}MnO_2 + 4H_2O$$

Subsequent addition of an excess of permanganate in the suspension and heating at 95°C for 24 hours provokes water oxidation with alkalinization and precipitation of birnessite, δ-MnO_2, according to

$$4\left[MnO_4\right]^- + 18H_2O \rightarrow 4\left[Mn(OH)_4(OH_2)_2\right]^0 + 3O_2 + 4\,HO^-$$

$$\left[Mn(OH)_4(OH_2)_2\right]^0 \rightarrow \delta\text{-}MnO_2 + 4H_2O$$

The crystallization of birnessite, which usually forms platelets with (001) basal faces, occurs on the facets of the cryptomelane wires in forming a set of four lateral wings perpendicular to each other, as arrow fins (Fig. 4.53 b) [132]. Epitaxy of birnessite onto cryptomelane may be understood by the similarity in spacing of {001} planes in δ-MnO_2 (d_{001} = 0.70 nm) and {010} planes in α-MnO_2 (d_{110} = 0.69 nm) (Fig. 4.53c, d). Growth of birnessite then occurs from the borders of platelets perpendicular to the faces of the cryptomelane wires. The phenomenon may also be favored because it involves a decrease in surface energy of the lateral faces of the cryptomelane wires.

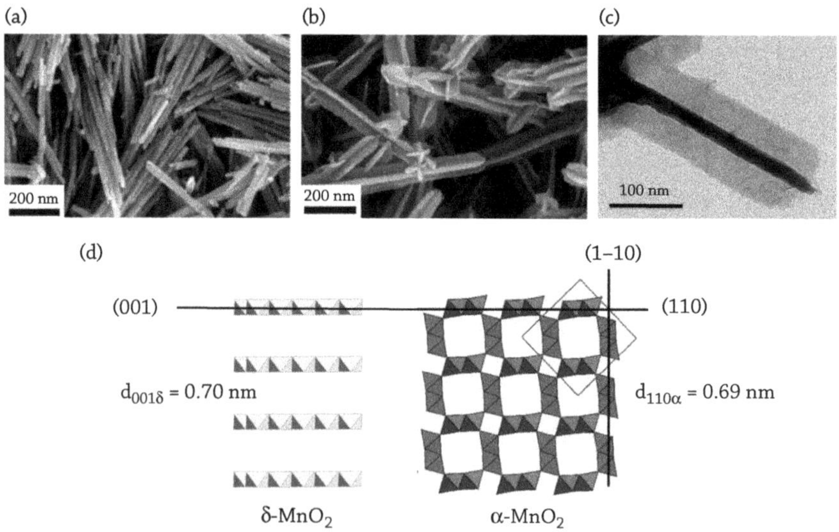

FIGURE 4.53 (a) SEM image of cryptomelane, α-MnO_2, nanowires. (b) SEM image and (c) HRTEM image of the cryptomelane nanowires on which birnessite, δ-MnO_2, has crystallized. (d) Scheme of the crystal structures of α- and δ-MnO_2 showing the epitaxy relationship (reproduced from [132] with permission of The Royal Society of Chemistry).

b) Ordered Aggregation or Oriented Attachment

Crystal growth more frequently proceeds by progressive condensation of complexes on a surface. However, high-resolution electron microscopy studies have revealed that particles may sometimes be formed through growth by aggregation of primary objects along a common crystallographic direction. That leads to side-by-side arrangements of particles at planar interfaces, with water elimination and interfacial reaction forming metal–oxygen bonds between the aggregated objects and allowing the crystal coherency (Fig. 4.54a), even if the mechanism includes defects such as stacking faults, twins, disorientations, and impurity inclusions [133]. The driving force of this assembling is still the elimination of the most reactive crystal faces as well as the lowering of the free enthalpy of the system. The morphology of the particles resulting from ordered aggregation is unusual, and it includes the monocrystalline chains of anatase crystals (Fig. 4.54b) and hematite particles (Figs. 7.26–7.28), together with the following examples.

The process may occur with various materials such as oxides, sulfides, carbonates, and phosphates [134–136]). More frequently, the ordered aggregation is accompanied by the dissolution–crystallization phenomenon, which complicates study of the kinetics of growth and does not allow the process to be modeled [137, 138]. In the current state of knowledge, it is difficult to know why ordered aggregation occurs, on what factors it depends. and what is its kinetics. It is also difficult to pinpoint the influence of this kind of growth on the microstructure, shape, and crystal phases of the final particles.

In addition to the examples given in the above-mentioned reviews, some insight into crystal growth by ordered aggregation may be found in the chemistry of manganese.

The reduction of permanganate by thiosulfate in solution forms diverse solid phases (Chap. 8). Among them, feitknechtite β-MnOOH is transitory, having been formed during the evolution at 60°C and pH 11 of mixtures with the composition

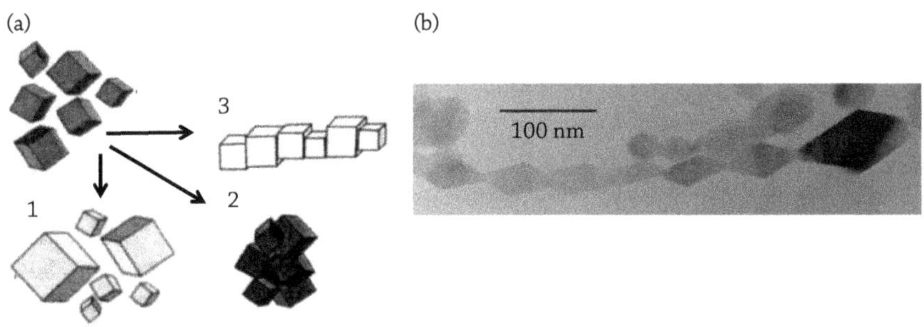

FIGURE 4.54 (a) Scheme illustrating the crystal growth by (1) enlarging the particles (Ostwald's ripening), (2) aggregation, and (3) ordered aggregation; (b) HRTEM image showing a monocrystal of TiO_2 anatase obtained under hydrothermal conditions and resulting from ordered aggregation of primary particles (reproduced from [Penn R.L., Banfield J.F., Morphology development and crystal growth in nanocrystalline aggregates under hydrothermal conditions: Insights from titania, Geochim. Cosmochim Acta 1999, 63, 1549–1557] with permission of Elsevier).

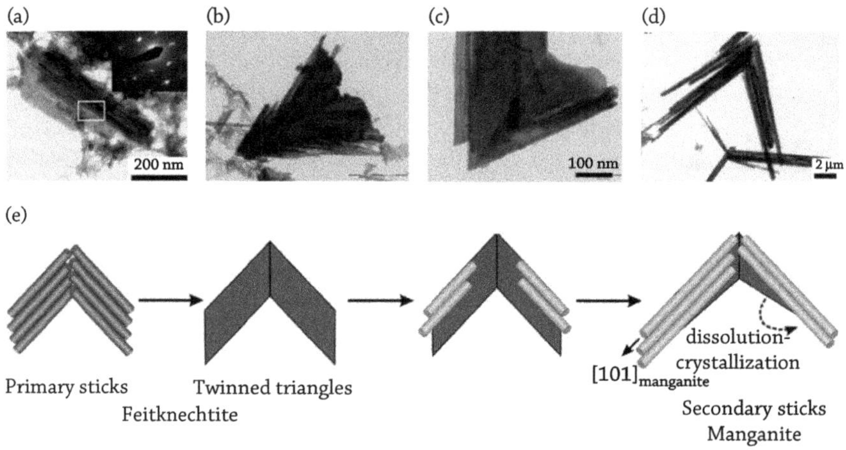

FIGURE 4.55 (a) TEM images of feitknechtite, β-MnOOH, obtained by reduction of permanganate with thiosulfate after heating for (a) 17 hours; (b) 18 hours; (c) 2 days; (d) manganite, γ-MnOOH, particles obtained by heating for 14 days; (e) scheme of the process involved in the formation of triangles of feitknechtite and their transformation into manganite (reproduced with permission from [139] Copyright 2010 American Chemical Society).

$S_2O_3/Mn = 3$ [139]. This metastable phase appears in suspension as attached wires (Fig. 4.55a) forming triangles (Fig. 4.55b). Aging permits elimination of joins and defects and smoothing of the surface of particles (Fig. 4.55c). Subsequent heating of suspension transforms feitknechite into manganite, γ-MnOOH. This last phase also exists as wires formed by heterogeneous nucleation and growth along the edges of the triangles of feitknechtite. The central part of the triangles progressively disappears to feed the growth of manganite wires according to a dissolution–crystallization process. It is remarkable that the assembling of wires in the form of a rafter or compass (Fig. 4.55d) retains the shape of the initial triangle (Fig. 4.55e).

This example illustrates the succession of processes that may be involved in the precipitation phenomenon: homogeneous nucleation of a metastable phase, ordered aggregation, dissolution–crystallization, heterogeneous nucleation of a thermodynamically more stable phase, and a new dissolution–crystallization step. Another example of a specific morphology resulting from ordered aggregation may be found in the chemistry of manganese (Chap. 8). Hollow double cones of ramsdellite, γ-MnO_2 (Fig. 4.56), are obtained by oxidation of manganous ions by permanganate in solution ($Mn^{VII}/Mn^{II} = 1/4$, 95°C, 1 day, initial pH 9.2, final pH 3.7) [140, 141], according to a complex mechanism involving different crystal phases.

From the start of the reaction, X-ray diffraction and electron microscopy allow identification of more or less crystalline nuclei of ramsdellite coexisting with rods of groutite, α-MnOOH, resulting from oxidation in air of the excess Mn^{II} ions. This oxyhydroxide transforms into ramsdellite, which is an isostructural phase. Electron microscopy clearly shows that the first nuclei of ramsdellite, themselves pyramidal shaped, are supports for ordered aggregation of rods of groutite, which transform in place by air oxidation into ramsdellite (Fig. 4.57). The final shape

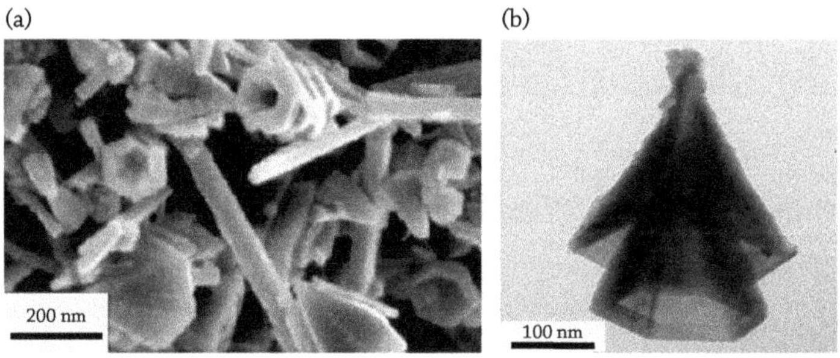

FIGURE 4.56 SEM (a) and TEM (b) images of cones of ramsdellite, γ-MnO_2, resulting from thermolysis of the mixture of permanganate and manganous ions (reproduced with permission from [140] Copyright 2009 American Chemical Society).

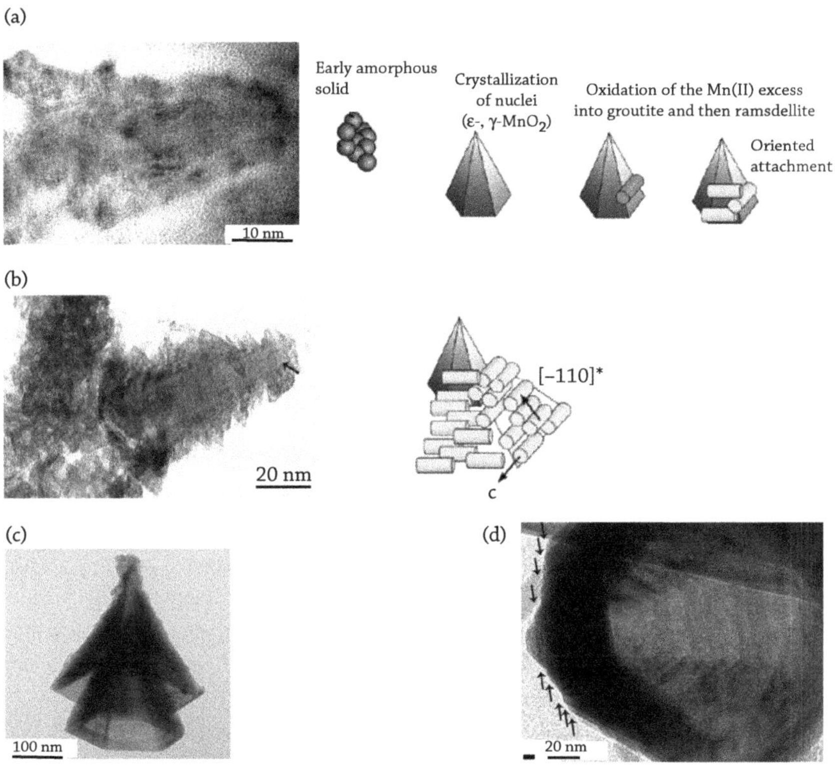

FIGURE 4.57 Possible stages in the formation of cones of ramsdellite, γ-MnO_2: (a) crystallization of nuclei in the early amorphous precipitate; (b) attachment of rods of groutite/ramsdellite resulting from aerial oxidation of the excess of Mn^{II} ions in solution; (c) smoothing and twinning of the cones by dissolution–crystallization. (d) The arrows show the steps on the faces of the cone resulting from the aggregation of rods (courtesy of D. Portehault, Univ. P. & M. Curie, Paris, France).

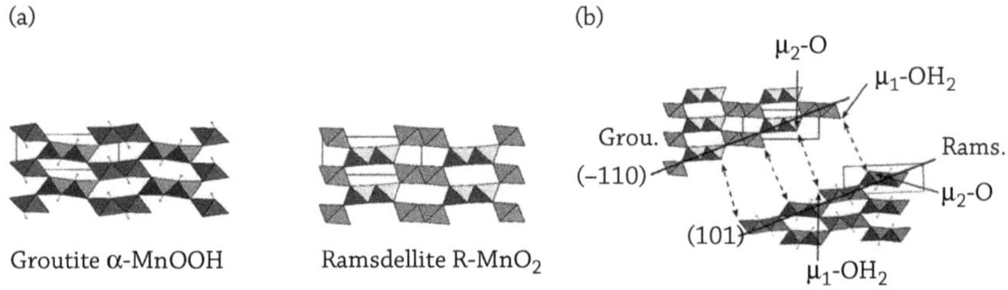

FIGURE 4.58 (a) Structures of groutite and ramsdellite. (b) Assembling of the (101) face of groutite and the (−110) face of ramsdellite.

of the hollow cones results from dissolution–crystallization, with smoothing and twinning of the six lateral faces.

The easy transformation of groutite, α-MnOOH, into ramsdellite, γ-MnO_2, may be explained by the relatively low stability of Mn^{III} compounds compared to those of Mn^{IV} and by the similarity of the two structures (Fig. 4.58a), allowing the shape of particles to be retained. The growth of the faces of cones proceeds by ordered aggregation involving the {101} faces of the rods of groutite and the {−110} faces of ramsdellite. These faces, which are practically not charged at the pH of the synthesis, may give rise to crystal continuity because of their quasi iso-structure (Fig. 4.58b). The six faces are joined by {021} twinning planes, forming angles between them and allowing the quasi-ideal hexagonal shape of the cone.

4.2.4 EFFECT OF MICROWAVE HEATING ON CRYSTALLIZATION IN SOLUTION

The synthesis of nanostructures in solution occurs at relatively low temperatures, from ambient to a few hundred degrees, depending on the technique used: precipitation, thermohydrolysis, or hydrothermal conditions. The solutions are usually heated by convection with electric ovens. Microwave heating, which appeared some years ago in the laboratory after its domestic uses were introduced, is an interesting technique for reducing the duration of the thermal treatment of solutions [142–144]. Microwave heating may also have specific nonthermal effects.

Microwave heating is based on the electromagnetic frequency of 2.45 GHz (wavelength 12.25 cm). The other frequencies of the microwave spectrum, 0.3–300 GHz (wavelength from 1 mm to 1m), are reserved for civil and military telecommunications. Two main mechanisms are involved in heating any material containing mobile electric charges: dipolar polarization, which involves the change in orientation of polar molecules in a liquid, and ionic conduction in a solvent or in a solid under the rapidly changing alternating electric field. Heat is generated by friction in the rotation of molecules or in the oscillating back-and-forth motion of ions. In conducting or semiconducting material, heating results from the Joule effect due to the motion of mobile electrons. The efficiency of microwave heating is dependent on the ratio of the dielectric loss of the substance (efficiency of the conversion of the electric field energy into heat) to its dielectric constant (ability of the substance to be polarized by the electric field). A solvent with no permanent dipolar moment is transparent to microwaves. The energy of the microwave photon is not high enough to break the chemical bond, so that microwaves induce heating but not chemical reactions as in photochemical processes. Microwave heating is meant to be rapid and homogeneous. It occurs in the whole volume of the sample simultaneously without thermal gradient, in contrast to the conventional heating by convection through the reactor walls. Reactions requiring hours or days with conventional heating may be achieved by some minutes of microwave heating. The conditions of implementation, however, are determinative: power density, volume and nature of the sample, geometry of the cavity that has an important impact on the distribution in the electromagnetic field throughout the sample, presence of protruding objects, and so on.

Microwave heating offers advantages over conventional heating, especially in accelerating reactions; in selectively heating a strongly absorbent material in a low polar solvent; and in avoiding the wall effect on the crystallization of solids. Microwave heating does not seem to give rise to particular nonthermal effects, in spite of different results that are sometimes obtained in "similar" conditions of conventional and microwave heating. An interesting demonstration has been given by C. O. Kappe's group [143] from the Karl Franzens University, Graz, Austria, in carrying out same experiences using vessels made of pyrex glass and silicium carbide of the same geometry. SiC, chemically inert, absorbs microwaves strongly, inducing a flow of electrons in the semiconducting material that heats the sample through electrical resistance, but because of the high microwave absorptivity of SiC, the contents of the vessel are efficiently shielded from the electromagnetic field. That the same results were obtained in both cases suggests that only thermal effects induce differences in the reaction rate and that the electromagnetic field does not have direct influence on the chemical pathway, especially on the crystallization process of oxides in aqueous solution. These same conclusions have been obtained for the synthesis of TiO_2 nanoparticles [145].

References

[1] Wells A. F. *Structural Inorganic Chemistry*, 5th ed. Oxford: Oxford University Press, 1991.

[2] Pottier A. Précipitation de cations métalliques en solution aqueuse. Influence de l'acidité sur les caractéristiques des particules d'oxyde. Thèse, Université P. et M. Curie, Paris, France, 1999.

[3] Defontaine G., Michot L.J., Bihannic I., Ghanbaja J., Briois V. Synthesis of NiGa layered double hydroxides. A combined extended X-ray absorption fine structure, small-angle X-ray scattering, and transmission electron microscopy study. 1. Hydrolysis in the pure Ni^{2+} system. *Langmuir* 2003, 19, 10588–10600.

[4] Le Bihan S., Figlarz M. Croissance de l'hydroxyde de nickel ($Ni(OH)_2$) à partir d'un hydroxyde de nickel turbostratique. *J. Cryst. Growth* 1972, 13–14, 458–461.

[5] Pagès T. Application de la chimie colloïdale en phase aqueuse à la précipitation de catalyseurs métalliques supportés: contrôle de la taille et de l'état d'agrégation des particules. Thèse, Université Pierre et Marie Curie, Paris, France, 1998.
[6] Baes C.F., Mesmer R.E. *The Hydrolysis of Cations*. New York: Wiley, 1976.
[7] Nicholas N.J., Franks G.W., Ducker W.A. The mechanism for hydrothermal growth of zinc oxide. *CrystEngComm*. 2012, 1232–1240.
[8] Rai P., Yu Y.T. Citrate-assisted hydrothermal synthesis of single crystalline ZnO nanoparticles for gas sensor application. *Sens. Actuators B Chem*. 2012, 173, 58–65.
[9] McBride R.A., Kelly J.M., McCormack D.E. Growth of well-defined ZnO microparticles by hydroxide ion hydrolysis of zinc salts. *J. Mater. Chem*. 2003, 13, 1–7.
[10] Meagley K.L., Garcia S.P. Chemical control of crystal growth with multidentate carboxylate ligands: Effect of ligand denticity on zinc oxide crystal shape. *Cryst. Growth Des*. 2012, 12, 707–713.
[11] Palumbo M., Lutz T., Giusca C.E., Shiozawa H., Stolojan V., Cox D.C., Wilson R.M., Henley S.J., Silva S.R.P. From stems (and stars) to roses: Shape-controlled synthesis of zinc oxide crystals. *Cryst. Growth Des*. 2009, 9, 3432–3437.
[12] Cavani F., Trifiro F., Vaccari A. Hydrotalcite-type anionic clays: preparation, properties and applications. *Catalysis Today* 1991, 11, 173–301.
[13] Miyata S. Anion-exchange properties of hydrotalcite-like compounds. *Clays Clay Miner.* 1983, 31, 305–311.
[14] Costantino U., Casciola M., Massinelli L., Nocchetti M., Vivani R. 1997. Intercalation and grafting of hydrogen phosphates and phosphonates into synthetic hydrotalcites and a.c.-conductivity of the compounds thereby obtained. *Solid State Ionics* 97, 203–212.
[15] Reichle W.T. Synthesis of anionic clay minerals (mixed metal hydroxides, hydrotalcite). *Solid States Ionics* 1986, 22, 135–141.
[16] Ogawa M., Kaiho H. Homogeneous precipitation of uniform hydrotalcite particles. *Langmuir* 2002, 18, 4240–4242.
[17] Adachi-Pagano M., Forano C., Besse J.P. Synthesis of Al-rich hydrotalcite-like compounds by using the urea hydrolysis reaction, control of size and morphology. *J. Mater. Chem*. 2003, 13, 1988–1993.
[18] Boclair J.W., Braterman P.S. Layered double hydroxide stability. 1. Relative stabilities of layered double hydroxides and their simple counterparts. *Chem. Mater*. 1999, 11, 298–302.
[19] Defontaine G., Michot L.J., Bihannic I., Ghanbaja J., Briois V. Synthesis of NiGa layered double hydroxides. A combined EXAFS, SAXS, and TEM study. 2. Hydrolysis of a Ni^{2+}/Ga^{3+} solution. *Langmuir* 2004, 20, 9834–9843.
[20] Defontaine G., Michot L.J., Bihannic I., Ghanbaja J., Briois V. Synthesis of NiGa layered double hydroxides. A combined EXAFS, SAXS, and TEM study. 3. Synthesis at constant pH. *Langmuir* 2004, 20, 11213–11222.
[21] Boclair J.W., Braterman P.S., Jiang J., Lou S., Yarberry F. Layered double hydroxide stability. 2. Formation of Cr(III)-containing layered double hydroxides directly from solution. *Chem. Mater*. 1999, 11, 303–307.
[22] Roussel H., Briois V., Elkaim E., de Roy A., Besse J.P., Jolivet J.P. Study of the formation of the layered double hydroxide [Zn-Cr-Cl]. *Chem. Mater*. 2001, 13, 329–337.
[23] Basolo F., Pearson R.G. *Mechanism of Inorganic Reactions*, 2nd ed. New York: Wiley, 1958.

[24] Spiccia L., Marty W. The fate of "active" chromium hydroxide, $Cr(OH)_3.3H_2O$, in aqueous suspension. Study of the chemical changes involved in its aging. *Inorg. Chem.* 1986, 25, 266–271.
[25] Giovanoli R., Stadelman W., Feitkneicht W. Über kristallines chrom(III)hydroxid. *Helv. Chim. Acta* 1973, 56, 839–847.
[26] Spiccia L., Marty W., Giovanoli R. Hydrolytic trimer of chromium (III). Synthesis through chromite cleavage and use in the preparation of the "active" trimer hydroxide. *Inorg. Chem.* 1988, 27, 2660–2666.
[27] Spiccia L., Stoeckli-Evans H., Marty W., Giovanoli R., A new "active" chromium (III) hydroxide: $Cr_2(\mu\ OH)_2\ (OH)_4\ (OH_2)_4.2H_2O$. Characterization and use in the preparation of salts of the $(H_2O)_4Cr(\mu\text{-}OH)_2Cr(OH_2)_4^{4+}$ ion. Crystal structure of $[(H_2O)_4Cr(\mu\ OH)_2Cr(OH_2)_4]\ [(H_3C)_3C_6H_2SO_3]_4.4H_2O$. *Inorg. Chem.* 1987, 26, 474–482.
[28] Singh K.K., Sarode P.R., Ganguly P. Evolution of α-Cr_2O_3 from hydrous chromium oxides: magnetic susceptibility and extended X-ray absorption fine structure studies. *J. Chem. Soc. Dalton Trans.* 1983, 1895–1899.
[29] Spiccia L., Marty W., Giovanoli R. Morphological and thermal investigations of the 'active' dimer hydroxide of chromium(III). *Helv. Chim. Acta* 1987b, 70, 1737–1744.
[30] Saraswat I.P., Vajpei A.C. Characterization of chromium-oxide hydrate gel. *J. Mater. Sci. Lett.* 1984, 3, 515–517.
[31] Swayambunathan V., Liao Y.X., Meisel D. Stages in the evolution of colloidal chromium (III) oxide. *Langmuir* 1989, 5, 1423–1427.
[32] Baker B.R., Pearson R.M. Water content of pseudoboehmite: A new model for its structure. *J. Catal.* 1974, 33, 265–278.
[33] Yang J.J., Martens W.N., Frost R.L. Transition of chromium oxyhydroxide nanomaterials to chromium oxide: a hot-stage Raman spectroscopic study. *J. Raman Spectrosc.* 2011, 42, 1142–1146.
[34] Hamilton W.C., Ibers J.A. Structures of $HCrO_2$ and $DCrO_2$. *Acta Cryst.* 1963, 16, 1209–1212.
[35] Yang J.J., Cheng H., Martens W.N., Frost R.L. Transition of synthetic chromium oxide gel to crystalline chromium oxide: a hot-stage Raman spectroscopic study. *J. Raman Spectrosc.* 2011, 42, 1069–1074.
[36] Pernet M., Berthet-Colominas C., Alario Franco M. A., Christensen A. N. Etude par diffraction neutronique de l'oxyhydroxyde de chrome β-CrOOH. Phys. *Status solidi (a)* 1977, 43, 81–88.
[37] Christensen A.N. The crystal structure of a new polymorph of CrOOH. *Inorg. Chem.* 1966, 5, 1452–1453.
[38] Christensen A.N. Hydrothermal preparation and magnetic properties of α-CrOOH, β-CrOOH, and γ-CrOOH. *Acta Chem. Scand. A* 1976, 30, 133–136.
[39] Charlot G. *L'analyse qualitative et les réactions en solution*, 6ème éd. Paris: Masson, 1969.
[40] Iler R.J. *The Chemistry of Silica*. New York: Wiley, 1979.
[41] Tobler D.J, Shaw S., Benning L.G. Quantification of initial steps of nucleation and growth of silica nanoparticles: An in-situ SAXS and DLS study. *Geochim. Cosmochim. Acta* 2009, 73, 5377–5393.
[42] Coudurier M., Baudru B., Donnet J.B. 1971. Etude de la polycondensation de l'acide disilicique. *Bull. Soc. Chim. Fr.* 9, 3147–3165.

[43] Bradley D.C., Mehrotra R.C., Gaur D.P. *Metal Alkoxides*. London: Academic Press, 1978.
[44] Turova N.Y., Turevskaya E.P., Kessler V.G., Yanoskaya M.I. *The Chemistry of Metal Alkoxides*. Boston: Kluwer, 2002.
[45] Brinker C.J., Scherer G.W. Sol-gel science. *The Physics and Chemistry of Sol-gel Processing*. San Diego: Academic Press, 1990.
[46] Stöber W., Fink A., Bohn E. Controlled growth of monodisperse silica spheres in the micron size range. *J. Colloid Interface Sc.* 1968, 26, 62–69.
[47] Green D.L., Jayasundara S., Lam Y.F., Harris M.T. Chemical reaction kinetics leading to the first Stober silica nanoparticles—NMR and SAXS investigation. *J. Non-Cryst. Solids* 2003, 315, 166–179.
[48] Fouilloux S., Désert A., Taché O., Spalla O., Daillant J., Thill A. SAXS exploration of the synthesis of ultra monodisperse silica nanoparticles and quantitative nucleation growth modeling. *J. Colloid Interface Sci.* 2010, 346, 79–86.
[49] Fouilloux S., Taché O., Spalla O., Thill A. Nucleation of silica nanoparticles measured in situ during controlled supersaturation increase. Restructuring toward a monodisperse nonspherical shape. *Langmuir* 2011, 27, 12304–12311.
[50] Gaillou E., Fritsch E., Aguilar-Reyes B., Rondeau B., Post J., Barreau A., Ostroumov M. Common gem opal: an investigation of micro- to nano-structure. *Amer. Miner.* 2008, 93, 1865–1873.
[51] Lefebvre J., Maria H. Etude des équilibres dans les solutions récentes de polyantomoniates. *Paris: C. R. Acad. Sci.*, 1963, 256, 3121–3124.
[52] Abe M., Ito T. Synthetic inorganic ion exchange materials. X. Preparation and properties of so-called antimonic (V) acid. *Bull. Chem. Soc. Jpn* 1968, 41, 333–342.
[53] Lemerle J. Rev. Chim. Min. 1972, 9, 863; Rev. Chim. Min. 1973, 10, 431.
[54] Clearfield A. *Inorganic Ion Exchange Materials*. Boca Raton, FL: CRC Press, 1982.
[55] Chowdhry U., Barkley J.R., English A.D., Sleight A.W. New inorganic proton conductors. *Mat. Res. Bull.* 1982, 17, 917–935.
[56] England W.A., Cross M.G., Hamnett A., Wiseman P.J., Goodenough J.B. Fast proton conduction in inorganic ion-exchange compounds. *Solid State Ionics* 1980, 1, 231–249.
[57] Baestle L.H., Huys D. Structure and ion-exchange characteristics of polyantimonic acid. *J. Inorg. Nucl. Chem.* 1968, 30, 639–649.
[58] Jolivet J.P., Lefebvre J. Etude de quelques antimonates cristallisés échangeurs d'ions. I) caractéristiques de la réaction d'échange et analyse des phases obtenues. *Rev. Chim. Min*. 1972, 9, 845.
[59] Kepert D.L. Isopolytungstates. *Prog. Inorg. Chem.* 1962, 4, 199–274.
[60] Hazel J.F., Mc Nabb W.M., Santini R. The formation and titration of colloidal vanadic acid. *J. Phys. Chem.* 1953, 57, 681–683.
[61] Gharbi N., Sanchez C., Livage J., Lemerle J., Nejem L., Lefebvre J. Mixed-valence polyvanadic acid gels. *Inorg. Chem.* 1982, 21, 2758–2765.
[62] Legendre J.J., Livage J. Vanadium pentoxide gels: I. Structural study by electron diffraction. *J. Colloid Interface Sci.* 1983, 94, 75–83.
[63] Souchay P. *Thermodynamique chimique*, 3ème éd. Paris: Masson, 1968.
[64] Legendre J.J., Aldebert P., Baffier N., Livage J. Vanadium pentoxide gels: II. Structural study by X-ray diffraction. *J. Colloid Interface Sci.* 1983, 94, 84–89.

[65] Lemerle J., Nejem L., Lefebvre J. Condensation process in polyvanadic acid solutions. *J. Inorg. Nucl. Chem.* 1980, 42, 17–20.

[66] Poznarsky G.A., McCormick A.V. ^{51}V NMR and EPR study of reaction kinetics and mechanisms in V_2O_5 gelation by ion exchange of sodium metavanadate solutions. *Chem. Mater.* 1994, 6, 380–385.

[67] Chirayil T., Zavalij P.Y., Whittingham M.S. Hydrothermal synthesis of vanadium oxides. *Chem. Mater.* 1998, 10, 2629–2640.

[68] Sanchez C., Babonneau F., Morineau R., Livage J. Semiconducting properties of V_2O_5 gels. *Phil. Mag. B* 1983, 47, 279–290.

[69] Kodak-Pathé. French patents 2318442 (1977) and 2429252 (1979).

[70] Selbin J. The chemistry of oxovanadium (IV). *Chem. Rev.* 1965, 65, 153–175.

[71] Goulpeau J., Chanéac C., Tronc E., Jolivet J.P. Unpublished results, 2002.

[72] Evans Jr H.T., Post J.E., Ross D.R., Nelen J.A. The crystal structure and crystal chemistry of fernandinite and corvusite. *Canad. Miner.* 1994, 32, 339–351.

[73] Théobald F. Contribution à l'étude des oxydes et hydrates d'oxydes du vanadium aux degrés d'oxydation IV, V et intermédiaires. Thèse, Université de Besançon, France, 1975.

[74] Carpy A., Galy J. Affinement de la structure cristalline du bronze $NaV_2O_5\alpha$'. *Acta Cryst. B* 1975, 31, 1481–1482.

[75] Besnardière J., Petrissans X., Ribot F., Briois V., Sarcin C., Morcrette M., Buissette V., Le Mercier T., Cassaignon S., Portehault D. Nanoparticles of low valence vanadium oxyhydroxides: reaction mechanism and polymorphism control by low-temperature aqueous chemistry. *Inorg. Chem.* 2016, 55, 11502–11512.

[76] Wu C., Dai J., Zhang X., Yang J., Xie Y. Synthetic häggite $V_4O_6(OH)_4$ nanobelts: oxyhydroxide as a new catalog of smart electrical switch materials. *J. Am. Chem. Soc.* 2009, 131, 7218–7119.

[77] Besnardière J. Synthèse et caractérisation de nanoparticules d'oxydes sous-stoechiométriques: vers des applications dans le domaine de l'énergie. Thèse, Université Pierre et Marie Curie, Paris, France, 2013.

[78] Evans Jr H.T., Mrose M.E. A crystal chemical study of the vanadium oxide minerals, häggite and doloresite. *Amer. Miner.* 1960, 45, 1144–1166.

[79] Son J.H., Wei J., Cobden D., Cao G., Xia Y. Hydrothermal synthesis of monoclinic VO_2 micro- and nanocrystals in one step and their use in fabricating inverse opals. *Chem. Mater.* 2010, 22, 3043–3050.

[80] Théobald F., Cabala R., Bernard J. Essai sur la structure de VO_2(B). *J. Solid State Chem.* 1976, 17, 431–438.

[81] Burdett J.K. *Chemical Bonding in Solids*. New York: Oxford University Press, 1995.

[82] Figlarz M. Filiation structurale et métastabilité en chimie douce du solide: exemple de la chimie de WO_3 hexagonal. *Rev. Chim. Min.* 1985, 22, 177–194.

[83] Figlarz M. New oxides in the WO_3-MoO_3 system. *Prog. Solid State Chem.* 1989, 19, 1–46.

[84] Kepert D.L., Kyle J.M. Transient intermediates in the polymerisation of tungstate. *J. Chem. Soc. Dalton Trans II* 1978, 133–137.

[85] Chemseddine A., Henry M., Livage J. Sol-gel derived electrochromic layers. *Rev. Chim. Min.* 1984, 21, 487–495.

[86] Chemseddine A., Babonneau F., Livage J. Anisotropic $WO_3 \cdot nH_2O$ layers deposited from gels. *J. Non Cryst. Solids* 1987, 91, 271–278.

[87] Freedman M.L. The tungstic acids. *J. Am. Chem. Soc.* 1959, 81, 3834–3839.

[88] Zhou L., Zou J., Yu M., Lu P., Wei J., Qian Y., Wang Y., Yu C. Green synthesis of hexagonal-shaped $WO_3{\cdot}0.33H_2O$ nanodiscs composed of nanosheets. *Cryst. Growth Des.* 2008, 8, 3993–3998.

[89] Gérand B., Nowogrocki G., Guenot J., Figlarz M. Structural study of a new hexagonal form of tungsten trioxide. *J. Solid State Chem.* 1979, 29, 429–434.

[90] Pope M.T. *Heteropoly and Isopolyoxometallates*. Berlin: Springer-Verlag, 1983.

[91] Marcilly Ch., Courty Ph., Delmon B. Preparation of highly dispersed mixed oxides and oxide solid solutions by pyrolysis of amorphous organic precursors. *J. Am. Ceram. Soc.* 1970, 53, 56–57.

[92] Pechini M.P. Method of preparing lead and alkaline earth titanates and niobates and coating method using the same to form a capacitor. US Patent Office n°3, 330, 697, 11 July 1967.

[93] Cooper J.K., Gul S., Toma F.M., Chen L., Glans P.A., Guo J., Ager J.W., Yano J., Sharp I.D. Electronic structure of monoclinic $BiVO_4$. *Chem. Mater.* 2014, 26, 5365–5373.

[94] Saison T., Chemin N., Chanéac C., Durupthy O., Ruaux V., Mariey L., Maugé F., Beaunier P., Jolivet J.P. Bi_2O_3, $BiVO_4$, and Bi_2WO_6: Impact of surface properties on photocatalytic activity under visible light. *J. Phys. Chem. C*, 2011, 115, 5657–5666.

[95] Yamashita H., Li H. *Nanostructured Photocatalysts. Advanced Functional Materials, Nanostructure Science and Technology*. Basel, Switzerland: Springer International Publishing, 2016.

[96] Saison T. Oxydes mixtes de bismuth pour la photocatalyse dans le visible. Thèse, Université P. et M. Curie, Paris, France, 2011.

[97] Saison T., Chemin N., Chanéac C., Durupthy O., Mariey L., Maugé F., Brezova V., Jolivet J.P. New insights into $BiVO_4$ properties as visible light photocatalyst. *J. Phys. Chem. C* 2015, 119, 12967–12977.

[98] Ren L., Jin L., Wang J.B, Yang F., Qiu M.Q., Yu Y. Template-free synthesis of $BiVO_4$ nanostructures: I. Nanotubes with hexagonal cross sections by oriented attachment and their photocatalytic property for water splitting under visible light. *Nanotechnology* 2009, 20,115603–115612.

[99] Meng X., Zhang L., Dai H., Zhao Z., Zhang R., Liu Y. Surfactant-assisted hydrothermal fabrication and visible-light-driven photocatalytic degradation of methylene blue over multiple morphological $BiVO_4$ single-crystallites. *Mater. Chem. Phys.* 2011, 125, 59–65.

[100] Saison T., Gras P., Chemin N., Chanéac C., Durupthy O., Brezova V., Colbeau-Justin C., Jolivet J.P. New insights into Bi_2WO_6 properties as a visible light photocatalyst. *J. Phys. Chem. C* 2013, 117, 22656–22666.

[101] Haruta M., Delmon B. Preparation of homodisperse solids. *J. Chim. Phys.* 1986, 83, 859–868.

[102] Sugimoto T. Preparation of monodispersed colloidal particles. *Adv. Colloid Interface Sci.* 1987, 28, 65–108.

[103] La Mer V.K., Dinegar R.H. Theory, production and mechanism of formation of monodispersed hydrosols. *J. Am. Chem. Soc.* 1950, 72, 4847–4854.

[104] Fouilloux S., Taché O., Spalla O., Thill A. Nucleation of silica nanoparticles measured in situ during controlled supersaturation increase. Restructuring toward a monodisperse nonspherical shape. *Langmuir* 2011, 27, 12304–12311.

[105] Nielsen A.E. *Kinetics of Precipitation*. Oxford: Pergamon Press, 1964.

[106] Kruger H. Techniques for the kinetic study of fast reactions in solution. *Chem. Soc. Rev.* 1982, 11, 227–255.
[107] Dhont J.K.G., Smits C., Lekkerkerker H.N.W. A time resolved static light scattering study on nucleation and crystallization in a colloidal system. *J. Colloid Interface Sci.* 1992, 152, 386–401.
[108] Tobler D.J, Shaw S., Benning L.G. Quantification of initial steps of nucleation and growth of silica nanoparticles: An in-situ SAXS and DLS study. *Geochim. Cosmochim. Acta* 2009, 73, 5377–5393.
[109] Mersmann A. Calculation of interfacial tensions. *J. Cryst. Growth* 1990, 102, 841–847.
[110] Bennema P., Söhnel O. Interfacial surface tension for crystallization and precipitation from aqueous solutions. *J. Cryst. Growth* 1990, 102, 547–556.
[111] Stumm W., Morgan J.J. *Aquatic Chemistry*. New York: Wiley-Interscience, 1970.
[112] Markov I.V. *Crystal Growth for Beginners: Fundamentals of Nucleation, Crystal Growth and Epitaxy*. Singapore: World Scientific, 1995.
[113] Noguera C., Fritz B., Clément A., Baronnet A. Nucleation, growth and ageing scenarios in closed systems I: A unified mathematical framework for precipitation, condensation and crystallization. *J. Crystal Growth* 2006, 297, 180–186.
[114] Noguera C., Fritz B., Clément A., Baronnet A. Nucleation, growth and ageing scenarios in closed systems II: Dynamics of a new phase formation. *J. Crystal Growth* 2006, 297, 187–198.
[115] Fritz B., Clément A., Amal Y., Noguera C. Simulation of the nucleation and growth of simple clay minerals in weathering processes: The NANOKIN code. *Geochim. Cosmochim. Acta* 2009, 73, 1340–1358.
[116] Fritz B., Noguera C. Mineral precipitation kinetics. Reviews in Mineralogy and *Geochemistry* 2009, 70, 371–410.
[117] Rothbaum H.P., Rohde A.G. Kinetic of silica polymerization and deposition from dilute solutions between 5 and 180°C. *J. Colloid Interface Sci.* 1979, 71, 533–559.
[118] Noguera C., Fritz B., Clément A. Precipitation mechanism of amorphous silica nanoparticles: a simulation approach. *J. Colloid Interface Sci.* 2015, 448, 553–563.
[119] Nielsen A.E. *Crystal Growth*. H.S. Peiser Ed. London: Pergamon Press, 1967.
[120] Dunning W. J. *Particle Growth in Suspension*. A.L. Smith Ed. London: Academic Press, 1973.
[121] Stanley H.E., Ostrowsky N., Eds. *On Growth and Form, Fractal and Non-fractal Patterns in Physics*, NATO Series E: Applied Sciences n° 100. Dordrecht: Martinus Nijhoff Publishing, 1986.
[122] Pulcinelli S.H., Santilli C.V., Jolivet J.P., Tronc E. Morphological characterization of aqueous tin oxyhydroxide gel. *J. Non-Cryst. Solids* 1994, 170, 21–26.
[123] Bailey J.K., Brinker C.J., Mecartney M.L. Growth mechanisms of iron oxide particles of differing morphologies from the forced hydrolysis of ferric chloride solutions. *J. Colloid Interface Sci.* 1993, 157, 1–13.
[124] Matijevic E., Bell A. *Particle Growth in Suspensions*. A.L. Smith, Ed. London: Academic Press, 1973.
[125] Sugimoto T., Muramatsu A., Sakata K., Shindo D. Characterization of hematite particles of different shapes. *J. Colloid Interface Sci.* 1993, 158, 420–428.
[126] Ocana M., Morales M.P., Serna C.J. The growth mechanism of α-Fe_2O_3 ellipsoidal particles in solution. *J. Colloid Interface Sci.* 1995, 171, 85–91.

[127] Füredi-Milhofer H. Spontaneous precipitation from electrolytic solutions. *Pure Appl. Chem.* 1981, 53, 2041–2055.

[128] Penners N.H.G., Koopal L.K. Preparation and optical properties of homodisperse haematite hydrosols. *Colloids and Surf.* 1986, 19, 337–349.

[129] Kratohvil S., Matijevic E. Preparation and properties of coated, uniform inorganic colloidal particles: 1. Aluminum (hydrous) oxide on hematite, chromia and titania. *Adv. Ceramic Mater.* 1987, 2, 798–803.

[130] Matijevic E. Colloid science of ceramic powders. *Pure Appl. Chem.* 1988, 60, 1479–1491.

[131] Garg A., Matijevic E. Preparation and properties of coated uniform colloidal particles. 11. Chromium hydrous oxide on hematite. *Langmuir* 1988, 4, 38–44.

[132] Portehault D., Cassaignon S., Baudrin E., Jolivet J.P. Synthesis of a manganese oxide nano-composite through heteroepitaxy in aqueous mediu. *Chem. Commun.* 2009, 674–676.

[133] Banfield J.F., Welch S.A., Zhang H., Ebert T.T., Penn R.L. Aggregation-based crystal growth and microstructure development in natural iron oxyhydroxide biomineralization products. *Science* 2000, 289, 751–754.

[134] Cölfen H., Antonietti M. *Mesocrystals and Nonclassical Crystallization*. Chichester, UK: Wiley, 2008.

[135] Niederberger M., Cölfen H. Oriented attachment and mesocrystals: non-classical crystallization mechanisms based on naoparticle assembly. *Phys. Chem. Chem. Phys.* 2006, 8, 3271–3287.

[136] Song R.Q., Cölfen H. Mesocrystals—Ordered nanoparticle superstructures. *Adv. Mater.* 2010, 22, 1301–1330.

[137] Burrows N.D., Yuwono V.M., Penn R.L. Quantifying the kinetics of crystal growth by oriented aggregation. *MRS Bull.* 2010, 35, 133–137.

[138] Zhang J., Huang F., Zhang L. Progress of nanocrystalline growth kinetics based on oriented attachment. *Nanoscale* 2010, 2, 18–34.

[139] Portehault D., Cassaignon S., Baudrin E., Jolivet J.P. Evolution of nanostructured manganese (oxyhydr)oxides in water through MnO_4^- reduction. *Cryst. Growth Des.* 2010, 10, 2168–2173.

[140] Portehault D., Cassaignon S., Baudrin E., Jolivet J.P. Twinning driven growth of manganese oxide hollow cones through self-assembly of nanorods in water. *Cryst. Growth Des.* 2009, 9, 2562–2565.

[141] Portehault D., Cassaignon S., Baudrin E., Jolivet J.P. Selective heterogeneous oriented attachment of manganese oxide nanorods in water: toward 3D nanoarchitectures. *J. Mater. Chem.* 2009, 19, 7947–7954.

[142] Bilecka I., Niederberger M. Microwave chemistry for inorganic nanomaterials synthesis. *Nanoscale* 2010, 2, 1358–1374.

[143] Baghbanzadeh M., Carbone L., Cozzoli P.D., Kappe C.O. Microwave-assisted synthesis of colloidal inorganic nanocrystals. *Angew. Chem. Int. Ed.* 2011, 50, 11312–11359.

[144] Zhu Y.J., Chen F. Microwave-assisted preparation of inorganic nanostructures in liquid phase. *Chem. Rev.* 2014, 114, 6462–6555.

[145] Dufour F., Cassaignon S., Durupthy O., Colbeau-Justin C., Chanéac C. Do TiO_2 nanoparticles really taste better when cooked in a microwave oven? *Eur. J. Inorg. Chem.* 2012, 2707–2715.

FIGURE 1.1 (c) Light emission under UV irradiation of CdSe nanoparticles dispersed in hexane. The particle size varies from 1.5 nm (blue) to 7 nm (red) (courtesy of B. Dubertret, ESPCI, Paris, France).

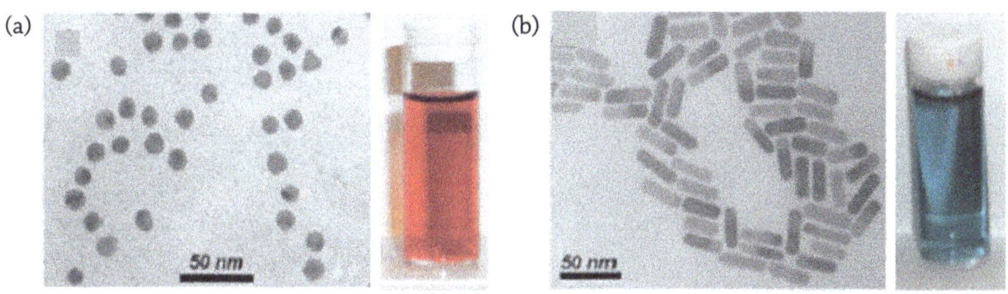

FIGURE 1.2 TEM images of aqueous dispersions of gold nanospheres (a) and nanorods (b) and colors of these dispersions (reproduced from Liz-Marzan L.M., Nanometals: formation and color, Materials Today 2004, 26–31 with permission of Elsevier).

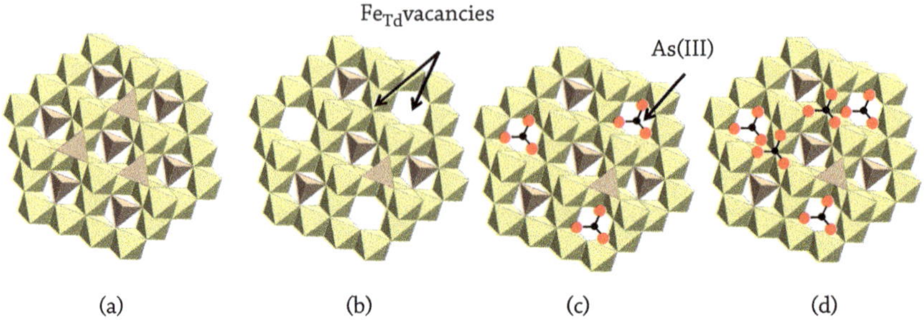

FIGURE 1.7 Scheme of adsorption sites for As^{III} on nanoparticles 6 nm in diameter based on structural information deduced from X-ray diffraction and As K-edge XAS analyses. (a) Idealized structure of {111} faces of maghemite, γ-Fe_2O_3 (octahedra and tetrahedra are FeO_6 and FeO_4, respectively). (b) Nanoparticle exhibiting Fe^{III} vacancies. (c) Filling of the most reactive sites by arsenic at low coverage rate. (d) Adsorption on lattice positions at higher surface coverage allowing a decrease of surface energy.

FIGURE 3.37 Scheme of the structure of $[Mo_{132}O_{372}(CH_3COO)_{30}(H_2O)_{72}]^{42-}$, including 12 (Mo_6) units (dark gray) linked by (Mo_2) bridges (light gray).

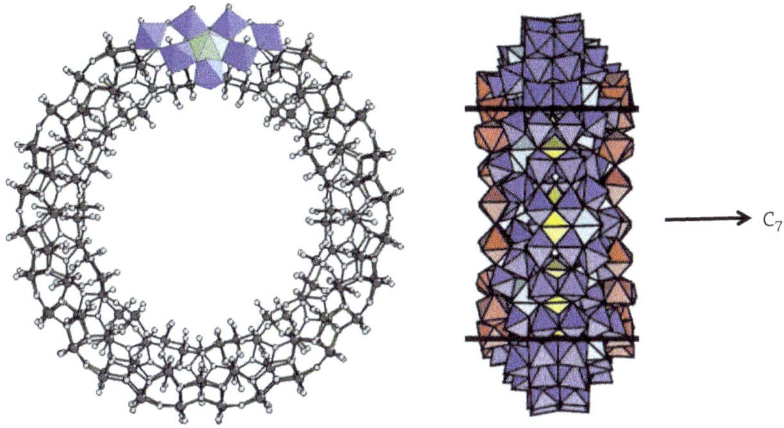

FIGURE 3.38 Scheme of the structure of the (Mo_{154}) wheel. Left, view perpendicular to the C_7 axis. Polyhedral representation of a (Mo_6) unit with the linkage MoO_6 octahedra. Right, polyhedral representation along the C_7 axis (reproduced from [Müller A., Kögerler P., Kuhlmann C., A variety of combinatorially linkable units as disposition: from a giant icosahedral keplerate to multi-functional metal–oxide based network structures, Chem. Commun. 1999, 1347–1358] with permission of The Royal Society of Chemistry).

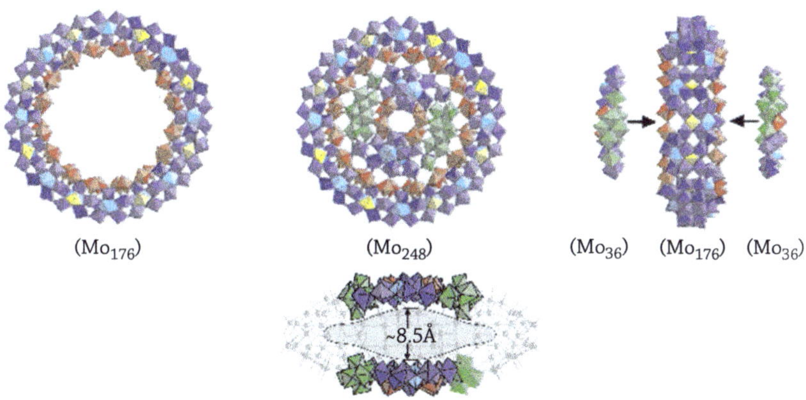

FIGURE 3.39 Schematic view of (Mo_{176}) and (Mo_{248}) (the MoO_7 polyhedra are in Turkish blue). View of the cavity in (Mo_{248}) and sketch of the association of the two (Mo_{36}) fragments with (Mo_{172}) (reproduced from [107] with permission of the Nature Publishing Group).

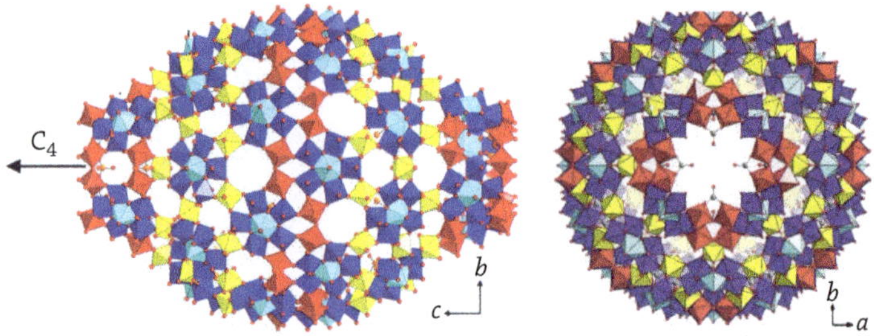

FIGURE 3.40 Scheme of the structure of (Mo_{368}) along th C_4 axis. The MoO_7 polyhedra are in Turkish blue (reproduced from [101] with permission of Elsevier).

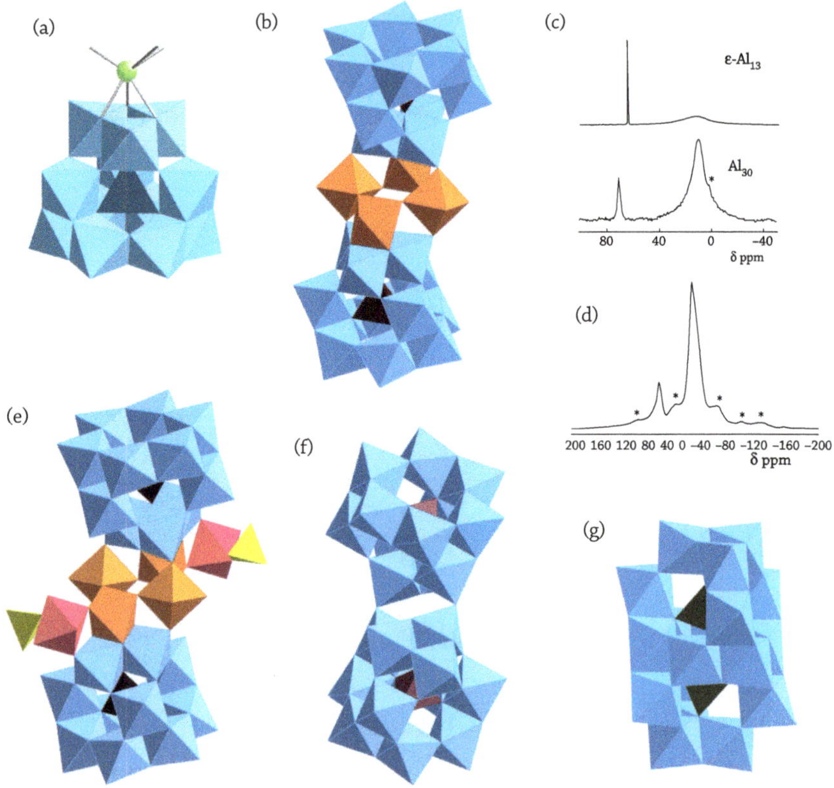

FIGURE 6.5 Structures of (a) $Na[Al_{13}O_4(OH)_{24}(H_2O)_{12}]^{8+}$ (δ-Al_{13}) and (b) $[Al_{30}O_8(OH)_{56}(H_2O)_{24}]^{18+}$ (Al_{30}); (c) ^{27}Al-NMR spectra of ε-Al_{13} and Al_{30} species in solution (reproduced with permission from [22] Copyright 2000 American Chemical Society); (d) ^{27}Al-NMR MAS spectrum of the sulfate salt $Al_{30}(SO_4)_9{\cdot}xH_2O$ (130.3 MHz, rot 5 kHz) (*rotation peaks) (reproduced with permission from [23] Copyright 2000 American Chemical Society); (e) structure of $[Al_{32}O_8(OH)_{60}(H_2O)_{28}(SO_4)_2]^{16+}$ (Al_{32}); (f) structure of $[Al_{26}O_8(OH)_{50}(H_2O)_{20}]^{12+}$ (Al_{26}); (g) structure of $[Ga_2Al_{18}O_8(OH)_{36}(H_2O)_{12}]^{8+}$ (Ga_2Al_{18}).

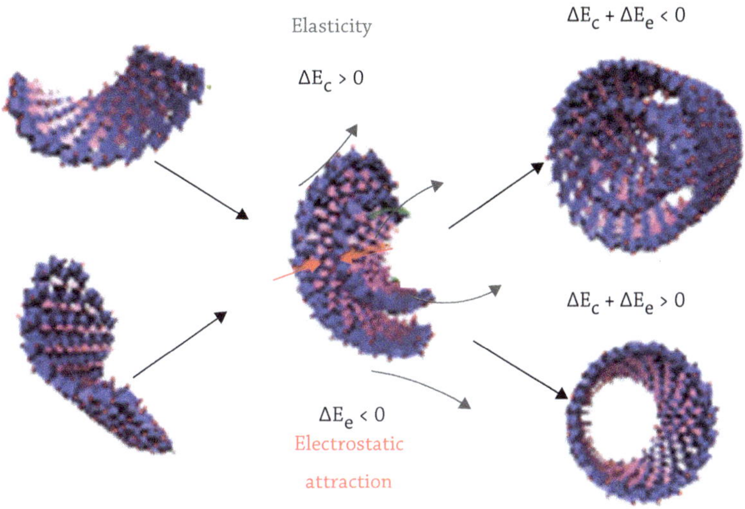

FIGURE 6.29 Schematic sketch for the proto-imogolite transformation into single- or double-wall imogolite nanotubes, depending on the energetic balance between electrostatic interaction and curvature change in the sheet (reproduced with permission from [102] Copyright 2012 American Chemical Society).

FIGURE 7.1 (a) Landscape of Luberon (southwestern France) showing a deposit of ochre (goethite and hematite).

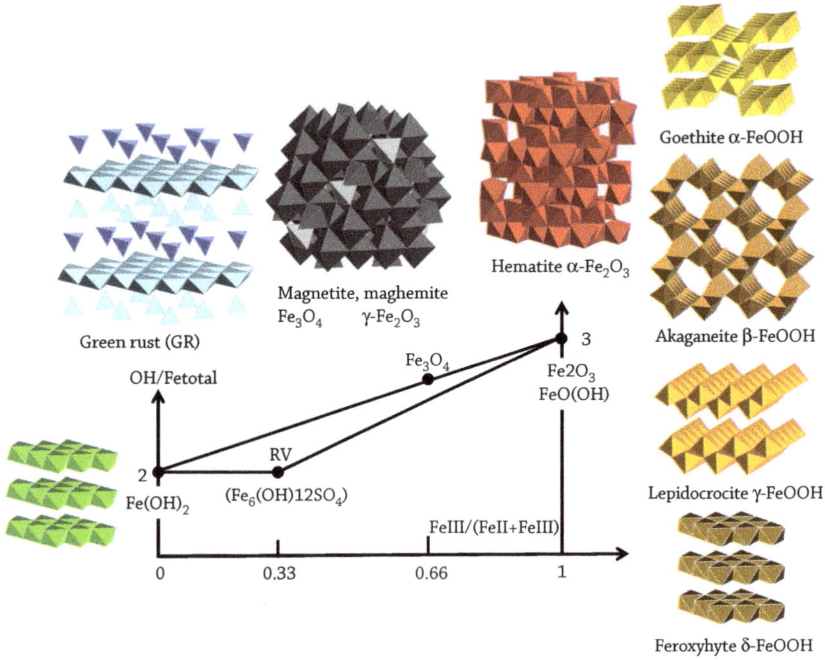

FIGURE 7.2 Main structural types of iron oxy(hydroxi)des as a function of the ferric-ferrous composition Fe^{III}/Fe_{total} and the hydroxylation ratio OH/Fe_{total}.

FIGURE 7.12 (a) Coyote Buttes, Paria Wilderness Area, Utah–Arizona border (USA). (b) Overview of moqui marbles in this area (courtesy of M. A. Chan, Univ. of Utah, USA). (c) Cross section of a broken moqui marble (image by Shinishi Kato with permission of rocksandminerals.com).

FIGURE 7.33 (a) Soil profile in a forest near Fougères (France). Under the surface overlying an organo-mineral horizon, the bluish-green fougerite stays in a reductomorphic zone (15–30 cm). Fougerite quickly turns to ochre upon contact with oxygen, after opening the profile (courtesy of Dr F. Trolard, INRA, Avignon, France).

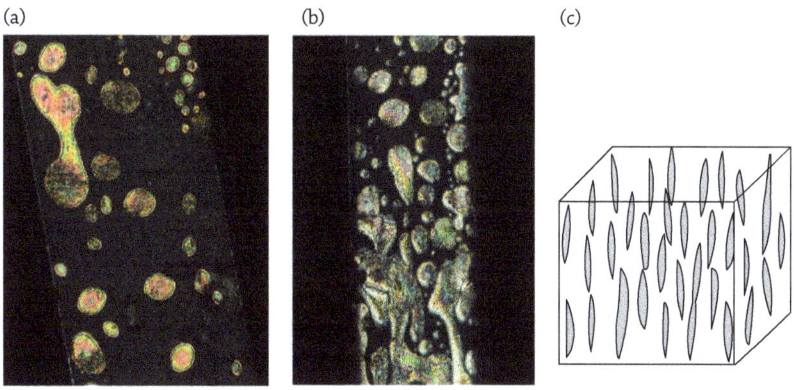

FIGURE 8.13 Images in optical microscopy between cross-polarizing crystals of suspensions of TiO_2 rutile particles (mean size $200 \times 11 - 15$ nm, L/D = 15, pH 2) with volumic fraction $\approx 6\%$ (a), $\approx 10\%$ (b). The colored zones are nematic domains (c) spontaneously formed in the isotropic phase.

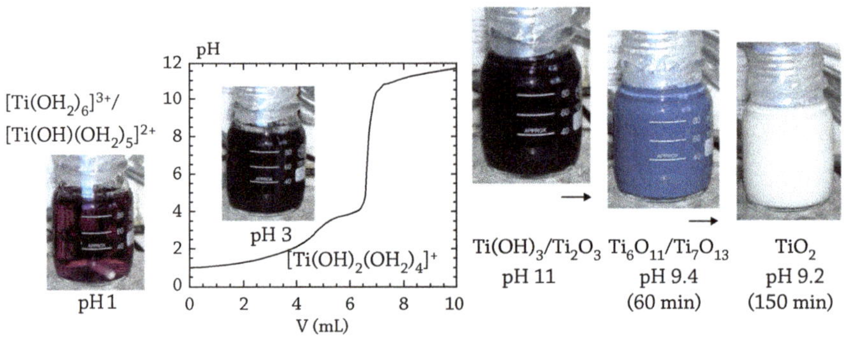

FIGURE 8.26 Titration curve by NaOH of a $TiCl_3$/HCl solution. From the left to right, images and compositions of solutions and suspensions at pH 1, 3, and 11 and suspensions aged at room temperature for 60 and 150 minutes in which the pH is decreasing with oxidation [42].

5 }

Surface Chemistry and Physicochemistry of Oxides

Oxide particles resulting from precipitation have at least one dimension less than a few nanometers. Therefore, as their specific area (surface-to-mass ratio) may reach several hundred square meters per gram, the behavior of these particles is closely related to their surface physical-chemical characteristics.

Thus, the *dispersion state* of particles in solution is dependent on attractive and repulsive forces *between surfaces*. The balance control of these forces limits the aggregation of particles and promotes the formation of sols or gels, or, contrariwise, flocculates the particles and separates them from a suspension. The *divison state* of solids resulting from precipitation is ruled by forces that exert themselves *onto the surface* (interfacial—or surface—tension). They determine the extent of the surface area and, therefore, the particle size. *Adsorption* of ions or molecules within the dispersion depends on forces exerting *between soluble species and the surface.* These forces may be due to electrostatic charges on the surface. They may also be due to the ability of the surface cations to be coordinated by soluble species and/or the ability of surface oxygenated groups to coordinate cations from solution. The attachment of various species on the surface of oxide particles plays a major role in various fields—for instance, the transport of matter in natural or industrial waters, catalysis and corrosion phenomena, formation of stable and homogeneous dispersions.

It is somewhat difficult to characterize the surface of nanometer-sized objects from structural as well as chemical standpoints. The geometry of such small particles is not easily defined with precision, and the surface often includes defects such as steps, truncations, and stacking faults. These sites are difficult to recognize but exhibit largely variable chemical reactivities. In addition, the study of the oxide–solution interface is complicated because few of its physical quantities are experimentally accessible. These quantities are treated as fitting parameters in more or less complex modelings. The current state of the art, however, allows suitable interpretation of experimental data.

This chapter focuses on the basic concepts behind the physical chemistry of the oxide–aqueous solution interface. Among these concepts, the electrostatic surface charge appears to be the major factor governing the chemistry and physical chemistry of aqueous dispersions of small oxide particles.

5.1. The Oxide–Solution Interface

The contact between the surface of an oxide particle and water induces a perturbation in the properties of the two partners: the solid is strongly solvated, and the surface oxygen atoms express their acid–base character toward the water molecules in exchanging protons. Thus, the surface gets an electrostatic charge, whereas the physical-chemical characteristics (structure, mobility, dielectric constant) of the surrounding water, strongly polarized by the surface, are sensitively affected on distances that may reach several nanometers.

5.1.1 ORIGIN OF THE ELECTROSTATIC SURFACE CHARGE

The surface charge of oxides results from ionization at the contact of the water of the hydroxylated groups, forming a nearly complete surface monolayer [1, 2].

Within the bulk of an oxide, the anionic and cationic charges are totally and strictly balanced, so that the solid is electrically neutral. In assuming that the charge z (oxidation state) of a cation is equally distributed on the bonds with the neighboring anions, the formal bond valence, v, of the cation is defined as $v = z/CN$. CN stands for the coordination number of the cation [3]. For oxygen atoms in the bulk, bounded to n_b cations, the partial charge is thus $\delta(O)_{vol} = n_b v - 2 = 0$. For the surface oxygen atoms, approximately half of the cations are missing, so that their charge is no longer exactly balanced and $\delta(O)_{surf} = n_s v - 2 < 0$. Consequently, the surface oxide ions behave as a base, inducing dissociation of water molecules in contact and leading to surface hydroxylation (Fig. 5.1). For hydroxide or oxide particles formed by precipitation from an aqueous solution, the surface hydroxylated groups are naturally present because they originate from the coordination sphere of the last cations incorporated into the solid as zero-charged complexes.

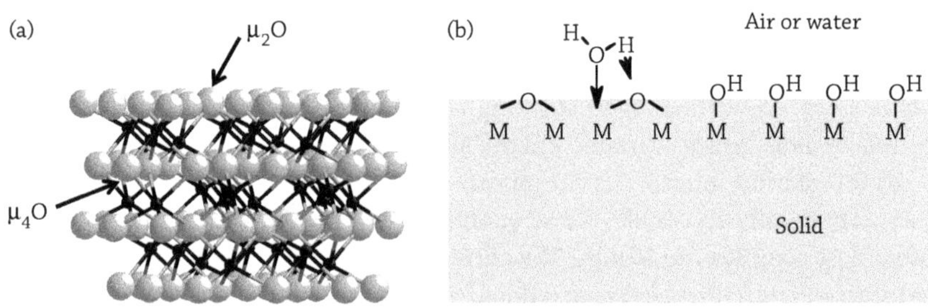

FIGURE 5.1 (a) Structure of oxygen ions in the bulk and at the surface of an oxide. (b) Scheme of hydroxylation of an oxide surface by water chemisorption.

The acidity, and therefore the electrostatic charge of the surface ligands, depend on two main factors: the characteristics of the underlying cation (oxidation state and coordination number) and the structure of the oxygenated group. Thus, the formal charge, $\delta(\text{OH})$, of singly coordinated OH ligands may be largely variable:

$$\begin{array}{llll} \text{Si}^{\text{IV}}\text{-OH} & \nu = 4/4 = 1 & \delta(\text{OH}) = 1 - 2 + 1 = 0 & \text{Si-OH}^{0} \\ \text{Ti}^{\text{IV}}\text{-OH} & \nu = 4/6 = 2/3 & \delta(\text{OH}) = (2/3) - 2 + 1 = -1/3 & \text{Ti-OH}^{-1/3} \\ \text{Al}^{\text{III}}\text{-OH} & \nu = 3/6 = 1/2 & \delta(\text{OH}) = 1/2 - 2 + 1 = -1/2 & \text{Al-OH}^{-1/2} \\ \text{Mg}^{\text{II}}\text{-OH} & \nu = 2/6 = 1/3 & \delta(\text{OH}) = (1/3) - 2 + 1 = -2/3 & \text{Mg-OH}^{-2/3} \end{array}$$

For doubly or triply coordinated OH groups, we have:

$$\begin{array}{llll} \text{Al}_2\text{-OH} & \nu = 1/2 & \delta(\text{OH}) = 2 \times (1/2) - 2 + 1 = 0 & \text{Al}_2\text{-OH}^{0} \\ \text{Al}_3\text{-OH} & \nu = 1/2 & \delta(\text{OH}) = 3 \times (1/2) - 2 + 1 = +1/2 & \text{Al}_3\text{-OH}^{+1/2} \end{array}$$

Thus, the surface of oxide particles in aqueous medium exhibits an acid–base behavior. The surface may bear an electrostatic charge, which is dependent on the nature and charge of the cation, on the structure of the surface oxygenated groups, and, of course, on the pH of the suspension. The acidity of an oxide is globally characterized by the point of zero charge (PZC) or by the isoelectric point (IEP). They correspond to the pH value at which the surface is globally noncharged: at pH < PZC, the net charge is positive, and at pH > PZC, it is negative. The PZC characterizes a surface on which there is no charged group at pH = PZC, whereas at an IEP, there are as many positive as negative charges. As will be seen later in this chapter, oxides should be characterized by an IEP, but the term "PZC" will be used hereafter. The main factors determining the value of the PZC (IEP) of oxides are:

1. *The coordination of the surface groups and their polarization by the cation.* The basicity of the surface oxygen atoms is weaker than their coordination is higher and the polarizing power of the cation stronger. The more the cation is small, charged, and of low coordination, the more important is the oxygen-to-metal electron transfer M←O and high is the acidity of the surface groups (Table 5.1).
2. *The crystal structure and morphology of particles.* These factors determine the type and relative extent of the different crystal faces and, therefore, the ratio of mono-, di-, or tricoordinated surface groups. As each type of surface group has its own acid–base properties, the morphology of the particles is an important factor influencing the PZC (IEP) value. This quantity thus does not appear to be an absolute characteristic of an oxide. The shift of the IEP of

TABLE 5.1 } Point of zero charge (PZC) of oxides of cations with different oxidation states, *z*. The coordination of the cation is indicated in brackets. Data from [1].

z	Oxide	PZC	*z*	Oxide	PZC
+II	MgO (6)	≈ 12.5	+IV	TiO_2 (6)	3.5 – 6.5
	ZnO (4)	9 – 10		SiO_2 (4)	2 – 4
+III	α-Fe_2O_3 (6)	5.5 – 9	+V	Sb_2O_5 (6)	≈ 0.5
	α-Al_2O_3(6)	6.5 – 10	+VI	WO_3 (6)	≈ 0.4

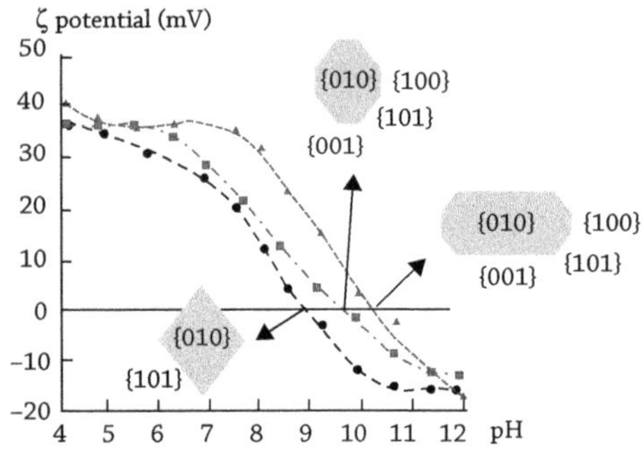

FIGURE 5.2 Experimental variations of zeta potential (§ 5.2.2) for suspensions of nanoplatelets of boehmite synthesized at different pHs (§ 6.3.1b). The IEP (PZC), defined as the pH for which the zeta potential is zero, varies from 8.9, 9.5 to 10.2 as the {001}/{101} lateral face ratio increases, from 0.02, 0.11 to 1.2 respectively. That clearly shows the difference in acidity of the μ_2–OH groups of these faces. The {010} basal faces are never charged (reproduced from [Chiche D., Chanéac C., Revel R., Jolivet J.P., Use of polyols as particle size and shape controllers: application to boehmite synthesis from sol-gel routes, Phys. Chem. Chem. Phys. 2011, 13, 6241–6248] with permission of The Royal Society of Chemistry).

boehmite, γ-AlOOH, as a function of the ratio of lateral faces of nanoplatelets, is a prime example (Fig. 5.2).

5.1.2 SURFACE ACIDITY: MULTISITE COMPLEXATION MODELING

The acidity of surface groups cannot be determined experimentally, and modeling is required to characterize the surface properties of oxides. A simple and successful model, the MUSIC model, was worked out by W. van Riemsdijk and coworkers at Wageningen, The Netherlands [4, 5]. This model is based on electrostatics, in connection with the chemistry of oxo- and hydroxocomplexes in solution.

As noted earlier, the surface oxygen atoms bear a noncompensated charge $\delta = nv - 2 + p$ (p is the number of protons bounded to the surface oxygen atom). One may thus consider the following protonation equilibria:

$$M_n - O^{(nv-2)} + H^+ \Leftrightarrow M_n - OH^{(nv-1)} \quad K_{n,1} = \frac{[M_nO^{(nv-2)}][H^+]}{[M_nOH^{(nv-1)}]} \qquad [5.1]$$

$$M_n - OH^{(nv-1)} + H^+ \Leftrightarrow M_n - OH_2^{(nv)} \quad K_{n,2} = \frac{[M_nOH^{(nv-1)}][H^+]}{[M_nOH_2^{(nv)}]} \qquad [5.2]$$

characterized by the dissociation constants $K_{n,1}$ and $K_{n,2}$. $[H^+]$ stands for the concentration of protons near the surface, related to the concentration of protons in the bulk of the solution by $[H^+]_{surf} = [H^+]_{sol}\exp(-e\psi_o/RT)$, ψ_o being the surface electrostatic potential (§ 5.2.2).

The free enthalpy change in the reactions [5.1] and [5.2] may be split in a purely electrostatic contribution $\Delta G°_{el}$, corresponding to the electrical energy involved in the approach of the proton toward the surface group and in other nonexplicit "chemical" contributions $\Delta G°_{ch}$:

$$\Delta G^O = \Delta G^O_{el} + \Delta G^O_{ch} \qquad [5.3]$$

The calculation principle consists of evaluating the electrostatic term in the constants $K_{n,p}$. Use of the formal bond valence concept allows us to simplify calculations in eliminating the influence of other surrounding ligands in the coordination sphere of the cation belonging to the surface group. Assuming point charges, the $\Delta G°_{el}$ term due to the approach of a proton and to O–H, OH–H, and M–H interactions can be expressed as

$$\Delta G^O_{el} = \frac{Z_H Z_{O(H)} Ne^2}{4\pi\varepsilon_1 r} + \frac{nZ_H vNe^2}{4\pi\varepsilon_2 L} \qquad [5.4]$$

L is the M–H distance, and r is the O–H distance (Fig. 5.3). $Z_H = 1$, $Z_O = -2$, $Z_{OH} = -1$. ε_1 and ε_2 are effective microscopic dielectric constants. ε_2 has no real physical significance because of the simplicity of the model. This single term replaces a sum of electrostatic interactions.

The equilibrium constant is expressed as

$$\Delta G^O = -RT\ln(1/K) = \Delta G^O_{el} + \Delta G^O_{ch} \qquad [5.5]$$

(the remaining K is a dissociation constant), and

$$-RT\ln(1/K) = \Delta G^O_{ch} + \frac{Z_H Z_{O(H)} Ne^2}{4\pi\varepsilon_1 r} + \frac{nZ_H vNe^2}{4\pi\varepsilon_2 L} \qquad [5.6]$$

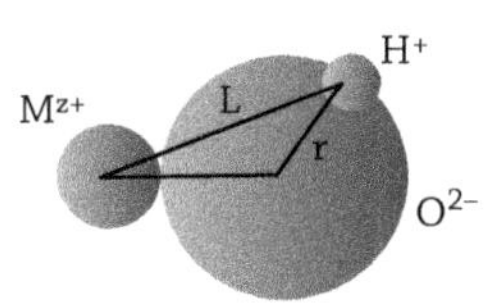

FIGURE 5.3 Geometry of surface groups -M-OH, as assumed through the MUSIC model (see text).

$$\text{or} \quad \log(1/K) = pK = A - \frac{Bnv}{L} \qquad [5.7]$$

$$\text{with} \quad A = -\frac{Z_H Z_{O(H)} N e^2}{2{,}3RT4\pi\varepsilon_1 r} - \frac{\Delta G_{ch}}{2{,}3RT} \quad B = \frac{Z_H N e^2}{2{,}3RT4\pi\varepsilon_2}$$

A and *B* are constants that can be evaluated in analogy with solution chemistry.

The protonation of soluble monomeric oxo- or hydroxocomplexes (labeled ^{m}):

$$M - O^{(v-2)} + H^+ \Leftrightarrow MOH^{(v-1)} \quad K^m_{1,1} \qquad [5.8]$$

$$M - OH^{(v-1)} + H^+ \Leftrightarrow MOH_2^{(v)} \quad K^m_{1,2} \qquad [5.9]$$

is characterized by equilibrium constants [6], which are well correlated to the *v/L* ratio for elements of similar electron structure in a wide range of formal charges (1 to 5) and coordination numbers (4 to 8) (Fig. 5.4). The intrinsic constants $K^m_{1,x}$ for protonation of monomers can therefore be written as

$$\log(1/K^m_{1,x}) = pK^m_{1,x} = A' - \frac{Bv}{L} \qquad [5.10]$$

which is analogous to relation [5.7] for $n = 1$.

The parallel straight lines in Figure 5.4 indicate that the successive protonation constants of oxo and hydroxo species are about 14 pK units apart. The difference is roughly the same for water (H_3O^+/H_2O, $pK = 0$; H_2O/HO^-, $pK = 14$). At the difference of polyacids for which the successive pKs are around 5 pK units apart [7], the successive protonation steps take place here on the same ligand involving strong H–H repulsions.

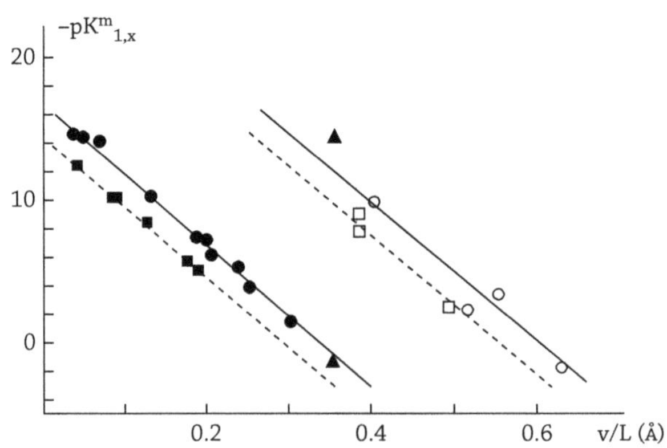

FIGURE 5.4 Protonation constants ($pK = -\log K$) of hydroxo- and oxomonomer complexes as a function of the formal bond valence to the M-H distance ratio. The symbol (●) is for protonation of neutral hydroxocomplexes of metals with a rare gas electron configuration ($pK^m_{1,2}$). The symbol (o) is for protonation of oxo forms of the same elements ($pK^m_{1,1}$). Constants $pK^m_{1,1}$ and $pK^m_{1,2}$ of d^{10} elements are labeled (■) and (□), respectively, and those of water (OH and H_2O) are labeled (▲). *L* values are estimated from ionic radii (reproduced from [4] with permission of Elsevier).

For surface reactions, the major difference with monomer species in solution is the term $\Delta G°_{ch}$ in Eq. [5.6], assuming that the gap between the values of $K_{n,1}$ and $K_{n,2}$ constants remains the same. For aluminum, the protonation constant of the monomer $Al(OH)_3(OH_2)_3$, $pK^m_{1,2}$, is 5.7 [6], whereas for gibbsite, $Al(OH)_3$, involving mainly reactive monocoordinated surface groups $-Al(OH)^{-1/2}$, the experimental value of $pK_{1,2}$ is 10 ± 0.5 [4] The difference between these values is due to structural and proximity constraints to which the protonable groups are subjected.

To calculate the p*K* of surface groups, the term *A′* in Eq. [5.10] has to be changed. It is obtained from the shift of the straight lines in Fig. 5.4 corresponding to the experimental value for protonation of monocoordinated Al–OH groups, which leads to average values $A = 34.06$ and 20.16 for the $K_{n,1}$ and $K_{n,2}$ steps, respectively, and $B = 52.7$ (*L* distances are in angströms). The protonation constants of groups with different structures (mono-, di-, or tricoordinated) can easily be calculated in introducing these values in Eq. [5.7], with the corresponding M–H (*L*) distances obtained from crystal data. A few examples follow.

Gibbsite $Al(OH)_3$ particles are hexagonal platelets in shape (Fig. 5.5). The basal {001} faces exhibit only doubly coordinated OH groups (13.8 groups/nm^2). The side {*hk*0} faces contain only singly and doubly OH groups (9.6 and 4.8 groups/nm^2, respectively). For singly coordinated groups, the $L_{(M-H)}$ distance is 2.59 Å [5]. Equation [5.7] gives $pK_{1,2} = 20.16 - 52.7 \times 0.5/2.59 = 9.98$, which corresponds to

$$-Al-OH^{-1/2}+H^{+} \Leftrightarrow -Al-OH_2^{+1/2} \quad pK_{1,2}=10$$

The p*K* of the $-AlO^-/-AlOH^{-1/2}$ couple ($pK_{1,1} = 23.88$) is very high, and singly coordinated groups are thus present only as $-Al\text{-}OH^{-1/2}$ or $-Al-OH_2^{+1/2}$ depending on pH.

For doubly coordinated groups ($L_{(M-H)} = 2.43$ Å), similar calculation gives:

$$pK_{2,1} = 34.06 - 52.7 \times 1/2.43 = 12.37$$

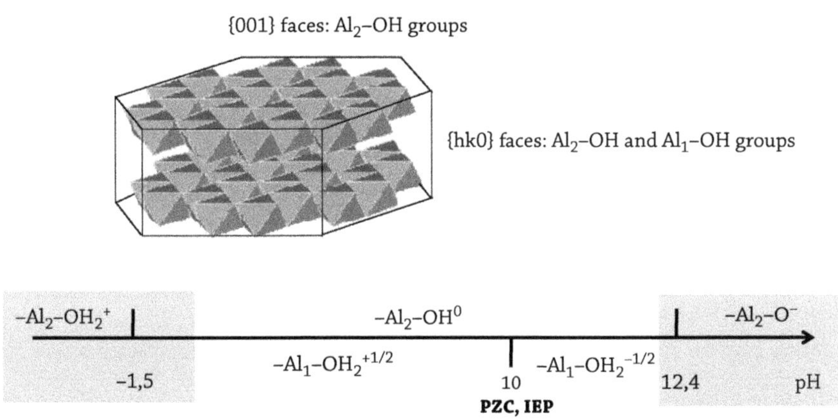

FIGURE 5.5 Scheme of a gibbsite $Al(OH)_3$ particle and protonation ranges of the surface groups. (The shaded areas are beyond the usual range of acidity in aqueous suspension.)

$$-Al_2-O^- + H^+ \Leftrightarrow -Al_2-OH \quad pK_{2,1}=12.4$$

and $pK_{2,2} = 20.16 - 52.7\times 1/2.43 = -1.53$

$$-Al_2-OH+H^+ \Leftrightarrow -Al_2-OH_2+ \quad pK_{2,2}=-1.5$$

Doubly coordinated groups exist only as the hydroxo Al_2-OH form within the usual pH range. The basal faces are therefore never charged, and the surface charge of gibbsite particles is exclusively borne by the lateral faces.

The values of constants calculated from the multisite complexation (MUSIC) model appear to be quite reasonable. They show that the acidity of the surface groups increases with their coordination number, as is also evidenced in the decrease of the calculated bond energy of the proton in hydroxo groups coordinated with the surface of model clusters [8]. Such a decrease in energy of the O–H bond is also reflected in the weakening of its infrared resonance frequency when the coordination number of the hydroxo ligand increases [9] (Table 5.2).

This example highlights several important points:

- The charge may be heterogeneously distributed on particles according to the structure of the exposed faces and then the structure of surface oxygenated groups. That helps us understand the variability in the reactivity of different faces of the same particle.
- The acidity of surface hydroxylated groups increases with their coordination number.
- The surface oxygenated groups involve the exchange of only one proton within the accessible range of pH in aqueous suspension because of mutual electrostatic repulsions. This means that no group exhibits an amphoteric behavior and that the *successive involvement of two protons by the same group* is completely unrealistic [10].
- The global charge of a particle is zero at the pH, equal to the pK of surface monocoordined groups, not because there is no charged group, but because there are as many positively as negatively charged groups bearing the same charge in absolute values. The oxide particles are characterized by an IEP rather than a PZC corresponding to the total absence of charged groups. Although the term "PZC" is most often used (in this book as well), it is important to keep in mind the meaning of these two terms.

TABLE 5.2 } Characteristics of hydroxylated groups on the surface of aluminum oxide Al_2O_3 (*from [7], ** from [8]).

Surface groups	$-Al-OH^{-1/2}$	$-Al_2-OH$	$-Al_3-OH^{+1/2}$
Calculated $pK_{n,1}$ (MUSIC)	24	12.3	1.6
Calculated bond energy of the proton (eV)*	21eV	19.4eV	17.7eV
Infrared absorption frequency (cm^{-1})**	3785–3800	3740–3745	3700–3710

An improvement in the MUSIC model gives important insight into the structural details and hydration effect of the surface [11]. For that purpose, the constants of Eqs. [5.1] and [5.2] are expressed as a function of the charge borne by a surface oxygen atom as $\mathrm{Ln}K_{n,x} = -A\delta(\mathrm{O})$. A is a constant, and $\delta(\mathrm{O})$ is defined by

$$\delta(O) = -2 + \sum S_M + xS_H.$$

$S_M = \exp[(r_o - r)/B]$ is the effective bond valence, as defined by Brown and Altermatt for solid-state chemistry; r is the metal–oxygen distance, and r_o and B are constants empirically determined from structural data [12]; and S_H is the effective bond valence of the proton, with a charge shared for 80% in the bond with oxygen in the

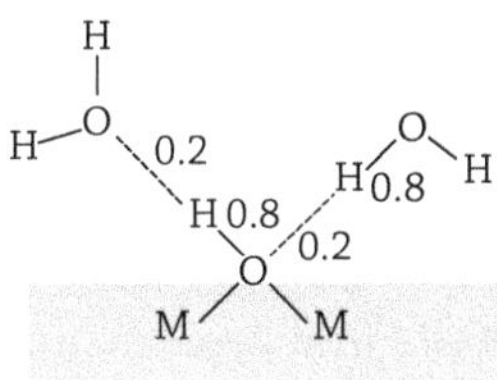

SCHEME 5.1

surface hydroxo group and for 20% within an hydrogen bond with the oxygen atom of a solvation water molecule, as indicated in Scheme 5.1:

Thus, an oxygen atom receives an 0.8 charge per proton in an OH bond and gives a 0.2 charge to another proton within a hydrogen bond and $S_H = 0.8$. The balance of charges is then: $\delta(\mathrm{O}) = -2 + \Sigma S_M + pS_H + m(1 - S_H)$, with $p + m = 2$ for a monocoordinated OH group, $p + m = 1$ or 2 for a dicoordinated group, and $p + m = 1$ for a tricoordinated OH group. Finally, the pK may be calculated from the relation $\mathrm{Ln}K_{n,x} = -A(\Sigma S_j - 2 + m)$ with the constant $A = 19.8$.

An application of this new version of the MUSIC model is given next for hematite, α–Fe_2O_3. Only the oxygenated groups belonging to the frequently observed crystal faces {001}, {100}, {110} and {104} (cf. Chap. 7) (Fig. 5.6) have been considered.

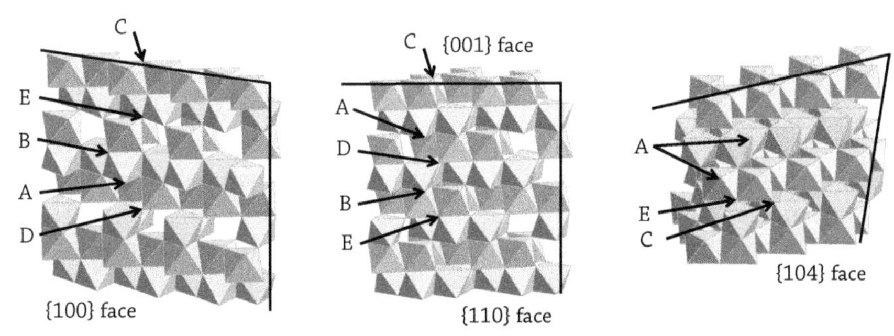

FIGURE 5.6 Scheme of the crystal structure of different faces of hematite, α-Fe_2O_3, particles. The different surface oxygenated groups are labeled A to E (see text and Fig. 5.7).

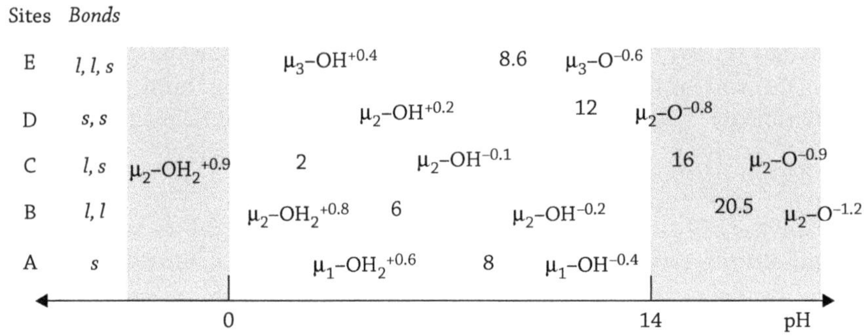

FIGURE 5.7 Acidity of the different surface oxygenated groups of hematite (see Fig. 5.6) calculated from the MUSIC model. The p*K* values are written in bold. (The shaded areas are beyond the usual range of acidity in aqueous suspension.)

For Fe^{3+} ions, r_o = 1.759Å and B = 0.37 [12]. In hematite, the face sharing provokes strong distortions of the iron coordination polyhedron, which exhibits short (d = 1.946 Å, S_{Fe} = 0.603) and long Fe–O bonds (d = 2.116 Å, S_{Fe} = 0.381). On the considered faces, there are five types of groups with long (labeled *l*) and short (labeled *s*) bonds, each having its own p*K* value (Fig. 5.7). So, the effective bond valence concept allows a refined approach of complex surfaces but, of course, without consideration of phenomena such as surface reconstruction or relaxation.

This example also shows that each surface group involves only one protonation/deprotonation equilibrium on the whole accessible pH range. Also, surface charge reversal is possible, and negatively charged groups may exist even if the surface is positively charged globally. That helps us understand why the adsorption of cations may occur in acidic medium, where generally surfaces are globally positively charged (§ 5.4). In addition, this example shows that the global positive and negative charges on both sides of the PZC (IEP) have no reason to be equal in absolute value because each group has a particular charge and the number of the different groups may be largely variable according to the structure of the surface.

The net surface charge density, σ_0, is defined as

$$\sigma_0 = (F/A)\,(\Sigma\,\delta^+[\mathrm{M_n-OH_x}^{\delta+}]\,-\Sigma\,\delta^-[\mathrm{M_n-OH_y}^{\delta-}]) \qquad [5.11]$$

A is the total particle surface area ($m^2.l^{-1}$), and F is the Faraday constant (96500 $C.mol^{-1}$); the quantities in brackets are the concentration ($mol.l^{-1}$) of charged surface groups. σ_0 may reach 0.2 to 0.4 $C.m^{-2}$ according to the nature of the oxide and the crystal structure of the exposed faces (Fig. 5.8). That quantity is measurable by potentiometric titration. The experimentally measured charge density is always notably lower than that calculated from previous parameters (structure, charge, p*K*, and density of surface groups) because of the electrostatic repulsive forces between nearby groups on the surface. These repulsions do not allow all chargeable groups to be effectively charged at a given pH. This phenomenon is highlighted by the ionic strength effect. As ionic strength allows electrostatic screening between surface groups themselves and between surface groups

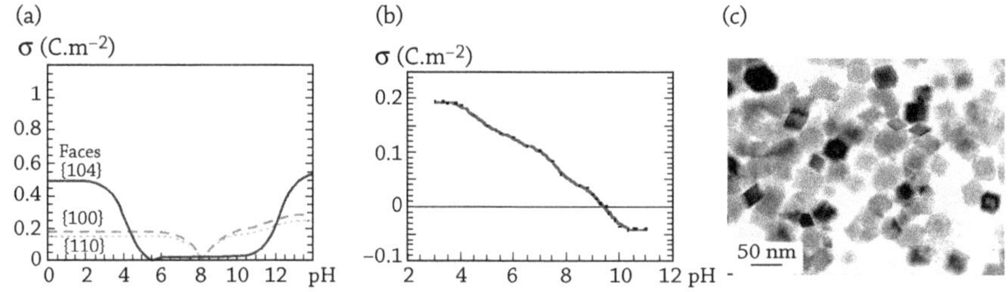

FIGURE 5.8 (a) MUSIC simulation of the variation in surface charge density with pH for different crystal faces of hematite (the surface charges are indicated in absolute value, positive at pH < 8, negative at pH > 8); (b) surface charge density experimentally determined by titration of hematite particles shown in MET image (c).

and ions in solution, an increase in ionic strength leads to an increase in surface charge density at constant pH (§ 5.2.2). It should therefore not be forgotten that the values of charges borne by oxygenated groups evaluated with the previous models do not simply allow a reasonable surface charge density to be calculated, as shown in Figures 5.8a, b.

5.2 Solvation and Structure of the Solid–Solution Interface

As regards ions in solution, the polarity and ionization of surface groups induce their solvation in the form of a hydration layer that is strongly bound to the surface of particles. This layer plays a major role in the structuration of the interfacial zone and in the physical chemistry of aqueous dispersions. It is involved in electrostatic screening effects on surface charges, in the extent of the diffuse layer around the particles, and in the interparticle interactions and stability of dispersions. Next we examine a descriptive and qualitative aspect of these phenomena, describing their possible occurrence in various events. For more extensive study of the quantitative aspect, see other reference books [13–15].

5.2.1 SOLVATION OF PARTICLES

Some properties of liquid water are due to the hydrogen bond network, but its local structure may be strongly modified by the presence of ions (Chap. 2). The polarized and charged surface groups also develop strong interactions with water, and the solvation of an oxide surface may induce perturbation of the structure of the liquid up to several molecular layers [16].

Measurements of immersion and adsorption heats, as well as dielectric measurements on α-Fe_2O_3, ThO_2, and SiO_2, show that the two or three physisorbed water layers, adjacent to the first chemisorbed water layer responsible for surface hydroxylation, are immobilized by pairs of hydrogen bonds with an order similar to that in ice and persisting at room temperature (Fig. 5.9) [2, 17–19]. The ordering

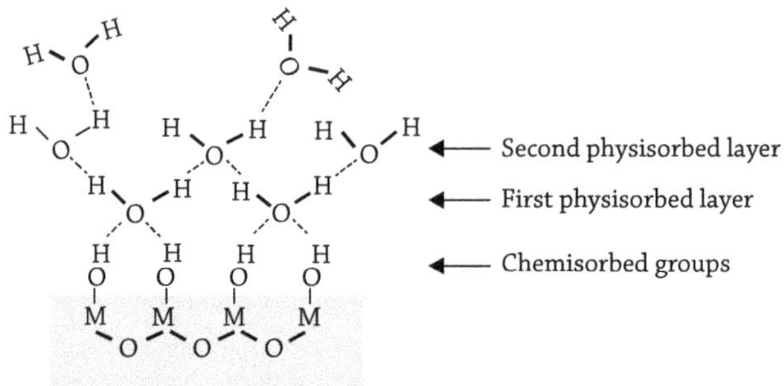

FIGURE 5.9 Scheme of water adsorption on an oxide surface.

of solvation layers progressively disappears from the surface. Beyond a distance of a few to tens of angströms, depending on the nature of oxide, water recovers its liquid characteristics [16, 20]. Molecular dynamics calculations on water molecules in the neighborhood of the {101} face of boehmite, α-AlOOH, globally noncharged but bearing μ_1–$OH^{-0,5}$ and μ_1–$OH_2^{+0,5}$ groups, show that a dense layer (0.1 molecule/Å^3) and well-organized water molecules exist until a distance of around 6 Å, after which the density resembles that of liquid water (0.033 molecule/Å^3) [21].

The thickness of the structured hydration layer increases with increasing surface polarity and increasing surface charge density. This solvation water constitutes the Stern compact layer. Its characteristics (thickness, dielectric constant) are almost impossible to determine experimentally, but its presence allows the differentiation of interactions between the surface and various electrolytes such as protons, hydroxyl ions, and inactive ions from an acid–base standpoint (§ 5.2.2).

Usually, the hydrodynamic size of nanoparticles (measured for instance by quasi-elastic light scattering) is largely higher than the size determined by electron microscopy or X-ray diffraction. This indicates that the oxide nanoparticles in aqueous dispersion are wrapped with an important solvation layer, especially when the surface charge density is high. Clearly, all this water is not really adsorbed, but, as is the case for ions in solution, the structuring electrostatic effect of the surface takes place on a considerable distance. A remarkable example is given by iron oxide γ-Fe_2O_3 nanoparticles (Chap. 7). Their mean diameter measured by electron microscopy or X-ray diffraction is 8 nm, but in suspension at pH ≈ 2 (at this pH, the positively charged particles bear a surface charge density of around 0.3 C.m^{-2}, equivalent to 2 electron charges per nm^2), their mean hydrodynamic diameter measured from quasi-elastic light scattering is 14 nm. So, a water layer of around 3 nm in thickness is carried along in the Brownian motion of particles. When these particles are separated by ultrafiltration from the sol at pH 2 and then dried at room temperature under reduced pressure, they form a powder that appears to be completely dry, somewhat similar to the powder obtained from the same particles flocculated at the PZC by addition of a base (the surface charge is canceled out) and then recovered according to the same process. The near-infrared absorption spectra of the two samples, however, highlight the difference in

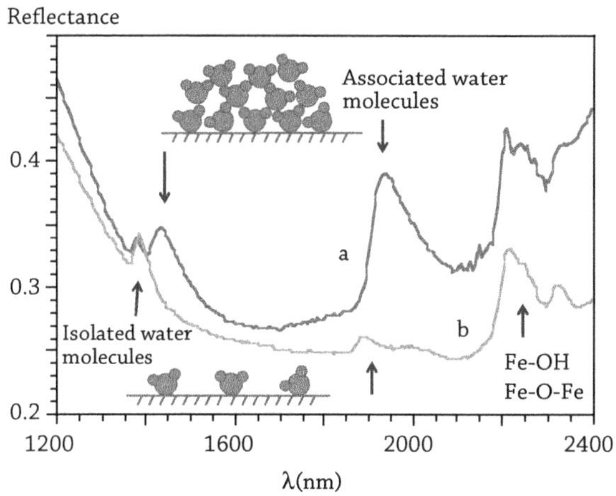

FIGURE 5.10 Near-infrared absorption spectra of γ-Fe_2O_3 nanoparticles 8 nm in mean diameter, ultrafiltered and then dried under reduced pressure from suspension at pH 2 (a) and at the PZC (b). The spectrum of particles dried at the PZC indicates adsorption of isolated water molecules, whereas for the particles dried at pH 2, absorption is essentially due to associated water molecules within the hydration layer. Absorption beyond 2300 nm is due to Fe-OH and Fe-O-Fe vibrators.

hydration (Fig. 5.10). The amounts of water remaining in these powders, evaluated by thermogravimetry, correspond to 5 and 1 layers of molecules per particle, respectively. In addition, the particles isolated at pH 2 spontaneously disperse in water, which is not the case for particles dried at the PZC (§ 7.2.3d). So, that shows the effect of the surface charge on surface hydration of oxide nanoparticles.

The transition zone between the ordered water layers, the first of which are strongly bounded to the surface, and "liquid" water is characterized by the destructuration of the solvent. Similarly to ion solvation, the water molecules within this zone are subject to competing forces: the forces inducing a collective behavior due to neighboring molecules in the solution and the forces inducing a particular orientation imposed by the surface of particles. This zone, characterized by a minimum viscosity of the solvent, corresponds to the glide or shearing plane of the solvent through the motion of particles in suspension [20]. Also, the structure induced in the solvation layer influences the forces exerting at short distance between the hydrated surfaces [14, 22].

5.2.2 SURFACE-ELECTROLYTE INTERACTIONS

The network of hydrogen bonds within the whole of the liquid, including the particle solvation layer, allows the easy diffusion of H^+ and HO^- ions from the bulk of solution toward the oxide surface. These ions react with hydroxylated surface groups without the special structuration of water in the proximity of the surface that hinders their motion. Only protons are truly chemisorbed to form OH or OH_2 ligands. HO^- ions are not really adsorbed because they can only deprotonate

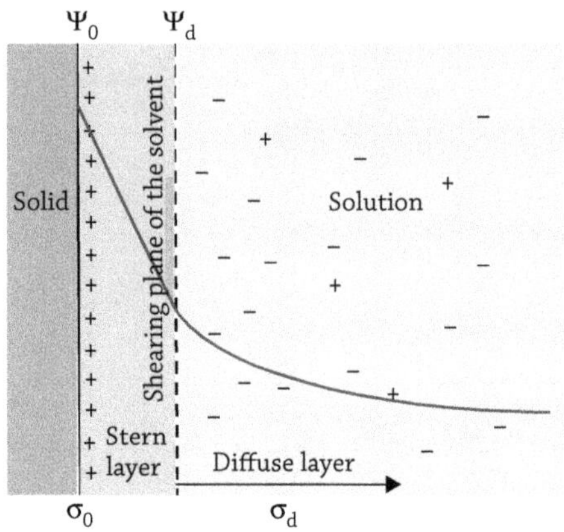

FIGURE 5.11 Scheme of the electrical double layer at the oxide–aqueous solution interface. In the scheme, the surface charge σ_o is positive. The curve represents the variation in electrostatic potential ψ_o from the surface. ψ_d is the potential at the beginning of the diffuse layer, at the position of the shearing plane of the solvent. It corresponds to zeta potential. Within the Stern layer, the decrease in potential is assumed linear.

surface ligands in forming water. These ions bring about the surface charge, σ_o, and then the electrostatic surface potential resulting from this charge (Fig. 5.11). Protons and hydroxyl ions are so-called potential-determining ions (PDIs).

The nonspecific electrostatic forces caused by the surface charge attract ions in solution having a charge of opposite sign (counterions) and repel ions of the same sign of charge (co-ions). Depending on the nature of counterions, their interaction with the surface may be more or less strong. The PDIs are, of course, submitted to the same constraints. That is why the concentrations of protons in the proximity of the surface, $[H^+]_{surf}$, taking place in the expression of the surface acid–base equilibrium constants, are related to the measurable proton concentration, $[H^+]_{sol}$, corresponding to the pH in the electrically nonperturbed zone of suspension, in using a Maxwell–Boltzman term leading to the relationship: $[H^+]_{surf} = [H^+]_{sol} \exp(-e\psi_o/RT)$. e is the electron charge, F the Faraday constant (96500 C.mol^{-1}), R the gas constant, and T the temperature.

Some ions electrostatically attracted by the surface have a high affinity for the solvation layer of particles. With oxides, the *structuring ions* penetrate most easily within the ordered layer of solvent because they preserve the local ordering in this zone. Alkaline ions are adsorbed on negatively charged oxide surfaces according to the sequence $Li^+ > Na^+ >> Cs^+$. This series results from the entropy loss of less mobile water molecules near the surface than in solution and is compensated by an increase in their bond energy with the most strongly hydrated cations [16]. On the other hand, AgI nanoparticles more specifically adsorb *breaking ions* because Ag^+ and I^-, themselves breaking ions, do not cause high solvation of the surface [23]. The mobility of the adsorbed ions within the solvation layer is limited because of the high structuration of this layer and of the electrostatic forces near the surface. This finding was clearly evidenced from NMR measurements of relaxation time of ^{23}Na adsorbed on silica [24]. The relaxation time, longer than that of Na^+ ions in solution, excludes the formation of ion pairs or complexes with negatively charged sites of the surface. The solvation layers of adsorbed Na^+ ions do not appear to be affected, and these ions submitted to nonspecific electrostatic forces within the hydrogen bonds network are only constrained not to diffuse outside the particle solvation layer.

The *breaking ions* and *type II structuring ions*, having a relatively low charge density such as Cs^+, ClO_4^-, and $N(CH_3)_4^+$ (Chap. 2), cannot penetrate the strongly structured water layer in contact with the surface. Their minimal distance of approach corresponds to the thickness of the compact Stern layer. It is identified with the solvent glide or shearing plane in the electrokinetic measurements (Fig. 5.11).

When subjected to electrostatic attraction and to thermal agitation, the counterions form a diffuse layer in solution beyond the Stern layer, in which the concentration gradient decreases with the distance. The electrostatic potential at the limit of the Stern layer, ψ_d, is defined as the electrokinetic zeta potential, ζ, calculated from the electrophoretic mobility [25]. This is the only measurable potential for oxide particles. The charge of the diffuse layer, σ_d, represents the countercharge of the particle (Fig. 5.11). It balances the surface charge, and electroneutrality imposes $\sigma_o + \sigma_d = 0$.

a) Effect of Ionic Strength on the Surface Charge Density

The counterions not only maintain the electroneutrality of the dispersion. These ions, in the proximity of the surface, also shield or screen the electrostatic repulsions between the charged groups and then modify the surface charge [23, 26]. At a given pH far enough from the PZC, the amount of surface-charged groups is limited by their mutual repulsive electrostatic interactions. The screening of these interactions by adsorbed counterions inside the solvation layer then allows an increase in the surface charge (Fig. 5.12a). Therefore, for a given pH, the charge depends on the concentration of electrolyte and hence on the screening effect. This is the "salt effect" used in the experimental determination of the PZC. At a given pH, the net surface charge increases with increasing concentration of the electrolyte, as shown by the shift of the curves in Figure 5.12b. Ions of electrolyte accumulating toward the surface and penetrating further into the Stern layer (cations if the surface charge is negative, anions if it is positive) have a screening effect on the repulsions between charged-surface groups themselves and also on repulsions between the surface and PDIs, protons, or hydroxyl ions, which are less repelled by the surface. More surface groups are ionized with electrolyte addition, and then the surface charge density increases. At PZC, the electrolyte has no effect. Therefore, the charge density–pH curves have a common intersection corresponding to the PZC [27].

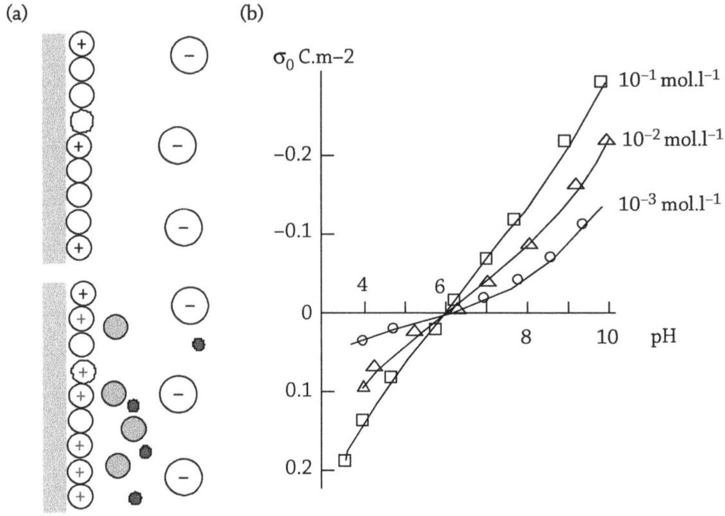

FIGURE 5.12 (a) Scheme of screening of electrostatic interactions between surface groups by ions of an electrolyte (anions are the large filled spheres and cations the small ones); (b) variation with pH in the surface charge density for TiO_2 anatase for several NaCl concentrations (reproduced from [Sprycha R., Surface charge and adsorption of background electrolyte ions at anatase/electrolyte interface, J. Colloid Interface Sci. 1984, 102, 173–185] with permission of Elsevier).

b) Effect of Ionic Strength on the Extent of the Diffuse Layer

In water at 25°C in the presence of a symmetrical [*z:z*] electrolyte of concentration *C*, the charge of the diffuse layer is given by Grahame's relation:

$$\sigma_d = -0.1173\sqrt{C}\sinh(19.48\, z\,\psi_d) \qquad [5.12]$$

with σ_d in $C.m^{-2}$, *C* in $mol.l^{-1}$ and ψ_d in volts. For low potentials ($z\psi_d << 50$ mV), Grahame's relation becomes [13, 28]:

$$\sigma_d = -\varepsilon\kappa\psi_d \qquad [5.13]$$

with $\kappa = (2F^2I10^3/\varepsilon RT)^{1/2}$ where *I* stands for the ionic strength, *I*, defined as $I = (1/2)\Sigma z_i^2 C_i$.

The variation of the potential at a distance *r* from the surface is given by

$$\psi_r = \psi_d \exp[-k(r-d)]$$

In water at 25°C, $\kappa = 0.329\sqrt{I}$ Å^{-1}. The term κ has the dimension of the reciprocal of length, and κ^{-1} (Debye–Hückel's length) usually represents the "thickness" of the diffuse layer. That is a misnomer because over the κ^{-1} distance, the potential decreases only by $\psi_d/\exp(1)=\psi_d/2.7$. However, in the low potential approximation, the diffuse layer is often treated as a parallel-plate capacitor, with plates separated by distance κ^{-1}, having the capacity $C_d = \varepsilon\kappa$ (Eq. [5.13]). The variation in potential

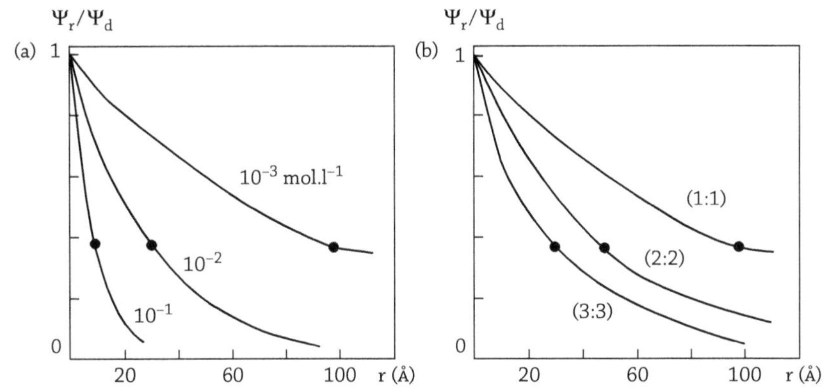

FIGURE 5.13 Variation in potential within the diffuse layer as a function of the distance from the Stern layer (low potential approximation). Influence of (a) the concentration of a (1:1) electrolyte, and (b) the charge of ions in (z:z) electrolytes at 10^{-3} mol.l^{-1}concentration. The marker (●) indicates the distance κ^{-1} (adapted from [12]).

in the solution as a function of the distance from the surface depends on the concentration and charge of the ions of electrolyte (Fig. 5.13).
The main observations deduced from Figure 5.13 are as follows:

- The electrically perturbed zone extends over distances that are of the same order of magnitude as the size of nanometric particles. For instance, $\kappa^{-1} = 30$ Å for a [1:1] electrolyte of concentration $C = 10^{-2}$ mol.l^{-1}.
- The thickness of the perturbed zone is sensitive to the concentration and charge of ions in the electrolyte, that is, to the ionic strength. As it increases, the diffuse layer becomes more compressed (κ^{-1} small).
- The response of the double layer to a change in ionic strength is a change in surface charge and/or potential.

5.3 Stability of Nanoparticle Dispersions Against Aggregation

The surface charge of nanoparticles considerably influences their mutual interactions. These particles are sensitive to Brownian motion and on the one hand may form homogeneous and stable aqueous suspensions if they are repelling and away from each other, spontaneously dispersing in water. On the other hand, if the particles are attracting each other and aggregate to form more or less compact agglomerates, they separate from the liquid and flocculate. The stability of a dispersion requires repulsive forces between particles in order for a *kinetic* energetic barrier to prevent their mutual approach. These forces may result from the presence of electrostatic charges on particle surfaces. The pH of the suspension and the nature and concentration of electrolytes are in this case the parameters to control. The stability of suspension may also be due to adsorption of macromolecules that exert a sterical hindrance between the surfaces. The strength of the surface–macromolecule interaction and the coverage rate of particles are in this case the factors ruling their dispersability in an aqueous or nonaqueous medium.

The stability of dilute suspensions of charged particles against aggregation and flocculation is usually described by the DLVO (Dejarguin, Landau, Verwey, Overbeek) theory [29]. It is based on the balance of the forces acting between particles, and it allows calculation of the height of the energetic barrier to avoid a permanent contact between particles. That is what is meant by the "kinetic" stability of the dispersions. The involved forces are of two kinds: attractive forces (London–Van der Waals forces), universal and omnipresent, depending only on the nature of the materials involved; and repulsive forces resulting from the interaction between electrical double layers, depending largely on pH and ionic strength.

How are such forces involved as the particles come together during Brownian collisions? This is indeed a complex question because the interaction energies are dependent on the transient behavior of double layers during the collisions. A quantitative treatment of this question is largely beyond the scope of this book. It is useful, however, to present a qualitative and brief description of the phenomena involved.

5.3.1 VAN DER WAALS FORCES

Van der Waals forces are always present and always attract particles of the same nature. They result from fluctuations of dipolar interactions at the molecular level [13, 14, 26]. The potential energy of interaction varies with the separation distance of dipoles, r, according to an r^{-6} function. The sum of interactions between macroscopic objects (regarding molecular dimensions) yields an expression of the interaction energy varying as r^{-2}. In the case of two identical spheres of radius a, with surfaces distant from D (center-to-center distance $R = D + 2a$), the potential energy of interaction is [30]:

$$V_A = \frac{A}{6}\left(\frac{2a^2}{R^2 - 4a^2} + \frac{2a^2}{R^2} + Ln\frac{R^2 - 4a^2}{R^2}\right) \quad [5.14]$$

A is the effective Hamaker constant for the system. It depends on the Hamaker constants of particles and the dispersion medium. For interacting objects in vacuum, the Hamaker constant is defined as $A = \pi^2 C\rho_1\rho_2$ [14]. ρ_1 and ρ_2 are the number of atoms per unit volume in both objects, and C is the coefficient in the potential interaction between pairs of atoms ($V = -Cr^{-6}$). For particles **1** in a dispersion medium **2**, the effective Hamaker constant is approximated by

$$A \approx \left(\sqrt{A_1} - \sqrt{A_2}\right)^2 \quad [5.15]$$

Typical values of A are usually a few kT (3.5 to 8.10^{-20} J for oxides and $3.7.10^{-20}$ J for water).

For small separation distances ($D/a << 1$), an approximation of the interaction energy is

$$V_A = -\frac{A}{12}\left(\frac{L}{D} + 2ln\frac{D}{L}\right) \qquad [5.16]$$

where $L = a + 3D/4$. The interaction is attractive because V_A is negative.

5.3.2 ELECTROSTATIC FORCES

In the encounter of two charged particles, the interpenetration of the diffuse part of double layers causes the increase in electrostatic potential within the separation space of surfaces. Locally, that causes an increase in the concentration of counterions and consequently of the ionic strength, I. The diffuse layers of identical counterions are repelling, but, even if the repulsive electrostatic ranges, κ^{-1}, decrease because κ is proportional to $I^{1/2}$, the osmotic pressure effect of counterions tends to repel the particles [14, 26, 31]. The approach of particles also allows the Stern layers to interact, which may lead to the discharge of surfaces. So, the repulsive electrostatic forces are governed in a complex manner by the physical chemistry of the system and by the pH and ionic strength of the suspension on which the surface charge and the extent of the diffuse layer are dependent.

5.3.3 TOTAL POTENTIAL ENERGY OF THE INTERACTION

The sum of attraction and repulsion energies is $V_T = V_A + V_R$ (Fig. 5.14).

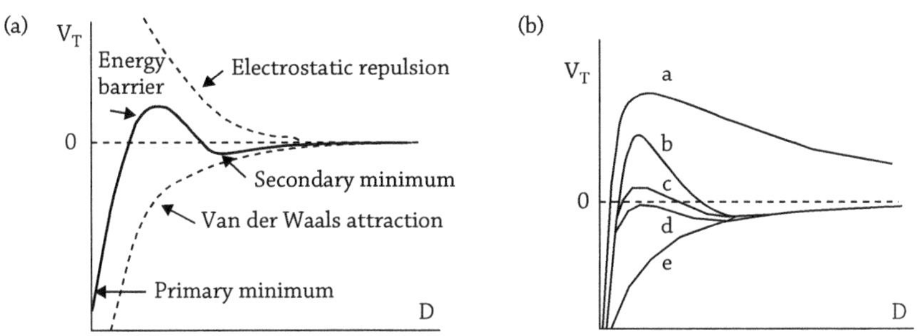

FIGURE 5.14 (a) Scheme of interaction energy $V_T = V_A + V_R$ (plain line) as a function of the distance between surfaces. (b) Different situations according to the relative importance of attractive and repulsive energies: a. the surfaces repel strongly; b. the surfaces are in a stable equilibrium in the secondary minimum; c,d. the surfaces are in an unstable equilibrium and give rise to a limited aggregation and a slow flocculation of particles; e. the surfaces strongly attract and the particles flocculate rapidly.

The repulsive energy V_R varies exponentially with the distance between the surfaces. It depends on pH and ionic strength through the potential ψ_o (or ψ_d) and κ. V_A in a first approximation is not sensitive to the physical-chemical conditions of the medium and varies as an inverse power of the distance. V_A is always predominating at short distances of approach (1 to 2 nm), which should induce a steep drop in the interaction energy. At very short distances between surfaces, the repulsive forces due to the interpenetration of the solvent layers are strongly oriented (structural forces) and due to the overlap of electron clouds (Born forces) exceed the van der Waals forces [14, 22, 32], so that a minimum (primary minimum) exists in the variation of V_T with distance (Fig. 5.14).

According to the physical-chemical conditions of the dispersion (pH, ionic strength), several different situations have to be considered. For highly charged surfaces in the presence of a diluted electrolyte (large values of Debye length κ^{-1}), the particles push away all the more strongly as they get closer. The energy barrier corresponding to the maximum of V_T (Fig. 5.14a) reaches some tens to hundred of kT that avoids aggregation of particles, which are maintained in a kinetically stable dispersion state. When the surface charge is low and ionic strength is high (Fig. 5.14e), the system is attractive, regardless of the distance between particles, and their flocculation is fast. For highly charged particles in the presence of a concentrated electrolyte, there is a weak minimum (secondary minimum) in the variation of the total energy (Fig. 5.14b). The particles are repelling at short distances, but they may stay in a pseudo-equilibrium at the distance of the secondary minimum without flocculation. Such a situation is characteristic of physical gels in which aggregates of particles form a loose network in which the trapped solvent prevents the collapse of aggregates and flocculation. That forms a semirigid system with special rheological properties [33]. As the energy barrier falls, the dispersion becomes less stable and aggregation of particles is possible or even favored. The *c* and *d* cases in Figure 5.14 are typical of a kinetically unstable critical state where coagulation is slow. The sol may appear to be kinetically stable because in temporary stationary states, aggregation is partially balanced by dispersion. The stationary states allow the formation of aggregates of various sizes without complete flocculation. It is the regime of kinetically limited aggregation [34].

An electrolyte is all the more flocculating as its concentration, C, and the charge of its ions, z, are important (Schulze and Hardy rule [35]). These two factors contribute to the decrease in Debye length κ^{-1} (κ is proportional to $zC^{1/2}$) and to the lowering of V_R. That explains the flocculation of muds transported by rivers (fresh water) when they reach the sea (salt water). It is also the reason lime is found in clayey grounds. The bivalent Ca^{2+} ions cause agglomeration of clay particles in the form of large grains that make the ground less compact, allowing better water circulation. To form a stable sol increasing the surface charge of particles by the addition of concentrated acid or base, the inverse effect may be obtained because the increase in ionic strength of the medium induces flocculation.

The interaction energy diagrams presented in Figure 5.14 show that if the particles approach some of the others at short distance by lowering the surface charge or adding electrolytes, the system will reach a deep well of energy and

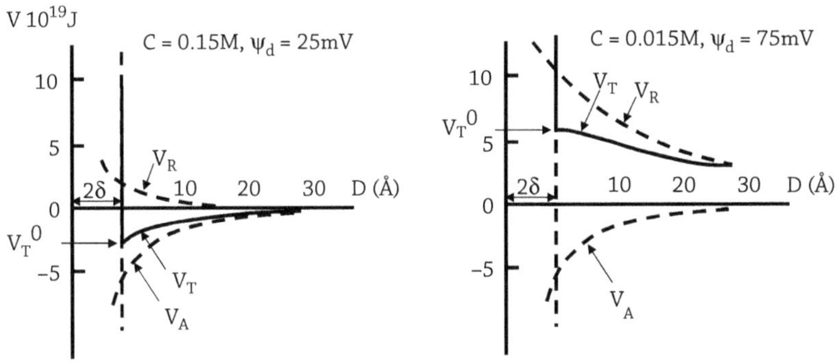

(a) Flocculation (b) Peptization

FIGURE 5.15 Diagrams of interaction energy corresponding to (a) flocculation and (b) peptization. Thickness of the Stern layer $\delta = 2$ Å; particle radius $a = 500$ Å; effective Hamaker constant $A = 5.10^{-20}$ J; the concentration c of a [1:1] electrolyte and the potential ψ_d of the diffuse layer are indicated on the figures (reproduced from [35] with permission of Elsevier).

dispersion will no longer be possible. The elimination of electrolytes by dialysis, for instance, and/or the increase in the surface charge, avoiding coagulation when particles are far from some of the others, no longer allows the dispersion of the flocculate. In order for dispersion (peptization) to occur, the surface charge must be balanced by counterions compressed on the surface and not canceled. In this way, the surfaces retain their strongly structured hydration layers, and the minimal distance of approach is at least two times the Stern layer thickness [30, 36, 37]. Direct contact between surfaces must be avoided to restrict the van der Waals forces and to avoid the possible condensation reactions between oxygenated groups on surfaces in contact. The diagrams corresponding to flocculation and peptization of particles (Fig. 5.15) are different from those in Figure 5.14 by the absence of a deep minimum because of a larger minimal distance of approach.

The texture of aggregates is usually dependent on the flocculation rate. When coagulation is slow, the aggregates are compact and often nondispersible (irreversible) because desolvation of surfaces and double-layer relaxation have time to occur. The compacity is high because the van der Waals attractive energy may be reinforced by a multiplication of the contact points between particles. If the flocculation is fast, however, the aggregates are usually bulky and of loose structure because a large amount of solvent is trapped between disordered clusters of particles. Such a precipitate is easily peptizable as long as the double layer relaxation does not occur. After a long enough period, the aggregates progressively become irreversible (coagulate). The particle surface charge is no longer screened (balanced), but it tends to decrease and even to cancel. For this reason, peptization of small particles has to be made rapidly after their flocculation.

5.4 Surface Reactivity: Adsorption

Ions or molecules attracted by electrostatic forces toward the surface may enter and remain in the Stern layer. They also may bind chemically to the surface. Such species are usually complexing anions, easily hydroxylable cations, or neutral molecules forming real coordination complexes with surface groups. Depending on the strength of the interaction between the adsorbed species and the surface, outer- or inner-sphere complexes are formed, and these species are so-called physi- or chemisorbed, respectively [38, 39]. Chemisorption, which is most often triggered by acid–base phenomena, may be confined on surfaces and form a more or less thick layer of material. It also may be relieved by redox phenomena within the core of particles exhibiting ionic or electronic mobility in the solid framework, giving rise to ionic and/or electronic exchanges at the solid–solution interface.

Adsorption is involved in various fields, such as the environment, for example, with respect to the transport of matter in natural or industrial waters and geochemistry. It also has many industrial uses, such as catalysis, stabilization of dispersions against aggregation, development of probes for medical diagnosis, and therapy. Adsorption of polymers onto oxide particles is one of various strategies available for the elaboration of organic-inorganic composite materials. Adsorption (or sorption if the amount of adsorbed matter exceeds the monolayer) is an important phenomenon, and it is useful to know the mechanisms of the interaction involved between the surface of oxides and various species in solution.

5.4.1 ELECTROSTATIC INTERACTIONS, OUTER-SPHERE SURFACE COMPLEXES

Nonhydroxylable and structuring cations such as Li^+ and Na^+ adsorb on oxides when the surface is negatively charged (pH > PZC). These ions are attracted to the surface by nonspecific forces such as electrostatic constraints and/or preferential solvation in the highly structured hydration layer of the surface (§ 5.2). The same kind of interaction is also involved with stable complexes of various cations (Cu^{2+}, Ni^{2+}, Co^{2+}). At high ammonia concentration, the hexaammine complex $[M(NH_3)_6]^{2+}$ is adsorbed on different substrates [40]. The UV-visible absorption band characteristic of the ammine complex of nickel is not shifted by adsorption on silica [41]. The negatively charged surface of oxides indeed behaves as the counterion of the cationic complex in forming, at best, ion pairs.

Some aquo cations may also be nonspecifically adsorbed, as Cu^{2+} ions in an acidic medium (pH < 3) impregnate silica particles with small pores (< 6 nm). The electron spin resonance (ESR) signal of copper is characteristic of the hexaaquo $[Cu(OH_2)_6]^{2+}$ complex in axial symmetry [42], but with reduced mobility, rotation, and reorientation, 1.5 times less than in aqueous solution. The mobility is strongly perturbed by the structuration of water, and the complex seems to be trapped in the eight to ten water layers that form a glassy matrix filling the small pores of silica. In a silica with larger pores (≥ 10 nm), all the filling water is not so structured, and a second ESR signal, still characteristic of $[Cu(OH_2)_6]^{2+}$ ions but in aqueous solution, is superimposed. Oxides such as MgO and Al_2O_3, which are more basic than silica, give rise to higher interactions with $[Cu(OH_2)_6]^{2+}$ ions, and the ESR spectrum looks like that of a solid [42]. Cu^{2+} ions are not desolvated, and the hexaaquo complexes are in fact immobilized onto the surface. It is the same when Cu^{2+} ions are adsorbed at pH 2.7 on TiO_2 [43], that is, largely down the PZC of titania located at pH 6.2. Although these adsorbed ions lead to the increased

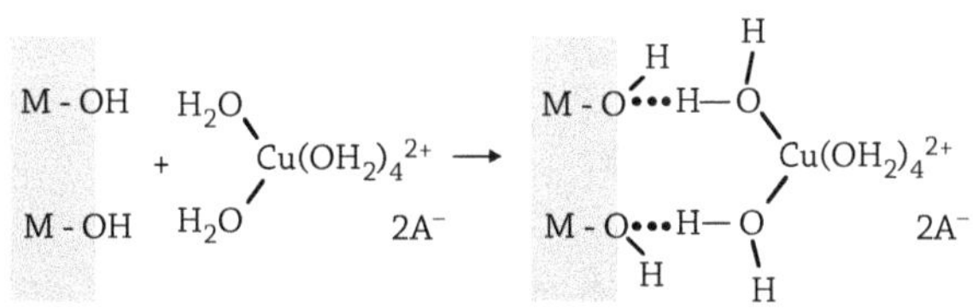

SCHEME 5.2

electrophoretic mobility of oxide particles (see farther) because they increase the surface charge, their "static" ESR spectrum is typical of hydrated copper(II) ions in a distorted octahedral coordination. That means the axis of the tetragonal distortion does not change in orientation at the time scale of the ESR spectroscopy.

This kind of interaction does not modify the coordination sphere of the adsorbed cations. They are indeed better solvated by the hydration layer of the surface than by "liquid" water, in bulk solution. In terms of coordination chemistry, the copper(II) ions form outer-sphere complexes with the surface (Scheme 5.2), likely involving hydrogen bonds between surface hydroxylated groups and the water molecules coordinated to the cation through the bridging $[H_3O_2]^-$ ligand [44, 45]. In the "physisorption" phenomenon, the surface plays the role of solvating agent of cations.

5.4.2 SPECIFIC INTERACTIONS: INNER-SPHERE SURFACE COMPLEXES

The easily hydroxylable cations such as Zn^{II}, Cd^{II}, Hg^{II}, Pb^{II}, Al^{III}, and the transition metals have a strong affinity for oxide surfaces, and their adsorption may occur against the electrostatic forces, at pH much lower than the PZC of the oxide. Adsorption of these cations, however, is strongly dependent on pH because the adsorption rate (the saturation rate of the surface) increases from 0 to 100% on a narrow range of pH, usually of one to two units (Fig. 5.16). However, the pH for the onset of the reaction is always lower than that corresponding to hydroxylation and precipitation of the cations alone in solution.

Adsorption of Co^{II} on goethite, α-FeOOH, occurs between pH 6 and 7 [46]. Co^{II} alone in solution forms around pH 7 the $[Co(OH)(OH_2)_5]^+$ complex and

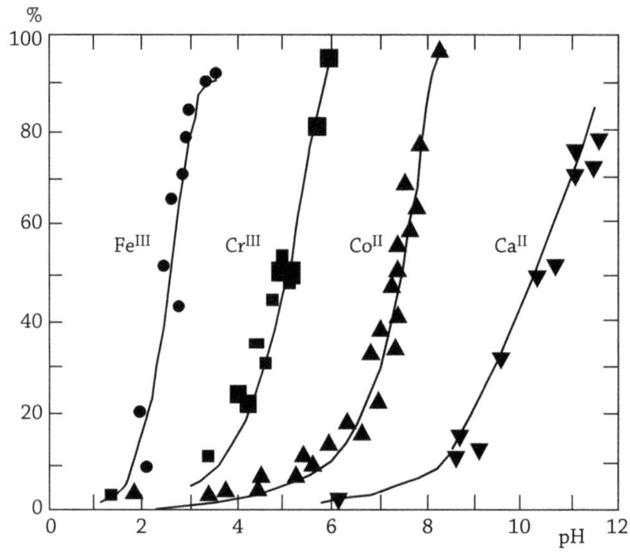

FIGURE 5.16 Adsorption isotherms at 25 °C of Fe^{III} ($1.2.10^{-4}$ mol.l^{-1}), Cr^{III} (2.10^{-4} mol.l^{-1}), Co^{II} ($1.2.10^{-4}$ mol.l^{-1}) and Ca^{II} ($1.4.10^{-4}$ mol.l^{-1}) on SiO_2 as a function of pH (reproduced from [James R.O., Healy T.W., Adsorption of hydrolyzable metal ions at the oxide-water Interface I. Co(II) adsorption on SiO_2 and TiO_2 as model systems, J. Colloid Interface Sci. 1972, 40, 42–52] with permission of Elsevier).

at concentration of 0.1 mol l^{-1}, $Co(OH)_2$ precipitates around pH 8. X-ray photoelectron spectroscopy (XPS) shows that adsorbed Co^{II} exists as very different species according to the pH at which the reaction occurs [46]. At pH < 6.5, cobalt is adsorbed as a hexaaquo complex. Since the PZC of the surface is 6.5, the aquo ions form outer-sphere complexes. The adsorption density is low because the surface is globally positively charged. From pH 6.5 to 7.5, the XPS data of adsorbed Co are similar to those of hydroxide $Co(OH)_2$. Since in solution at the same concentration Co^{II} alone is hydroxylated only at pH > 7.4, it is likely that hydroxylation of Co^{II} ions is induced by surface groups rather than by the solution. Thus, the surface hydroxylated groups act as hydroxylating ligands, and chemisorption of cobalt results in inner-sphere complex formation. When adsorption occurs at pH > 8, the XPS data indicate the presence of $Co(OH)_2$ and CoO(OH) on the surface. These compounds are also detected by X-ray diffraction and XPS in the precipitates obtained with Co alone at the same pH. It therefore seems that the precipitation of cobalt occurs in solution and that the particles are then physisorbed on goethite.

The hydroxylating and condensing role of the surface of oxides is also nicely shown by the study of $^{57}Fe^{III}$ ion adsorption on hematite α-$^{56}Fe_2O_3$ [47]. As the ^{56}Fe isotope is silent in Mössbauer spectroscopy, the spectrum of adsorbed ferric ions is not obfuscated by that of particles. At pH 2.5, the Mössbauer spectrum at RT of the suspension consists of a sextet (Fig. 5.17a), indicating that all ferric ions are in magnetic interaction between them and those of the substrate. The characteristics

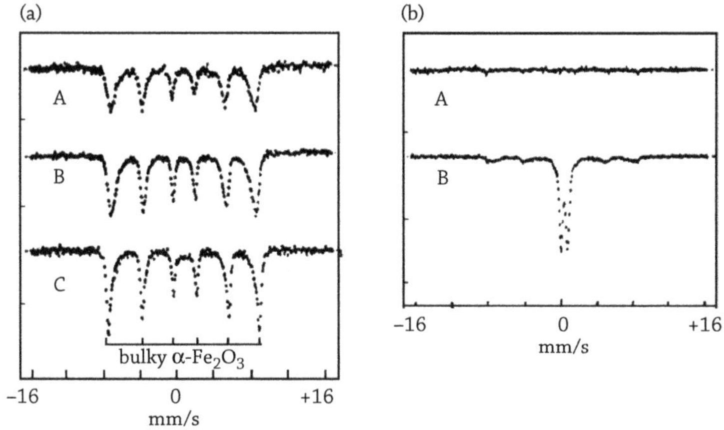

FIGURE 5.17 Mössbauer spectra (298 K) of $^{57}Fe^{III}$ ions adsorbed on hematite α-$^{56}Fe_2O_3$ particles at pH 2.5 (a) [A: suspension; B: after drying; C: after heating 6 at 600°C for 6 hours]; at pH 6 (b) [A: suspension; B: after drying] (reproduced with permission from [46] Copyright 1990 American Chemical Society).

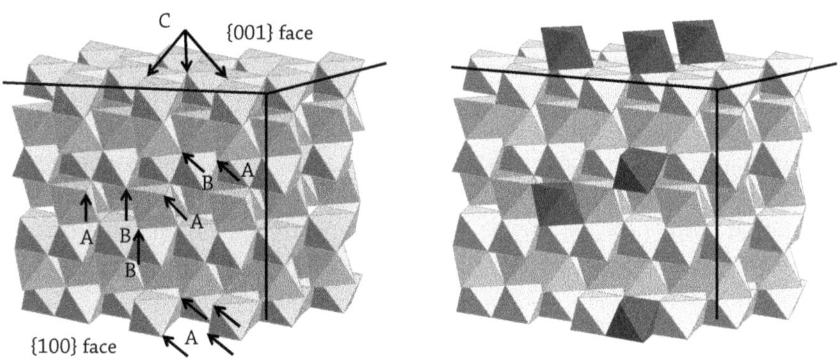

FIGURE 5.18 Scheme of crystal faces of hematite particles showing sites of possible coordination of ferric complexes from solution (see also Figs. 5.6 and 5.7).

of adsorbed ions (isomer shift, hyperfine magnetic field) are similar to those of bulk hematite. The adsorbed ferric ions occupy the typical crystal sites of hematite, and the adsorbed layer forms an extension of the corundum-type crystal framework of particles. At this acidity, the ferric ions are unable to nucleate a solid and certainly not in these conditions of the corundum framework. Clearly, these ions are hydroxylated by the surface groups and chemisorbed.

At pH 2.5, the ferric ions in diluted solution (10^{-3} mol l^{-1}) exist as $[Fe(OH_2)_6]^{3+}$ (70%), $[Fe(OH)(OH_2)_5]^{2+}$ (20%), $[Fe(OH)_2(OH_2)_4]^{+}$ (3%), and $[Fe_2(OH)_2(OH_2)_8]^{4+}$ (6%). At the same pH, the hematite particles are globally positively charged, but it exists surface groups negatively charged of *C*-type (Fig. 5.6 and 5.18) and then nucleophilic. That explains why the surface of hematite can behave as hydroxylating and condensing ligand toward adsorbed aquo ferric complexes. In addition, the groups of *A*- and *B*-type carry strongly polarized water molecules that can be easily substituted by hydroxo ligands of ferric complexes. Also, the structural details of the surface show that adsorption occurs in an ordered manner, allowing adsorbed ferric ions to occupy crystal sites corresponding to the corundum-type framework.

When adsorption takes place at pH 6, the wet suspension exhibits no Mössbauer signal. After drying, an intense quadrupole doublet in the presence of a weak magnetic component is observed (Fig. 5.17b). At this acidity, ferric ions in solution form a gelatinous solid that is physisorbed by the substrate. The absence of Mössbauer signal by the wet solid is due to the very weak recoilless absorption fraction of γ-rays by the hydroxide gel loosely bound to the surface of hematite particles. Dehydration of the sample at RT decreases the mobility of the physisorbed gelatinous solid, which then gives a Mössbauer spectrum similar to that of amorphous ferric gels. Adsorption at pH 4 involves both mechanisms.

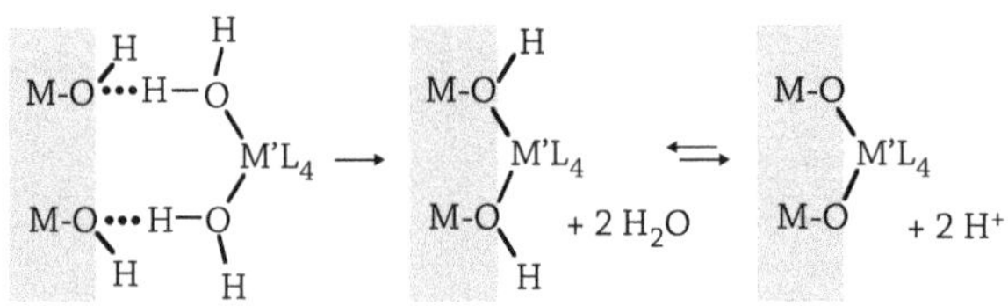

SCHEME 5.3

Adsorption of hydroxylable cations proceeds according to different mechanisms in which hydroxylation of cations by the nucleophilic groups of the surface or by the solution itself (HO^- ions, water molecules) are in competition. As long as the pH of hydrolysis and precipitation of the cations in solution is not reached, the

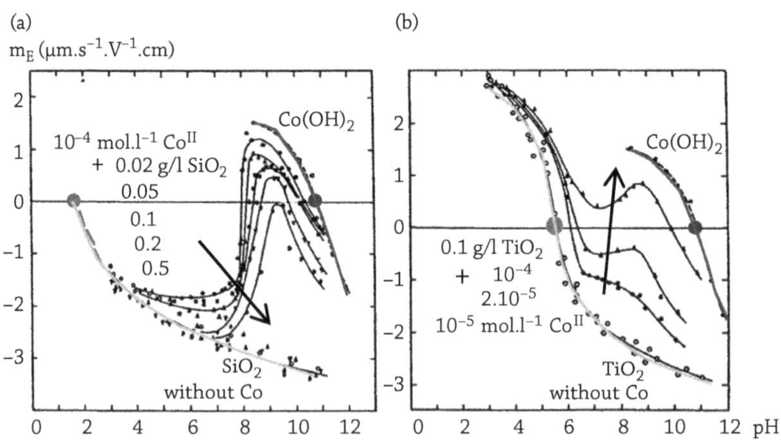

FIGURE 5.19 Electrophoretic mobility of various (a) SiO_2 amounts and (b) TiO_2 in the presence of $Co(NO_3)_2$ (reproduced from [47] with permission of Elsevier). The arrows indicate increasing concentration of SiO_2 in (a) and increasing concentration of Co^{II} ions in (b).

surface may behave as a ligand of soluble ions and chemisorb them by a real substitution and coordination reaction similar to olation or oxolation reactions involved between hydroxylated species in solution (Scheme 5.3).

The different regimes of interaction between ions in solution and the surface as depending on pH (chemisorption of cations, physisorption of a hydroxide or an oxide) may be highlighted by the variations in electrophoretic mobility of the suspensions. The mobility, m_E, of the particles is defined as $m_E = v_E/E$. v_E is the velocity of particles under the electric field E [28]. Depending on the sign of the particle charge, the particles move in the direction of the field or in the opposite direction. At pH = PZC, the mobility is zero. Adsorption of a hydroxylable cation at a large enough concentration that allows entire covering of the available surface area in suspension changes the electrophoretic behavior of particles. Three charge reversals (CRs) take place as the pH increases (Fig. 5.19). The passage at the PZC (or IEP) corresponds to the CR_1 (+/−). It is close to pH 2 for SiO_2 and to pH 5.5 for TiO_2 (Fig. 5.19). Adsorption of positively charged species (aquo ions, hydroxylated complexes) progressively balances the negative surface charge, and a CR_2 (−/+) takes place around pH 8. As pH increases, the positive charge of the adsorbed species decreases, and a CR_3 indicates the PZC of the adsorbed layer, at pH 11. The positions of CR_2 and CR_3 depend on the amount of adsorbed ions. When coverage of the surface is complete, CR_3 is equal to the PZC of the hydroxide or oxide formed by the adsorbed cation itself. CR_2 is lower than the pH of precipitation of the cation alone in solution. It marks the onset of precipitation on the almost completely covered surface [48, 49]. Such a variation in electrophoretic mobility in suspension reflects the progressive change in the nature of the surface during adsorption, but that implies a homogeneous distribution on adsorbed species on the surface. This behavior is not general because the adsorption mechanism depends on the nature of the surface as well as on that of the adsorbed ions.

The adsorption of anions and weak acids may also be *specific* and proceed by substitution of aquo or hydroxo ligands from the oxide surface [50]. Infrared spectroscopy shows that various ligands such as oxalate, benzoate, phosphate, and silicate substitute hydroxo monocoordinated groups on the surface of various oxy(hydroxi)des of iron and aluminum. These ligands generally behave as bridging ligands at a low coverage rate and around pH 3–4 [51–54]. The di- and tricoordinated surface groups are more difficult to substitute because with underlying cations they form more covalent bonds than monocoordinated surface groups. Specific adsorption of anions is a surface complexation reaction, and, as a rule, anions are more efficiently adsorbed than they have a high complexing power in solution [38, 55]. The chelating or bridging effect, which reinforces the complexing power in solution, also favors anion adsorption. Thus, ethylenediaminetetraacetic acid (EDTA) and aminocarboxylates, which may form soluble complexes that are more stable than oxide, allow dissolution of small oxide particles [56]. Such ligands are used, for instance, for cleaning the corrosion products of industrial equipment.

Anion adsorption on magnetic nanoparticles of maghemite, $\gamma\text{-}Fe_2O_3$ (Chap. 7), highlights the coordination of surface cations and the change in their

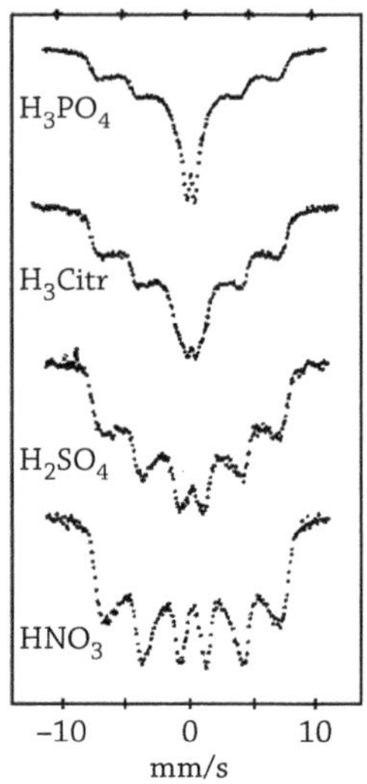

FIGURE 5.20 Mössbauer spectra (290 K) of γ-Fe_2O_3 particles flocculated from aqueous suspensions by addition of various acids.

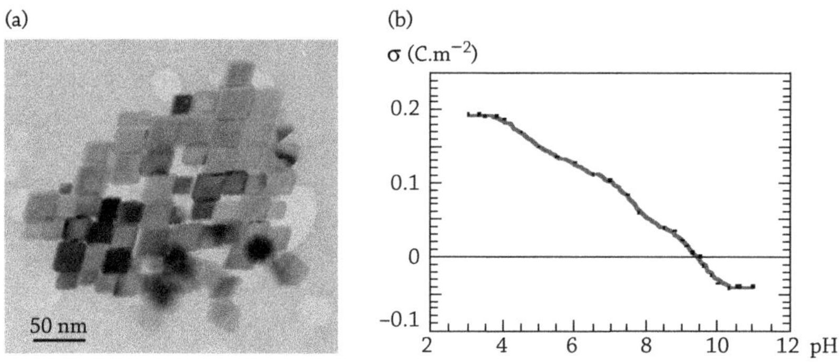

FIGURE 5.21 (a) TEM image of rhombohedral hematite nanoparticles and (b) their surface charge density as a function of pH.

magnetic properties (§ 1.1.1). In spinel crystal nanostructures, ferric cations are ferrimagnetically coupled, as long as no surface effect occurs, as exemplified in Mössbauer spectroscopy of the powder of particles flocculated with nitric acid addition in suspension (Fig. 5.20). Nitrate ions act only as counterions of the electrostatic surface charge. As affinity of adsorbed species for the surface increases (NO_3^- << SO_4^{2-} < $HCitr^{2-}$ < HPO_4^{2-}), the magnetic coupling of the surface Fe^{3+} cations decreases. With the most strongly complexing anions (citrate and phosphate), the magnetic coupling vanishes, and the surface ferric ions form paramagnetic complexes, which appear on Mössbauer spectra as a quadrupole doublet superimposed onto the magnetic component from the core of nanoparticles [57].

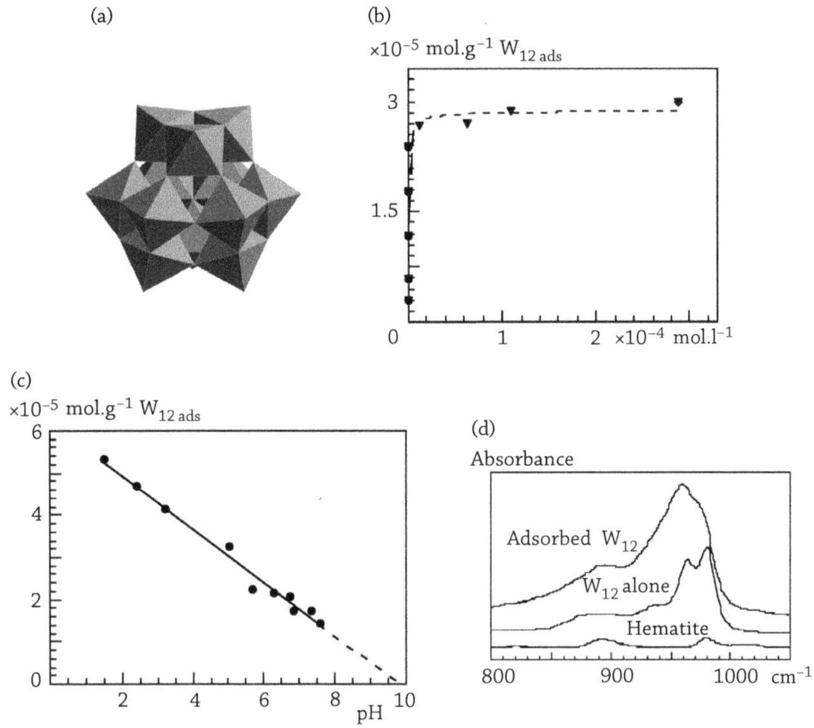

FIGURE 5.22 (a) Structure of $[H_2W_{12}O_{40}]^{6-}$ (W_{12}), (b) adsorption isotherm on hematite at pH 5, (c) amount of adsorbed polyanion as a function of pH and (d) infrared spectrum of the polyanion alone (sodium salt) and adsorbed on hematite, and hematite alone.

It is interesting to compare the adsorption of α-$[H_2W_{12}O_{40}]^{6-}$ polyanion to that of $[Al_{13}O_4(OH)_{24}(OH_2)_{12}]^{7+}$ (ε-Al_{13}) polycation on hematite nanoparticles (Fig. 5.21) [58] because these entities, which geometrically are very similar, spherical, and around 1 nm in diameter (Figs. 5.22a and 5.23a), bear charges of opposite signs but are quasi-equivalent in absolute value. These two structures are derivatives of Keggin's polyanion (§ 6.2 and § 3.3.2).

On the range $2 \leq pH \leq 8$, the polyanion is stable and its charge does not change. Its adsorption on positively charged hematite particles is rapid and leads immediately to flocculation. Infrared spectroscopy indicates that the structure of the adsorbed polyanion is preserved (Fig. 5.22d). The adsorbed amount decreases as the pH increases, that is, with the decreases in surface charge of particles. The adsorption is reversible: the polyanions adsorbed at pH 2 are quickly and almost totally desorbed when the pH increases. The adsorbed amount as a function of pH extrapolates to zero at pH close to 10, a value corresponding to the PZC of hematite (Fig. 5.22c). The interaction of the polyanion with hematite is clearly of an electrostatic nature. It only involves the balance of charge between the two partners.

Al_{13} polycation adsorbs on positively charged hematite nanoparticles, and pH 5 is the optimal value for stability in solution of the polycation. The saturation

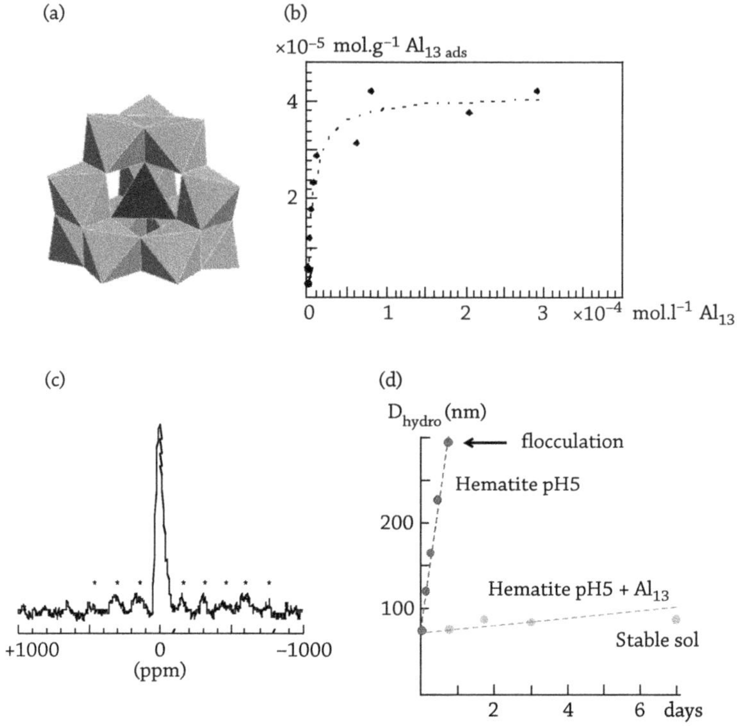

FIGURE 5.23 (a) Structure of $[Al_{13}O_4(OH)_{24}(OH_2)_{12}]^{7+}$ (Al_{13}), (b) adsorption isotherm of Al_{13} on hematite at pH5, (c) ^{27}Al NMR spectrum of adsorbed polycation, and (d) variation of hydrodynamical size of hematite particles at pH 5 and of particles bearing adsorbed polycation.

corresponds to a coverage rate of around 60% of the total surface area (Fig. 5.23b), but adsorption destroys the polycation. The solid ^{27}Al NMR spectrum of hematite containing adsorbed Al_{13} exhibits only aluminum in octahedral coordination (Fig. 5.23c). The signal characteristic of tetrahedral aluminum, which is the signature of Al_{13}, has totally disappeared (Fig. 6.2). On the other hand, the stability of hematite nanoparticles against aggregation at pH 5 is considerably reinforced, as shown by the evolution of the hydrodynamical size of particles (Fig. 5.23d). When the hematite particles in suspension at pH 5 aggregate with a cluster size increasing rapidly in one day, the particles, having adsorbed Al_{13}, remain stable, and the dispersion state has practically not evolved during one week. At pH 5, the electrokinetic potential ζ of hematite, around 10 mV, is not high enough to avoid aggregation, and flocculation is completed after 2 days, whereas at pH 2, with a ζ potential of 32 mV, the dispersion is stable during a very long period. In the presence of Al_{13} at pH 5, the ζ potential of the particles is close to 30 mV, and the suspension is stable during at least 1 week, indicating that the polycation adsorption increased the electrostatic surface charge. That shows that adsorption of a cation on a globally positively charged oxide surface is possible and that the reaction results from the presence of surface nucleophilic oxygenated groups (Figs. 5.6 and 5.7). Even the polycation is destroyed by coordination on the surface (see § 6.2 for study of isomers of Al_{13}), its fragments remain adsorbed, and their electrostatic charge is gained by the surface of particles.

Adsorption of organic macromolecules is likely the oldest method used to stabilize dispersions of small particles. In ancient Egypt, stable inks were made by dispersing carbon nanoparticles (carbon black) into aqueous solutions of acacia gum (arabic gum). Polymer adsorption may help avoid aggregation and adhesion of particles during rehydration of dispersions kept dry or, on the other hand, be used as binder between particles for the elaboration of composite materials. The same polymer can play the one or the other role according to the adsorption density and the saturation rate of the particle surface. Various modes of interaction can be involved between hydroxylated and/or charged surface groups of the oxide and the type of polymer: hydrogen bonds with the polar functional groups distributed on the chain of nonionic macromolecules (ethylene polyoxide, polyacrylamide, polyvinylic alcohol) [59, 60]; electrostatic interactions with polyelectrolytes [61, 62]; and coordination or covalent grafting with polymer bearing reactive groups such as alkoxysilanes [63]. Some excellent reviews are available on the physical chemistry of such systems [61, 62, 63].

5.4.3 ADSORPTION AND TRANSFERS THROUGH THE OXIDE–SOLUTION INTERFACE

Cation chemisorption on many oxides involves interaction with the surface without implication of the core of particles. In this case, the interface is polarizable, which means no exchange takes place between the solid and the solution. For materials exhibiting ionic or electronic mobility, such as ion exchangers, semiconductors, or conductors, the surface acid–base phenomena that induce cation adsorption may involve a redox response from the core of particles appearing as ion and/or electron transfer through the solid–solution interface.

The interaction of ferric oxides in suspension with ferrous ions is likely the most studied redox system for iron. It highlights the enhancement of the reducing power of ferrous ions in solution [66–68] because the ferric ions produced on the surface can be structurally incorporated into the solid and thus strongly stabilized with respect to soluble oxidation products. Thus, the redox potential of the Fe^{III}-Fe^{II} couple involving soluble species can be notably lowered. Another key point is that adsorption of ferrous ions on ferric solids resulting from the acid–base reaction (coordination of the cation by surface hydroxo ligands) is accompanied by a redox phenomenon (Fe^{II}-Fe^{III} electron transfer), which leads to an atomic exchange between the solution and the solid. The phenomenon has been evidenced in using ^{57}Fe and ^{56}Fe isotopes that are active and silent, respectively, in Mössbauer spectroscopy. The interaction of $^{57}Fe^{II}$ ions in solution with $^{56}Fe^{III}$ oxides leads to solids, including $^{57}Fe^{III}$ ions and the release of $^{56}Fe^{II}$ ions in solution [69]. The kinetics and the exchange rate are dependent on the used solid—ferrihydrite, goethite, lepidocrocite, hematite, and magnetite—and on factors such as pH and concentration in solution [70, 71]. On the other hand, ferrous ion adsorption on TiO_2 particles does not involve Fe^{2+}–Ti^{4+} transfer, and the adsorbed ferrous ions remain in the ferrous state [69]. The Fe^{2+}–Ti^{4+} transfer is possible, however, at room temperature, for instance, being at the origin of the blue color of sapphire, but it is dependent on structural constraints that require the two cations Ti^{4+} and Fe^{2+} to occupy face-sharing octahedral sites in the corundum structure. Special redox effects may thus be due to ferrous ion adsorption on ferric oxides because it allows the electron injection into the solid. The crystal structure and electronic properties of the solid appear as the key points of the redox reactivity of iron oxide nanoparticles. Ferrous ion adsorption may so induce the crystallization of ferrihydrite into spinel [72, 73] and be involved in the transformation of maghemite (§ 7.2.3b).

Another illustrative example concerns the ferrous ion adsorption on hematite α-Fe_2O_3. It has previously been seen that ferric ion adsorption in acidic medium is the result of a real coordination reaction allowing the adsorbed cations to occupy sites of the underlying crystal lattice (Fig. 5.18). In similar acidity conditions and up to pH 7 (conditions in which Fe^{II} alone does not precipitate), the adsorption of ferrous ions on hematite involves specific morphological effects. These ions preferentially attach onto the {001} basal faces of synthetic or natural tabular hematite particles in forming pyramidal islands having the same crystal orientation as that of the support, whereas the {0*hk*} edge faces of particles are eroded by the release of Fe^{II} ions in solution (Fig. 5.24) [74, 75]. The phenomenon has been studied with Mössbauer spectroscopy within adsorption of $^{57}Fe^{II}$ ions on hematite α-$^{56}Fe_2O_3$ (only the adsorbed ions are detectable by Mössbauer spectroscopy): the spectrum of the solid phase indicates ferric ions with spectral data (isomeric shift, magnetic hyperfine field) very similar to those of cations in hematite, while the solution, which initially contained only $^{57}Fe^{II}$, is enriched in $^{56}Fe^{II}$ in a 1:1 ratio [75, 76]. The X-ray reflectivity measurements also show the adsorbed phase in epitaxy and the absence of ferrous ions on the surface [77]. It can be concluded that the oxidative adsorption occurring on the {001} basal face is accompanied by a reductive dissolution of the {0*hk*} edge faces.

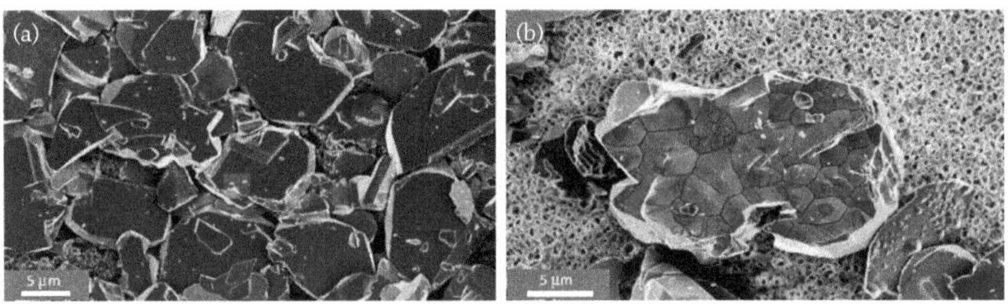

FIGURE 5.24 MEB image of tabular platelets of synthetic hematite before (a) and after (b) reaction at 75°C for 24 hours with a solution of Fe^{II} ions and oxalic acid at pH 2.1 in anaerobic conditions. The growth of pyramidal islands on basal (001) faces is clearly visible (reproduced with permission from [74] Copyright 2010 American Chemical Society).

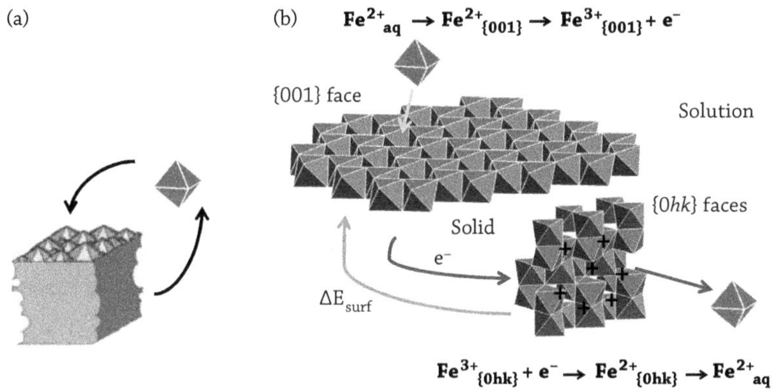

FIGURE 5.25 (a) Scheme of the morphological transformation of tabular hematite particles during Fe^{II} ion adsorption. (b) Scheme of the coupling of electron transfers on different faces of the particles (dark octahedra). The gradient in surface electrostatic potential within the crystal, chemically induced by ferrous ion adsorption (light octahedra), involves an electric current within the bulk. The net balance corresponds to the injection of electrons on the {001} faces and their elimination from the {ohk} faces.

Modeling hematite particles (§ 5.1.2) shows that in acidic medium (pH < 7), the {*hk*0} faces are globally positively charged, while the {001} basal faces are practically not charged or only weakly negatively. That explains why ferrous ion adsorption preferentially occurs on the {001} faces with oxidation into ferric ions increasing the interaction energy. As the material is semiconducting, the diffusion of the electronic charges within the bulk lowers the charge density in the zones, with high potential involving reduction of surface ferric ions and dissolution of the ferrous species thereby generated (Fig. 5.25).

Thus, the accumulation of ionic charges on certain faces of particles and the diffusability of electronic charges through the bulk create conditions under which the interfacial charge transfer on one face is coupled to the transfer on other faces by an electric current in the bulk. In other words, the gradient of potential (some tenths of volt at pH 2), chemically induced by adsorption on hematite particles, is high enough and the resistivity of the bulk low enough that the growth of certain faces is bound to the simultaneous dissolution of other faces. Thus, a real redox cycle takes place without a net balance of reduction of iron with two coupled interfacial processes:

1. Growth of basal faces by oxidative sorption: $Fe^{2+}_{aq} \rightarrow Fe^{2+}_{001} \rightarrow Fe^{3+}_{001} + e^-$
2. Reductive dissolution of edge faces: $Fe^{3+}_{hk0} + e^- \rightarrow Fe^{2+}_{hk0} \rightarrow Fe^{2+}_{aq}$.

The coupling of these two reactions is mediated by charge transport through the crystal. The process is sufficiently facile at room temperature to support a small current.

Much smaller hematite particles (30 m^2g^{-1}), synthesized by precipitation in aqueous solution, behave differently in giving rise to an obvious size effect. Adsorption of ferrous ions still induces an interfacial Fe^{II}-Fe^{III} electron transfer and exchange of Fe^{II} ions with the solution, but saturation occurs beyond the monolayer. Subsequent sorption of Fe^{II} ions forms ferrous hydroxide [76]. It seems that the Fe^{II}-Fe^{III} electron transfer does not occur beyond the first layer of adsorbed iron ions. This limitation may be due to surface disorder and defects that prevent epitactic crystallization of hematite [75, 76]. Surface defects and geometric constraints may also hinder the orbital overlapping required for the Fe^{II}-Fe^{III} charge transfer [78]. With nanoparticles, the various crystal faces may also be less well differentiated, so that the gradient in the electric field through the particle is too weak to serve as a driving force in the process.

The fate of electrons injected into the solid is strongly dependent on the electronic properties of the material. Magnetite, which is a semimetal, may behave as a reductant toward various reactives in forming nonstoichiometric spinel phases that can in turn be reduced through injection of electrons: with maghemite γ-Fe_2O_3, the Fe^{II} ions are sorbed in building an epitactic spinel oxide layer (§ 7.3.2b), the injected electrons being delocalized in the conduction band. With a semiconductor such as hematite, the electron can partake in reductive dissolution as indicated earlier. Electrons can also be located in a trapping site such as a surface state created, for instance, by a strongly chelating ligand, which forms a complex with a strong surface–ligand electron coupling giving rise to the sensitization phenomenon (§ 1.1.3c). With an isolator such as goethite, the adsorbed iron (II) can reduce a species in solution at the proximity of the surface with an enhanced reducing power [68, 79], for instance, Cr^{VI} [80] or arsenic [81]. The fate of the injected electrons in various iron oxides by ferrous ion adsorption may be schemed in terms of electronic structure (Fig. 5.26) [71].

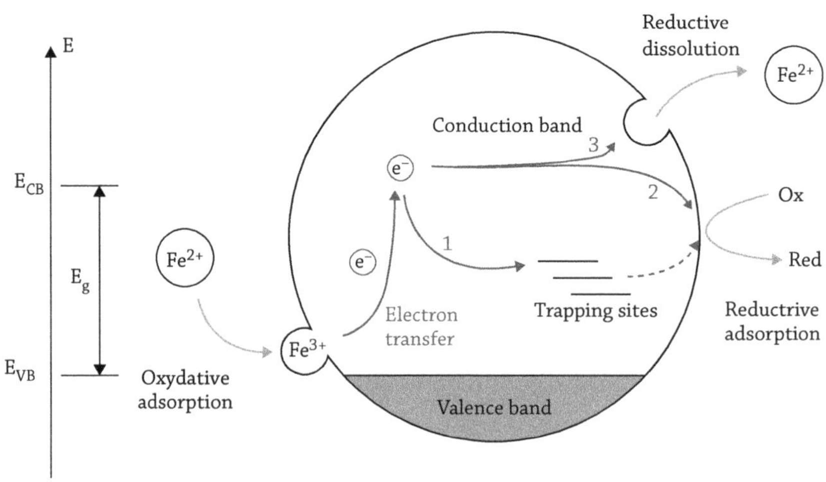

FIGURE 5.26 Scheme of possible fates of an injected electron in the conduction band of iron oxides by oxidative adsorption of ferrous ions in solution: 1. localization in a trap in the structure; 2. reduction of an oxidant (pollutant) in solution; 3. reductive dissolution of the solid (reproduced with permission from [70] Copyright 2011 American Chemical Society).

a) Redox Catalysis

Divalent cations other than Fe^{II} can be adsorbed on iron oxides and give rise to an enhanced redox reactivity. It is the case, for instance, for the oxidation of Mn^{II} ions by oxygen to form manganite according to

$$Mn^{2+} + 1/4\ O_2 + 3/2\ H_2O \rightarrow MnOOH + 2\ H^+$$

The reaction is extremely slow in a homogeneous phase at pH < 8.5.

In the presence of hematite, the reaction is accelerated even more because particles are small [82]. Normalized with respect to specific surface area, the reaction rate with particles 7 nm in mean diameter is one order of magnitude higher than that with 37 nm particles (the particles were synthesized by thermohydrolysis of acidic solutions of ferric nitrate, § 7.2.2). Acceleration of the reaction may be due to three main factors: (1) the adsorption of Mn^{II} increases its reducing power and decreases the free enthalpy of the reaction [67, 68]; (2) the presence of more surface defects on the smallest hematite particles favors stabilization of Mn^{III} ions strongly distorted by the Jahn–Teller effect [83, 84]; and (3) the increase in the Lewis base character of oxygen surface atoms as the particle size decreases [85] favors the interaction with adsorbed cations. The effect of catalysis results from a stronger adsorption of Mn^{II} ions on the smallest particles, which lowers the activation energy for electron transfer between reactives.

Surprisingly, the same oxidation of Mn^{II} ions in the presence of hematite appears to slow down when the particle size decreases [86]. In this study carried out according to the same process outlined above, particles of 7, 9, and 38 nm in diameter were obtained by thermolysis of acidic solutions of ferric nitrate and commercial particles 150 nm in size. The rate constants normalized with respect

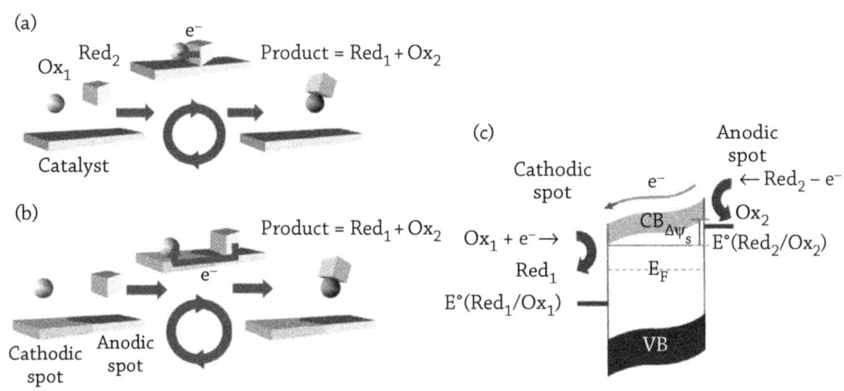

FIGURE 5.27 Schemes of the chemical (a) and electrochemical (b) mechanisms for heterogeneous catalysis of the redox reaction $Ox_1 + Red_2 \rightarrow Red_1 + Ox_2$. The chemical mechanism involves the charge transfer between the two adsorbates Ox_1 and Red_2 at proximity, while the electrochemical mechanism involves two distinct anodic and cathodic sites. (c) Scheme of the electrochemical mechanism for a *n*-type semiconductor (reproduced with permission from [86] Copyright 2011 American Chemical Society).

to specific surface area were practically the same in these two studies, but for the largest particles the results were different, probably because of differences in the experimental procedure and/or the treatment of kinetic data. It seems more likely that commercial and lab-synthesized particles exhibit differences in nature and in the characteristics of the exposed faces. Either way, the redox phenomenon can be interpreted according to an electrochemical model in which the two oxidation and reduction half-reactions are spatially separated and electrically coupled by a charge transfer within the bulk or in the surface of the solid [86, 87]. Thus, rather than a "chemical" paradigm based on the activated complex model assuming the proximity of the redox partners on the surface, one may consider that the change in the electron structure of the particle with size determines its (electro)chemical reactivity (Fig. 5.27).

In this model, the lowered reactivity results from widening the forbidden band with decreasing particle size (§ 1.1.3c). The estimated band gap is found to be 2.18, 2.55, and 2.95 eV for hematite nanoparticles of 120, 38, and 7 nm, respectively [88]. The widening of the band gap is attributed to the weakened hybridization of 3*d* and 2*p* orbitals of iron and oxygen, respectively, because of lattice distortions and the relaxation of the trigonal distortion of coordination polyhedra of ferric ions. Thus, the rise in energy of the conduction band hinders the electron injection and slows down the redox reaction.

These considerations again raise a question about the size effect on the electronic structure of semiconducting solids, especially hematite for which contradictory data have been published (§ 1.1.3c). Although a change in electronic structure with a decrease in particle size is theoretically expected, the observed diversity in behavior of certain nanoparticles could be due to nonoptimized synthesis and purification conditions, giving rise to nonreproducible results.

5.4.4 ADSORPTION AND SURFACE ENERGY: ROLE OF ACIDITY IN PARTICLE SIZE AND MORPHOLOGY

The effect of adsorption on the surface energy of particles dispersed in an electrolyte is expressed by the Gibbs equation [26]:

$$d\gamma = -(\Sigma n_i^s / A)d\mu_i = -\Sigma \Gamma_i d\mu_i \qquad [5.17]$$

where Γ_i is the adsorption density of the component *i* (mol.m^{-2}), *A* is the total surface area of particles, and n^s_i represents the number of entities *i* adsorbed under the chemical potential μ_i. γ is the surface energy or interfacial tension (solid–liquid interface). It is the energy required to reversibly and isothermically increase the surface area of a unit quantity [14, 26]. The surface tension of finely divided solids cannot be directly measured, at the difference of liquids, but it can be calculated [89–91]. For oxides, it varies from 50 to a few hundred mJ.m^{-2}. Recent works involving potentiometric measurements similar to concentration cells reported an estimated value of 0.35 J.m^{-2} for nanoparticles of nickel oxide [92].

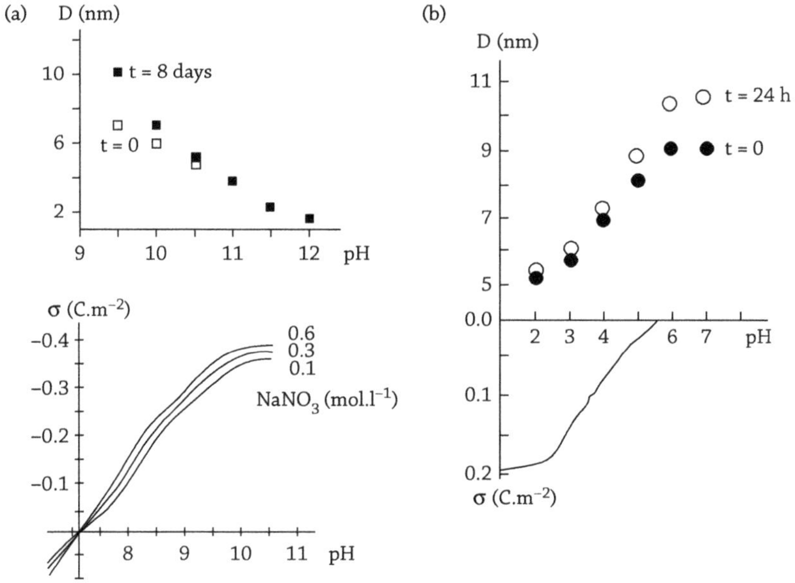

FIGURE 5.28 Effect of pH on the mean size and surface charge density of particles of (a) magnetite Fe_3O_4 and (b) TiO_2 anatase. The particle size of magnetite was measured just after precipitation and after aging for 8 days of suspension (25 °C, I = 0.5M). The surface charge density was determined by acid–base titration at various ionic strengths. The mean particle size of anatase was measured just after precipitation and after aging for 24 hours in suspension at 60°C.

In lowering the interfacial tension, adsorption allows the spontaneous extension of the interface. The phenomenon is easily observed with labile systems such as water–oil (toluene, cyclohexane) mixtures in which the change in dispersion state of a phase in another one does not involve the breaking of tight bonds. Water and oil are not miscible, and the two phases are separated by a planar interface of minimum area with an interfacial tension typically around 0.1 $J.m^{-2}$. The introduction into the mixture of an amphiphilic tensioactive agent such as sodium dodecylsulfate, which is adsorbed at the water–oil interface (the -SO_3^- polar head of the molecule is solvated by water, whereas the aliphatic chain interacts with oil), induces spontaneous and rapid emulsification. It results from the drop in interfacial tension up to around 10^{-3} to 10^{-4} $J.m^{-2}$ and from the ability of the interfacial area to extend without kinetic hindrance and with no increase in free enthalpy. That is the principle of microemulsion formation [14, 93]. With oxides, the effect of adsorption on already formed particles is not perceptible because the stability of the metal–oxygen bonds gives the crystal lattice a high chemical inertness. It is possible, however, to observe such an effect from the variation in size of particles precipitated at different pH and ionic strengths or in the presence of various additives (complexing anions, polymers, etc.). An abundance of articles in the literature report the use of more or less complexing ligands in reducing the particle size or favoring the growth of peculiar crystal faces. The acidity of the precipitation medium may play a role similar to that of complexation in control of size or the morphology of particles. This is the case, for instance, for magnetite, Fe_3O_4, and titanium dioxide, TiO_2 (Fig. 5.28), or also boehmite, γ-AlOOH (§ 6.3.1b). Acidity and/or ionic strength in the precipitation medium govern the protonation/deprotonation rate of oxide surfaces, and a variation in these factors modifies the chemical composition and the charge density of surfaces and thus the surface energy. This effect of thermodynamic origin can help control the particle size and even avoid Ostwald ripening [94–96]. It is possible to semiquantitatively evaluate the lowering in interfacial tension and to explain some morphological changes resulting exclusively from acidity [96].

Protonation of an oxide surface (pH < PZC) causes the variation in the oxide–solution interfacial tension as indicated by the Gibbs equation [5.17]:

$$d\gamma = -\Gamma_H d\mu_H \qquad [5.18]$$

In a similar way, the effect of deprotonation of the surface in basic medium (pH > PZC) is expressed as

$$d\gamma = +\Gamma_H d\mu_H \qquad [5.19]$$

(The + sign is due to desorption of protons. In basic medium, adsorption of hydroxyl ions could be considered, but these species are not really bound to the surface because they form water with the neutralized protons.)

To evaluate the density of surface-charged sites as a function of acidity, at minimum two different kinds of acid–base groups, M_1–OH and M_2–OH, are considered in the charge reversal on each side of the PZC (§ 5.1.2). The surface equilibria are written as

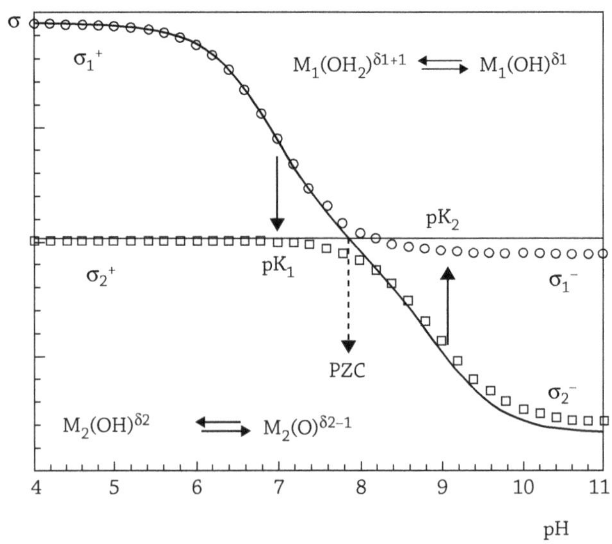

FIGURE 5.29 Surface charge density as a function of pH (plain line) for two surface acid–base couples.

$$(M_1 - OH_2)^{\delta_1 + 1} \Leftrightarrow (M_1 - OH)^{\delta_1} + H^+ \qquad [5.20]$$

$$(M_2 - OH)^{\delta_2} + OH^- \Leftrightarrow (M_2 - O)^{\delta_2 - 1} + H_2O \qquad [5.21]$$

The surface charge generated by the M_1 and M_2 sites is drawn in Figure 5.29. The net surface charge density is given as

$$\sigma_0 = (F/A)[(\delta_1 + 1)n_{M_1OH_2} + \delta_1 n_{M_1OH} + \delta_2 n_{M_2OH} + (\delta_2 - 1)n_{M_2O}] \qquad [5.22]$$

where n_x is the number of sites of x-type per area unit.

Consider the case of pH ≥ PZC and the pKs of equilibria [5.20] and [5.21] which are distinct enough, being different by two to three units of pH. For simplicity, consider the negative surface charge at pH > PZC due solely to deprotonation of the M_2–OH groups ($\delta_1 << 1$, as represented in Figure 5.29). The charge density is then given by

$$\sigma_0 = (F/A)[\delta_2 n_{M_2OH} + (\delta_2 - 1)n_{M_2O}] \qquad [5.23]$$

N_2 is the total number of active groups ($N_2 = n_{M_2OH} + n_{M_2O}$). The adsorption density, Γ_H, is

$$\Gamma_H = n_{M_2O}/A = (\delta_2 N_2/A) - (\sigma_0/F) = (1/F)(\sigma_2^+ - \sigma_0) \qquad [5.24]$$

where σ_2^+ is the maximal positive charge density acquired in acidic medium, at pH << pK_2, σ_2^- that was acquired at pH >> pK_2.

At equilibrium, the chemical potential of adsorbed protons, equal to that of protons in solution, is given by

$$\mu_H^{surface} = \mu_H^{solution} = \mu_0 + RT\ln a_H = \mu_0 + RT\ln[H^+] \qquad [5.25]$$

where a_H is the activity of proton in solution. From the constant of Eq. [5.21] we obtain:

$$[H^+] = K_2(n_{M_2OH}/n_{M_2O})\exp(F\psi_0/RT) \qquad [5.26]$$

where ψ_0 is the surface electrostatic potential.

From the relations $n_{M_2OH} = (A/F)(\sigma_0 - \sigma_2^-)$ and $n_{M_2O} = (A/F)(\sigma_0^+ - \sigma_0)$, we obtain:

$$[H^+] = K_2[(\sigma_0 - \sigma_2^-)/(\sigma_2^+ - \sigma_0)]\exp(F\psi_0/RT) \qquad [5.27]$$

and:

$$d\mu_H = RTd\ln[(\sigma_0 - \sigma_2^-)/(\sigma_2^+ - \sigma_0)] + Fd\psi_0 \qquad [5.28]$$

Then the introduction of the relations [4.24] and [4.28] into Eq. [4.19] gives :

$$d\gamma = [(\sigma_2^+ - \sigma_0)/F][Fd\psi_0 + RTd\ln[(\sigma_0 - \sigma_2^-)/(\sigma_2^+ - \sigma_0)]$$

$$= (\sigma_2^+ - \sigma_0)d\psi_0 + (RT/F)(\sigma_2^+ - \sigma_0)d\sigma_0/(\sigma_0 - \sigma_2^-) \qquad [5.29]$$

Integration of Eq. [5.29] requires a relation between surface potential and charge. As Nernst's law cannot be applied to oxides, Grahame's equation is used (Eq. [5.12]):

$$-\sigma_d = \sigma_0 = p\sinh(F\psi_0 / 2RT) \text{ with } p = 0.0131(\varepsilon I)^{1/2} \quad [5.30]$$

ε is the dielectric constant of water in the interfacial zone, and $I = (1/2)\sum c_i z_i^2$ is the ionic strength. By derivation of this equation, we obtain:

$$d\psi_0 = (2RT / F)d\sigma_0 / \sqrt{p^2 + \sigma_0^2} \quad [5.31]$$

Integration of Eq. [5.29] with $\gamma = \gamma_0$ for $\sigma_0 = 0$ (pH = PZC) gives:

$$\begin{aligned}\Delta\gamma &= \gamma - \gamma_0 \\ &= (RT / F)(2p + 2\sigma_2^+ \sinh^{-1}(\sigma_0 / p) - 2\sqrt{p^2 + \sigma_0^2} \\ &\quad + (\sigma_2^+ - \sigma_2^-)\ln[1 - (\sigma_0 / \sigma_2^-)]\end{aligned} \quad [5.32]$$

A similar expression is obtained for proton adsorption at pH < PZC when the surface is globally positively charged:

$$\begin{aligned}\Delta\gamma &= \gamma - \gamma_0 \\ &= (RT / F)(2p + 2\sigma_1^- \sinh^{-1}(\sigma_0 / p) - 2\sqrt{p^2 + \sigma_0^2} \\ &\quad + (\sigma_1^+ - \sigma_1^-)\ln[1 - (\sigma_0 / \sigma_1^+)]\end{aligned} \quad [5.33]$$

Equations [5.32] and [5.33] show that the interfacial tension is at maximum at the PZC, where the surface is not charged. Graphical representation for different crystal faces of AlOOH nanoparticles are drawn in Fig. 6.15b. It must be kept in mind that this approach is however simplified and that other factors should also

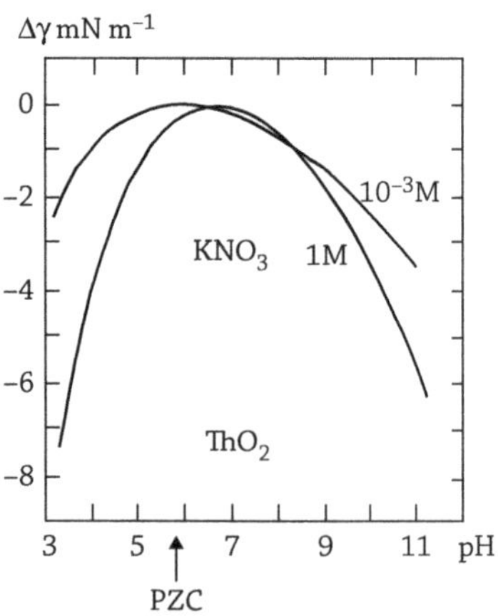

FIGURE 5.30 Variation in surface energy as a function of pH, calculated from the experimental σ = f(pH) curves at different ionic strengths for ThO_2 (reproduced with permission from [89] Copyright 1966 Canadian Science Publishing) and for Al_2O_3 (reproduced with permission from [90] Copyright 1969 American Chemical Society).

be taken into account, especially the strong solvation of the charged oxide surfaces and its variation with the charge density, the presence of counterions near the surface, the influence of the surface charge on the metal–oxygen bond length, the geometry of the surface, and its curvature radius. Unfortunately, it is practically impossible to evaluate such effects. Even so, the model allows clarification of the influence of the acidity as a major factor in the synthesis of oxide nanoparticles.

The *variation* in interfacial energy can also be more easily calculated from integration of the experimental $\sigma = f(\text{pH})$ curves at different ionic strengths (Fig. 5.30) [90, 91]. The variation appears to be similar to that calculated in using the above equations.

A strong decrease in the oxide–solution interfacial tension, which can reach zero or even become negative, is expected at high ionic strength and at a pH far from the PZC. (A negative surface tension should correspond to acidity conditions involving the oxide dissolution.) In such conditions, the decrease in particle size is closely correlated to the increase in surface charge density, as observed, for instance, for magnetite Fe_3O_4 [97] and anatase TiO_2 [98] (Fig. 5.28). In addition, when the surface is electrostatically saturated, Ostwald ripening, which causes the reduction in surface area by growth of particles (§ 4.2.3), no longer occurs. It should also be noted that the lowering of the surface tension has a kinetic consequence favoring the nucleation of the solid during precipitation (Eq. [4.4]) and contributing to the decrease in particle size.

The particles of magnetite and anatase formed under the conditions just described are isotropic. The size of their quasi-spherical shape is practically not modified because all the exposed faces have more or less the same physicochemical and energetic features. It is not the same for anisotropic particles, which have faces with different physicochemistries and energies. This effect is well evidenced with boehmite, γ-AlOOH, which forms platelets because of its two-dimensional crystal structure. According to the pH of formation, the particles exhibit not only different sizes, but also different ratios in the edge faces extension (§ 6.3.1b). The modeling of the acid–base properties of the different crystal faces of boehmite, in using Eqs. [5.32] and [5.33], helps us to evaluate the variations in energy of the different faces (Fig. 6.15) and to account for the morphological change of particles [95].

Thus, the precise adjustment of precipitation conditions (pH, nature, and concentration of the ionic strength) allows some control of the size and shape of particles over the nanorange. Such controlled conditions of synthesis may also involve the adsorption of ligands on specific particle faces (§ 6.3.1b) and of polymers, even if they are the more often less selectively adsorbed on oxides.

References

[1] Parks G.A. The isoelectric points of solid oxides, solid hydroxides and aqueous hydroxo complex systems. *Chem. Rev.* 1965, 65, 177–198.

[2] Micale F.J., Kiernan D., Zettlemoyer A.C. Characterization of the surface properties of iron oxides. *J. Colloid Interface Sci.* 1985, 105, 570–576.
[3] Pauling L. The principles determining structure of complex ionic crystal. *J. Am. Chem. Soc.* 1929, 51, 1010–1026.
[4] Hiemstra T., van Riemsdijk W.H., Bolt G.H. Multisite proton adsorption modeling at the solid/solution interface of (hydr)oxides: A new approach I. Model description .and evaluation of intrinsic reaction constants. *J. Colloid Interface Sci.* 1989, 133, 91–104.
[5] Hiemstra T., De Wit J.C.M., van Riemsdijk W.H. Multisite proton adsorption modeling at the solid/solution interface of (hydr)oxides: a new approach II. Application to various important (hydr)oxides. *J. Colloid Interface Sci.* 1989, 133, 105–117.
[6] Baes C.F., Mesmer R.E. *The Hydrolysis of Cations*. New York: Wiley, 1976.
[7] Phillips C.S.G., Williams R.J.P. *Inorganic chemistry, Vol.1*. London: Oxford University Press, 1965.
[8] Kawakami H., Yoshida S. Quantum-chemical studies of alumina. Part 2. Lewis acidity. *J. Chem. Soc. Faraday Trans II* 1985, 81, 1117–1127.
[9] Knözinger H., Ratnasamy P. Catalytic alumina surface models and characterization of surface sites. *Catal. Rev. Sci. Eng.* 1978, 17, 31–70.
[10] Healy T.W., White L.R. Ionizable surface group models of aqueous interface. *Adv. Coll. Interface Sci.* 1978, 9, 303–345.
[11] Hiemstra T., Venema P., van Riemsdijk W. H. Intrinsic proton affinity of reactive surface groups of metal (hydr)oxides: the bond valence principle. *J. Colloid Interface Sci.* 1996, 184, 680–692.
[12] Brown I.D., Altermatt D. Bond-valence parameters obtained from a systematic analysis of the inorganic crystal structure database. *Acta Cryst. B* 1985, 41, 244–247.
[13] Hiemenz P.C. *Principles of Colloid and Surface Chemistry*. New York: M. Dekker, 1977.
[14] Israelachvili J. *Intermolecular and Surface Forces*, 2nd ed. London: Academic Press, 1992.
[15] Evans D.F., Wennerström H. *The Colloidal Domain Where Physics, Chemistry, Biology and Technology Meet*. New York: VCH Publishers, 1994.
[16] Michot L.J., Villiéras F., François M., Bihannic I., Pelletier M., Cases J.M. Water organisation at the solid–aqueous solution interface. *C. R. Geoscience* 2002, 334, 611–631.
[17] Hair M.L., Hertl W. Adsorption on hydroxylated silica surfaces. *J. Phys. Chem.* 1969, 73, 4269–4276.
[18] Holmes H.F., Fuller E.L., Secoy C.H. Heats of immersion in the thorium oxide-water system. IV. Variation of the net differential heat of adsorption with specific surface area. *J. Phys. Chem.* 1968, 72, 2095–2100.
[19] McCafferty E., Zettlermoyer A.C. Adsorption of water vapor on α-Fe_2O_3. *Disc. Faraday Soc.* 1971, 52, 239–263.
[20] Bérubé Y.G., De Bruyn P.L. Adsorption at the rutile-solution interface. *J. Colloid Interface Sci.* 1968, 28, 92–105.
[21] Motta A., Gaigeot M.P., Costa D. Ab initio molecular dynamics study of the AlOOH boehmite/water interface: role of steps in interfacial Grotthus proton transfers. *J. Phys. Chem. C* 2012, 116, 12514–12524.
[22] Ninham B.W. Surface forces—The last 30 Å. *Pure Appl. Chem.* 1981, 53, 2135–2147.

[23] Lyklema J. Progress in interfacial chemistry in relation to colloidal stability. *Pure Appl. Chem.* 1981, 53, 2199–2209.

[24] Jang H.M., Fuerstenau D.W. The nature of simple monovalent cation-silica interaction as reflected in the spin-lattice relaxation time of ^{23}Na. *Langmuir* 1987, 3, 1114–1118.

[25] Doane T.L., Chuang C.-H., Hill R.J., Burda C. Nanoparticle ζ -potentials. *Acc. Chem. Res.* 2012, 45, 317–326.

[26] Hunter R.J. *Foundations of Colloid Science*. Vol.1. Oxford: Clarendon Press, 1987.

[27] Lyklema J. Points of zero charge in the presence of specific adsorption. *J. Colloid Interface Sci.* 1984, 99, 109–117.

[28] Shaw D.J. *Introduction to Colloid and Surface Chemistry*, 3rd ed. London Butterworths, 1980.

[29] Overbeek J. Th. Strong and weak points in the interpretation of colloidal stability. *Adv. Coll. Interf. Sci.* 1982, 16, 17–30.

[30] Overbeek J. Th. Recent developments in the understanding of colloid stability. *J. Colloid Interface Sci.* 1977, 58, 408–422.

[31] Healy T.W., Chan D., White L.R. Colloidal behavior of materials with ionizable group surfaces. *Pure Appl. Chem.* 1980, 52,1207–1219.

[32] Feke D.L., Prabhu N.D., Adin Mann Jr J., Adin Mann III J. A formulation of the short-range repulsion between spherical colloidal particles. *J. Phys. Chem.* 1984, 88, 5735–5739.

[33] Brinker C.J., Scherer G.W. *Sol-gel Science. The Physics and Chemistry of Sol-gel Processing*. San Diego: Academic Press, 1990.

[34] Derjaguin B.V. Some new aspects and conclusions on theory of stability of colloids and their experimental verification. *Disc. Faraday Soc.* 1978, 65, 306–312.

[35] Overbeek J. Th. The rule of Schulze and Hardy. *Pure Appl. Chem.* 1980, 52, 1151–1161.

[36] Frens G., Overbeek J. Th. Repeptization and the theory of electrocratic colloids. *J. Colloid Interface Sci.* 1972, 38, 376–387.

[37] Frens G. On coagulation in the primary minimum. *Discuss. Faraday Soc.* 1978, 65, 146–155.

[38] Schindler P.W., Stumm W. *Aquatic Surface Chemistry, Chemical Processes at the Particle-Water Interface*. W. Stumm, Ed. New York: Wiley, 1987.

[39] Haworth A. A review of the modelling of sorption from aqueous solution. *Adv. Colloid Interface Sci.* 1990, 32, 43–78.

[40] Fuerstenau D.W., K. Osseo-Asare. Adsorption of copper, nickel and cobalt by oxide adsorbents from aqueous ammoniacal solutions. *J. Colloid Interface Sci.* 1987, 118, 524–542.

[41] Bonneviot L., Clause O., Che M., Monceau A., Dexpert H. EXAFS characterization of the adsorption sites of nickel ammine and ethylenediamine complexes on a silica surface. *Catalysis Today* 1989, 6, 39–46.

[42] Bassetti V., Burlamacchi L., Martini G. Use of paramagnetic probes for the study of liquid adsorbed on porous supports. Copper(II) in water solution. *J. Am. Chem. Soc.* 1979, 101, 5471–5477.

[43] Bleam W.F., McBride M.B. The chemistry of adsorbed Cu(II) and Mn(II) in aqueous titanium dioxide suspensions. *J. Colloid Interface Sci.* 1986, 110, 335–346.

[44] Ardon M., Bino A. A new aspect of hydrolysis of metal ions: the hydrogen-oxide bridging ligand $H_3O_2^-$. Structure and Bonding 65, 1–28, Springer Verlag, Berlin, 1987.

[45] Che M., Bonneviot L. Role of oxide surface in coordination chemistry of transition metal ions in catalytic systems. *Pure Appl. Chem.* 1988, 60, 1369–1378.
[46] Schenk C.V., Dillard J.G., Murray J.W. Surface analysis and the adsorption of Co(ll) on goethite. *J. Colloid Interface Sci.* 1983, 95, 398–409.
[47] Ambe S., Ambe F. Mossbauer study of ferric ions adsorbed at α-ferric oxide/aqueous solution interface. *Langmuir* 1990, 6, 644–649.
[48] James R.O., Healy T.W. Adsorption of hydrolyzable metal ions at the oxide-water interface II. Charge reversal of SiO_2 and TiO_2 colloids by adsorbed Co(ll), La(lll), and Th(IV) as model systems. *J. Colloid Interface Sci.* 1972, 40, 53–64.
[49] Tewari P.H., Lee W. Adsorption of Co(II) at the oxide-water interface. *J. Colloid Interface Sci.* 1975, 52, 77–88.
[50] Blesa M.A., Weisz A.D., Morando P.J., Salfity J.A., Magaz G.E., Regazzoni A.E. The interaction of metal oxide surfaces with complexing agents dissolved in water. *Coord. Chem. Rev.* 2000, 196, 31–63.
[51] Parfitt R.L., Fraser A.R., Russel J.D., Farmer V.C. Adsorption on hydrous oxides. II. Oxalate, benzoate, and phosphate on gibbsite. *J. Soil Sci.* 1977, 28, 40–47.
[52] Parfitt, R. L., van der Gast, S. J., Childs, C. W. A structural model for natural siliceous ferrihydrite. *Clays Clay Miner.* 1992, 40, 675–681.
[53] Atkinson R.J., Parfitt R.L., Smart R.S.C. Infra-red study of phosphate adsorption on goethite. *J. Chem. Soc. Faraday Trans.* I 1974, 70, 1472–1479.
[54] Cradwick P.D.G., Farmer V.C., Russel J.D., Masson C.R., Wada K., Yoshinaga N. Imogolite, a hydrated aluminium silicate of tubular structure. *Nature Phys. Sci.* 1972, 240, 187–189.
[55] Stumm W., Wehrli B., Wieland E. Surface complexation and its impact on geochemical kinetics. *Croat. Chem. Acta* 1987, 60, 429–456.
[56] Chang H.C., Matijevic E. Interactions of metal hydrous oxides with chelating agents IV. Dissolution of hematite. *J. Coll. Interface Sci.* 1983, 92, 479–488.
[57] Tronc E., Jolivet J.P. Interactions between charged particles. In *Magnetic Properties of Fine Particles*, Dormann J.L., Fiorani D., Eds. Amsterdam: North Holland, 1992.
[58] Hernandez J. Synthèse de nanoparticules d'oxydes de fer et d'aluminium pour l'étude de l'adsorption d'entités inorganiques polycondensées. Thèse, Université Pierre et Marie Curie, Paris, France, 1998.
[59] Lee L.T., Somasundaran P. Adsorption of polyacrylamide on oxide minerals. *Langmuir* 1989, 5, 854–860.
[60] Dickinson E., Ericsson L. Particle flocculation by adsorbing polymers. *Adv. Coll. Interface Sci.* 1991, 34, 1–29.
[61] Cohen Stuart M.A., Fleer G.J., Lyklema J., Norde W., Scheutjens J.M.H.M. Adsorption of ions, polyelectrolytes and proteins. *Adv. Coll. Interface Sci.* 1991, 34, 477–535.
[62] Wang T.K., Audebert R. Adsorption of cationic copolymers of acrylamide at the silica-water interface: hydrodynamic layer thickness measurements. *J. Colloid Interface Sci.* 1991, 121, 32–41.
[63] Laible R., Hamann K. Formation of chemically bound polymer layers on oxide surfaces and their role in colloid stability. *Adv. Coll. Interface Sci.* 1980, 13, 65–99.
[64] Napper D.H. *Polymeric Stabilization of Colloidal Dispersions*. London: Academic Press, 1983.

[65] De Gennes P.G. Polymers at an interface; a simplified view. *Adv. Coll. Interface Sci.* 1987, 27, 189–209.
[66] Mulvaney P., Cooper R., Grieser F., Meisel D. Charge trapping in the reductive dissolution of colloidal suspensions of iron(III) oxides. *Langmuir* 1988, 4, 1206–1211.
[67] Werhli B., Sulzberger B., Stumm W. Redox processes catalysed by hydrous oxide surfaces. *Chem. Geol.* 1989, 78, 167–179.
[68] Silvester E., Charlet L., Tournassat C., Géhin A., Grenèche J.M., Liger E. Redox potential measurements and Mössbauer spectrometry of FeII adsorbed onto FeIII (oxyhydr) oxides. *Geochim. Cosmochim. Acta* 2005, 69, 4801–4815.
[69] Williams A.G.B., Scherer M.M. Spectroscopic evidence for Fe(II)-Fe(III) electron transfer at the iron oxide-water interface. *Environ. Sci. Technol.* 2004, 38, 4782–4790.
[70] Jeon B.H., Dempsey B.A., Burgos W.D. Kinetics and mechanisms for reactions of Fe(II) with iron(III) oxides. *Environ. Sci. Technol.* 2003, 37, 3309–3315.
[71] Gorski C.A., Scherer M.M. Fe^{2+} Sorption at the Fe oxide-water interface: a revised conceptual framework, 315–343. In *Aquatic Redox Chemistry*, Tratnyek P. et al. Eds. ACS Symposium Series. Washington, DC: American Chemical Society, 2011.
[72] Jolivet J.P., Belleville P., Tronc E., Livage J. Influence of Fe(II) ions on the formation of spinel iron oxide in aqueous medium. *Clays Clay Miner.* 1992, 40, 531–540.
[73] Pedersen H.D., Postma D., Jakobsen R., Larsen O. Fast transformation of iron oxyhydroxides by the catalytic action of aqueous Fe(II). *Geochim. Cosmochim. Acta* 2005, 69, 3967–3977.
[74] Yanina S.V., Rosso K.M. Linked reactivity at mineral-water interfaces through bulk crystal conduction. *Science* 2008, 320, 218–222.
[75] Rosso K.M., Yanina S.V., Gorski C.A., Larese-Casanova Ph., Scherer M.M. Connecting observations of hematite (α-Fe_2O_3) growth catalyzed by Fe(II). *Environ. Sci. Technol.* 2010, 44, 61–67.
[76] Larese-Casanova Ph., Scherer M.M. Fe(II) sorption on hematite: new insights based on spectroscopic measurements. *Environ. Sci. Technol.* 2007, 41, 471–477.
[77] Catalano J.G., Fenter P., Park C., Zhang Z., Rosso K.M. Structure and oxidation state of hematite surfaces reacted with aqueous Fe(II) at acidic and neutral pH. *Geochim. Cosmochim. Acta* 2010, 74, 1498–1512.
[78] Sherman D.M. Molecular orbital (SCF-Xa-SW) theory of metal-metal charge transfer processes in minerals II. Application to $Fe^{2+} \rightarrow Ti^{4+}$ charge transfer transitions in oxides and silicates. *Phys. Chem. Minerals* 1987, 14, 364–367.
[79] Handler R.M., Beard B.L., Johnson C.M., Scherer M.M. Atom exchange between aqueous Fe(II) and goethite: an Fe isotope tracer study. *Environ. Sci. Technol.* 2009, 43, 1102–1107.
[80] Deng B., Stone A.T. Surface-catalyzed chromium(VI) reduction: reactivity comparisons of different organic reductants and different oxide surfaces. *Environ. Sci. Technol.* 1996, 30, 2484–2494.
[81] Amstaetter K., Borsch T., Larese-Casanova Ph., Kappler A. Redox transformation of arsenic by Fe(II)-activated goethite (α-FeOOH). *Environ. Sci. Technol.* 2010, 44, 102–108.

[82] Madden A.S., Hochella Jr. M. F. A test of geochemical reactivity as a function of mineral size: Manganese oxidation promoted by hematite nanoparticles. *Geochim. Cosmochim. Acta* 2005, 69, 389–398.
[83] Madden A.S., Hochella Jr. M. F., Luxton T. P. Insights for size-dependent reactivity of hematite nanomineral surfaces through Cu^{2+} sorption. *Geochim. Cosmochim. Acta* 2006, 70, 4095–4104.
[84] Junta J.L., Hochella Jr M.F. Manganese (II) oxidation at mineral surfaces: A microscopic and spectroscopic study. *Geochim. Cosmochim. Acta* 1994, 58, 4985–4999.
[85] Noguera C., Pojani A., Casek P., Finocchi F. Electron redistribution in low-dimensional oxide structures. *Surf. Sci.* 2002, 507–510, 245–255.
[86] Chernyshova I. V., Ponnurangam S., Somasundaran P. Effect of nanosize on catalytic properties of ferric (hydr)oxides in water: Mechanistic insights. *J. Catalysis* 2011, 282, 25–34.
[87] Chernyshova I. V., Ponnurangam S., Somasundaran P. Tailoring (bio)chemical activity of semiconducting nanoparticles: Critical role of deposition and aggregation. *J. Am. Chem. Soc.* 2011, 133, 9536–9544.
[88] Chernyshova I. V., Ponnurangam S., Somasundaran P. On the origin of an unusual dependence of (bio)chemical reactivity of ferric hydroxides on nanoparticle size. *Phys. Chem. Chem. Phys.* 2010, 12, 14045–14056.
[89] Dalal E.N. Calculation of solid surface tensions. Langmuir 1987, 3, 1009–1015.
[90] Ahmed S.M. Studies of the dissociation of oxide surfaces at the liquid-solid interface. *Can. J. Chem.* 1966, 44, 1663–1670.
[91] Ahmed S.M. Studies of the double layer at oxide-solution interface. *J. Phys. Chem.* 1969, 73, 3546–3555.
[92] Croué K., Jolivet J.P., Larcher D. Direct determination of oxide surface free energy through potentiometric measurements. Electrochem. *Solid-State Lett.* 2012, 15, F8-F11.
[93] Aveyard R. Ultralow tensions and microemulsions. *Chem. Ind.* 1987, 14, 474–478.
[94] Overbeek J. Th. Microemulsions, a field at the border between lyophobic and lyophilic colloids. *Faraday Disc. Chem. Soc.* 1978, 65, 7–19.
[95] Stol R.J., De Bruyn P.L. Thermodynamic stabilization of colloids. *J. Colloid Interface Sci.* 1980, 75, 185–198.
[96] Jolivet J.P., Froidefond C., Pottier A., Chanéac C., Cassaignon S., Tronc E., Euzen P. Size tailoring of oxide nanoparticles by precipitation in aqueous medium. A semi-quantitative modeling. *J. Mater. Chem.* 2004, 14, 3281–3288.
[97] Vayssières L., Chanéac C., Tronc E., Jolivet J. P. Size tailoring of magnetite particles formed by aqueous precipitation an example of thermodynamic stability of nanometric oxide particles. *J. Colloid Interface Sci.* 1998, 205, 205–212.
[98] Pottier A., Cassaignon S., Chanéac C., Villain F., Tronc E., Jolivet J.P. Size tailoring of TiO_2 anatase nanoparticles in aqueous medium and synthesis of nanocomposites. Characterization by Raman spectroscopy. *J. Mater. Chem.* 2003, 13, 877–882.

6 }

Aluminum Oxides

ALUMINA AND ALUMINOSILICATES

Aluminum is the third most abundant element in Earth's crust (8.3% in mass), behind oxygen (45.5%) and silicon (27.2%). It forms in nature various oxygenated mineral phases: hydroxides $Al(OH)_3$, oxyhydroxides AlOOH, of which bauxite is the main ore, and oxides, Al_2O_3, alumina. Corundum, α-Al_2O_3, is the component of many gems: sapphire (pure Al_2O_3, perfectly colorless), ruby (red colored due to the presence of Cr^{3+} ions), and blue sapphire (blue colored by the presence of Ti^{4+} and Fe^{2+} ions), among many others. The content of foreign elements substituted for Al^{3+} ions in these phases accounts for only a small percentage of the total. Aluminum also forms many natural phases in combination with various elements, especially silicon in aluminosilicates, such as feldspars, clays, zeolites, allophanes, and imogolites. The biochemical cycling of the elements involves many soluble complexes of aluminum in natural waters [1, 2].

6.1 Introduction

Aluminum oxides and oxy(hydroxi)des are important materials and nanomaterials used in many fields: for instance, as active phase for adsorption in water treatment; as inert support and active phase in catalysis; as active phase in flame-retardant polymers; as refractory material for laboratory tools and in the ceramics industry; and as abrasives [3, 4]. Alumina Al_2O_3 is produced in various forms (tubes, balls, fibers, and powders) for numerous industrial uses (laboratory tools, filtration membranes, ball bearings, fine powders as catalysis supports, etc.).

The structural chemistry of aluminum oxy(hydroxi)des is rich. There are various hydroxides, $Al(OH)_3$ (gibbsite, also named hydrargillite, bayerite, and some other polytypes such as nordstrandite and doyleite), oxyhydroxides, AlOOH (boehmite and diaspore), and a series of oxides, Al_2O_3, so-called transition aluminas. These last phases have different degrees of hydration and different degrees of order of the Al^{3+} cations within the cubic close packing of oxygen atoms according to the temperature at which they have been submitted. They belong to various structural types (γ, δ, θ, η, κ, etc.). These aluminas of huge specific surface areas are usually used in catalysis, especially γ-alumina of spinel crystal structure. α-alumina (corundum) is the most thermodynamically stable phase among aluminum oxides. Its structure is based on a hexagonal closest packing of oxygen atoms. β-aluminas are based on a structure that is closely related to that of spinel stabilized by incorporation of alkali ions such as Na^+ or K^+ [5].

This chapter presents the main features of the chemistry of aluminum in aqueous solution involving condensation phenomena and the formation of polycations and nanoparticles, including the influence of many factors on the size and shape of particles. The formation of aluminosilicates is also briefly examined.

6.2 Hydroxylation and Condensation in Solution: Polycations

In acidic medium, Al^{3+} ions form the octahedral aquocomplex $[Al(OH_2)_6]^{3+}$. Because of polarization by the cation of the coordinated water molecules, the complex behaves as an acid and forms a series of hydroxylated complexes $[Al(OH)_h(OH_2)_{6-h}]^{(3-h)+}$, with h varying up to 4 according to the pH of the medium (Fig. 6.1a) [6].

The hexaaquo complex $[Al(OH_2)_6]^{3+}$, which is stable up to pH ~ 4, is less acid than its ferric homologue. With the same charge (+3) and in spite of a larger size ($r_{Fe3+} = 0.64$Å *vs* $r_{Al3+} = 0.53$Å [7]), however, the ferric ions are more strongly polarizing because iron has higher electronegativity than aluminum (1.72 vs. 1.47 on Allred and Rochow's scale). In the aluminate complex corresponding to $h = 4$, $[Al(OH)_4]^-$, which is formed in strongly alkaline medium, aluminum is tetracoordinated, as is clearly shown by the ^{27}Al-NMR at 80 ppm (Fig. 6.1b). The reduction in coordination

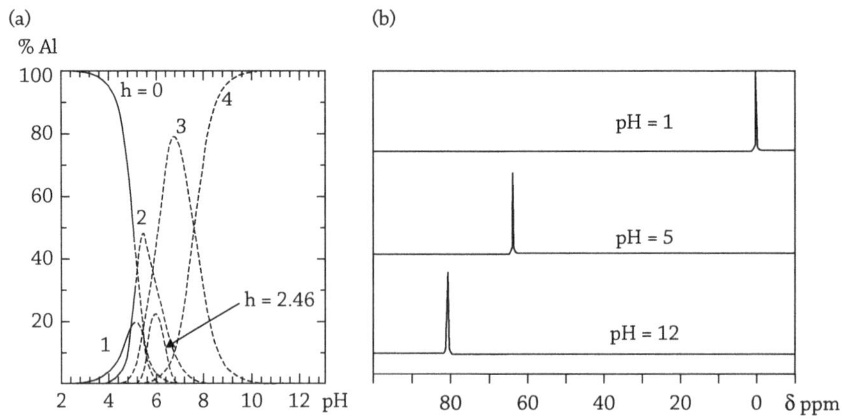

FIGURE 6.1 (a) Speciation diagram of aluminum in dilute solution (adapted from [6]). The dashed lines correspond to supersaturation with respect to gibbsite α-$Al(OH)_3$. (b) ^{27}Al-NMR spectra in solution of some complexes of aluminum: $[Al(OH_2)_6]^{3+}$ at pH 1, $[Al_{13}O_4(OH)_{24}(OH_2)_{12}]^{7+}$ ("ε-Al_{13}") at pH 5 and $[Al(OH)_4]^-$ at pH 12.

of Al^{3+} ion, which occurs with the increase in the hydroxylation rate, can be explained by its relatively small size and the increase of the hydroxo ligand to metal electron transfer. The electrophilic demand of the cation and consequently the polarization of water molecules in the hypothetical hexacoordinated complex $[Al(OH)_4(OH_2)_2]^-$ are thus lowered, so that the coordinated water molecules become solvation molecules (§ 2.2, Table 2.3). Al^{3+} ions have also tended to easily adopt the tetrahedral coordination in solids such as mixed oxides with the spinel structure, particularly in solids formed in alkaline medium such as zeolites and clays. Ferric ions can also adopt the tetrahedral coordination in some solids, for instance, magnetite and maghemite, but less extensively. In solution, the tetrahedral complex $[FeO_4]^{5-}$ exists only in strongly alkaline medium (Chap. 7).

The cationic hydroxylated complexes of aluminum exhibit a marked tendency to condense in forming soluble polycations by olation. In contrast, aluminate $[Al(OH)_4]^-$, as a tetrahydroxo anionic complex, does not easily condense by oxolation, although the hydroxo ligands are highly nucleophilic (basic). The cation aluminum, strongly stabilized in its hydroxo neighborhood by intense ligand to metal transfers, does not suffer nucleophilic attack, and the complex remains monomer, except at very high concentration ($\approx$ 6 mol.l^{-1}) where a dimer has been identified by infrared, Raman, and ^{27}Al-NMR spectroscopies [8].

Progressive addition of a base to a solution of aluminum chloride or nitrate at room temperature over the range $0 < h \leq 2.46$ (pH $\leq$ 4) forms aquohydroxo complexes that condense by edge sharing of octahedra in forming a series of polycations (Fig. 6.2a) [9, 10].

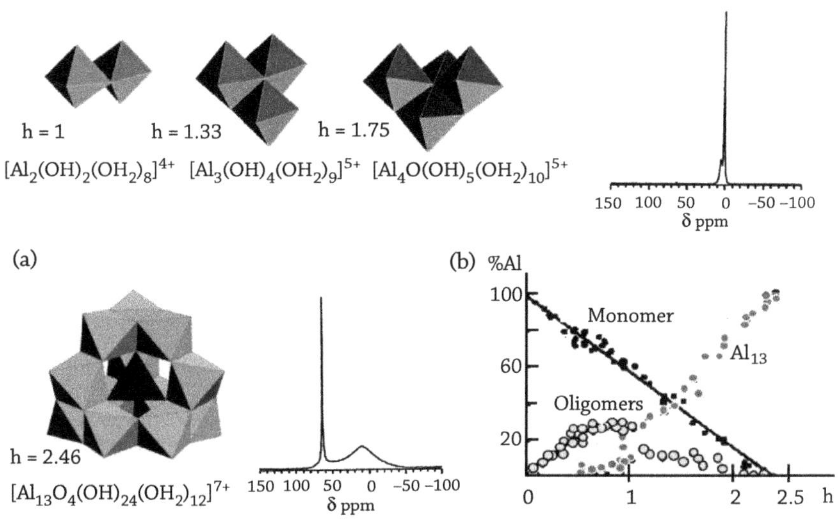

FIGURE 6.2 (a) Structures of polycations of aluminum with corresponding ^{27}Al-NMR spectra in solution (reproduced from [10] with permission of The Royal Society of Chemistry); (b) distribution of species in solution (reproduced from [9] with permission of The Royal Society of Chemistry).

A dimer, a trimer, and a tetramer corresponding to the formal values of $h = 1$, 1.33, and 1.76, respectively, are together characterized by ^{27}Al-NMR resonance around 4 ppm. The relative amount of these oligomers varies with the hydroxylation rate and is at maximum around $h = 1$ (Fig. 6.2b). Only the trimer $[Al_3(OH)_4(OH_2)_{12}]^{5+}$ has been characterized by potentiometric titration [11]. ^{1}H, ^{17}O, and ^{27}Al-NMR studies have shown that the di-μ_2 hydroxo dimer does not form in any detectable amount after neutralization of acid solutions [9, 10]. A sulfate of the dimer, characterized by X-ray diffraction, crystallizes from sulfuric solutions [12], but ^{27}Al-NMR shows that after dissolution of crystals in pure water, the dimer disappears in forming the monomer $[Al(OH_2)_6]^{3+}$ and a trimeric polycation of unknown structure [9].

Beyond $h = 1$, the amount of these oligomers decreases to the benefit of the polycation $[Al_{13}O_4(OH)_{24}(OH_2)_{12}]^{7+}$ ("ε-Al_{13}") (Fig. 6.2b). This species, globally spherical in shape, is made of four trimers linked by edges, including a thirteenth aluminum atom in tetrahedral coordination. It corresponds to the ε isomer of the so-called Keggin's structure (§ 3.3.2) (Fig. 6.2a) [9, 13]. This polycation, quantitatively formed for $h = 2.46$, has a signature at 63 ppm on the ^{27}Al-NMR spectrum due to the tetrahedral central aluminum atom. Another rather wide peak at around 9 ppm is characteristic of the strongly distorted 6-coordinated aluminum atoms constituting the shell of the polycation. The areas of these peaks are in the 12 to 1 ratio. Such M_{13} polycation is only formed by Al^{3+} and Ga^{3+} ions, which easily adopt tetra- and octahedral coordination [14, 15]. Analogous structures with Fe^{3+} and Cr^{3+} cations can be formed but never in similar physicochemical conditions [16]. The Fe_{13} cluster has been synthesized in the presence of strongly complexing ligands [17] and in nonaqueous medium [18]. The peculiar behavior of Al^{3+} ion in comparison with 3*d* elements like Cr^{3+} and Fe^{3+} by which it forms such a polycation can be explained by the electronic structure of these three elements (§ 4.1.3).

The formation of the Al_{13} polycation can be understood from the planar trimeric species $[Al_3(OH)_4(OH_2)_9]^{5+}$ reacting with the hexaaquo monomer, which coexists in solution at this hydroxylation rate. The trimeric structure is encountered in many polycondensed species because of its geometry, which allows minimizing electrostatic repulsions between the cations. It may constitute an intermediate species for subsequent condensation. In this species, the oxygen atom of the μ_3–OH bridge establishes three bonds with the underlayer cations and the fourth bond with a proton. As the oxygen atom of the μ_3–OH bridge is more highly polarized by three small Al^{3+} cations than by those of the μ_2–OH bridges, the μ_3–OH bridge is the most acidic and dissociates as

$$\left[Al_3(OH)_4(OH_2)_9\right]^{5+} + H_2O \rightarrow \left[Al_3O(OH)_3(OH_2)_9\right]^{4+} + H_3O^+$$

The polarization effect also makes the water molecules in the *cis* position from the μ_3–O bridge more acidic than the water molecules in the *trans* position (Fig. 6.3a). The formation of the $[H_3O_2]^-$ ligand can thus induce in moderated hydrolysis conditions ($h < 2.6$) the deprotonation of *cis* aquo ligands according to

$$\left[Al_3O(OH)_3(OH_2)_9\right]^{4+} + 3\,OH^- \rightarrow \left[Al_3O(OH)_3(O_2H_3)_3(OH_2)_9\right]^{+} + 3H_2O$$

The μ_3–O bridge in the trimer is then able to act as a nucleophilic ligand toward a coexisting monomer $h = 0$ in solution, but, owing to steric hindrance, the central aluminum atom adopts the tetrahedral coordination in the complex:

$$\left[Al(OH_2)_6\right]^{3+} + 4\left[Al_3O(OH)_3(O_2H_3)_3(OH_2)_9\right]^{+} \rightarrow$$
$$\left[Al\{Al_3O(OH)_3(O_2H_3)_3(OH_2)_{3+}\}_4\right]^{7+} + 6\,H_2O$$

The four coordinated trimers to the central aluminum atom can then condense between them by olation with elimination of the *cis*-water molecules:

$$\left[Al\left\{Al_3O(OH)_6(OH_2)_6\right\}_4\right]^{7+} \rightarrow \left[Al_{13}O_4(OH)_{24}(OH_2)_{12}\right]^{7+} + 12H_2O$$

This last step helps ensure the stability of the polycation. It is favored by heating at 80°C of the aluminum solution hydrolyzed at $h = 2.6$. It is indeed in these conditions that the ^{27}Al-NMR signals characteristic of the polycation are detected [10]. The hexacoordinated aluminum ions of the shell are strongly distorted, as indicated by the wide signal observed in the MAS-NMR spectrum of crystals of the sulfate salt of the polycation (Fig. 6.3b, c). This signal is very different from that observed on spectra of species in solution.

This pathway appears to be in accordance with ^{1}H, ^{17}O, and ^{27}Al-NMR studies of moderately hydrolyzed aluminum solutions [9]. These studies show that only small oligomers, likely trimers, and the monomer directly condense without the

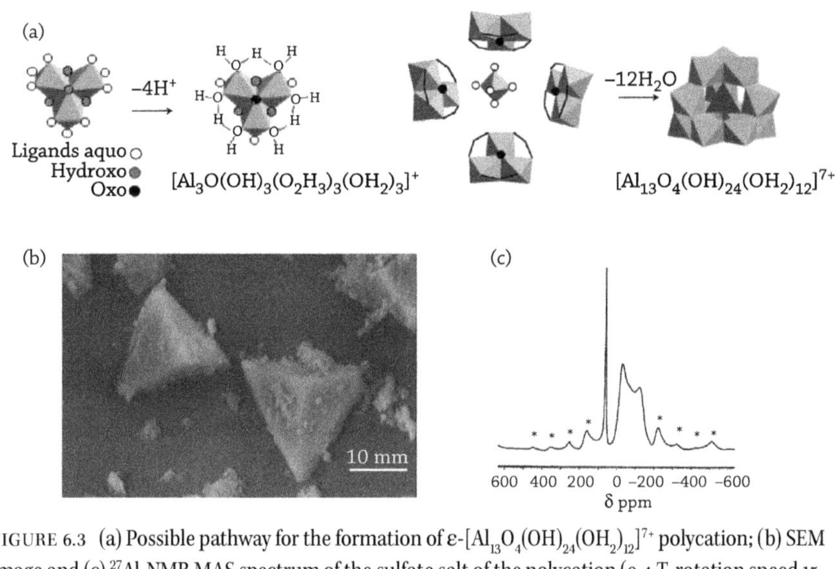

FIGURE 6.3 (a) Possible pathway for the formation of ε-$[Al_{13}O_4(OH)_{24}(OH_2)_{12}]^{7+}$ polycation; (b) SEM image and (c) ^{27}Al-NMR MAS spectrum of the sulfate salt of the polycation (9.4 T, rotation speed 15 kHz; *rotation peaks).

appearance of other intermediate species during Al_{13} formation. Indeed, the speciation diagram shows that the hexaaquo complex coexists with oligomers on the whole hydroxylation range. Formation of the isostructural Ga_{13} polycation studied in using ^{71}Ga-NMR and EXAFS at Ga K-edge during hydroxylation of gallium chloride or nitrate seems to occur according to a slightly different mechanism [19]. The temporal and structural evolution of Ga species in solution clearly indicates that Ga_{13} is built up from direct association of three planar trimers and a tetramer. Such a mechanism appears to be in accordance with the speciation of gallium. So, the two cations behave differently in solution because gallium, which is much more electronegative than aluminum (1.82 vs. 1.47 on Allred and Rochow's scale), is more easily hydrolyzed. The $[Ga(OH_2)_6]^{3+}$ complex ceases to exist beyond $h = 1$ [6, 19], whereas the aluminum hexaaquo complex is still present in solutions beyond $h = 2$. That explains why Ga_{13} forms only from oligomers existing in solution.

The oxygenated sites in the ε-Al_{13} polycation exhibit quite different reactivities to oxygen exchange with ^{17}O. This is the finding of the ^{17}O-NMR study of dissolution of the sulfate salt of the polycation in ^{17}O-enriched water in the presence of $BaCl_2$ [20, 21]. The 12 water molecules are instantaneously exchangeable (characteristic exchange time is $\tau_{298} \approx 10^{-3}$ s): the corresponding peak on the ^{17}O-NMR spectrum (20 ppm) appears instantaneously, and intensity remains constant. Conversely, the μ_4–O oxygen atoms are inert to substitution. Exchange occurs only if the polycation is destroyed by acidification in ^{17}O-enriched water and reformed by addition of base. The characteristic peak is then located at 55 ppm. Exchange on the μ_2–OH bridges is characterized by a peak at 35 ppm, with increasing intensity according to a distinctly bi-exponential rate. This peak was assigned to two resonances nonresolved from one another and corresponding to two sets of kinetically distinct μ_2–OH bridges (Fig. 6.4a). The 12 μ_2–OH^b bridges in the trimer groups are slowly exchangeable ($\tau_{298} \approx 17$ h), and the 12 others, μ_2–OH^a, which link the trimer groups, are rapidly exchangeable ($\tau_{298} \approx 1$mn). As the four μ_4–O sites remain unexchanged, the exchange on all other oxygenated groups occurs with no complete dissociation of the polycation according to the scheme shown in Figure 6.4.

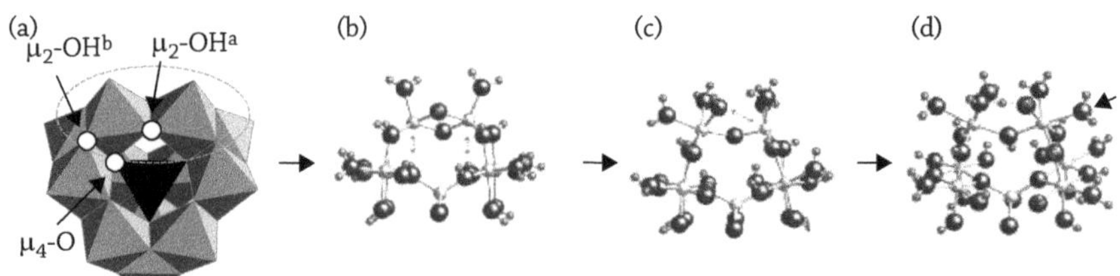

FIGURE 6.4 Stable form of the ε-Al_{13} polycation (a) and steps of the isotopic exchange of oxygen of the μ_2–OH^a bridges shown in the zone limited by dotted line: (b) dissociation of bonds with μ_4–O bridge; (c) μ_2–OH bridge rupture; (d) $H_3O_2^-$ bridge formation with a bulk water molecule. Dehydration of the bridge leads to exchange, with a probability of 50% (reproduced with permission from [21] Copyright 2006 American Chemical Society).

Some other polycationic species stem from heating at 80°C or more during long periods of time the ε-Al_{13} solution or aluminum chloride solutions, which are partially hydrolyzed with NaOH ($2.25 \le h \le 2.5$).

Thermolysis of an $h = 2.25$ $AlCl_3$ solution at 80°C for several days provokes the isomerization of ε-Al_{13} into δ-Al_{13}. This form is the δ-isomer of Keggin's structure in which three trimeric groups are connected by edge sharing and the fourth, having rotated by 60°, is connected by corner sharing (Fig. 3.28 and Fig. 6.5a). In the crystallized sulfate salt, $Na[Al_{13}O_4(OH)_{24}(OH_2)_{12}](SO_4)_4{\cdot}19H_2O$, the δ-$Al_{13}$ units seem to be stabilized thanks to the coordination of a Na^+ ion by three bridging OH ligands from the trimeric group having rotate and by three water molecules in forming an extremely deformed octahedral polyhedron. The ^{27}Al-NMR resonance around 64.5 ppm is the signature of the δ-Al_{13} unit in solution [22, 23]. In fact, the main species in solution in these conditions is a bigger polycation, $[Al_{30}O_8(OH)_{56}(H_2O)_{24}]^{18+}$ ("Al_{30}") (Fig. 6.5b). This species, whose existence had been suspected for a long time [22], has now been structurally characterized [23, 24].

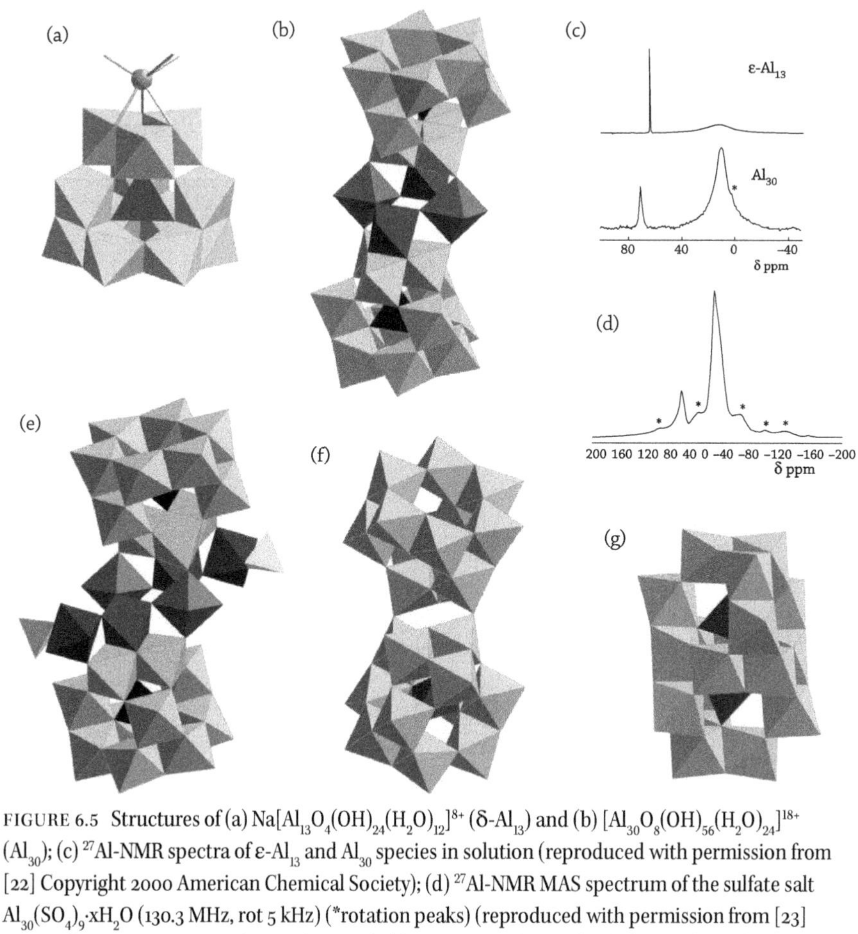

FIGURE 6.5 Structures of (a) $Na[Al_{13}O_4(OH)_{24}(H_2O)_{12}]^{8+}$ (δ-Al_{13}) and (b) $[Al_{30}O_8(OH)_{56}(H_2O)_{24}]^{18+}$ (Al_{30}); (c) ^{27}Al-NMR spectra of ε-Al_{13} and Al_{30} species in solution (reproduced with permission from [22] Copyright 2000 American Chemical Society); (d) ^{27}Al-NMR MAS spectrum of the sulfate salt $Al_{30}(SO_4)_9{\cdot}xH_2O$ (130.3 MHz, rot 5 kHz) (*rotation peaks) (reproduced with permission from [23] Copyright 2000 American Chemical Society); (e) structure of $[Al_{32}O_8(OH)_{60}(H_2O)_{28}(SO_4)_2]^{16+}$ (Al_{32}); (f) structure of $[Al_{26}O_8(OH)_{50}(H_2O)_{20}]^{12+}$ (Al_{26}); (g) structure of $[Ga_2Al_{18}O_8(OH)_{36}(H_2O)_{12}]^{8+}$ (Ga_2Al_{18}).

This polycation is made up of two δ-Al_{13} moieties connected by a ring of four AlO_6 octahedra linked between them by corners. Two of these four monomers of the central ring share three edges with the groups having rotate in each δ-Al_{13} moities, while the two others monomers of the central ring share two corners with the two δ-Al_{13} moities. (Fig. 6.5b). Al_{30} has a characteristic ^{27}Al-NMR at 70 ppm, corresponding to the two tetrahedrally coordinated aluminum ions (Fig. 6.5c, d). The formation of Al_{30} (h = 2.25) by thermolysis of the ε-Al_{13} solution (h = 2.45) seems to result from the acidification of the medium caused by thermolysis. That may explain why partial decondensation occurs transitorily, yielding some monomers and simultaneously allowing the rearrangement of the polycation in preserving the tetrahedral sites during the process. This mechanism has been established by ^{27}Al-NMR during the kinetics of transformation of an Al_{13} solution by heating [22, 25, 26] (Fig. 6.6). The fact that crystals of δ-Al_{13} have been observed as a by-product of Al_{30} synthesis [23] also supports this formation pathway.

Another polycation, Al_{32}, was obtained from thermolysis at 70°C for 36 hours of an $AlCl_3$ solution partially neutralized with NaOH (h = 2.5) and then heated at 150°C for 8 hours in autoclave. After addition of sodium sulfate (SO_4/Al = 0.30) and aging 1 week at room temperature, crystals of $[Al_{32}O_8(OH)_{60}(H_2O)_{28}(SO_4)_2]^{16+}$ were obtained [27]. The structure is similar to that of Al_{30}, with two supplementary $[Al(OH)_2(OH_2)_3(SO_4)]^-$ groups belonging to the central ring linking the two Al_{13} units (Fig. 6.5e). The same Al_{32} polycation may also be obtained in which sulfate groups are replaced by the bidentate iminodiacetate ligand [28].

The δ-Al_{13} unit may dimerize during crystallization in the presence of 2,6-naphtalene disulfonate (2,6-NDS) to form an Al_{26} species [29] (Fig. 6.5f). The crystals are obtained by heating at 100°C for 16 hours a 0.25 mol.l^{-1} $AlCl_3$ (h = 2.4) solution preliminarily heated at 80°C and then cooled at room temperature. After cooling, a solution of 2,6-NDS is added, and the mixture is left undisturbed for 21 days. A few crystals of $[Al_{26}O_8(OH)_{50}(OH_2)_{20}\cdot(2,6\text{-NDS})_6\cdot(H_2O)_{12.4}]$ are obtained. In the polycation, the two δ-Al_{13} units are linked by corner sharing through the

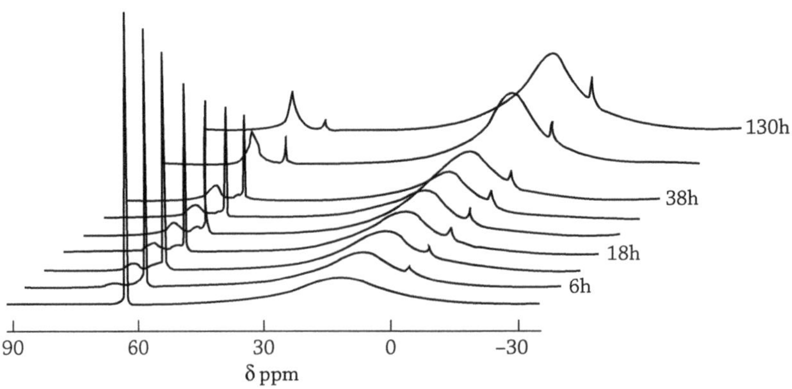

FIGURE 6.6 ^{27}Al-NMR spectra (130.3 MHz) showing thermal evolution of a solution of $Al_{13}Cl_7$ (0.035 mol.l^{-1}) at 85°C (reproduced with permission from [22] Copyright 2000 American Chemical Society).

rotated trimers of each unit. In the crystal, the Al_{26} polycations are arranged into two-dimensional sheets, their charge being balanced by 2,6-NDS anions. This organosulfonate molecule appears to be an interesting crystallization controller because of the affinity of the sulfonate groups to aluminum polycations and of π-π intermolecular interactions between the naphthalene rings.

Surprisingly and interestingly, too, a germanium–aluminum heteropolycation Ga_2Al_{18}, $[Ga_2Al_{18}O_8(OH)_{36}(H_2O)_{12}]^{8+}$ (Fig. 6.5g), similar to P_2W_{18}, the Dawson's polyoxometalate (Fig. 3.31), has been synthesized by heating at 80°C for 24 hours the partially hydrolyzed Ga-$AlCl_3$ aqueous mixture ((Ga/Al = 1/2) [30]. Crystallization occurred in the presence of 2,7-NDS by evaporation at room temperature. The polycation is made of fragments of ε-Al_{13} in which Ga is tetrahedrally coordinated, at the difference of the P_2W_{18} Dawson's structure, which is based on α-Keggin's fragments. Even if the Ga^{3+} ion considerably stabilizes the ε-$GaAl_{12}$ structure in avoiding its transformation into equivalent Al_{30} [21], through heating, it allows partial reconstruction, leading to the Ga_2Al_{18} unit. Many other heteroatom derivatives of Al polycations have been obtained [21].

Another Al_{13} isomer, α-Al_{13}, in which the four trimers are corner-shared, has been observed in only one mineral named zunyite, $Al_{13}Si_5O_{20}(OH,F)_{18}Cl$ (Fig. 6.7), discovered in the Zuni Mine in Colorado. This compound is a sorosilicate present in highly aluminous shales, resulting from hydrothermally altered volcanic rocks. In hydrothermal conditions (350–400°C, 0.1 kbar), it can be synthesized from the mixture of gibbsite, fluoride, aluminum chloride, and silica gel [31]. The presence of at least 2 fluorine atoms in substitution of hydroxyl ions by Al_{13} unit is essential [32, 33]. In the crystal, the α-Al_{13} units ($Al_{13}O_{16}OH_{24}$) are linked by silicic pentamers (Si_5O_{16}) to the μ_2 bridges of trimers and by Al-O-Al bridges from terminal Al-OH bonds of the same trimers [34] (Fig. 6.7b). This species has a ^{27}Al-NMR signature at 71.4 ppm, characteristic of the tetrahedral central aluminum atom, and at 7.7 ppm for the hexacoordinated aluminum atoms [35] (Fig. 6.7c).

In the three ε-, δ-, α-Al_{13} species, the ^{27}Al-NMR resonance characteristic of the tetrahedrally coordinated aluminum increases from 63 to 71.4 ppm as the mean Al-O distance decreases: 63 ppm in ε-Al_{13} ($d_{Al\text{-}O}$ 1.833Å), 64.5 ppm in δ-Al_{13}

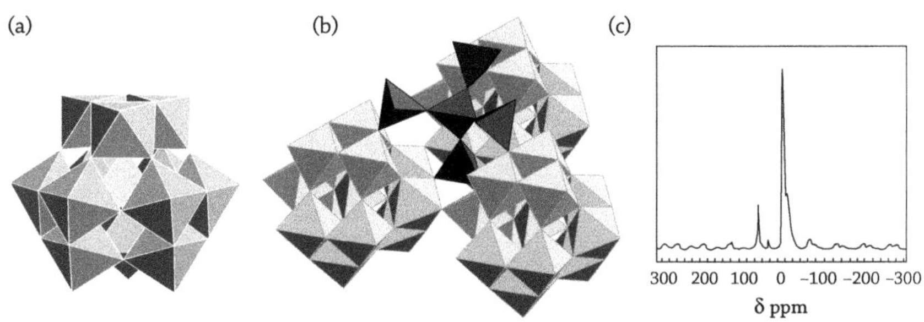

FIGURE 6.7 Structure of the α-Al_{13} group in zunyite (a); connection of the Al_{13} units in the crystal by silicic pentamers and Al-O-Al bridges; (c) ^{27}Al-NMR MAS spectrum (11.7T) of zunyite (reproduced from [35] with permission of The Mineralogical Society of America).

($d_{Al\text{-}O\ mean}$ 1.812Å), and 71.4 ppm in α-Al_{13} ($d_{Al\text{-}O}$ 1.800Å). This effect appears to be correlated to the number of corner-shared trimeric groups around the tetrahedral aluminum. It seems that the corner sharing by a trimer with its neighbors is allowed as a result of its own coordination to other entities such as a sodium cation in δ-Al_{13}, aluminic octahedra for the connection between two δ-Al_{13} units in the Al_{30} polycation or silicic tetrahedra in zunyite crystal. Instead, the ε-Al_{13} units in the sulfate salt are linked to sulfate anions only by electrostatic interactions. Then, it seems that the coordination of any species to a trimer imposes its bonding to the other trimers, in the Al_{13} group, by corners in order to ensure the stability of the Al_{13} polycation. This kind of bonding likely allows the relaxation of the distortion of the coordination polyhedron of octahedral Al^{3+} ions in the "complexed" trimer. It is also to be noted that the adsorption of the Al_{13} polycation on hematite particles leads to its decomposition (§ 5.4.2), which shows the fragility of the ε-Al_{13} polycation toward coordination.

Some other polycations of aluminum can be obtained in unusual synthesis conditions. They are the octamer $[Al_8(OH)_{14}(H_2O)_{18}]^{10+}$ ($h = 1.75$) (Fig. 6.8a), formed by crystallization of the sulfate salt after attack of metallic aluminum by sulfuric acid in the presence of a small amount of mercury [36], and the planar polycation $[Al_{13}(OH)_{24}(H_2O)_{24}]^{15+}$ (h = 1.84) (Fig. 6.8b), including 13 hexacoordinated

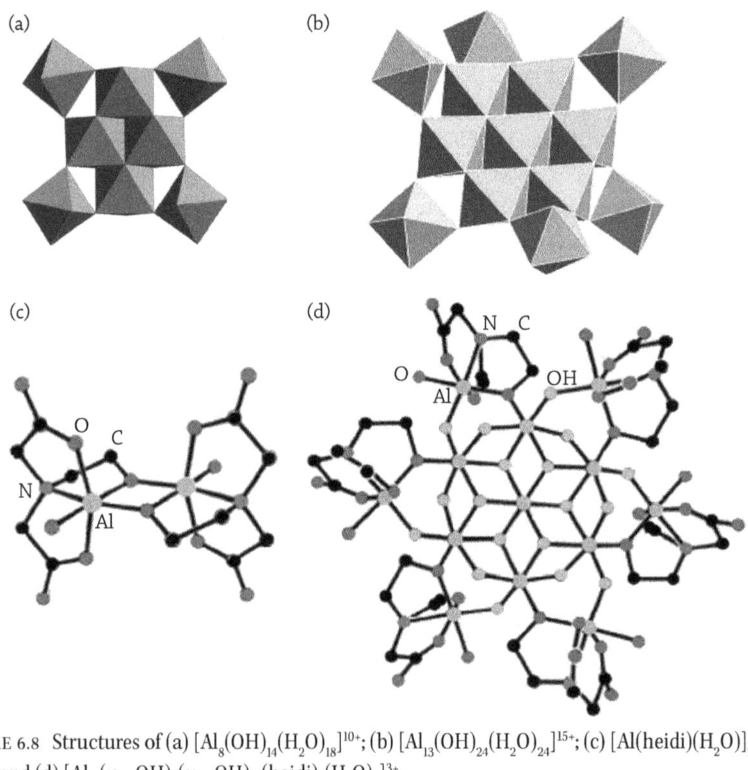

FIGURE 6.8 Structures of (a) $[Al_8(OH)_{14}(H_2O)_{18}]^{10+}$; (b) $[Al_{13}(OH)_{24}(H_2O)_{24}]^{15+}$; (c) $[Al(heidi)(H_2O)]_2$, $2H_2O$; and (d) $[Al_{13}(\mu_3\text{-}OH)_6(\mu_2\text{-}OH)_{12}(heidi)_6(H_2O)_6]^{3+}$.

octahedral Al^{3+} ions edge and corner sharing [37]. This polycation, crystallized as hydrated chloride salt, $[Al_{13}(OH)_{24}(H_2O)_{24}]Cl_{15}\cdot 13H_2O$, is obtained by aging a concentrated solution of an aluminium basic chloride for several days [38, 39]. Aminocarboxylate ligands, especially *N*-(2-hydroxyethyl) iminodiacetic acid, $N(CH_2CO_2H)_2(CH_2CH_2OH)$, the so-called "heidi" ligand, allows formation of a dimer and the planar tridecamer with Al/heidi molar ratios of 1/2 at pH 4.3 and 2/1 at pH 5, respectively [40]. The tetradentate heidi ligand ensures, by itself, condensation of the Al^{3+} ions in the dimer in forming the μ_3–O bridges (Fig. 6.8c). At higher pH, heidi competing with hydroxo ligands is coordinated only at the periphery of the planar Al_{13} polycation, whose core is structurally similar to aluminum trihydroxide (Fig. 6.8d).

6.3 Formation of Solid Phases

The formation of solids involves noncharged species corresponding formally to the $h = 3$ complex of aluminum. Because of equilibria between species in solution, the solid forms from the global hydroxylation rate of around $h = 2.5$, just beyond the existence range of the Al_{13} polycation. The precursors of solid phases are always aquohydroxo or oxohydroxo complexes, but other ligands can also be included in the coordination sphere of the cation. If such ligands are strongly complexing (basic) ligands, they can be incorporated in the solid with well-defined stoichiometry and crystallographic positions, in forming a so-called *basic salt*. That is typically the case of aluminosilicates (§ 6.3.2). A more weakly complexing ligand will not be present in the final solid, but it may play an important role in the formation of the solid during the transient steps of precipitation. The processes involved in precipitation, for instance, direct alkalinization (addition of a base to the solution of the cation), thermolysis or hydrothermal conditions, and microwave heating, also have an important influence on the obtained solid. Because of the technological interest in aluminas in many fields (mineralogy and geochemistry, ceramics, pharmacy, heterogeneous catalysis, etc.), the aluminic gel obtained under various conditions gave rise to numerous studies of characterization and an extensive literature in the years 1940–1980. The development of absorption and diffusion of X-ray, especially of ^{27}Al-NMR spectroscopy, has led to considerable progress since then.

6.3.1 ALUMINUM HYDROXIDES, OXYHYDROXIDES, AND OXIDES

a) Hydroxylation by Addition of Base at Room Temperature

Alkalinization with sodium hydroxide or ammonia of a solution of an aluminum salt, chloride, or nitrate, up to pH 6 to 8, quasi-instantaneously forms a transparent gel amorphous to X-ray diffraction (§ 4.1.3) Natural clays may also contain other. The ^{27}Al-NMR spectrum exhibits a broad signal at ~ 0 ppm relative to 6-coordinated aluminum ions without any signal detectable at 63 ppm characteristic

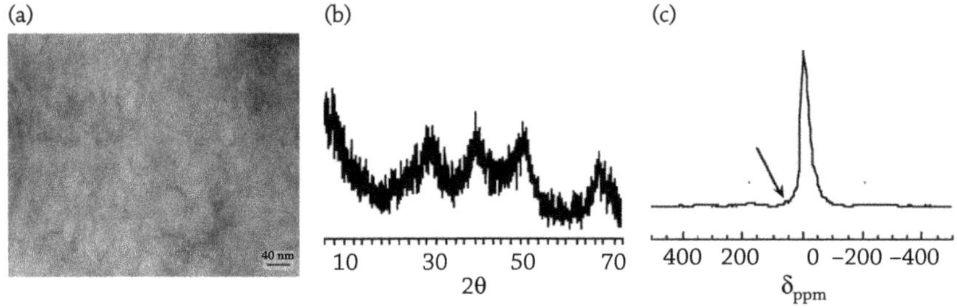

FIGURE 6.9 (a) SEM image, (b) X-ray diffraction pattern and (c) ^{27}Al-NMR spectrum of the gel freshly synthesized from alkalinization at pH 6 of an aluminum nitrate solution. (The arrow on the NMR spectrum indicates the position of the peak relative to the tetrahedral aluminum atom in the Al_{13} polycation.)

of the Al_{13} species [41] (Fig. 6.9). Moreover, the study by SAXS and ^{27}Al-NMR of the gel obtained by alkalinization of a solution of previously synthesized Al_{13} shows that it is composed of aggregates of clusters that progressively densify until all trace of the polycation is lost [42]. So, the polycations transiently formed during progressive alkalinization of the solution are rapidly destroyed beyond their acidity range of formation and stability, because of their high reactivity. Thus, as a rule, these species cannot be viewed as building blocks in the formation of the solid.

The aluminic gel is highly metastable and evolves according to different pathways, depending on the acidity during aging:

- At $5 < pH < 8$, the gel transforms slowly but spontaneously into small particles of boehmite, γ-AlOOH (Fig. 6.10). As the solubility of aluminum is very low at this acidity range (around 10^{-6} mol.l^{-1}), the transformation proceeds in situ by dehydration and local structuration in solid state. That explains the very small particle size—because the structuration does not spread on long distances—and thus the very high specific surface area of the solid, around 300 m^{2}g^{-1}. It also explains why the obtained phase is not the most thermodynamically stable phase at room temperature.
- At $pH < 5$ and $pH > 8$, the solubility of aluminum is higher, and the gel can consequently transform according to a dissolution–crystallization process leading to hydroxide $Al(OH)_3$, the thermodynamically stable phase at room temperature (Fig. 6.10). Its structure is made of layers of $Al(OH)_6$ octahedra sharing three of their edges (Figs. 6.10 and 6.11c). The formation of such a structure can be justified in considering condensation of octahedral $[Al(OH)_3(OH_2)_3]^0$ or tetrahedral $[Al(OH)_3(OH_2)]^0$ complexes in equilibrium with other charged complexes released in solution from the amorphous gel during aging at $pH < 6$ or at $pH > 8$, respectively. Dimers resulting from the first step of condensation can be formed by olation or, in basic medium, by nucleophilic addition and extension of coordination (Fig. 6.11a,b).

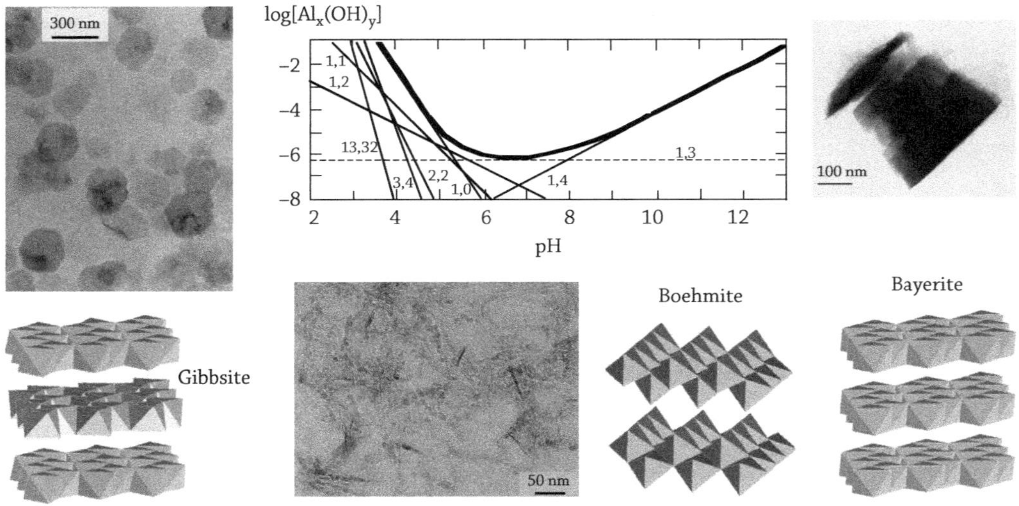

FIGURE 6.10 Solubility diagram of aluminum (soluble species are labeled (x,y) corresponding to the composition $[Al_x(OH)_y]$; the line in bold represents the sum of all soluble species). TEM images and crystal structures of the phases obtained after aging of the aluminic gel at room temperature and at different pHs (gibbsite at pH < 5, boehmite at 5 < pH < 8, bayerite at pH > 8).

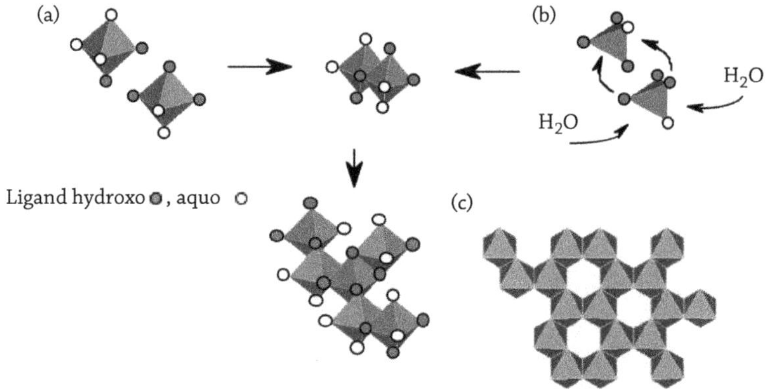

FIGURE 6.11 Possible pathway in formation of $Al(OH)_3$ layers during dissolution–crystallization from the amorphous aluminic gel.

> Their subsequent condensation, which always takes place by olation, can then directly form the characteristic planes of the structure. The process is similar to that described in the formation of brucite layers (Chap. 4), but with cationic vacancies because of the stoichiometry of the solid, $M(OH)_3$ instead of $M(OH)_2$. The stacking of layers and the cohesion of the structure result from interactions by hydrogen bonds. Aging of the gel thus forms bayerite at pH > 8 and gibbsite at pH < 5. The precipitation of these two phases is all the slower as pH is high (bayerite) or low (gibbsite) because of the effect of acidity on the amount of the zero-charged complexes. However, bayerite precipitates much quicker than gibbsite (1 week at pH 10 and room temperature for bayerite; several months at pH 4.5 at room temperature; or 2 weeks at 60°C for gibbsite) [41].

Gibbsite is thermodynamically a little more stable at room temperature than bayerite [43, 44]. These two crystalline phases correspond to different stacking of the same $Al(OH)_3$ layers (Fig. 6.10). In bayerite, the successive layers have the same orientation (ABAB stacking of the planes of oxygen atoms), while in gibbsite, two successive layers are offset by a rotation of 30° (ABBA stacking of the planes of oxygen atoms). Precipitation with ammonia or ethylene diamine leads to nordstrandite, whose structure, with the ABABBABA stacking of the oxygenated planes, is intermediate between that of gibbsite and bayerite. The platelets of bayerite are highly aggregated in forming different morphologies—usually cylinders and cones, depending on synthesis conditions (pH, rate of addition of reactives, duration of precipitation) [45]—whereas the particles of gibbsite, which are not or quite a little aggregated, form stable dispersions by aging the aluminic gel. This difference in morphology is not yet clearly understood [46]. Although the same octahedra layers are involved in the two cases, their different stacking may result from a peculiar system of intra- and interlayer hydrogen bonds according to the pH [43]. That could also be the case for nordstrandite and doyleite

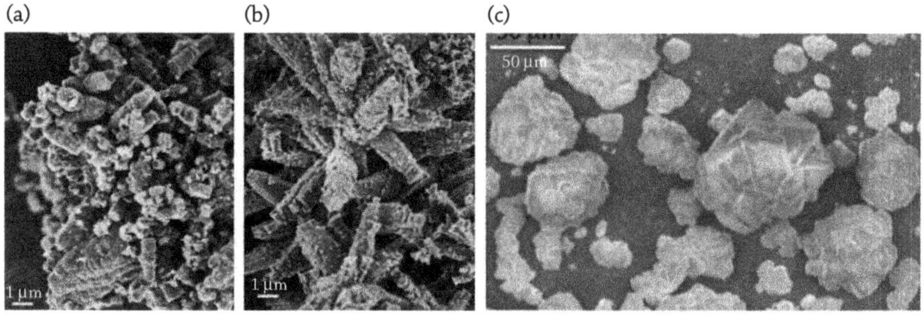

FIGURE 6.12 SEM images of (a) bayerite formed at pH 9.4 and (b) at pH 10.4 (reproduced with permission from [45] Copyright 2003 American Chemical Society), (c) gibbsite forming disordered agglomerates obtained by the Bayer process (courtesy of M. Siebentritt, ENSCP, Paris, France).

(a polytype intermediate between gibbsite and nordstrandite), which are more rarely encountered $Al(OH)_3$ phases [47].

Stabilization of gibbsite particles against aggregation at pH ≈ 5 could be due to the adsorption of polycations, especially Al_{13}, which coexist in the acidity conditions corresponding to partial dissolution of the gel. It has indeed been observed than iron oxide, goethite, and hematite can be stabilized in aqueous suspension by adsorption of the Al_{13} polycation at pH 5 (§ 5.4.2) [41]. Although the polycation is destroyed by adsorption, its fragments remain adsorbed and retain their electrostatic charge, thereby stabilizing the dispersion and avoiding particle aggregation. A similar mechanism could occur with gibbsite particles, but it is obviously difficult to identify and characterize aluminum on a surface of aluminum hydroxide. That is the reason for which aluminum adsorption has been studied on iron oxides.

Acidification of aluminate $[Al(OH)_4]^-$ also forms $Al(OH)_3$. In diluted aluminum solution (< 1 mol.l^{-1}) and with sodium hydroxide (pH ≤ 14), bayerite is obtained, whereas in concentrated medium, gibbsite is formed [48, 49] (Fig. 6.12). Bayerite, rarely observed in nature [46, 50], precipitates only at pH 8–12. Because of its slightly lower thermodynamic stability, it transforms into gibbsite by aging in suspension.

The industrial Bayer process is used to produce gibbsite. The process is based on the treatment of bauxites in strongly alkaline medium (concentration in sodium hydroxide of 3 to 6 mol.l^{-1}) at a temperature between 140 and 300°C in a stainless steel autoclave [51, 52]. Aluminum is solubilized as aluminate $[Al(OH)_4]^-$, which allows separation of insoluble impurities, especially iron oxide, at the origin of the red color of the ore. The solution is then cooled around 70–45°C, diluted, and seeded to precipitate gibbsite. The morphology of particles formed in the Bayer process is quite different from that obtained in acidic medium around pH 4–5, as previously indicated. The seeding is intended to form large particles with a morphology that can be varied. It depends on the alkali cations present, as well as on synthesis conditions (e.g., basicity, aluminate concentration, temperature). Some morphologies seem to result from various assembling of small hexagons

FIGURE 6.13 SEM images of gibbsite crystals obtained in the Bayer process: (a) faceted hexagons of a few micrometers in size and (b) of some tens of micrometers, (c) block-shaped crystals, (d) crystals with large-chamfered faces (adapted from [Sweegers C., On the diversity of gibbsite crystal morphologies, shapes defects and kinetics, PhD Thesis, University of Nijmegen, The Netherlands, 2001).

or diamond-shaped particles forming large twinned prismatic particles involving various formation mechanisms: oriented attachment, aggregation, dissolution–crystallization [53, 54] (Fig. 6.13).

b) Thermohydrolysis

Thermolysis of an aluminum nitrate solution at 95°C forms boehmite, γ-AlOOH, the thermodynamically stable phase at this temperature on the whole range of acidity [55] (Fig. 6.14). The morphology and size of particles are closely dependent on the acidity [56, 57]. The particles formed under such conditions (specific surface area of around 100 $m^2.g^{-1}$) are, however, larger than those resulting from the aging of aluminic gel at pH around 8 and room temperature (specific surface area of around 300 $m^2.g^{-1}$). Electron microscopy and modeling of X-ray diffraction patterns [57] help us to characterize platelets with different extensions of edge facets (Fig. 6.15).

The boehmite particle's change in shape can be explained by the effect of acidity on surface energy. Basically, protonation/deprotonation of surface-oxygenated surface groups corresponds to a change in the chemical composition of the surface and consequently modifies the surface energy (§ 5.4.4). So, the electrostatic surface charge density is the expression of the change in protonation/deprotonation of the

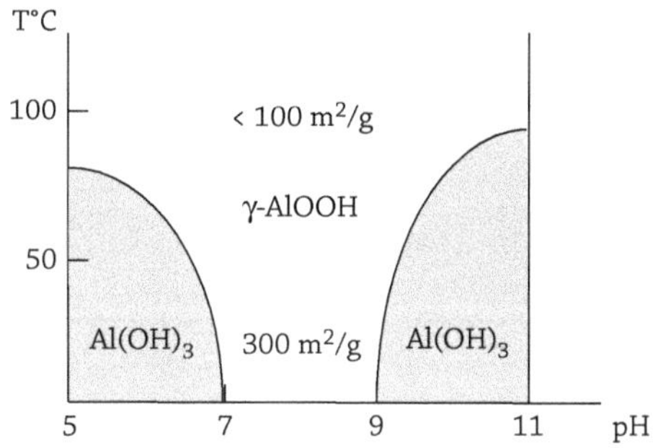

FIGURE 6.14 Schematic phase diagram of aluminum (oxy)hydroxides.

surface by reference to the point of zero charge (pH ≈ 8 for boehmite). It follows that the higher the surface charge is (in absolute value), the farther the pH from the PZC, the more important the lowering of surface energy, and the more energetically favored the extension of the surface. For boehmite, the electrostatic charge is borne by the edge faces of platelets, the {001}, {101} and {100} faces that have the highest energy but are variable, whereas the basal {010} faces, never charged, have the lowest surface energy independent of pH (Fig. 6.15b). Consequently, as the pH in synthesis is far from the PZC, the particles are formed under a decreasing edge surface energy that favors the relative extension of the {101} faces. The increase in the global particle size of boehmite is allowed because the basal faces having the lowest surface energy can extend themselves without incurring any penalty in energy [56].

Organic molecules such as polyols and carboxylic acids are also commonly used to control the shape and/or the size of boehmite nanoparticles. However, these molecules do not have to possess an overabundance of complexing strength in order to not form basic salts or complex structures such as alumoxanes, which are organic-inorganic hybrid compounds resulting from the reaction of boehmite with carboxylic acids [58].

The effect of a polyol such as xylitol $C_5H_7(OH)_5$ on boehmite nanoparticles has been finely characterized [59]. The presence of the polyol (10^{-2} mol.l^{-1}) in the solution of aluminum nitrate (10^{-1} mol.l^{-1}) at pH 11.5, thermolyzed at 95°C for a week, leads to a decrease in particle size without producing a significant change in morphology but showing a marked increase in the area of {101} edge facets compared to {010} basal faces (Fig. 6.16). The edge-to-basal area ratio is near enough to double passing from 0.9 to 1.7 for syntheses without and with xylitol, respectively. In these conditions, the adsorbed xylitol covers around 6% of the surface area of particles.

The adsorption isotherm of xylitol on boehmite indicates an adsorption constant that is rather low, suggesting an adsorption of molecular type that is without coordination and ligand substitution. Adsorption of xylitol results from a combination in interactions of intra- and intermolecular hydrogen bonds with surface μ_2-OH ligands, as indicated by density functional theory (DFT) modeling of surface energy (Fig. 6.17). It shows that adsorption is advantaged on the {010} and

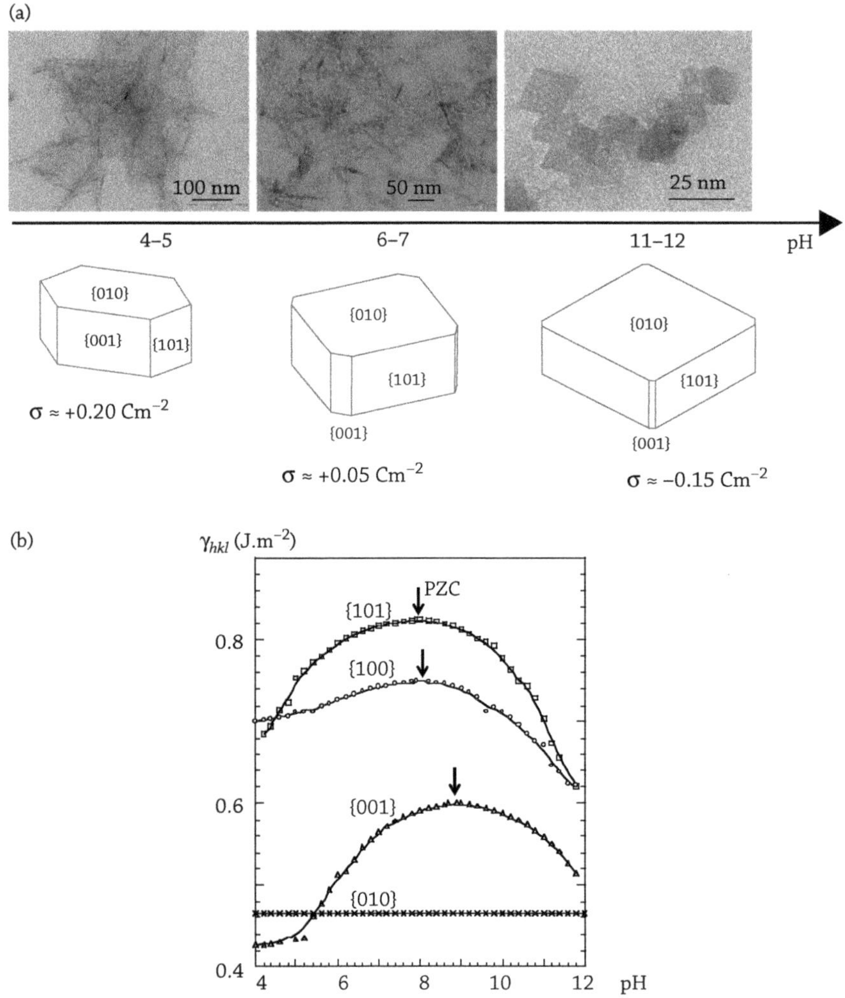

FIGURE 6.15 (a) TEM images of boehmite nanoparticles with different morphologies produced by thermolysis at 95°C of aluminum nitrate at different pH values: fibers formed at pH 4 constituted of piles of small platelets, chains of small platelets formed at pH 6, and large platelets formed at pH 11. The particle shapes are deduced from simulations of X-ray diffraction patterns (reproduced with permission from [57] Copyright 2008 American Chemical Society). The electrostatic surface charge densities measured from acid–base titrations are indicated for each type of particle. (b) Calculated variations in surface energy with pH of different facets of boehmite particles from Eqs. [5.32] and [5.33] (§ 5.4.4). These curves are positioned in energy from the γ_0 values of noncharged surfaces deduced from *ab-initio* simulation [Raybaud P., Digne M., Iftimie R., Wellens W., Euzen P., Toulhoat H., Morphology and surface properties of boehmite (γ -AlOOH): a density functional theory study, J. Catal. 2001, 201, 236–246]. The maximum on these curves (indicated by arrows) corresponds to the point of zero charge (reproduced from [56] with permission of The Royal Society of Chemistry).

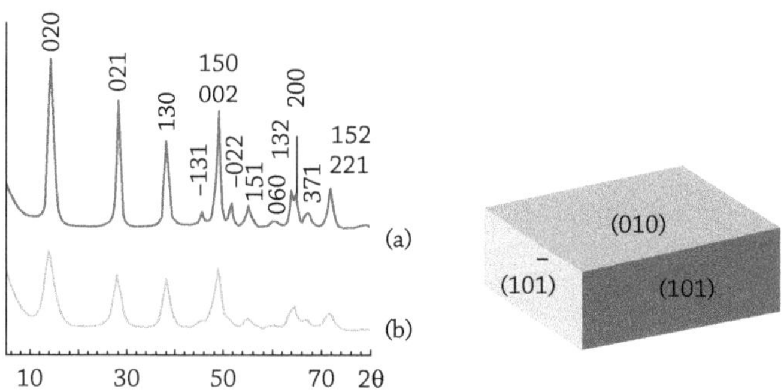

FIGURE 6.16 X-ray diffraction pattern of boehmite particles synthesized at pH 11.5 without (a) and with (b) xylitol, and scheme of morphology of particles (reproduced from [59] with permission of The Royal Society of Chemistry).

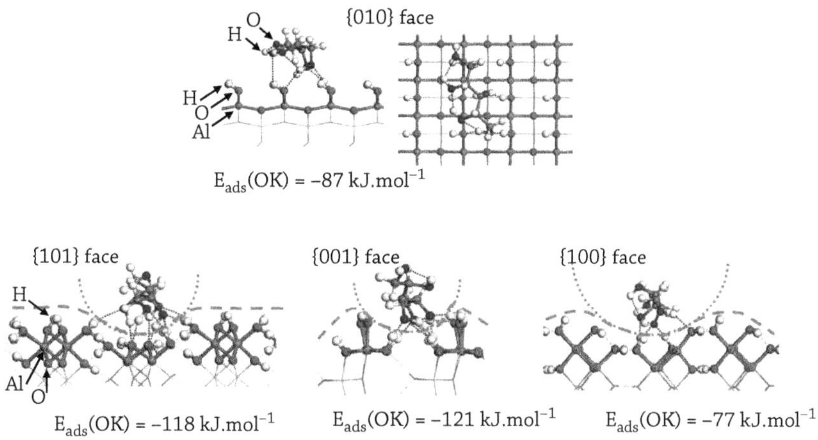

FIGURE 6.17 Representations of the most stable calculated configurations of xylitol molecule adsorbed on the different faces of boehmite platelets. The dashed lines define the cavities on the edge faces of boehmite, and the dotted lines indicate the local curvature (reproduced from [59] with permission of The Royal Society of Chemistry).

{001} edge faces of particles compared to basal faces. The edge faces exhibit cavities in which the molecule of xylitol is better stabilized in establishing hydrogen bonds inside the curvature of the surface than on flatter basal surfaces where H–bond interactions can only occur on a plane (Fig. 6.17). That shows why xylitol is preferentially adsorbed on edge faces of the platelets of boehmite in favoring their extent.

The effect of polyol is more important when its molecular size increases. So, with linear molecules from C_2 (ethyleneglycol) to C_6 (dulcitol), the specific surface area increases from 180 to 350 m^2g^{-1}, respectively, which corresponds to platelets from 12 nm wide and 6 nm thick to 6 nm wide and 4 nm thick [60] (Fig. 6.18). Without polyol, the mean size of particles is 15×6.5 nm. In all cases, the {101}

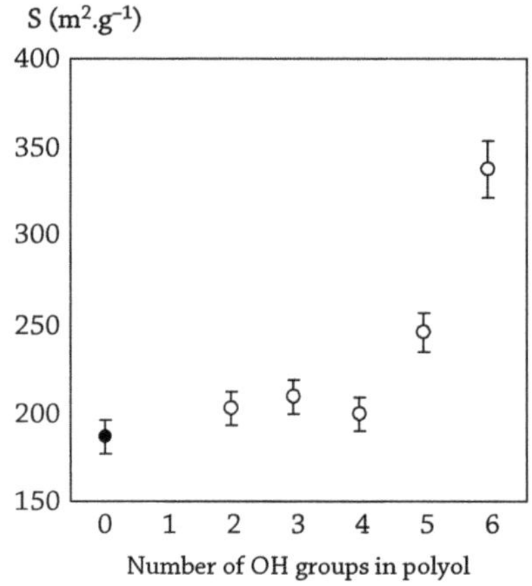

FIGURE 6.18 Specific surface area of boehmite particles synthesized by thermolysis for one week at 95°C and pH 11.5 in the absence and presence of polyols $C_nH_{n+2}(OH)_n$ of increasing length, from C_2 to C_6 (reproduced from [60] with permission of The Royal Society of Chemistry).

edge faces of the platelets are dominant, and their proportion increases with the polyol size.

It is also important to consider the molecular structure of the polyol because *threo* configurations (the OH groups lie on the same side of the molecule; Scheme 6.1), which can behave as a tridentate ligand, are more efficient than *erythro* configurations in which the OH groups lie on both sides of the carbonated chain of the molecule. Similar effects to those due to polyols have been observed with gibbsite formed by the Bayer process [61].

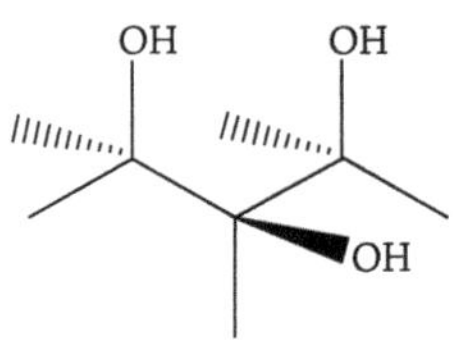

SCHEME 6.1

Hydrothermal treatment of suspensions of aluminum salts in the presence of surfactants forms other morphologies of boehmite nanoparticles: nanofibers by heating at 120°C in the presence of polyethylene oxide (PEO) of the solid resulting from the precipitation of aluminate with acetic acid [62]; nanorods and nanoplatelets by heating at 200°C of aluminum nitrate solutions of pH adjusted at 5 and 10, respectively, with hydrazine [63]; and nanotubes by heating at 120°C of aluminum chloride in alkaline medium in the presence of CTAB (N-cetyl-N,N,N-trimethylammonium bromide) [64] (Fig. 6.19).

The interest in the morphological control of boehmite nanoparticles lies in their transformation by heating around 450–500°C in gamma alumina, γ-Al_2O_3, a widely used catalyst or catalyst support in the petroleum industry [4].

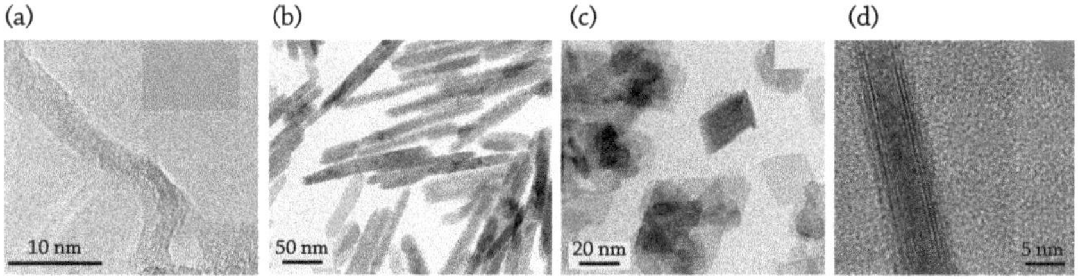

FIGURE 6.19 TEM images of various morphologies of boehmite nanoparticles: (a) fibers (reproduced with permission from [62] Copyright 2004 American Chemical Society), (b) rods and (c) platelets (reproduced from [63] with permission of Elsevier), (d) tubes (reproduced from [64] with permission of The Royal Society of Chemistry).

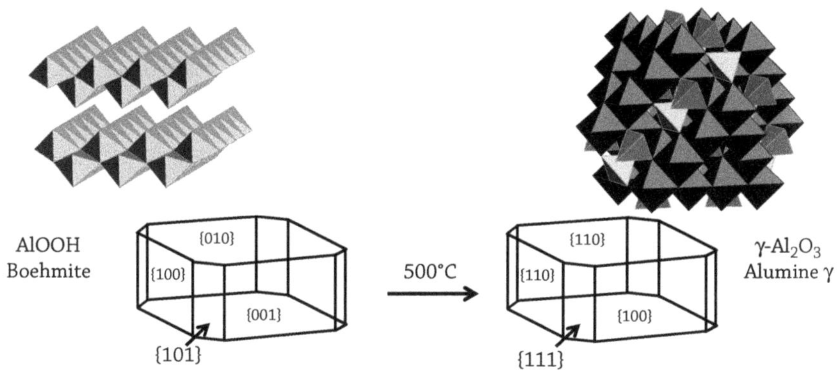

FIGURE 6.20 Structural relationship in the topotactic transformation of boehmite in γ-alumina having spinel crystal structure.

As the thermal transformation of boehmite is topotactic, the shape and size of nanoparticles are preserved after transformation [43, 65] (Figs. 6.20, 6.21). It is possible to optimize the performance of gamma alumina for a given reaction by the promotion of certain crystalline faces of the nanoparticles.

Compared to ferric ions, the behavior of Al^{3+} ions appears to be governed by a lower polarization of the oxygenated ligands, which gives better stability to the hydroxo ligands in the aluminum compounds. The aluminum hydroxides are stable phases, while the $Fe(OH)_3$ trihydroxide does not exist. Oxyhydroxides MOOH are formed by the two metals, but the isomorphous compounds are obtained under different conditions: at the difference of goethite, α-FeOOH, which is the phase most easily formed by precipitation of ferric ions, the isostructural aluminic phase, diaspore α-AlOOH, is obtained only by hydrothermal treatment at 380°C under 6 kbars of $Al(OH)_3$ or γ-Al_2O_3 in the presence of sodium carbonate [66] (Fig. 6.22). Boehmite is easily obtained through precipitation of an aluminum salt, but the ferric isostructural phase, lepidocrocite, γ-FeOOH, can only be formed by rapid oxidation of ferrous hydroxide in suspension (Chap. 7). Formation of aluminas, α- as well as γ-Al_2O_3, never occurs in aqueous suspension. It requires heating of

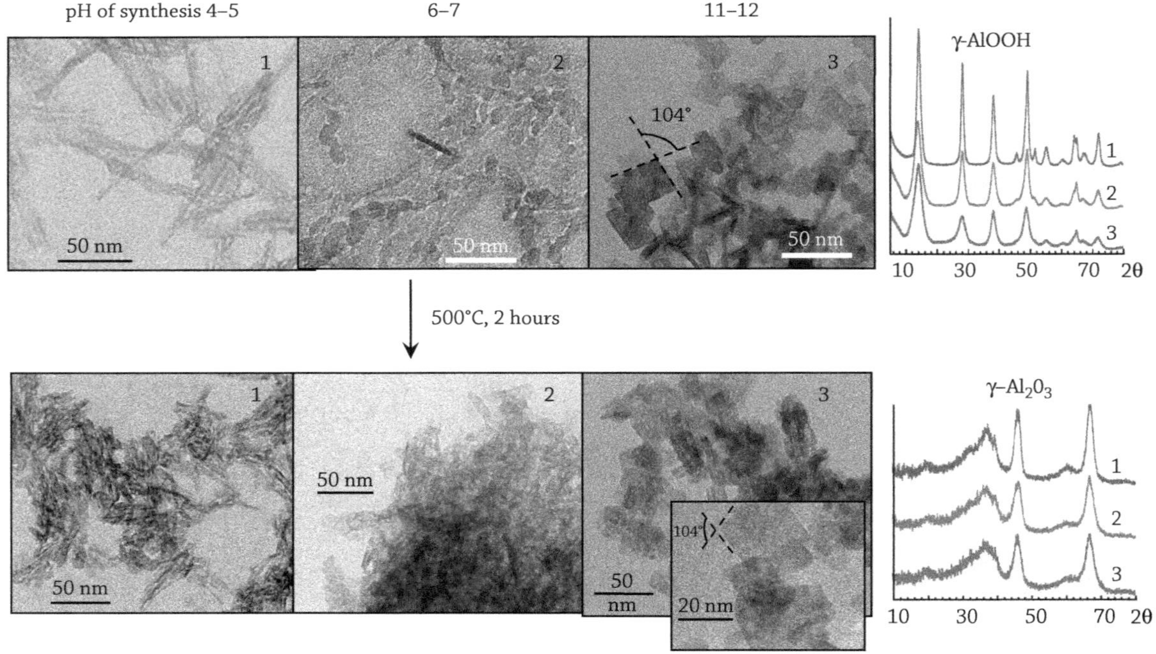

FIGURE 6.21 TEM images and X-ray diffraction diagrams of boehmite with various morphologies formed at different pHs (Fig. 6.16) and of corresponding γ-aluminas resulting from heating at 500°C for 2 hours.

FIGURE 6.22 SEM images of particles of diaspore α-AlOOH obtained by hydrothermal treatment (380°C, 6 kbars) (a) from hydroxide $Al(OH)_3$ (Na_2CO_3 1 mol.l^{-1}) and (b) from γ-alumina γ-Al_2O_3 (Na_2CO_3 0.5 mol.l^{-1}) [68]. (c) Crystal structure of diaspore.

hydroxides in solid state, whereas hematite α-Fe_2O_3 and magnetite Fe_3O_4 are easily precipitated from solution.

6.3.2 ALUMINOSILICATES

Aluminosilicates form a huge family of minerals, with extremely variable structures and compositions [5]. These compounds can be regarded as basic salts or as mixed complexes of the Al^{3+} cation and silicate with hydroxo ligands. Silicates $[SiO_{4-h}(OH)_h]^{(4-h)-}$ ($h < 4$) are basic entities that are highly nucleophilic and able to condense by oxolation to form anionic polysilicates having a large structural diversity evidenced by ^{29}Si-NMR [67. 68] (§ 3.3.1). Silicic acid, $Si(OH)_4$, the noncharged entity ($h = 4$), leads to amorphous hydrated silica with a morphology that is highly dependent on pH (§ 4.1.4): optically transparent polymeric gels formed of long chains of SiO_4 tetrahedra with a few branching points and many dangling Si–OH bonds in acidic medium, and strongly diffusing gels containing nanoparticles formed of highly branched chains in neutral medium [69]. Silica and silicates can react with aluminum, according to the different pathways briefly reviewed in the following section.

6.3.2.1 Adsorption of Aluminum on Amorphous Silica

Amorphous hydrated silica reacts with soluble aluminum by adsorption involving different kinds of interactions. The polar and charged oxygenated surface groups may induce nonspecific electrostatic interactions with molecules or metal complexes in forming outer-sphere surface complexes within the solvation layer of particles according to a physisorption process. This type of interaction may have various consequences: the restriction in mobility of soluble species near the surface as shown by ESR spectroscopy of the hexaaquo-copper(II) paramagnetic complex adsorbed on fully hydrated silica at pH 3 [70] (§ 5.4.1); the electrostatic screening leading to destabilization of dispersions and flocculation of silica particles by aluminum-based flocculants [71] (§ 5.3); and the transport of pollutants such as molecules and heavy metals partially hydrolyzed in fresh waters.

Adsorption may also involve specific interactions in a chemisorption process. Polar or electrostatically charged oxygenated surface groups act as hydroxylating ligands toward hydroxylable cations in solution, allowing their coordination as surface inner-sphere complexes (§ 5.4.2). The role of the surface is highlighted by the observation that adsorption occurs at a pH lower than that corresponding to hydroxylation of the cation alone in solution [72] This behavior is exhibited by many metal oxides, but amorphous silica with a high porosity and high surface reactivity due to many strongly nucleophilic dangling silanol bonds can not only coordinate but also incorporate Al^{3+} ions into its framework because of the ability of the cation to adopt the fourfold coordination. This was observed with colloidal silica [71] and also with biogenic silica from diatom frustules [73, 74]. Through incubation of silica from postmortem diatom frustules free from organic matter with dilute aluminum in seawater at pH ~ 8, aluminum is incorporated mainly with tetrahedral coordination, causing an unexpected slight decrease in pore size distribution, suggesting the formation of a superficial aluminosilicate layer [74]. In contrast, with living diatoms, aluminum is exclusively incorporated with tetrahedral coordination inside the silica framework involving a biosynthetic mechanism leading to the decreased solubility of silica in the marine environment [73]. This incorporation of aluminum likely stabilizes silica against hydrolysis and dissolution because of reinforcement of the silica network by eliminating dangling silanol bonds and by filling and flattening the silica surface.

6.3.2.2 Crystalline Aluminosilicates: Zeolites, Clays, Imogolite

Like phosphates, silicates $[SiO_{4-h}(OH)_h]^{(4-h)-}$ behave as strong ligands toward soluble aluminum, but silicates have a special role because, unlike phosphates which always remain monomer in solution, they can condense. This difference explains the richness of the crystal chemistry of silicates compared to that of phosphates. Even if the crystal chemistry of aluminosilicates appears to be complicated and regardless of their specific formation conditions, the acidity of the medium may be the major factor in the formation of the main aluminosilicic structural types. Indeed, the acidity controls the relative reactivities of aluminum and silicates toward condensation. In basic medium (pH around 12), silicates can condense and simultaneously complex aluminate, which is monomeric in these conditions. That corresponds to the formation of zeolites, which are made of a tridimensional framework of AlO_4 and SiO_4 tetrahedra corner sharing by Si-O-Si and Si-O-Al bridges, but without any Al-O-Al bond. In weakly basic to slightly acid media ($9 \geq pH \geq 5$), both silicates and octahedrally coordinated aluminum simultaneously condense. Aluminum forms by itself octahedral layers of gibbsite type, but the strong complexing ability of silicates toward aluminum controls the condensation in forming polysilicic layers adjacent to the aluminic layers. Various kinds of such sandwich structures form clays. In more acidic medium ($5 \geq pH \geq 3$), aluminum still forms similar layers of gibbsite type, but the tendency of silicic acid to condense is rather low, as indicated by the structure of silicic acid gels. Silicate is then coordinated as a monomeric ligand to aluminic layers involving three OH groups per tetrahedron. This kind of bonding results in the curvature of the gibbsite-type

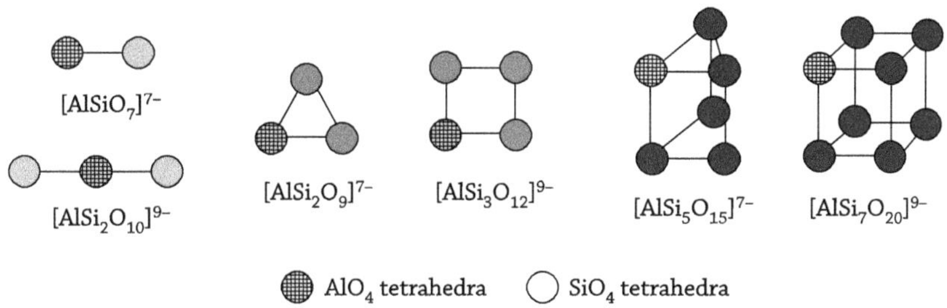

FIGURE 6.23 Scheme of aluminosilicic polyanions formed at pH >12 in the presence of Na^+ or K^+ cations (the eventual protonation of oxo ligands is not taken into account). The polyhedral species are formed in the presence of tetralkylammonium NR_4^+ cations.

layers and the formation of tubes of imogolite. More detailed insight is presented in the following section.

A) ZEOLITES

In strongly basic medium, small polysilicate units condense with tetrahedral aluminate $[Al(OH)_4]^-$. It results in a rather complex chemistry of aluminosilicic polyanions that structurally have been well characterized by ^{29}Si and ^{27}Al NMR (Fig. 6.23) [75–78].

These mixed species, which are highly labile and never contain Al–O–Al bonds because aluminate cannot self-condense, are believed to be building blocks for crystallization of zeolites. These aluminosilicic compounds are made of a three-dimensional, largely open anionic framework of SiO_4 and AlO_4 corner-sharing tetrahedra forming cavities and channels of variable size and shape in which water molecules and various countercations are included (Fig. 6.24) [79, 80]. Many natural compounds as well as synthetic phases exist. They are obtained mainly by (hydro) thermal treatment of suspensions of aluminum hydroxide or aluminate and silica or silicates, strongly alkalinized by sodium or tetraalkylammonium hydroxide NR_4OH and containing a template such as amine or crown-ether. Under such conditions, the thermal treatment of suspensions allows the formation of soluble precursors, permitting the growth of crystals. Many factors control the crystal structure and composition, such as Si/Al ratio, pH, nature of cations (e.g., Na^+, NR_4^+) and the surfactant used as template, temperature, pressure, and so on. The intimate mechanism of formation of zeolites is still poorly understood, even if their synthesis is technically efficient and well controlled [81]. However, there is no obvious correlation between the geometries of the organic template molecules and the structures they promote. In addition, there is no direct correlation between the structures of precursor solutes and zeolite products [75, 82, 83]. Contrary to the traditional picture of zeolite formation from structured "secondary building units" preexisting in solution, cyclic and cage-like aluminosilicate solute species are not directly involved in the crystallization of zeolites from aqueous solutions. Rather, they appear simply to serve as reservoirs for small, active, acyclic species responsible for crystal growth [84].

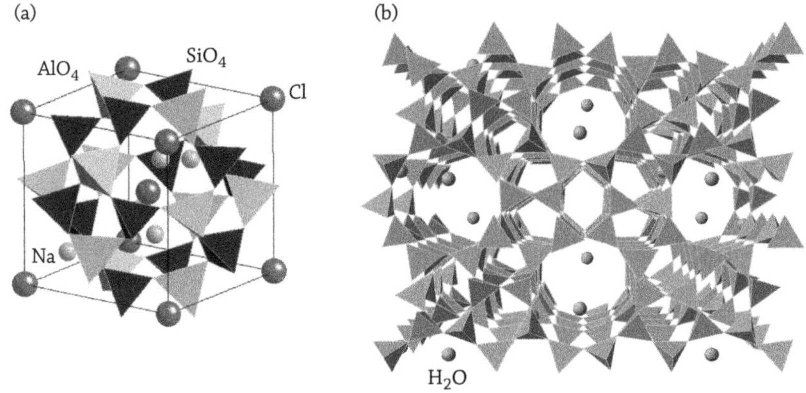

FIGURE 6.24 Representation of (a) the sodalite cavity ($Na_4(AlO_2)_3(SiO_2)_3Cl$), which is a basic structural unit of natural zeolites and of (b) the synthetic ZSM-5 (Zeolite Socony Mobil), $[Al_nSi_{96-n}O_{192}]\cdot Na^+_n\cdot (H_2O)_{16}$ with $n < 27$.

B) CLAYS

Clays form a wide family of aluminosilicates characterized by different two-dimensional structures, the phyllosilicates. They basically result from the stacking of layers of edge-sharing AlO_6 octahedra (O) and layers of corner-sharing SiO_4 tetrahedra (T). Many stacking sequences of such layers occur, for instance, T-O T-O in kaolinite, $Si_2O_2Al_2(OH)_4$, or T-O-T T-O-T in pyrophyllite, $Si_4O_{10}Al_2(OH)_2$. The chemical composition of the layers is highly variable because of many substitutions inside each layer—for instance, Al^{3+} or Fe^{3+} to Si^{4+} in T-layers as in muscovite or mica, and Al^{3+} to Si^{4+} in T-layers and Mg^{2+} to Al^{3+} in O-layers as in montmorillonite. Resulting from these substitutions is an excess of negative charges in the layers, which is balanced by intercalation of cations (e.g., K^+, Ca^{2+}) inside the interlayer spaces and also of water molecules as for smectites (Fig. 6.25). Natural clays may also contain other substituting cations, depending on their place of formation [85, 86].

In the laboratory, smectites can be synthesized according to largely variable precipitation conditions, at low temperature (< 100°C) or under hydrothermal conditions [87]. Coprecipitation of Al^{3+}, M^{2+}, and Si^{4+} is carried out in the presence of a base such as $Ca(OH)_2$, NaOH, or KOH, at pH between 5 and 9. The requirements for synthesis are a basic to neutral medium, an especially important dilution of silicate (concentration in the range 10^{-4} to 10^{-3} mol.l^{-1}), and a slow precipitation [88–90]. Under these conditions, one may suggest a possible schematic mechanism for the formation of clays in which the fastest step of the process is the formation of O-type layers resulting from condensation by olation of M^{2+} and Al^{3+} cations in brucite-type ($Mg(OH)_2$) or gibbsite-type ($Al(OH)_3$) mixed layers (Fig. 6.26a).

Because of its low concentration, silicate as $SiO(OH)^-$ or silicic acid $Si(OH)_4$ self-condenses very slowly. These species, whether monomeric or oligomeric, are

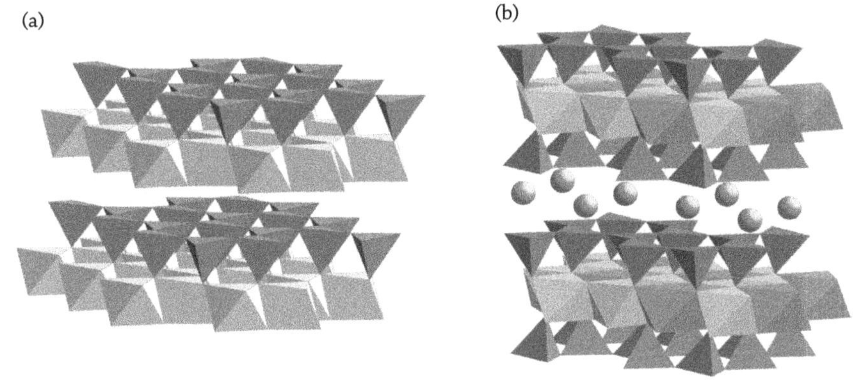

FIGURE 6.25 Idealized structures of (a) kaolinite $Si_2O_2Al_2(OH)_4$ and (b) montmorillonite $[Si_4O_{10}Al_{2-x}M_{2+x}(OH)_2]C_x{\cdot}nH_2O$.

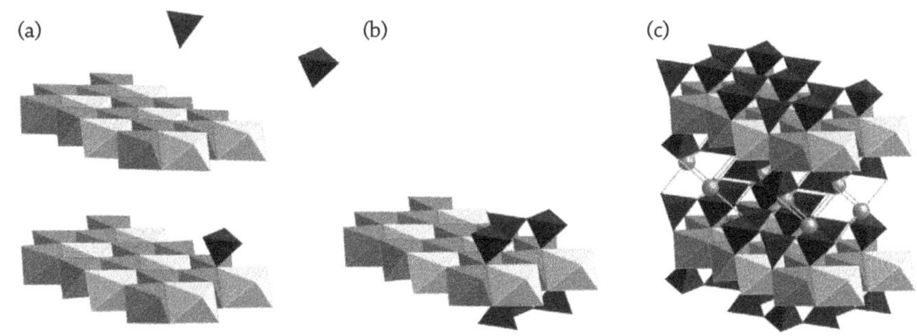

FIGURE 6.26 Scheme of a possible formation mechanism of clay minerals highlighting the complexation of brucite-type octahedral layers by silicic groups forming tetrahedral layers (see text).

highly nucleophilic and strongly complexing. They can thus fix on each side of the O-layers and then condense between them to form rings of corner-shared tetrahedra characteristic of the structure of the T-layers (Fig. 6.26b). So, because of the trend toward complexation and simultaneously toward condensation of the silicic species, the T- and O-layers are thereby connected by corners of SiO_4 tetrahedra linked to μ_2OH–bridges in the aluminic layers. This pathway shows how the Al-based layers are at the origin of the bidimensionality of the clay mineral structure.

C) IMOGOLITES

Imogolite is another interesting crystalline aluminosilicate. It is a tubular nano-object of around 2 nm in diameter and several micrometers in length with the composition $Al_2SiO_3(OH)_4$. The nanotubes are generally clustered in bundles around 30 nm in diameter [91]. As is true of many other natural aluminosilicates, imogolite results from the alteration of volcanic glasses and ashes. The crystal structure of imogolite can be described by a gibbsite-type layer of AlO_6 octahedra

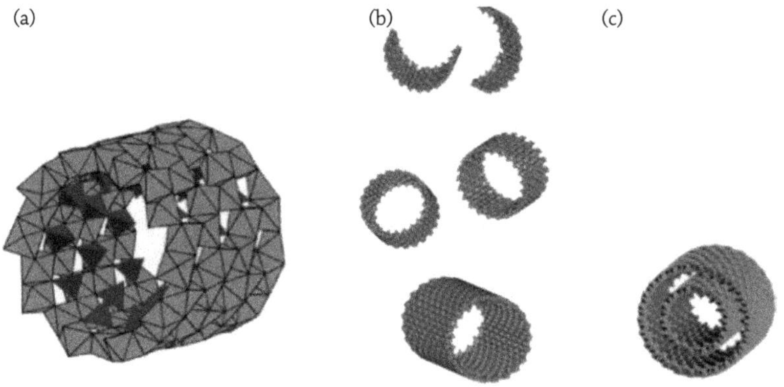

FIGURE 6.27 (a) Schematic structure of imogolite showing SiO_4 tetrahedra linked on the internal face of the wrapped gibbsite-type layer. (b) Building fragments of imogolite nanotubes detected by NMR and X-ray absorption spectroscopies. (c) Scheme of a double-wall imogolite nanotube.

wrapped up to form a cylinder. Isolated SiO_4 tetrahedra are grafted by three corners on the internal surface of the tube at the upright of the vacancies in the gibbsite layer (Figs. 6.27a, 6.28a) [92]. Other aluminosilicic compounds close to imogolite, such as allophanes, have a rather large variability in composition ($1 \leq Al/Si \leq 2$) and exhibit a structural short-distance order with different morphologies such as spheroids and spherical caps [91, 93].

Imogolite is usually synthesized by heating dilute solutions ($\sim 10^{-3}$ mol.l^{-1}) of the OH/Al = 2 hydroxylated Al^{3+} complex for 5 days, in the presence of silicate (tetraethylorthosilicate, $Si(OCH_2CH_3)_4$, Al/Si = 2) in slightly acid medium (pH on the range 3.5–4.5) [94]. These conditions are somewhat similar to those involved in the formation of clays, but for imogolite, the solution is more acidic, the Si/Al ratio being lower and dilution higher. These factors promote condensation of aluminum and formation of aluminosilicic complexes in avoiding self-condensation of silicic acid $Si(OH)_4$ resulting from hydrolysis of the alkoxide (§ 4.1.4a). It has been shown, however, that imogolite can be obtained from more concentrated media (decimolar concentrations in Al) in order to increase the yield of the reaction [95]. The growth of imogolite under such conditions is notably delayed inasmuch as it requires around 1 month of heating at 95°C. It has been shown through ^{29}Si and ^{27}Al NMR investigations that the growth of imogolite from aluminum perchlorate and tetraethylorthosilicate is in fact limited by the decondensation rate of polysilicic species that are concomitantly formed. It is also likely that complexation of aluminum by monomeric silicic species inhibits the formation of boehmite, as is the case for the formation of clays.

The linkage of SiO_4 tetrahedra by three Si-O-Al bonds per tetrahedron, instead of by only one Si-O-Al bond as in clays, induces the curvature of the gibbsite layer because the O-O distance in SiO_4 tetrahedra is shorter than that in gibbsite layers. Such an effect has been proved by molecular modeling and has been understood as resulting from an equilibrium between the mechanical stiffness of the aluminosilicic layer and the dissymmetry in its interfacial tension [96]. This effect was also nicely evidenced in using germanate analogues of imogolite similarly synthesized from tetraethylorthogermanate [97, 98]. With a better electronic contrast between Al and Ge than between Al and Si, it was shown, using X-ray spectroscopy, X-ray scattering, and NMR spectroscopy, that the curvature of the gibbsite-like fragments was due to the coordination of the GeO_4 tetrahedra and that the maximum in curvature was obtained when all vacancies in the gibbsite layer were coordinated with tetrahedra. The diameter of Ge-Al imogolites (2.5 nm) is larger, however, than that of Si-Al imogolites (2 nm) because the GeO_4 tetrahedron, larger than the SiO_4 tetrahedron, induces less stress in the gibbsite layer and thus less force in the curvature.

It has also been shown that imogolite precursors were curved, rooftile-shaped fragments of around 5 nm in size, assembled edge to edge to form short cylindrical sections (Fig. 6.27b) [96]. A kinetic model for the growth of nanotubes [99] is in accordance with a two-step process: very fast formation of curved roof tiles (proto-imogolite) with edge-to-edge assembling to form short tubular units and then association of these units by oriented aggregation to form the nanotube. At an adequate concentration in aluminum (0.25 mol.l^{-1}), germanium allows the formation of double-wall imogolite nanotubes [100] (Figs. 6.27c, 6.28). This structure has been finely characterized using SAXS and IR spectroscopy [100, 101].

The formation of double-wall tubes results from a subtle equilibrium between the adhesion energy of proto-imogolite fragments and their flexibility. Indeed, the tiles of proto-imogolite bear, at pH 4–5, a positive electrostatic charge due to Al_2-OH groups on the aluminic side and a negative charge on the other face resulting from the Ge-OH (or Si-OH) groups. Two fragments can thus attract and superpose because of the electrostatic attraction between the positive charge on the external face of a sheet and the negative charge on the internal face of another sheet, with, in addition, the establishment of a hydrogen bond network between these solvated surfaces (Fig. 6.29). The flexibility of these fragments allows adaptation of their curvature. Thus, during the formation of a double-wall nanotube, the external fragment decreases its curvature, whereas the internal fragment increases

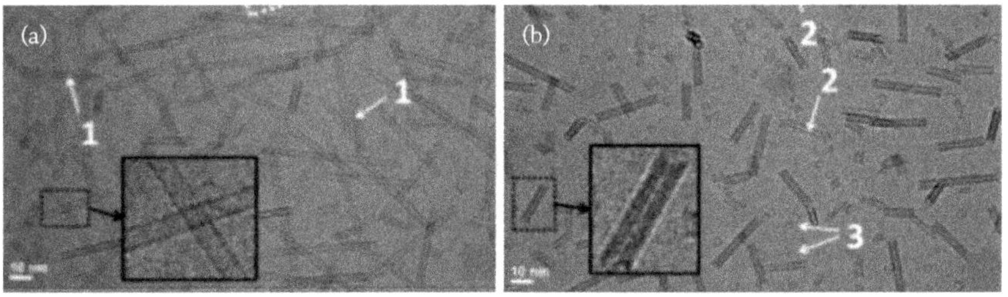

FIGURE 6.28 Cryo-TEM images of Al-Ge imogolites formed at aluminum concentration of 0.5 mol.l^{-1} (a) and 0.25 mol.l^{-1} (b). Two types of nanotubes are visible: double-wall tubes labeled 1 and single-wall tubes labeled 2. The small objects labeled 3 are precursors of nanotubes. The inserts are enlargements of selected zones in dotted lines (reproduced with permission from [100] Copyright 2010 American Chemical Society).

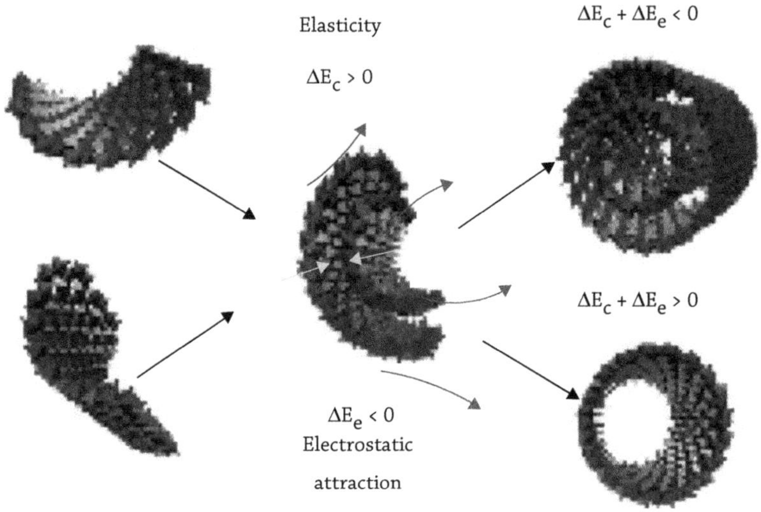

FIGURE 6.29 Schematic sketch for the proto-imogolite transformation into single- or double-wall imogolite nanotubes, depending on the energetic balance between electrostatic interaction and curvature change in the sheet (reproduced with permission from [102] Copyright 2012 American Chemical Society).

it, thereby increasing the global elastic energy from the spontaneous curvature of the two elements. The change in curvature shifts some IR absorption peaks. Thus, two possibilities exist: if the attractive electrostatic energy exceeds the increase in elastic energy, double-wall tubes are formed; otherwise, superposition of two fragments is more difficult, and single-wall tubes form (Fig. 6.29) [101, 102].

In the series of $Al_2Si_xGe_{1-x}O_3(OH)_4$ compounds, the SiO_4 tetrahedra reinforce the rigidity of the gibbsite-like sheet so that only single-wall nanotubes are formed ($x = 1$), while, with germanium, there are fewer mechanical constraints and double-wall nanotubes are obtained ($x = 0$). The structural transition is progressive, which means that the two structures are coexisting for $x \leq 0.25$. Only single-wall nanotubes are formed by the compounds richer in silicon ($x \geq 0.25$) [101]. The double-wall nanotubes are larger (diameter of 3 ± 0.5 nm) and shorter (length of 20 ± 15 nm) than single-wall ones (diameter of 2.5 ± 0.5 nm, length of 50 nm). The kinetic growth model [99] shows that the same amount of proto-imogolite allows the formation of less tubular units and that collisions leading to the growth of tubes from double-wall units are around five times less efficient than those allowing the formation of single-wall nanotubes.

References

[1] Pédro G.G. Cycles biogéochimiques et ecosystèmes continentaux. Rapport sur la Science et la Technologie n° 27, Académie des Sciences, EDP Sciences, Les Ulis, France, 2007.

[2] Sposito G. *The Environmental Chemistry of Aluminium*. Boca Raton, FL: CRC Press, 1989.

[3] Castel B. *Les alumines et leurs applications*. Paris: Nathan, 1990.

[4] Euzen P., Raybaud P., Krokidis X., Toulhoat H., Le Loarer J.L., Jolivet J.P., Froidefond C. Alumina. In *Handbook of Porous Solids*, F. Schüth, K. Sing and J. Weitkamp, Eds., Vol. 3, 1591. Weinheim, Germany: Wiley-VCH Verlag GmbH, 2002.

[5] Wells A.F. *Structural Inorganic Chemistry*, 5th ed. Oxford: Oxford University Press, 1991.

[6] Baes C.F., Mesmer R.E. *The Hydrolysis of Cations*. New York: Wiley, 1976.

[7] Shannon R.D. Revised effective ionic radii and systematic studies of interatomic distances of halides and chalcogenides. *Acta Cryst. A* 1976, 32, 751–767.

[8] Moolenaar R. J., Evans J. C., McKeever L. D. The structure of the aluminate ion in solutions at high pH. *J. Phys. Chem.* 1970, 74, 3629–3636.

[9] Akitt, J.W., Elders, J.M. Multinuclear magnetic resonance studies of the hydrolysis of aluminium(III). Part 8. Base hydrolysis monitored at very high magnetic field. *J. Chem. Soc. Dalton Trans.* 1988, 1347–1355.

[10] Thompson A.R., Kunwar A.C., Gutowsky H.S., Oldfield, E. Oxygen-17 and aluminium-27 NMR spectroscopic investigations of aluminium(III) hydrolysis products. *J. Chem. Soc. Dalton Trans.* 1987, 2317–2322.

[11] Brown P.L., Sylva R.N., Batley G.E., Ellis J. The hydrolysis of metal ions. Part 8. Aluminium(III). *J. Chem. Dalton Trans.* 1985, 1967–1970.

[12] Johansson G. The crystal structures of $[Al_2(OH)_2(H_2O)_8](SO_4)_2.2H_2O$ and $[Al_2(OH)_2(H_2O)_8](SeO_4)_2.2H_2O$. *Acta Chem. Scand.* 1962, 16, 403–420.

[13] Johansson G. On the crystal structure of the basic aluminum sulfate $13Al_2O_3.6SO_3.xH_2O$. *Arkiv. Kemi* 1963, 20, 305–321.

[14] Bradley S. M., Kydd R. A., Yamdagni R. Comparison of the hydrolyses of gallium(III) and aluminium(III) solutions by nuclear magnetic resonance spectroscopy. *J. Chem. Soc., Dalton Trans.* 1990, 2653–2656.

[15] Bradley S. M., Kydd R. A., Yamdagni R. Detection of a new polymeric species formed through the hydrolysis of gallium(III) salt solutions. *J. Chem. Soc. Dalton Trans.* 1990, 413–417.

[16] Bradley S. M., Kydd R. A. Comparison of the species formed upon base hydrolyses of gallium(III) and iron(III) aqueous solutions: The possibility of existence of an $[FeO_4Fe_{12}(OH)_{24}(H_2O)_{12}]^{7+}$ polyoxocation. *J. Chem. Soc. Dalton Trans.* 1993, 2407–2413.

[17] Murugesu M., Clérac R., Wernsdorfer W., Anson C.E., Powell A.K. Hierarchical assembly of $\{Fe_{13}\}$ oxygen-bridged clusters into a close-packed superstructure. *Angew. Chem. Int. Ed.* 2005, 44, 6678–6682.

[18] Bino A., Ardon M., Lee D., Springler B., Lippard S.J. Synthesis and structure of $[Fe_{13}O_4F_{24}(OMe)_{12}]^{5-}$: The first open-shell Keggin ion. *J. Am. Chem. Soc.* 2002, 124, 4578–4579.

[19] Michot L.J., Montargès-Pelletier E., Lartiges B.S., d'Espinose de la Caillerie J.B., Briois V. Formation mechanism of the Ga_{13} Keggin ion: A combined EXAFS and NMR study. *J. Am. Chem. Soc.* 2000, 122, 6048–6056.

[20] Rustad J.R., Loring J.S., Casey W.H. Oxygen-exchange pathways in aluminum polyoxocations. *Geochim. Cosmochim.* 2004, 68, 3011–3017.

[21] Casey W. H. Large aqueous aluminum hydroxide molecules. *Chem. Rev.* 2006, 106, 1–16.

[22] Fu G., Nazar L. F., Bain A. D. Aging processes of alumina sol-gels: characterization of new aluminum polyoxycations by ^{27}Al NMR spectroscopy. *Chem. Mater.* 1991, 3, 602–610.

[23] Rowsell J., Nazar L. F. Speciation and thermal transformation in alumina sols: structures of the polyhydroxy-oxoaluminum cluster $[Al_{30}O_8(OH)_{56}(H_2O)_{26}]^{18+}$ and its α-Keggin moiete. *J. Am. Chem. Soc.* 2000, 122, 3777–3778.

[24] Allouche L., Gérardin C., Loiseau T., Férey G., Taulelle F. Al_{30}: A giant aluminum polycation. *Angew. Chem. Int. Ed.* 2000, 39, 511–514.

[25] Shafran K.L., Perry C.C. A systematic investigation of aluminium ion speciation at high temperature. Part 1. Solution studies. Dalton Trans. 2005, 2098–2105.

[26] Shafran K.L., Deschaume O., Perry C.C. The static anion exchange method for generation of high purity aluminium polyoxocations and monodisperse aluminium hydroxide nanoparticles. *J. Mater. Chem.* 2005, 15, 3415–3423.

[27] Sun Z., Wang H., Tong H., Sun S. A giant polyaluminum species S-Al_{32} and two aluminum polyoxocations involving coordination by sulfate ions S-Al_{32} and S-K-$Al_{13.}$ *Inorg. Chem.* 2011, 50, 559–564.

[28] Abeysinghe S., Unruh D.K., Forbes T.Z. Surface modification of Al_{30} Keggin-type polyaluminum molecular clusters. *Inorg. Chem.* 2013, 52, 5991–5999.

[29] Abeysinghe S., Unruh D.K., Forbes T.Z. Crystallization of Keggin-type polyaluminum species by supramolecular interactions with disulfonate anions. *Cryst. Growth Des.* 2012, 12, 2044–2051.

[30] Fairley M., Corum K.W., Johns A., Unruh D.K., Basile M., de Groot J., Mason S.E., Forbes T.Z. Isolation and characterization of the $[Ga_2Al_{18}O_8(OH)_{36}(H_2O)_{12}]^{8+}$ cluster: cationic variations on the Wells-Dawson topology. *Chem. Commun.* 2015, 51, 12467–12469.

[31] Baumer A., Gimenez H., Caruba R., Turco G. Remplacements de regroupements atomiques dans la structure de la zunyite. *Bull. Soc. Fr. Minéral. Cristallogr.* 1974, 97, 271–277.

[32] Baumer A., Turco G. Essais de synthèse de la zunyite exempte de fluor par voie hydrothermale. *C.R. Acad. Sci. Paris, série D, T* 1968, 267, 467–470.

[33] Konta J., Mraz L. Dillnite and its relation to zunyite. *Amer. Miner.* 1961, 46, 629–636.

[34] Louisnathan S. J., Gibbs G. V. Aluminum-silicon distribution in zunyite. *Amer. Miner.* 1972, 57, 1089–1108.

[35] Dirken P. J., Kentgens A. P. M., Nachtegaal G. H., Van Der Eerden M. J., Jansen J. B. H. Solid-state MAS NMR study of pentameric aluminosilicate groups with 180° intertetrahedral AI-Q-Si angles in zunyite and harkerite. *Amer. Miner.* 1995, 80, 39–45.

[36] Casey W. H., Olmstead M. M., Phillips B. L. A new aluminum hydroxide octamer, $[Al_8(OH)_{14}(H_2O)_{18}](SO_4)_5.16H_2O$. *Inorg. Chem.* 2005, 44, 4888–4890.

[37] Seichter W., Mögel H.J., Brand P., Salah D. Crystal structure and formation of the aluminium hydroxide chloride $[Al_{13}(OH)_{24}(H_2O)_{24}]Cl_{15}.13\ H_2O$. Eur. *J. Inorg. Chem.* 1998, 795–797.

[38] Breuil H. Sur les chlorures et bromures basiques d'aluminium. *Ann. Chim.* 1965, 10, 467–493.

[39] Walter-Levy, L., Breuil, H. C. R. *Acad. Sci.* 1962, 255,1744–1746.

[40] Jordan P.A., Clayden N.J., Heath S.L., Moore G.R., Powell A.K., Tapparo, A. Defining speciation profiles of $A1^{3+}$ complexed with small organic ligands: the Al^{3+}-heidi system. *Coord. Chem. Rev.* 1996, 149, 281–309.

[41] Hernandez J. Synthèse de nanoparticules d'oxydes de fer et d'aluminium pour l'étude de l'adsorption d'entités inorganiques polycondensées. Thèse, Université Pierre et Marie Curie, Paris, France, 1998.
[42] Bottero, J.Y., Axelos, M., Tchoubar, D., Cases, J.M., Fripiat, J.J., Fiessinger, F. Mechanism of formation of aluminum trihydroxide from Keggin Al_{13} polymers. *J. Colloid Interface Sci.* 1987, 117, 47–57.
[43] Digne M., Sautet P., Raybaud P., Toulhoat H., Artacho E. Structure and stability of aluminum hydroxides: a theoretical study. *J. Phys. Chem. B* 2002, 106, 5155–5162.
[44] Gale J.D., Rohl A.L., Milman V., Warren M.C. An ab initio study of the structure and properties of aluminum hydroxide: gibbsite and bayerite. *J. Phys. Chem. B* 2001, 105, 10236–10242.
[45] Lefèvre G., Pichot V., Fédoroff M. Controlling particle morphology during growth of bayerite in aluminate solutions. *Chem. Mater.* 2003, 15, 2584–2592
[46] Schoen R., Roberson C.E. Structures of aluminum hydroxide and geochemical implications. *Amer. Miner.* 1970, 55, 43–77.
[47] Demichelis R., Catti M., Dovesi R. Structure and stability of the $Al(OH)_3$ polymorphs doyleite and nordstrandite: a quantum mechanical ab initio study with the CRYSTAL06 code. *J. Phys. Chem C* 2009, 113, 6785–6791.
[48] Phambu N. Préparation d'hydroxydes d'aluminium. Caractérisation structurale, morphologique et superficielle. Application à l'étude d'une couche de passivation d'aluminium. Thèse, Université de Nancy 1, France, 1996.
[49] Li H., Addai-Mensah J., Thomas J. C., Gerson A. R. The crystallization mechanism of $Al(OH)_3$ from sodium aluminate solutions. *J. Cryst. Growth* 2005, 279, 508–520.
[50] Bentor Y.K., Gross H., Heller L. Some unusual minerals from the « mottled zone » complex. Israel. *Amer. Mineral.* 1963, 48, 924–930.
[51] Hind A.R., Bhargava S. K., Grocott S. C. The surface chemistry of Bayer process solids: a review. *Coll. Surf. A* 1999, 146, 359–374.
[52] Misra C., White E.T. Crystallisation of Bayer aluminium trihydroxide. *J. Cryst. Growth* 1971, 8, 172–178.
[53] Sweegers C., de Coninck H.C., Meekes H., van Enckevort W.J.P., Hiralal I.D.K., Rijkeboer A. Morphology, evolution and other characteristics of gibbsite crystals grown from pure and impure aqueous sodium aluminate solutions. *J. Crystal Growth* 2001, 233, 567–582.
[54] Freij S. J., Parkinson G. M. Surface morphology and crystal growth mechanism of gibbsite in industrial Bayer liquors. *Hydrometallurgy* 2005, 78, 246–255.
[55] Poisson R., Brunelle J.P., Nortier P. *Catalyst Supports and Supported Catalysts: Theoretical and Applied Concepts*. Boston: A.B. Stiles, Butterworths, 1987.
[56] Jolivet J.P., Froidefond C., Pottier A., Chanéac C., Cassaignon S., Tronc E., Euzen P. Size tailoring of oxide nanoparticles by precipitation in aqueous medium. A semi-quantitative modeling. *J. Mater. Chem.* 2004, 14, 3281–3288.
[57] Chiche D., Digne M., Revel R., Chanéac C., Jolivet J.P. Accurate determination of oxide nanoparticle size and shape based on X-ray powder pattern simulation: Application to boehmite AlOOH. *J. Phys. Chem. C* 2008, 112, 8524–8533.
[58] Callender R.L., Harlan C.J., Shapiro N.M., Jones C.D., Callahan D.L., Wiesner M.R, MacQueen D.B., Cook R., Barron A.R. Aqueous synthesis of water-soluble alumoxanes: environmentally benign precursors to alumina and aluminum-based ceramics. *Chem. Mater.* 1997, 9, 2418–2433.

[59] Chiche D., Chizallet C., Durupthy O., Chanéac C., Revel R., Raybaud P., Jolivet J.P. Growth of boehmite particles in the presence of xylitol: morphology oriented by the nest effect of hydrogen bonding. *Phys. Chem. Chem. Phys.* 2009, 11, 11310–11323.
[60] Chiche D., Chanéac C., Revel R., Jolivet J.P. Use of polyols as particle size and shape controllers: application to boehmite synthesis from sol-gel routes. *Phys. Chem. Chem. Phys.* 2011, 13, 6241–6248.
[61] Paulaine A.M., Seyssiecq I., Veesler S. The influence of organic additives on the crystallization and agglomeration of gibbsite. *Powder Technology* 2003, 130, 345–351.
[62] Zhu H.Y., Gao X.P., Song D.Y., Bai Y.Q., Ringer S.P., Gao Z., Xi Y.X., Martens W., Riches J.D., Frost R.L. Growth of boehmite nanofibers by assembling nanoparticles with surfactant micelles. *J. Phys. Chem. B* 2004, 108, 4245–4247.
[63] Chen X. Y., Zhang Z. J., Li X. L., Lee S. W. Controlled hydrothermal synthesis of colloidal boehmite (γ-AlOOH) nanorods and nanoflakes and their conversion into γ-Al_2O_3 nanocrystals. *Solid State Comm.* 2008, 145, 368–373
[64] Kuang D., Fang Y., Liu H., Frommen C., Fenske D. Fabrication of boehmite AlOOH and γ-Al_2O_3 nanotubes via a soft solution route. *J. Mater. Chem.* 2003, 13, 660–662.
[65] Chiche D. Elaboration d'alumines à faces fortement hydroxylées. Thèse, Université Pierre et Marie Curie, Paris, France, 2007.
[66] Dillinseger C. Modélisation expérimentale de la dissolution des oxy-hydroxydes d'aluminium et de fer dans une solution 5 M NaOH, à 200 et 250°C. Application au procédé Bayer de traitement de la bauxite. Thèse, Université d'Orléans, France, 1995.
[67] Knight C.T.C. A two-dimensional silicon-29 NMR spectroscopic study of the structure of the silicate anions present in an aqueous potassium silicate solution. *J. Chem. Soc. Dalton Trans.* 1988, 1457–1460.
[68] Knight C.T.C., Kirkpatrick R.J., Oldfield E. Silicon-29 Multiple quantum filtered NMR spectroscopic evidence for the presence of only six single site silicate anions in a concentrated potassium silicate solution. *J. Chem. Soc. Chem. Commun.* 1989, 919–921.
[69] Iler R.J. *The Chemistry of Silica*. New York: Wiley, 1979.
[70] Bassetti V., Burlamacchi L., Martini G. Use of paramagnetic probes for the study of liquid adsorbed on porous supports. Copper(II) in water solution. *J. Am. Chem. Soc.* 1979, 101, 5471–5477.
[71] Lartiges B. S., Bottero J. Y., Derrendinger L. S., Humbert B., Tekely, P., Suty, H. Flocculation of colloidal silica with hydrolyzed aluminum: an ^{27}Al solid state NMR investigation. *Langmuir* 1997, 13, 147–152.
[72] James R.O., Healy T.W. Adsorption of hydrolyzable metal ions at the oxide-water Interface I. Co(ll) adsorption on SiO_2 and TiO_2 as model systems. *J. Colloid Interface Sci.* 1972, 40, 42–52.
[73] Gehlen M., Beck L., Calas G., Flank A.M., Van Bennekom A. J., Van Beusekom J. E. E. Unraveling the atomic structure of biogenic silica: Evidence of the structural association of Al and Si in diatom frustules. *Geochim. Cosmochim. Acta* 2002, 66, 1601–1609.
[74] Koning E., Gehlen M., Flank A.-M., Calas G., Epping E. Rapid post-mortem incorporation of aluminum in diatom frustules: Evidence from chemical and structural analyses. *Mar. Chem.* 2007, 106, 208–222.

[75] Swaddle T.W., Salerno J., Tregloan P. A. Aqueous aluminates, silicates, and aluminosilicates. *Chem. Soc. Rev.* 1994, 319–325.

[76] Mortlock R.F., Bell A.T., Radke C.J. NMR Investigations of tetrapropylammonium aluminosilicate and borosilicate solutions. *J. Phys.Chem.* 1991, 95, 372–378.

[77] Mortlock R.F., Bell A.T., Chakraborty A.K., Radke C.J. Effect of silicate ratio on the distribution of silicate and aluminosilicate anions in TPA aluminosilicate solutions. *J. Phys.Chem.* 1991, 95, 4501–4506.

[78] Mortlock R.F., Bell A.T., Radke C.J. Incorporation of aluminum into silicate anions in aqueous and methanoic solutions of TMA silicates. *J. Phys.Chem.* 1991, 95, 7847–7851.

[79] Breck D.W. *Zeolite Molecular Sieves. Structure, Chemistry and Uses.* New York: Wiley, 1974.

[80] Barrer R.M. *Hydrothermal Chemistry of Zeolites.* London: Academic Press, 1982.

[81] Cundy C.S., Cox P.A. The hydrothermal synthesis of zeolites: precursors, intermediates and reaction mechanism. *Micropor. Mesopor. Mater.* 2005, 82, 1–78.

[82] Kinrade S.D., Knight C. T. G., Pole D.L., Syvitski R.T. Silicon-29 NMR studies of tetraalkylammonium silicate solutions. 1. Equilibria, ^{29}Si chemical shifts, and ^{29}Si relaxation. *Inorg. Chem.* 1998, 37, 4272–4277.

[83] Kinrade S.D., Knight C. T. G., Pole D.L., Syvitski R.T. Silicon-29 NMR studies of tetraalkylammonium silicate solutions. 2. Polymerization kinetics. *Inorg. Chem.* 1998, 37, 4278–4283.

[84] Swaddle T.W. Silicate complexes of aluminum(III) in aqueous systems. *Coord. Chem. Rev.* 2001, 219–221, 665–686.

[85] Berry L.G., Mason B. *Mineralogy, Concepts, Description, Determination*, 2nd ed. San Francisco: W.H. Freeman, 1972.

[86] Caillère S., Hénin S. Rautureau M., *Minéralogie des argiles,* Vols. 1 and 2. Paris: Masson, 1982.

[87] Kloprogge J.T., Komarneni S., Amonette J. E. Synthesis of smectite clay minerals; a critical review. *Clays Clay Miner.* 1999, 47, 529.

[88] Harder H. The role of magnesium in the formation of smectite minerals. *Chem. Geol.* 1972, 10, 31–39.

[89] Harder H. Clay mineral formation under lateritic weathering conditions. *Clay Miner.* 1977, 12, 281–288.

[90] Decarreau A. Experimental crystallogenesis of Mg-smectite: hectorite, stevensite. *Bull. Miner.* 1980, 103, 579.

[91] Levard C. Nanoparticules naturelles: imogolites et allophanes. Structure, mécanismes de croissance et capacité de rétention des éléments traces métalliques. Thèse, Université Aix-Marseille III, France, 2008.

[92] Cradwick P.D.G., Farmer V.C., Russel J.D., Masson C.R., Wada K., Yoshinaga N. Imogolite, a hydrated aluminium silicate of tubular structure. *Nature Phys. Sci.* 1972, 240, 187–189.

[93] Creton B., Bougeard D., Smirnov K.S., Guilment J., Poncelet O. Structural model and computer modeling study of allophone. *J. Phys. Chem. C* 2008, 112, 358–364.

[94] Farmer V.C., Fraser A.R., Tait J. M. Synthesis of imogolite: a tubular aluminum silicate polymer. *J. Chem. Soc. Chem. Commun.* 1977, 462–463.

[95] Levard C., Masion A., Rose J., Doelsch E., Borschneck D., Dominici C., Ziarelli F., Bottero J.Y. Synthesis of imogolite fibers from decimolar concentration at low

temperature and ambient pressure: a promising route for inexpensive nanotubes. *J. Am. Chem. Soc.* 2009, 131, 17080–17081.

[96] Guimarães L., Enyashin A.N., Frenzel J., Heine T., Duarte H.A., Seifert G. Imogolite nanotubes: stability, electronic and mechanical Properties. *ACS Nano* 2007, 1, 362–368.

[97] Levard C., Rose J., Thill A., Masion A., Doelsch E., Maillet P., Spalla O., Olivi L., Cognigni A., Ziarelli F., Bottero J.-Y. Formation and growth mechanisms of imogolite-like alumino-germanate nanotubes. *Chem. Mater.* 2010, 22, 2466–2473.

[98] Konduri S., Mukherjee S., Nair S. Controlling nanotube dimensions: correlation between composition, diameter and internal energy of single-walled mixed oxide nanotubes. *ACS Nano* 2007, 1, 393–402.

[99] Maillet P., Levard C., Spalla O., Masion A., Rose J., Thill A. Growth kinetic of single and double-walled aluminogermanate imogolite-like nanotubes: an experimental and modeling approach. *Phys. Chem. Chem. Phys.* 2011, 13, 2682–2691.

[100] Maillet P., Levard C., Larquet E., Mariet C., Spalla O., Menguy N., Masion A., Doelsch E., Rose J., Thill A. Evidence of double-walled Al-Ge imogolite-like nanotubes. A cryo-TEM and SAXS investigation. *J. Am. Chem. Soc.* 2010, 132, 1208–1209.

[101] Thill A., Guiose B., Bacia-Verloop M., Geertsen V., Belloni L. How the diameter and structure of $(OH)_3Al_2O_3Si_xGe_{1-x}OH$ imogolite nanotubes are controlled by an adhesion versus curvature competition. *J. Phys. Chem. C* 2012, 116, 26841–26849.

[102] Thill A., Maillet P., Guiose B., Spalla O., Belloni L., Chaurand P., Auffan M., Olivi L., Rose J. Physico-chemical control over the single- or double-wall structure of aluminogermanate imogolite-like nanotubes. *J. Am. Chem. Soc.* 2012, 134, 3780–3786.

7 }

Iron Oxides

AN EXAMPLE OF STRUCTURAL VERSATILITY

Iron is Earth's fourth most widespread element (6.2% in mass), behind oxygen, silicon, and aluminum. It exists mostly as ferric oxide and oxyhydroxide (Fig. 7.1a) and to a lesser extent as sulfide (pyrite), carbonate (siderite), and silicate (fayalite). Iron oxides are largely used in technological areas such as metallurgy, colored pigments, magnetic materials, and catalysts. They also play an important role in the environment because the dissolution of ferric oxides in natural waters, promoted by acid–base, redox, photochemical phenomena, and also microbial mediation [1, 2], allows iron to be involved in many biogeochemical processes [3–5].

Iron is present in many living organisms such as plants, bacteria, mollusks, animals, and humans in various forms:

- Porphyrinic complexes of iron, which are active centers of hemoglobin and several ferredoxins involved in biological functions, especially respiration mechanism and photosynthesis [6].
- Nanoparticles of amorphous ferric oxyhydroxides in animal and human organisms as ferritin, which allows regulation and storage of iron [7] and in various nanophases present in plants as phytoferritin [8].
- Crystalline iron oxy(hydroxi)des produced by biomineralization processes [9]. Goethite, lepidocrocite, and magnetite are the main constituents of radulas and the teeth of mollusks (limpets, chitons). Magnetite nanoparticles produced by magnetotactic bacteria (Fig. 7.1b), as well as by bees and pigeons, are used for purposes of orientation and guiding along the lines of force of the Earth's magnetic field [3, 10]. Green rusts are also ferric-ferrous compounds belonging to the biogeochemical cycle of iron [11].

The crystal chemistry of iron oxy(hydroxi)des is very rich. The ferric, ferrous, and mixed ferric-ferrous oxygenated compounds correspond to around a dozen crystal structural types (Fig. 7.2) [3, 12]. Most of these crystal phases can be synthesized from solutions in the laboratory, giving rise to a most diversified chemistry [13, 14]. They are also formed in nature because of the large variability

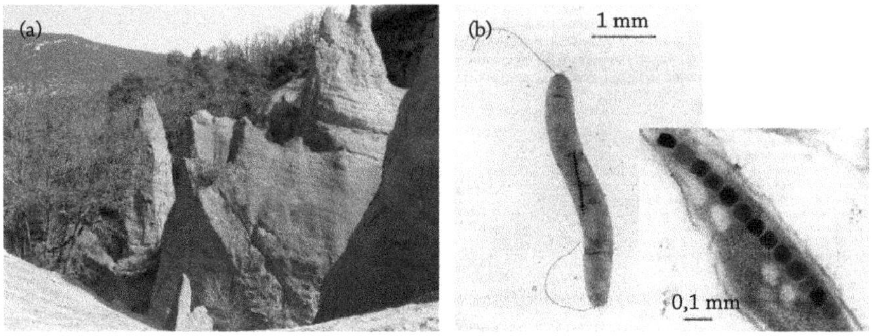

FIGURE 7.1 (a) Landscape of Luberon (southwestern France) showing a deposit of ochre (goethite and hematite). (b) Magnetotactic bacterium containing an alignment of magnetite particles (reproduced with permission from [Fukumori Y., Enzymes for magnetite synthesis in *Magnetosprillum magnetotacticum*, in Biomineralization, E. Bauerlein, Ed., 2nd ed., p. 75, Weinheim, Germany: Wiley-VCH, 2004] Copyright Wiley-VCH Verlag GmbH & Co. KGaA).

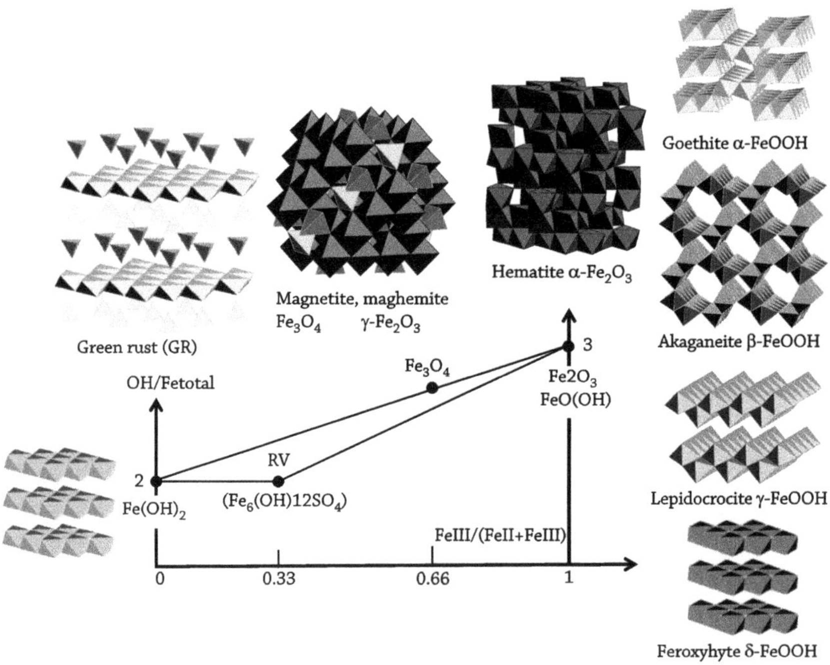

FIGURE 7.2 Main structural types of iron oxy(hydroxi)des as a function of the ferric-ferrous composition Fe^{III}/Fe_{total} and the hydroxylation ratio OH/Fe_{total}.

of physicochemical conditions: an acidity range from around pH 0 to 13; redox conditions from oxic to totally anoxic media; bacterial activity that can be extremely intense; salinity largely varying from almost pure waters to real brines; presence of many organic and inorganic ligands; and various photochemical processes.

The large versatility of iron aqueous chemistry results essentially from the existence of two oxidation states that are stable over a wide range of acidity and also from a high reactivity of the iron complexes toward redox and acid–base phenomena. In this chapter, we examine the formation of clusters and crystallization—in the laboratory—of ferrous, ferric, and mixed-valent compounds.

7.1 Speciation of Iron and Condensation in Aqueous Solution

Like many 3*d* transition elements in aqueous solution, iron in its most common oxidation states, +II and +III, forms hexaaquo complexes $[Fe(OH_2)_6]^{z+}$ in which polarization of the coordinated water molecules depends closely on the oxidation state and on the size of the cation. For these reasons, ferric ions ($z = 3$, $r_{ion} = 0.64$ Å) are much more acidic than ferrous ions ($z = 2$, $r_{ion} = 0.77$ Å). Thus, hydroxylation of complexes occurs along distinct acidity ranges: between pH 7 and pH 9 at room temperature for ferrous ions; and from pH 1 to pH 4 to 5 for ferric ions (Fig. 7.3). The iron aquohydroxo complexes $[Fe(OH)_h(OH_2)_{6-h}]^{(z-h)+}$ exhibit a strong trend toward condensation. The reaction, proceeding by olation, is rapid because of the high lability of coordinated water molecules.

The formation of ferrous polycationic species is poorly documented. As for other divalent cations (§ 4.1.1), a tetramer $[Fe_4(OH)_4(OH_2)_{12}]^{4+}$ seems to be formed near pH 6 (Fig. 7.4) from $h = 1$ hydroxylated complexes [15], but to our knowledge, no structural data have been reported. In principle, the $h = 1$ complex should form a dimer, but the hydroxo ligands, weakly polarized by divalent cations, remain highly basic in the dimer. Thus, these ligands tend to increase their coordination by association of dimers in forming a tetramer in which all hydroxo ligands are tricoordinated (Fig. 7.4).

The chemistry of ferric iron is rich and more complicated than that of other trivalent elements (Al, Cr) because of the high reactivity of Fe^{3+} ions to substitution and condensation reactions (§ 4.1.3). It is difficult to stop condensation in solution

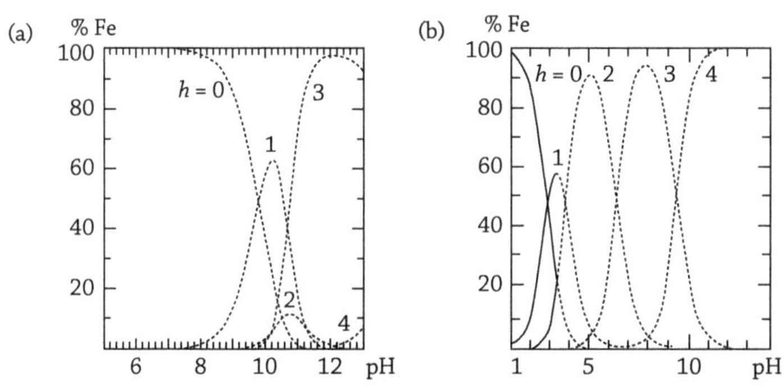

FIGURE 7.3 Speciation of ferrous (a) and ferric (b) ions in highly diluted solution at room temperature. The *h* values correspond to $[Fe(OH)_h(OH_2)_{6-h}]^{(z-h)+}$ complexes. The dotted lines indicate supersaturation with respect to $Fe(OH)_2$ (a) and to goethite α-FeOOH (b) (adapted from [15]).

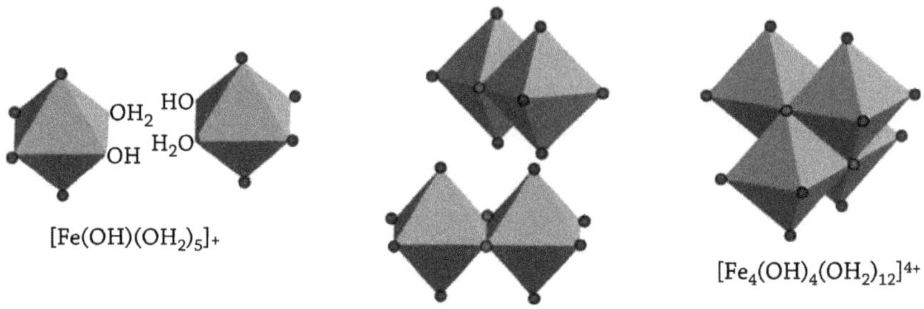

FIGURE 7.4 Ferrous species in solution and formation scheme of the $[Fe_4(OH)_4(OH_2)_{12}]^{4+}$ polycation from the $[Fe(OH)(OH_2)_5]^+$ complex.

of ferric complexes at a molecular polycationic stage, even when pH is carefully controlled, as can be done with Cr^{3+} and Al^{3+} ions. Although dimers and trimers have been evidenced using X-ray absorption techniques [16], once initiated, hydroxylation of ferric complexes likely becomes progressively easier and leads to their partition in solution in the form of highly hydroxylated complexes rapidly precipitated and others that are less hydroxylated, even if not at all. Such behavior has been encountered with Ni^{2+} ions in solution (§ 4.1.1). It has also been evidenced in recent studies of partially hydrolyzed ferric solutions ($C_{Fe} = 0.2$ mol.l^{-1}, $1.8 < pH < 3$, $0.5 < h < 2.5$) using in situ time-resolved synchrotron-based X-ray diffraction, quick-scanning EXAFS, rapid UV–vis, and quenched frozen-solution Mössbauer spectroscopic analyses [17, 18]. These studies show that $h \leq 0.5$ solutions contain mainly the hexaaquo $[Fe(OH_2)_6]^{3+}$ complex, the μ-oxo $[Fe_2O(OH_2)_{10}]^{4+}$ dimer (surprisingly instead of the μ_2-hydroxo $[Fe_2(OH)_2(OH_2)_8]^{4+}$ species; this point is discussed later in this chapter), and a small amount of $[Fe(OH)(OH_2)_5]^{2+}$. From $h = 0.7$, the same species are detected in solution with the presence of growing amounts of solid, ferrihydrite (§ 7.2.2), as h increases. The amount of dimer decreases, and that of the two other species, monomer and solid, become more abundant with increasing reaction time. This shows that polycationic entities with increasing size when hydroxylation ratio, h, increases do not form, as occurs with Al and Cr, because the growth of a solid is continuously fed.

Under these conditions, it is also possible that the lifetime of such species is too short to be detected. Thus, control of the reactivity of ferric ions requires the use of strongly complexing polydentate ligands and allows the chelation of the cation. In this way, a tetramer can be obtained at pH 4.5 (80°C, acetate buffer), in the reaction of ferric chloride and 1,3-diamino-2-hydroxypropane-N,N,N′,N′-tetraacetic acid (HPDTA) in the presence of alkali cation [19]. The HPDTA ligand (Scheme 7.1a) behaves as a heptadentate ligand able to coordinate two Fe^{3+} cations by carboxylate and amine functions (Scheme 7.1b; oxygen and nitrogen atoms in HPDTA are labeled 1 and 2, respectively). Two similar units are linked by two oxo or hydroxo bridges (labeled 3) to form the tetramer. The coordination of the ferric ions is completed by oxygen atoms (labeled 2) from acetate, carbonate, or water.

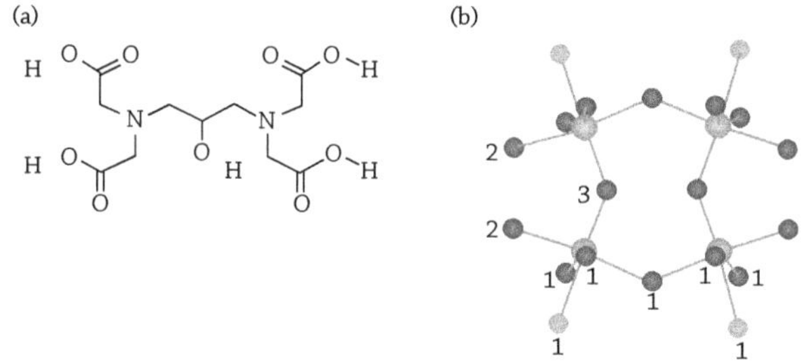

SCHEME 7.1

Many other polycationic ferric complexes exhibiting varied structures have been obtained in various aqueous and nonaqueous media with polycarboxylate or polyamino ligands [20–23] (Fig. 7.5). In reference to aluminum, a Fe_{13} cluster has Keggin's structure similar to that of the ε-Al_{13} polycation (Fig. 7.5) but formed under different conditions. The $[Fe_{13}(\mu_3\text{-}OH)_6(\mu_3\text{-}O)_6(OH_2)_6(Hntp)_8]^{5+}$ cluster is obtained in water in the presence of the chelating multipode nitrilotripropionate ligand [24], while the anionic cluster $[Fe_{13}O_4F_{24}(OCH_3)_{12}]^{5-}$, including fluoride and methoxy ligands, is formed by heating $FeF_3{\cdot}3H_2O$ in methanol in the presence of pyridine under anaerobic conditions [25].

In ferric polycationic species, Fe^{3+} ions are most often linked with oxo bridges rather than hydroxo bridges like Al^{3+} and Cr^{3+} ions lies in their respective polycations. It should also be noted that ferric hydroxide, $Fe(OH)_3$, does not exist (see later), whereas $M(OH)_3$ hydroxides of aluminum and chromium are stable phases. The bridging hydroxo ligand in ferric structures thus appears to be of rather low stability, likely because iron has a higher electronegativity than that of chromium and aluminum: the charge transfer from bridging oxygen to Fe^{3+} cations is more intense, and the higher acidity of the hydroxo bridge favors the formation of the oxo bridge. It is also possible that deprotonation of the hydroxo bridge favors orbital overlapping between iron and oxygen in order to optimize the superexchange magnetic interaction within the antiferromagnetic coupling of ferric ions in many molecular and solid structures.

In a strongly alkaline medium (NaOH > 10 $mol.l^{-1}$), the ferric iron, like Al^{3+} ions, is tetrahedrally coordinated in the $[FeO_4]^{5-}$ anionic complex (ferrate III). This tetraoxo species does not condense, and it exists in solid state as isolated entities in the compound Na_5FeO_4 [26]. Under similar conditions, the ferrous ions also are soluble in the form of anionic species $[Fe(OH)_3]^-$ and $[Fe(OH)_4]^{2-}$ with a coordination that is not yet clearly defined [27].

Ferrates can exist with unusual +IV, +V, +VI oxidation states (o.s.) for iron. They form anionic tetraoxo $[FeO_4]^{x-}$ species in which the Fe–O bonds shorten as the oxidation state of iron increases: $d_{Fe\text{-}O}$ = 1.889 Å (o.s. +III), 1.807 Å (o.s. +IV), 1.720 Å (o.s. +V), 1.647 Å (o.s. +VI) [28–30]. These species are formed in strongly alkaline medium under various conditions: heating of FeO in the presence of Na_2O_2 at 400°C to form $Na_4Fe^{IV}O_4$; heating the mixture $Ba(OH)_2$, KOH, and Fe_2O_3 at 750°C

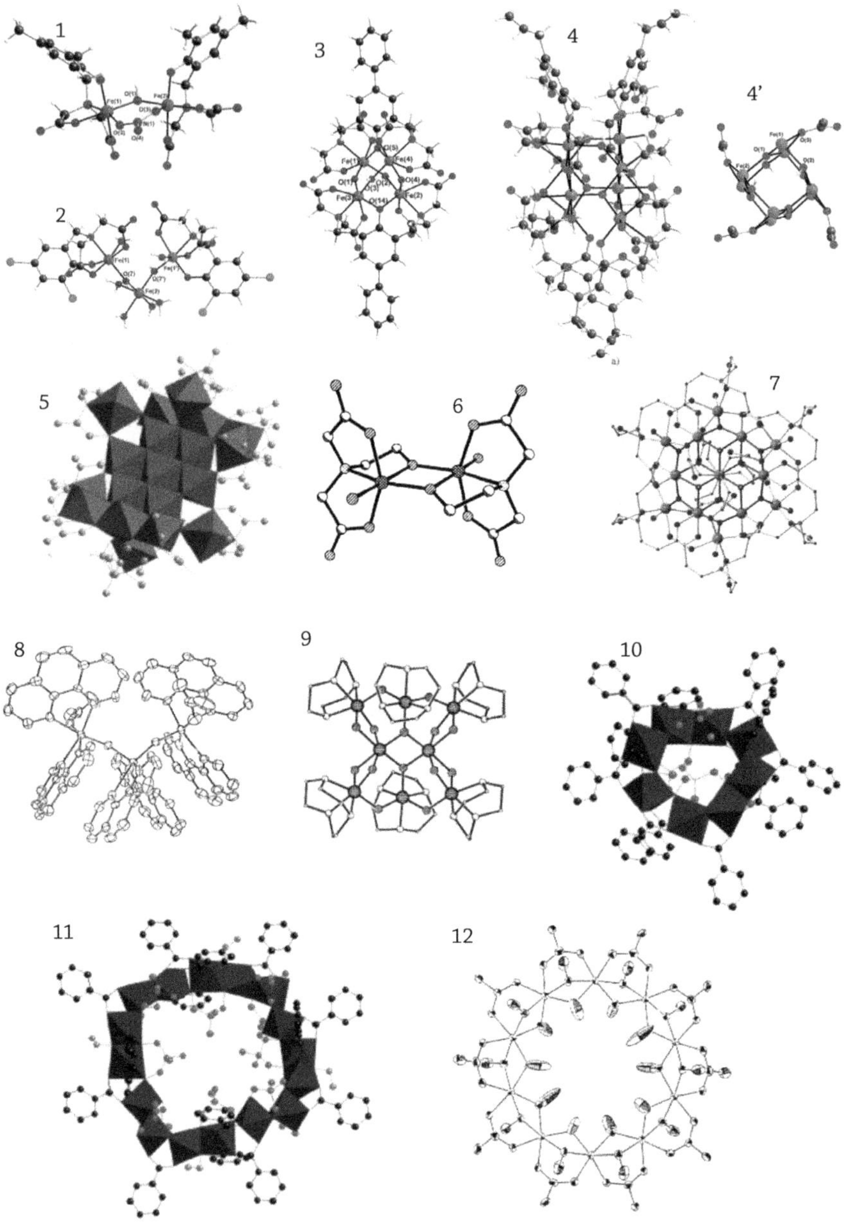

FIGURE 7.5 Structures of some ferric complexes formed in the presence of strongly complexing multidentate ligands. Iminodiacetate R-N(CH_2COO)$_2$ ligands: $[(Me_2hda)_2Fe_2(\mu\text{-}OH)(\mu\text{-}NO_3)]^{2-}$ **1**, $[(Cl_2hda)_2Fe_2(\mu\text{-}OH)_2Fe(OH_2)_4]^{+}$ **2**, $[(4\text{-Ph-hxta})_2Fe_4(\mu\text{-}O)(\mu\text{-}OH)_3]^{3-}$ **3**, $[(2\text{-allyl-4-OMe-hda})_4Fe_8(\mu_3\text{-}O)_4(\mu\text{-}OH)_4(en)_4]$ **4**, and view of the core $Fe_8(\mu_3\text{-}O)_4(\mu\text{-}OH)_4$ **4′** (reproduced from [23] with permission of Elsevier). Heidi (L) ($HOCH_2CH_2$)N(CH_2COOH)$_2$ ligand: $[Fe_{19}O_6(OH)_{14}L_{10}(H_2O)_{12}]^{+}$ **5**, $[FeL(H_2O)_2]_2$ **6** with L/Fe = 0.5 and 1, respectively (reproduced with permission from Heath S.L., Powell A.K. The trapping of iron hydroxide units by the ligand "heidi": two new hydroxo(oxo)iron clusters containing 19 and 17 iron atoms. Angew. Chem. Int. Ed. Eng. 1992, 31, 191–193] Copyright Wiley-VCH Verlag

to form $Ba_2Fe^{IV}O_4$; heating hematite Fe_2O_3 at around 350°C in the presence of sodium or potassium peroxide under oxygen stream to form $Na_2Fe^{VI}O_4$ [31]; and anodic dissolution of iron in NaOH or in a KOH-NaOH highly concentrated solution to form soluble $[Fe^{VI}O_4]^{2-}$ species or crystals of $K_2Fe^{VI}O_4$, respectively [32, 33]. All these compounds are unstable both in solid state and in acidic to neutral solution because they are powerful oxidizers [34–38]. Ferrate (VI), with a standard electrochemical potential $E°$ ($[Fe^{VI}O_4]^-/Fe^{III}$) of 2.20 V, is more oxidizing than ozone ($E°$ (O_3/O_2) = 2.07 V). It is therefore used for water treatment because of its oxidizing power as ferrate and also because of the flocculating ability of ferric oxyhydroxide particles produced by its reduction [39].

7.2 Formation of Solid Phases

The precursor of the solid is always a noncharged aquohydroxo complex. The starkly different crystal chemistry of the Fe^{II} and Fe^{III} cations can be understood not only from knowledge of the functionality of precursors, but also according to the process involved in precipitation—for instance, base addition in the solution of precursors or thermolysis of acid solutions in the temperature range of around 60–100°C. Some effects of hydrothermal treatment of solutions are examined later (§ 7.2.2 c).

7.2.1 FERROUS HYDROXIDE AND OXIDIZED DERIVATIVES: FEROXYHYTE AND LEPIDOCROCITE

The noncharged ferrous complex, $[Fe(OH)_2(OH_2)_4]^0$, formed at around pH 6 to 7, under strictly anaerobic conditions, leads to the hydroxide, $Fe(OH)_2$. Like the hydroxides of many divalent elements (§ 4.1.1), $Fe(OH)_2$ has the brucite-type structure, which is built by stacking sheets of edge-sharing octahedra (Fig. 7.6).

FIGURE 7.5 Continued

GmbH & Co. KGaA). Nitrilotripropionate ($H_3ntp = N(CH_2CH_2COOH)_3$) ligand: $[Fe_{13}(\mu_3\text{-}OH)_6(\mu_3\text{-}O)_6(H_2O)_6(Hntp)_8]^{5+}$ **7** (Reproduced with permission from [24] Copyright Wiley-VCH Verlag GmbH & Co. KGaA); $[Fe_3O_2(phen)_6(OOC)_2]^{3+}$ **8** (Reproduced with permission from [Vincent J.M., Menage S., Latour J.L., Bousseksou A., Tuchagues J.P., Decian A., Fontecave M. A new type of trinuclear oxoiron(III) cluster. Angew. Chem. Int. Ed. Engl. 1995, 34, 205–207] Copyright Wiley-VCH Verlag GmbH & Co. KGaA); $[Fe_8(\mu_3\text{-}O)_2(\mu_2\text{-}OH)_{12}(tacn)_6]^{8+}$ (tacn = 1,4,7-triazacyclononane) **9** (Reproduced with permission from [Wieghardt K., Pohl K., Jibril I., Huttner G. Hydrolysis products of the monomeric amine complex $(C_6H_{15}N_3)FeCl_3$: the structure of the octameric iron(III) cation of $\{[(C_6H_15N_3)_6Fe_8(\mu_3\text{-}O)_2(\mu_2\text{-}OH)_{12}]Br_7(H_2O)\}Br.8H_2O$. Angew. Chem. Int. Ed. Engl. 1984, 23, 77–78] Copyright Wiley-VCH Verlag GmbH & Co. KGaA); $[Fe_8(PhCOO)_{12}(thme)]_4$ 2 Et_2O **10** and $[Fe_{16}(EtO)_4(PhCOO)_{16}(Hthme)_{12}](NO_3)_4.4\ CH_2Cl_2.10\ H_2O$ **11** (H_3thme = 1,1,1-tris(hydroxymethyl)ethane (Jones 2002); $[Fe(OMe)_2(O_2CCH_2Cl)]_{10}$ **12** (reproduced with permission from [Taft K.L., Lippard S.J. Synthesis and structure of $[Fe(OMe)_2(O_2CCH_2Cl)]_{10}$, a molecular ferric wheel. J. Am. Chem. Soc. 1990, 112, 9629–9630] Copyright 1990 American Chemical Society).

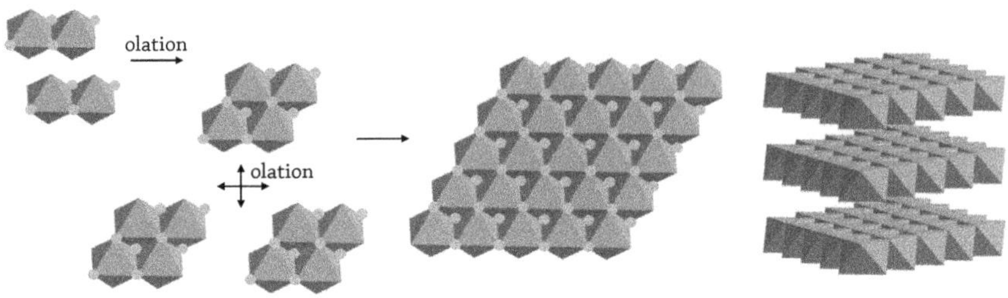

FIGURE 7.6 (a) Possible pathway for the formation of ferrous hydroxide $Fe(OH)_2$ from complexes in solution. (b) Scheme of the brucite structure.

There is obviously no structural relationship between the compact tetramer formed from the $h = 1$ complex (Fig. 7.4) and hydroxide (Fig. 7.6). In fact, the $h = 2$ noncharged precursor, bifunctional from the condensation standpoint, can form dimers leading readily by olation to planar tetramers $[M_4(OH)_8(OH_2)_8]^0$ containing tricoordinated hydroxo groups. Such a species could be right building units for the solid because they can assemble bidirectionally by olation to form the brucite-type sheets of the crystal structure (Fig. 7.6, § 4.1.1).

In solid state or in aqueous suspension, ferrous hydroxide is sensitive to oxidation, and different phases are formed depending on existing conditions, as follows.

- When oxidation proceeds rapidly with a strong oxidant such as hydrogen peroxide (H_2O_2) in a basic medium (pH ~ 10 to 12), feroxyhyte, δ-FeOOH, is formed (Fig. 7.2). This phase retains the brucite-type structure but with stacking faults and even the presence of cations in tetrahedral holes [40–42]. As feroxyhyte is never obtained by precipitation of ferric ions, its formation likely proceeds by in situ oxidation of ferrous hydroxide, with deprotonation of half of the hydroxo ligands. Feroxyhyte thus appears as the oxidized track of ferrous hydroxide.
- When oxidation of ferrous hydroxide occurs in softer conditions, and especially in air around pH 7, lepidocrocite, γ-FeOOH, isostructural with boehmite, γ-AlOOH, is obtained (Fig. 7.7). As for feroxyhyte, this phase does not directly form from precipitation of ferric ions. Since oxidation of ferrous ions occurs near pH 7, which corresponds to the minimum solubility of ferric ions (Fig. 7.8), the transformation likely results from a reaction in solid state. Probably because of the slowness of the process, it involves the local rearrangement of the brucite sheets in forming double chains of octahedra characteristic of the corrugated sheets of lepidocrocite [43, 44]. Such double chains are also present in goethite, α-FeOOH, and akaganeite, β-FeOOH, but with different linking (see later). The process induces fragmentation of the initial hydroxide particles to form slat-shaped nano-objects (Fig. 7.7a). Lepidocrocite can be transformed into hematite, α-Fe_2O_3, or into maghemite, γ-Fe_2O_3, by dehydration according to a topotactic process [45].

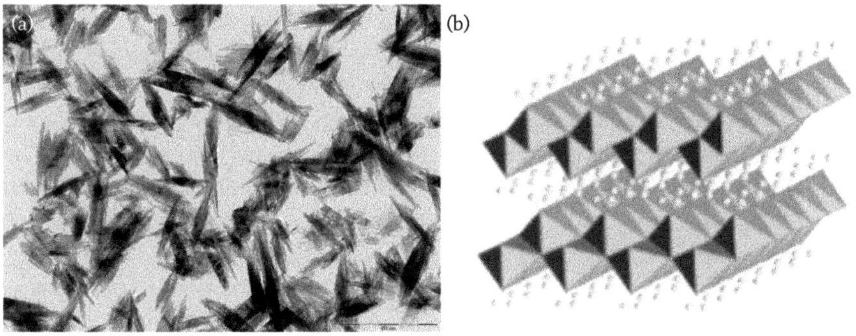

FIGURE 7.7 (a) TEM image of lepidocrocite, γ-FeOOH, particles resulting from aerial oxidation of ferrous hydroxide $Fe(OH)_2$ in aqueous suspension. (b) Crystal structure of lepidocrocite (the small spheres represent the hydrogen atoms of hydroxo ligands).

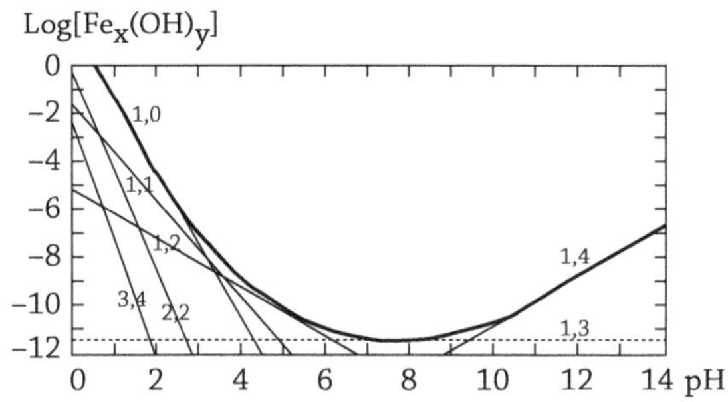

FIGURE 7.8 Solubility diagram of the amorphous ferric gel (adapted from [15]). Soluble species are labeled (x, y) in correspondence with the formal composition $[Fe_x(OH)_y]$.

- Incomplete oxidation of ferrous hydroxide may lead to other mixed-valent phases such as green rusts and magnetite (see § 7.2.3).

7.2.2 FERRIC COMPOUNDS: FERRIHYDRITE, GOETHITE, HEMATITE, AKAGANEITE

The behavior of ferric ions in solution is strongly dependent on the physicochemical conditions of hydroxylation (e.g., processing, temperature, presence of anions, or complexing ligands), which gives the structural chemistry of ferric oxy(hydroxi)des a high complexity that is increased by various surface and size effects (§ 1.1.3). Indeed, at the nanometric scale, metastable phases in bulk state may become thermodynamically stable, and they can be formed in syntheses in the laboratory as well as in natural conditions when they are not a priori expected [46–49].

a) Alkalinization of Solutions: Ferrihydrite, Goethite, Hematite

The addition of a base up to pH > 3 to a solution of ferric chloride or nitrate at room temperature forms quasi-instantaneously a poorly organized, heavily hydrated, and poorly defined composition solid phase made of very small particles around 2 nm in size. This solid, called "ferric gel" (§ 4.1.3) or ferrihydrite, exhibits an X-ray diffraction pattern containing only two wide peaks, hence the name "two-line ferrihydrite" [50, 51]. As previously mentioned, ferrihydrite forms from a low hydroxylation rate of the ferric solution ($h \geq 0.7$), in coexistence with the hexaaquo monomer and the μ-oxo dimer $[Fe_2O(OH_2)_{10}]^{4+}$ [18]. However, this dimer is not structurally compatible with any iron oxide because of the long Fe···Fe distance of about 3.6 Å [17]. In ferrihydrite, the Fe···Fe distances are 3.04 and 3.45 Å for edge- and corner-sharing octahedra, respectively [18]. Nucleation of the solid likely involves reconfiguration of this species into the μ_2-hydroxo dimer $[Fe_2(OH)_2(OH_2)_8]^{4+}$. Quantum chemical calculations show that their free enthalpies do not differ substantially [17]. This last species is probably immediately incorporated into ferrihydrite in releasing H^+ ions and thus is not detected in solution. The increase in pH, which causes a more abundant precipitation of ferrihydrite, could favor the formation of this μ_2-hydroxo dimer.

Many efforts have been and are still being expended in characterizing the structure of ferrihydrite. It is an ubiquitous material important in various ways, especially from an environmental standpoint because of its role in trapping metals and heavy elements, especially arsenic [52, 53]. It is also the inorganic constituent of ferritin, a protein that allows storage and regulation of iron in living organisms [7, 54]. Ferrihydrite's structure is very controversial, specifically with regard to the presence of tetrahedral ferric ions in coexistence with octahedral ferric ions. On the one hand, it is believed that the local ordering in ferrihydrite is similar to that existing in goethite and akaganeite in which iron is only hexacoordinated [42, 55–58], while on the other hand, it is claimed that an important proportion, up to 25 to 30%, of tetracoordinated iron is present in ferrihydrite according to the structural model of akdalaite $^{VI}Al_8{}^{IV}Al_2O_{14}(OH)_2$ [59–63]. Tetracoordinated iron in ferrihydrite is somewhat astonishing because, to our knowledge, it does exists neither in soluble ferric species (at the difference of aluminum and except in strongly alkaline medium (§ 7.1)), nor in solid phases other than magnetite and maghemite, which are of a spinel-type structure. In addition, the aging of ferrihydrite in suspension in the absence of ferrous and other divalent ions never leads to structured solids containing tetrahedral iron. Consequently, there is no reason for a spontaneous reduction in coordination of ferric ions within a particularly metastable solid phase in which the presence of tetrahedral iron thus appears highly doubtful. Taking account of tetrahedral iron in ferrihydrite could likely result from the fitting procedures of experimental data essentially provided by X-ray absorption spectroscopy [42, 64–66].

Because of its poor structural ordering and its extreme division state (the mean particle size deduced from electron microscopy observation is about 2–3 nm), the two-line ferrihydrite is thermodynamically unstable (§ 1.1.3a). In aqueous suspension, it transforms into various crystalline phases according to different pathways that are dependent on the acidity and presence of diverse cations and anions [3, 50]. In purely ferric systems at pH between around 5 and 8, ferrihydrite transforms slowly into very small particles of hematite, α-Fe_2O_3 [67]. As in these acidity conditions the solubility of iron is low, around 10^{-7} mol.l^{-1} (Fig. 7.8), the transformation proceeds by dehydration and local rearrangement. That explains the formation of an oxide under the form of very small particles because the structural ordering, proceeding only by water elimination and ion diffusion within the solid, cannot spread on large distances. The EXAFS study of aging of the ferric gel at pH 6.5 and 92°C [67] clearly shows the evolution of Fe-Fe distances: until 6 hours after the synthesis, distances of 3.40 and 3.05 Å corresponding to corner and edge sharing, respectively, between FeO_6 octahedra are observed. Beyond 6 hours, a new short Fe-Fe distance at 2.9 Å appears, corresponding to the face sharing of FeO_6 octahedra typical of the corundum-type structure of hematite (Fig. 7.9). After 130 hours in these conditions, crystallization is achieved.

When the solubility of the ferric gel is higher, approximately at pH lower than 4 and higher than 8, the two-line ferrihydrite transforms into goethite, α-FeOOH, via a dissolution–crystallization process. At the difference of aluminum which forms hydroxide $Al(OH)_3$ in similar synthesis conditions, ferric hydroxide, $Fe(OH)_3$, is never obtained, nor is it identified by X-ray diffraction. This difference in behavior is likely due to the higher electronegativity of iron, which is more polarizing for hydroxo ligands than aluminum and preferentially forms an oxyhydroxide. However, it exists only as a very rare mineral, bernalite, $Fe(OH)_3 \cdot nH_2O$ ($0 \leq n \leq 0.25$), found in the Broken Hill Mine in Australia, which has been structurally identified [68]. It is in fact a hydrated hydroxide with a perovskite-like structure made of $Fe(OH)_6$ octahedra linked by corners and

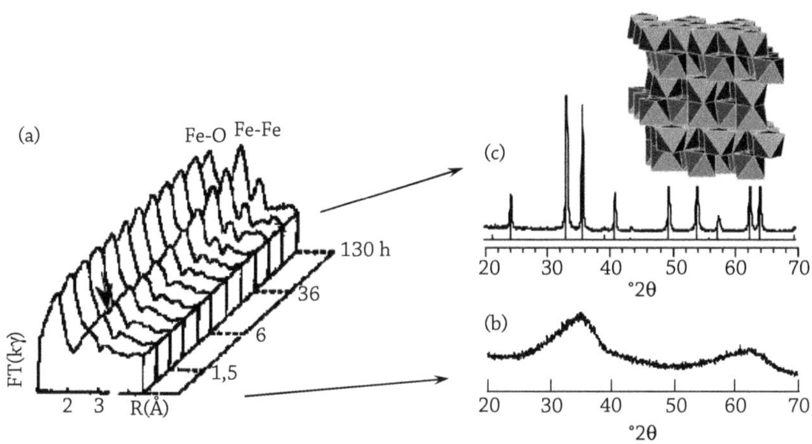

FIGURE 7.9 (a) Evolution of the EXAFS radial distribution function around Fe atoms during crystallization into hematite of a ferric gel in suspension at 92 °C (reproduced from [67] with permission of Elsevier). The arrow indicates the appearance of a shoulder characteristic of a Fe-Fe distance at 2.86 Å typical of face-sharing FeO_6 octahedra. X-ray diffraction patterns of (b) two-line ferrihydrite and (c) hematite with a scheme of the crystal structure.

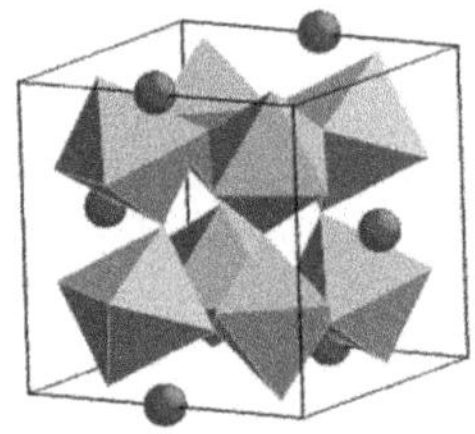

FIGURE 7.10 Crystal structure of bernalite, $Fe(OH)_3 \cdot 0.25H_2O$, including corner-shared $Fe(OH)_6$ octahedra. The water molecules included in the crystal are represented by spheres.

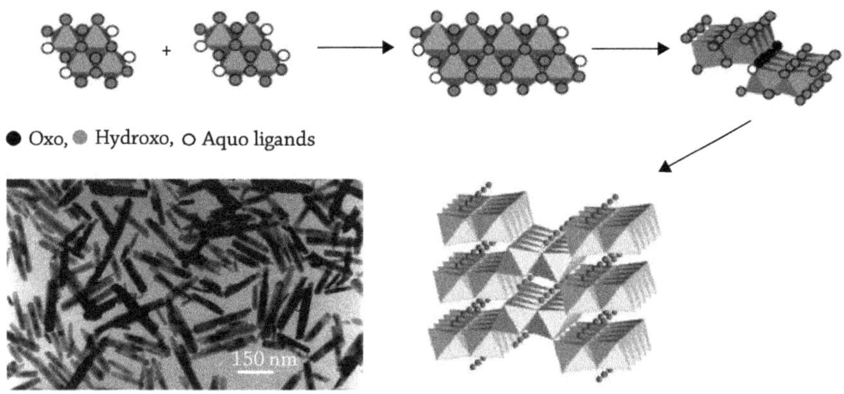

FIGURE 7.11 Possible pathway in the formation of goethite α-FeOOH by aging at pH 11 and room temperature of suspension of ferrihydrite and TEM image of the particles so obtained.

including water molecules (Fig. 7.10). This compound is not synthesizable in the laboratory, and the geological conditions of its formation are still mysterious, all the more so since in the deposit, the mineral is associated with goethite. This phase may be stabilized by various impurities such as silicon, zinc, lead, and carbonate existing in the deposit.

The growth of goethite from the ferric gel proceeds according to a dissolution–crystallization mechanism that can be understood to involve the planar tetramer $[Fe_4(OH)_{12}(OH_2)_4]^0$, which itself is formed from aquohydroxo dimers (§ 4.1.3). Its condensation by olation because of the presence of aquo ligands enables it to form embryos of the double chains of octahedra characteristic of the crystal structure of goethite (Fig. 7.11). Oxolation allowing the linkage between the double chains by μ_3–oxo bridges occurs then because of the relative kinetics of the two reactions. Also, a set of hydrogen bonds exists between oxo and hydroxo ligands within the solid. Such a schematic pathway accounts for a preferred growth direction leading to formation of nanorods rather monodispersed in size of about 150 to 200 nm in length, 25 nm in width, and 10 nm in thickness.

Goethite is used mainly as yellow pigment for various applications in the construction and cosmetics industries [3]. Because of the high anisotropy of their shape, the nanoparticles in concentrated aqueous dispersion exhibit a behavior of liquid crystal in forming nematic phases in which the particles are spontaneously aligned along a common direction [69]. Goethite nanorods may also be used as reinforcement to improve the mechanical properties of organic polymers, such as poly(hydroxyethylmethacrylate), involving strong interactions between the surface of particles and hydroxyl and carboxylate groups of the chains of polymer. The nanoparticles thus behave as reticulation points within the polymer [70].

Complexing ligands of iron, such as phosphate [71–73] or silicate [74, 75], delay or even avoid the transformation of ferrihydrite into crystalline phases. These ligands in the environment are thus allowed to retain the existence of ferrihydrite in the oldest soils (200,000 years) in New Zealand, for instance. It is also through complexation that ferric iron is reversibly stored as ferritin nanoparticles in living organisms. Divalent metal cations also have an important role in the transformation of ferrihydrite into crystal phases. At low content, many cations induce the transformation into goethite, hematite, or spinel ferrite (§ 7.2.3b), while at high content, they allow crystallization of various polymetallic phases [3, 50].

Strange natural ferric concretions named *moqui marbles* or *iron sandstone balls* are found in large amounts in the desert areas of Utah, Colorado, and Arizona in the American Southwest [76]. These balls, pea- to baseball-sized, are made of a sandstone core that is a concretion of cemented quartz sand grains. The core is covered by a more or less large thickness of hematite (Fig. 7.12b, c). The formation of such balls likely results from a geological process involved in the bleaching of highly ferruginous sandstone present in these states. Infiltration of rainwaters in conjunction with some reducing matter allows dissolution of ferric oxide and its transport in groundwater as ferrous species because of the much higher solubility of Fe^{II} than Fe^{III} ions near the neutrality [1]. Depending on the porosity and permeability of sandstone, the leaching on geological duration can produce alternating red and white zones from where iron oxide has been partially or fully removed, respectively, due to enormous amounts of water having flushed through the porous sandstone (Fig. 7.12a). From resurgence of reducing water loaded in iron or when it mixes with oxygenated water, oxidation of ferrous species is immediate and iron precipitates as ferrihydrite in the contact of small more or less spherical sandstone concretions. The precipitate forms a cement that seals the porosity of sandstone. Subsequent deposits of iron, likely induced by heterogeneous nucleation (§ 4.2.3 a) and according to a dehydratation/crystallization process in conditions close to neutrality, form a wrapping of compact hematite with a thickness of a few millimeters to centimeters. The rounded shape of the *moqui marbles* probably results from that of the core in which the deposit occurred. The erosion caused by water and wind during geological periods of several millions of years has also probably contributed to this shape. Similar concretions found on the planet Mars by NASA's robot Opportunity could be an indication that Mars had water in the past.

FIGURE 7.12 (a) Coyote Buttes, Paria Wilderness Area, Utah–Arizona border (USA). (b) Overview of moqui marbles in this area (courtesy of M. A. Chan, Univ. of Utah, USA). (c) Cross section of a broken moqui marble (image by Shinishi Kato with permission of rocksandminerals.com).

b) Thermolysis of Acidic Solutions: Hematite, Akaganeite

Heating of acidic ferric nitrate or perchlorate solutions ($1 \leq pH \leq 3$) at 90–100°C without any additive or complexing anion forms hematite [3, 50, 51, 77, 78]. Hematite crystallization seems to be due to several factors: notably, the acidity of the medium and the increase in temperature, which favors condensation by oxolation of hydroxylated complexes and weakens the kinetic limitation to form the most stable phase. It results in the formation of oxide, hematite Fe_2O_3, instead goethite FeOOH.

After heating for a few minutes, the ferric solution becomes blood red and contains small particles, only around 3 nm in diameter, exhibiting the X-ray diffraction pattern characteristic of 6-line ferrihydrite (Fig. 7.13a, b). This phase of a somewhat better crystal ordering than 2-line ferrihydrite directly transforms into hematite during thermolysis (Fig. 7.13c, d). After 24 hours (95°C, pH 2), perfect rhombohedra limited by {104}-type faces with a size of about 30 nm are obtained (Fig. 7.13e) [79, 80].

The stronger the acidity of the solution, the slower the growth of the particles but the larger their size, reaching 100 nanometers (Fig. 7.13f). Indeed, the stronger the acidity, the more the hydroxylation of cations and nucleation of the solid are restricted, but the more the aggregation is favored because of the increase in ionic strength (salt effect, § 5.3) and the more the dissolution–crystallization mechanism is at work to feed the growth of hematite particles. In less acidic medium ($pH > 3$), thermolysis of ferric solutions forms a mixture of hematite and goethite particles.

The nature of the anions present and the acidity of the medium during the thermolysis have a strong influence on the shape of the particle (Figs. 7.14, 7.15). The

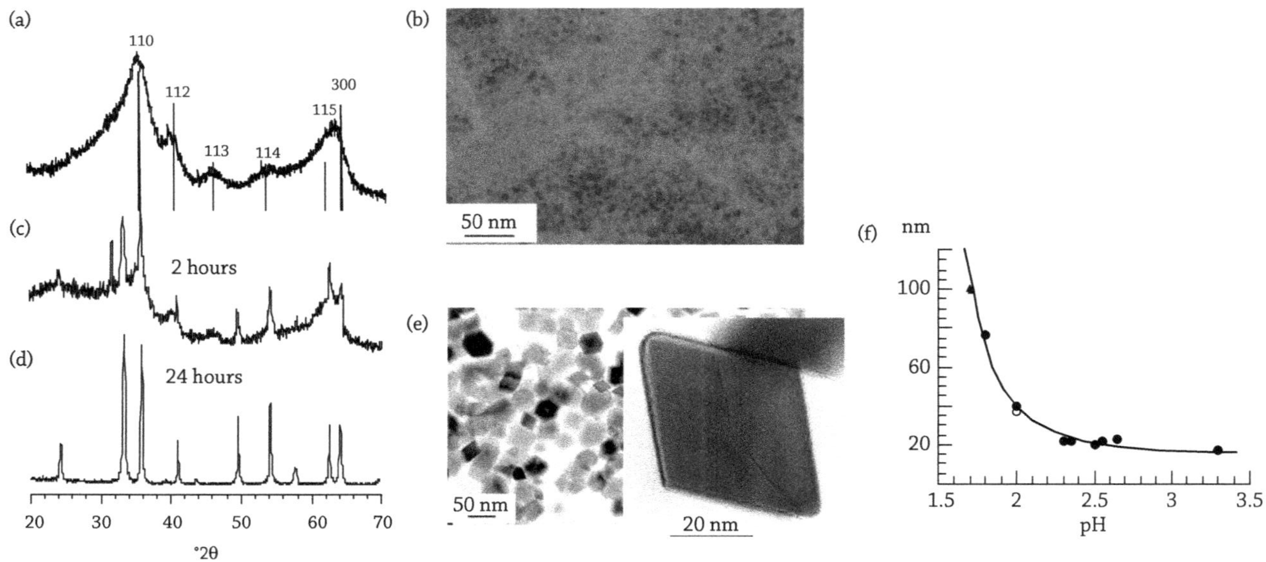

FIGURE 7.13 X-ray diffraction pattern (a) and TEM image (b) of 6-line ferrihydrite obtained after a few minutes by thermolysis of ferric nitrate ($C_{Fe} = 5.10^{-3}$ mol.l^{-1}, 95°C, pH ≈ 2); X-ray diffraction pattern of the solid aged 2 hours (c) and 24 hours (d) in suspension. TEM images of hematite particles formed after 24 hours (e). Mean size of hematite particles obtained by thermolysis of ferric nitrate solutions at various acidities (f) (from [79] with permission).

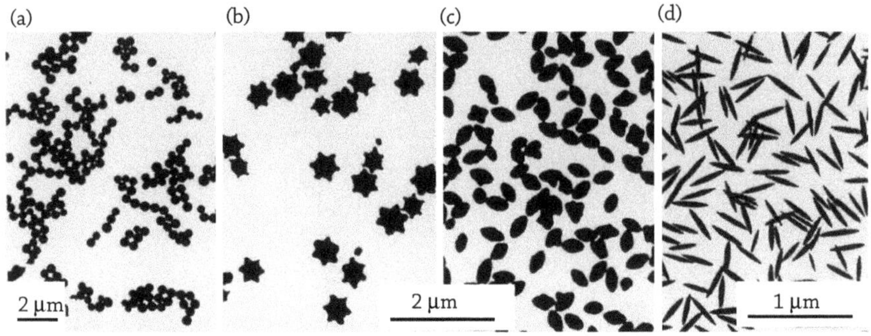

FIGURE 7.14 TEM images of hematite particles resulting from thermohydrolysis of acid ferric solutions (100°C, pH 0.5–2.5) in the presence of (a) Cl^-, (b) ClO_4^-, (c) NO_3^-, (d) $H_2PO_4^-$ (courtesy of E. Matijevic, 1996).

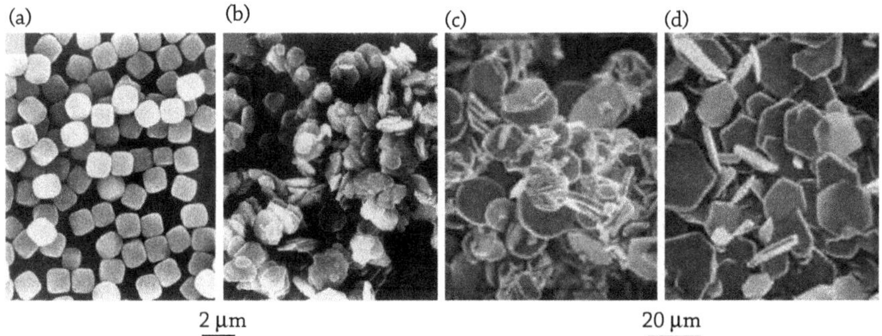

FIGURE 7.15 TEM images of hematite particles formed in alkaline medium: (a) pseudocubic particles obtained in the aqueous mixture $FeCl_3$ 1mol.l^{-1}+ NaOH ($OH^-/Fe = 5.4$) aged for 8 days at 100°C; (b) platelets obtained in the suspension $FeCl_3$ 1mol.l^{-1}+ NaOH ($OH^-/Fe = 8$) aged for 2 hours at 180 °C (reproduced from [Sugimoto T., Muramatsu A., Sakata K., Shindo D., Characterization of hematite particles of different shapes, J. Colloid Interface Sci. 1993, 158, 420–428] with permission of Elsevier); (c) platelets formed in the alkaline solution of Fe(III)-triethanolamine chelate (NaOH 1.2 mol.l^{-1}, 250°C) in the presence of ClO_4^- and H_2O_2 and (d) in the presence of NO_3^- and CH_3COO^- (adapted from [78]).

anisotropy of particles increases when the complexing ability of the anion toward ferric ions is marked [77]. But in addition, among the anions, chloride has a special role in the crystallization of ferric oxide.

In the presence of a high concentration of chloride (or with a concentration in ferric chloride higher than 4.10^{-2} mol.l^{-1}), the precipitation at room temperature or by thermolysis of acidic solutions leads first to the formation of akaganeite, β-FeOOH [3, 77, 81, 82]. The crystal structure of this phase is related to that of goethite because it contains the same double chains of edge-shared octahedra. However, the linking between them by corner sharing, different from that which is in goethite, forms an arrangement of channels with square cross sections characteristic of the hollandite-type structure (Fig. 7.16).

(a) (b)

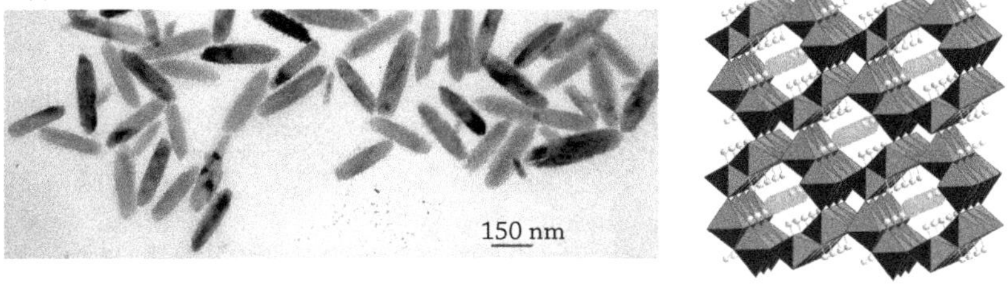

FIGURE 7.16 (a) TEM image of akaganeite, β-FeOOH, nanoparticles. (b) Scheme of the crystal structure. The large spheres in the center of channels indicate the positions of chloride ions. The small spheres represent the hydrogen atoms of hydroxo ligands.

Akaganeite always contains chloride anions and water molecules with an average composition of $FeOOH{\cdot}(HCl)_{0.12}{\cdot}(H_2O)_{0.25}$. Chloride ions can be removed from the solid by titration with a base and then reintroduced by acidification without change of the crystal structure. They can be exchanged by fluoride but not by too bulky perchlorate ions [83–85]. Chloride ions are located within the channels of the framework, at the center of prisms determined by eight hydrogen atoms belonging to hydroxo groups of the chains of octahedra. They occupy around two-thirds of these sites in balancing the positive charge of the protonated groups in the same channels [86]. The formation pathway of akaganeite can be understood as follows.

Chloride is a rather weak complexing species of ferric ions in solution [87]. At a low hydroxylation rate (h = HO−/Fe ≤ 0.3), EXAFS measurements indicate that two Cl^- ions in *trans* position (Fe-Cl distance of 2.31 Å) form, with four oxygen atoms (Fe-O distance of 2.01 Å), the coordination sphere of iron [56, 82]. Cl^- ions are progressively removed as the hydroxylation rate increases. In the first stages in the condensation of iron, Fe···Fe correlations at 3.45 Å and at 3.01 Å correspond to the linking of octahedra by corners, Fe-O-Fe, and by edges, $Fe\text{-}(OH)_2\text{-}Fe$, respectively. These correlations are the same as those in the presence of noncomplexing anions such as NO_3^- and ClO_4^-. For $h > 2$, chloride is no more detected in the coordination sphere of iron inside the polymers. However, Cl^- ions remain trapped in these polymers and on the chains during their formation because of the strong interactions forming outer-sphere complexes such as $(Fe^{3+})_3\text{-O-H}^{\delta+}\cdots Cl^-$. Interchain hydrogen bonds between OH groups can no longer be established as in goethite, and because of the large size of chloride ions (r_{Cl^-} = 1.81 Å), the ordering of the bonds around them forms another organization of the double chains of octahedra. Chloride ions would thus seem to act more as a template than as a real ligand of ferric ions. Also, the low complexing power of chloride toward ferric ions explains why oxychloride FeOCl, with a layered crystal structure of $CdCl_2$-type similar to that of lepidocrocite, γ-FeOOH, does not form by precipitation under conditions in which akaganeite, β-FeOOH, forms. Oxychloride is obtained in the reaction of hematite $\alpha\text{-}Fe_2O_3$ with $FeCl_3$ around 400°C in sealed tube [88].

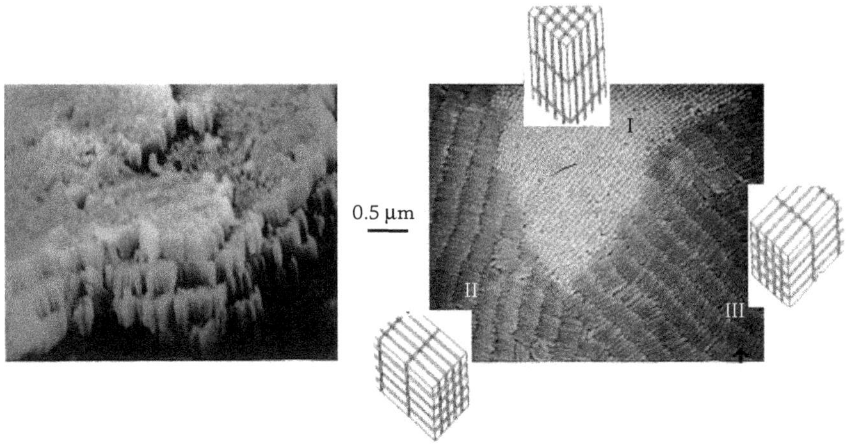

FIGURE 7.17 SEM images of smectic domains in dried sols of highly monodispersed akaganeite nanoparticles. In the domain I, the ends of particles forming horizontal layers are visible. In domains II and III, the particles form vertical layers, and only the faces of particles are visible (reproduced with permission from [89] Copyright 1996 American Chemical Society).

The high anisotropy of akaganeite particles and their homogeneity in size promotes their highly ordered aggregation in concentrated suspension or by drying. Such ordered structures are known as Schiller's layers (Fig. 7.17). They form smectic liquid crystals exhibiting wonderful iridescent colors due to bright interference phenomena within the ordered domains [89].

Akaganeite is a metastable phase. By dehydroxylation under vacuum at 170°C, it forms β-Fe_2O_3, an oxide of bixbyite β-Mn_2O_3-type crystal structure (Fig. 7.18). This phase does not exist in nature, but it has been recognized in some industrial processes of steel treatments. It can also be obtained by reaction in solid state between NaCl and $Fe_2(SO_4)_3$ at 400°C [90, 91] and by hydrothermal treatment at 150°C of $FeCl_3$ in the presence of urea at pH 9–10 [92]. This phase has been obtained only under the nanometric form. It transforms at 500°C into hematite.

In aqueous suspension, akaganeite transforms by heating into hematite, yielding various particle morphologies according to the experimental conditions and the nature of anions. Two mechanisms are involved in the transformation: dissolution–crystallization and/or ordered aggregation of primary particles. In most cases, the hematite particles obtained are highly homogeneous in size.

At iron concentration in the range 2.10^{-2} – 4.10^{-2} mol.l^{-1}, akaganeite transforms slowly at 100°C into spherical, ellipsoidal, or pseudocubic α-Fe_2O_3 particles [77, 81, 93, 94–96] (Fig. 7.19a,b). The size and shape of hematite particles are closely related to the acidity of the aging medium, as well as to the size and aggregation of the initial akaganeite particles. These can also be a support for heterogeneous nucleation of hematite. The mechanism of the transformation involved is clearly dissolution–crystallization, which is perfectly illustrated by the formation of double-ellipsoid-shaped hematite particles (Fig. 7.19e). In this case, the largest sized β-FeOOH initial

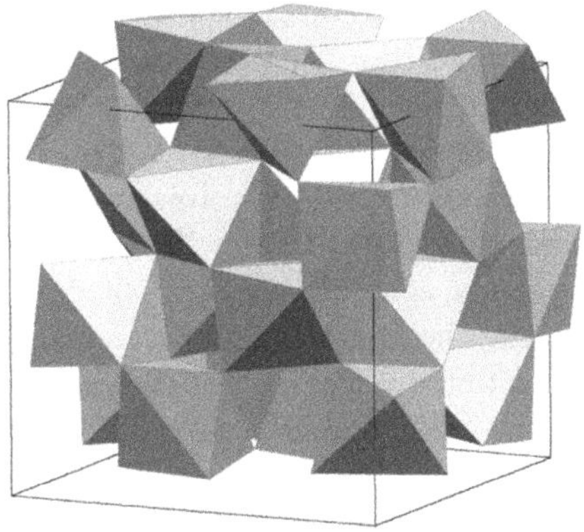

FIGURE 7.18 Ordering of FeO_6 octahedra in the bixbyite-type crystal structure of β-Fe_2O_3.

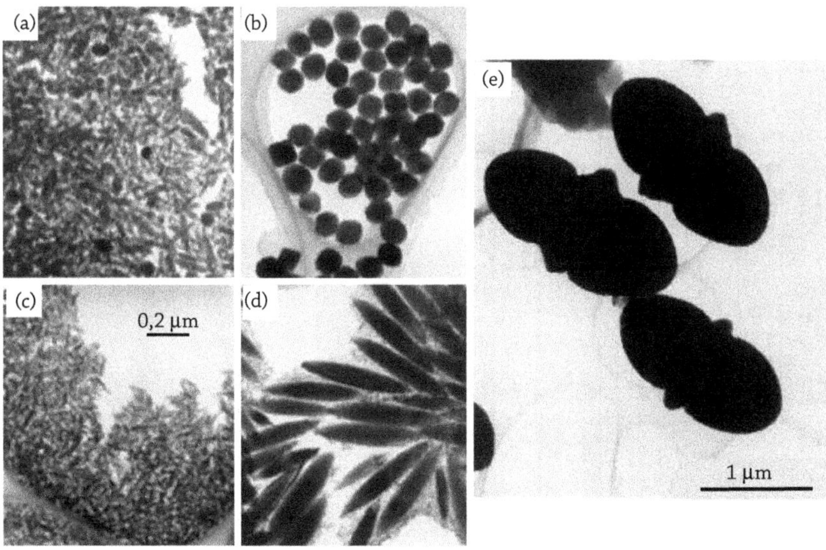

FIGURE 7.19 α-Fe_2O_3 particles obtained by thermohydrolysis at 100°C of $FeCl_3$ solutions (0.02 mol.l^{-1}) for 8 hours (a) and 2 days (b) of aging; α-Fe_2O_3 particles obtained in the presence of KH_2PO_4 (4.5.10^{-4} mol.l^{-1}) for 8 hours (c) and 4 days (d) of heating (reproduced from [96] with permission of Elsevier). (e) Double-ellipsoid-shaped α-Fe_2O_3 particles obtained from 0.018 mol.l^{-1} $FeCl_3$, 0.05 mol.l^{-1} HCl, 1 week at 100°C (reproduced from [94] with permission of Elsevier).

particles are used as the support for recrystallization of hematite by heterogeneous nucleation. However, in the presence of a high concentration in chloride, the dissolution–crystallization equilibria transform all of the akaganeite, so that the initial rod of akaganeite is still present at the end of the process [77, 94]. At higher concentration in chloride, iron complexation is too efficient, so that recrystallization of hematite can no longer occur and akaganeite becomes stabilized during the aging of suspensions [81].

The transformation of akaganeite during thermolysis in the presence of phosphate produces spindle-shaped hematite particles with a high anisotropy (Fig. 7.19c, d). The aspect ratio increases with the amount of phosphate and reaches the value of 6 for the particles shown in Figure 7.19d. The particle growth, which is highly directed by phosphate adsorption, involves an ordered aggregation mechanism of very small primary hematite particles. The phosphate adsorption cancels the electrostatic surface charge of the primary particles in allowing their aggregation along their major axis. The interaction between surfaces, their desolvation, and the acidity of the solution are factors that promote the slow desorption of phosphate and allows the transformation of polycrystalline aggregates into monocrystals [81, 93]. The intervention of the ordered aggregation phenomena has been demonstrated by using a new experimental procedure involving a valve-assisted pressure autoclave through which sampling and quenching of samples during the transformation of akaganeite are carried out in a hydrothermal treatment (Fig. 7.20) [97]. Such a technique allows nearly in situ snapshot experiments. The growth of α-Fe_2O_3 nanorods observed in the images in Figure 7.20 can be illustrated by the sequential pathway in Scheme 7.2: (1) precipitation of β-FeOOH

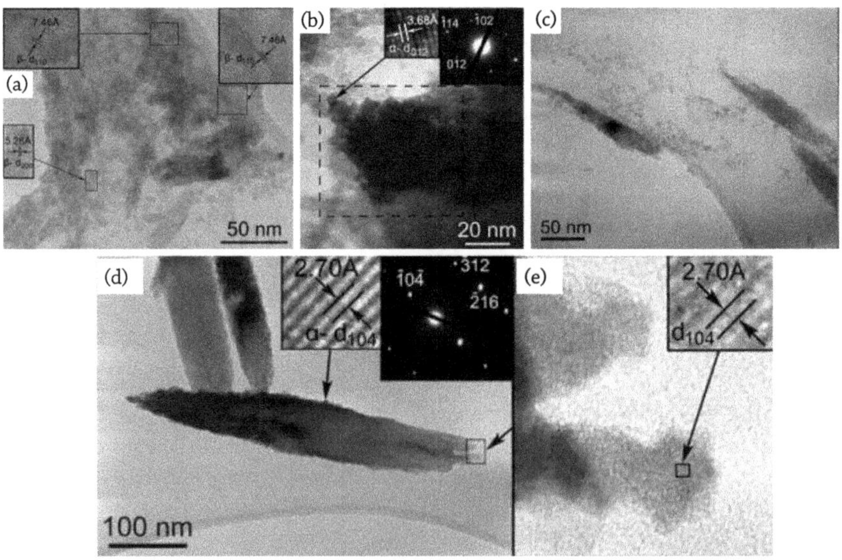

FIGURE 7.20 Bright field and phase contrast TEM images of quenched hydrothermal products in liquid nitrogen after 30 mn of processing of $FeCl_3$ in the presence of ammonium phosphate at 200°C. (a) β-FeOOH, as identified by lattice fringes (inset) and (b) an α-Fe_2O_3 nanoparticle (boxed), identified by lattice fringes and corresponding electron diffraction pattern (inset). (c) Larger α-Fe_2O_3 nanorod (~ 150 nm long, ~ 25 nm wide) among smaller α-Fe_2O_3 nanoparticles (< 10 nm) formed after 40 minutes of processing at 200°C. (d) Large ellipsoidal α-Fe_2O_3 nanorod obtained after 50 minutes of processing at 200°C identified by lattice fringes and corresponding electron diffraction pattern (inset), (e) filamentary feature at the tip of an α-Fe_2O_3 nanorod during formation (reproduced from [97] with permission of The Royal Society of Chemistry)

nanorods (gray); (2) nucleation of α-Fe_2O_3 nanoparticles (black dots); (3) growth of α-Fe_2O_3 nanoparticles along with β-FeOOH phase dissolution; and (4) when β-FeOOH has disappeared, oriented attachment, aggregation, and coarsening of α-Fe_2O_3 nanoparticles, giving the final shape mediated by the presence of phosphate acting as a surfactant [97].

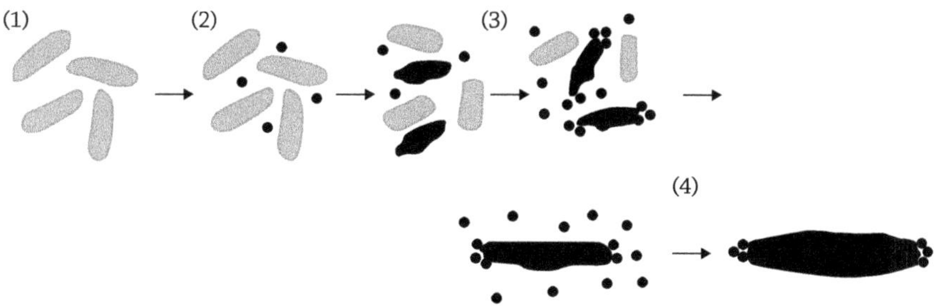

SCHEME 7.2 Pictorial pathway for the transformation of β-FeOOH into α-Fe_2O_3 nanorods in the presence of phosphate by the hydrothermal process (reproduced from [97] with permission of The Royal Society of Chemistry).

Various other complexing species of ferric ions, for instance, acid-alcohols such as citrate, amino acids, and phosphonates, as well as polymers such as polyvinylic alcohol, allow us to control the morphology of hematite particles (spheres, cubes, ellipsoids, etc.) during the thermolysis of acid solutions of ferric chloride [98–101]. As with phosphate, the aspect ratio of particles increases with the complexing ability of the ligand, but too high concentrations in citrate or the use of too strongly complexing ligand such as EDTA lead to the dissolution of hematite particles [102, 103].

In conditions of fast precipitation, when a concentrated and acidic solution of ferric chloride is introduced in boiling water, the early amorphous solid transforms directly into hematite without going through akaganeite formation [104]. When an important solvent evaporation occurs during heating and aging allowing a two- to fourfold increase in concentration, the aggregation of hematite primary particles forms large discs and tetrahedra (Fig. 7.21). The reasons for the formation of such objects remain unknown.

Hematite can also be obtained as nanotubes by anodic oxidation of an iron foil in ethylene glycol containing ammonium fluoride and water [105]. The process is similar to that used in the formation of tubular structures of titanium oxide (§ 8.2.3b, Fig. 8.31). The film of amorphous iron oxyhydroxide formed on the surface of the metal is structured in nanotubes 55 nm in diameter and 3 μm in length, with a wall thickness of around 10 nm (Fig. 7.22), but the size of tubes is highly dependent on temperature, anodization voltage, concentration of ammonium fluoride, and amount of water. The formation of nanotubes results from a delicate balance between three simultaneous processes: (1) electrochemical oxidation of iron, which forms the oxide layer with acidification of the solution, (2) dissolution and transport of ferric ions under the electric field within the

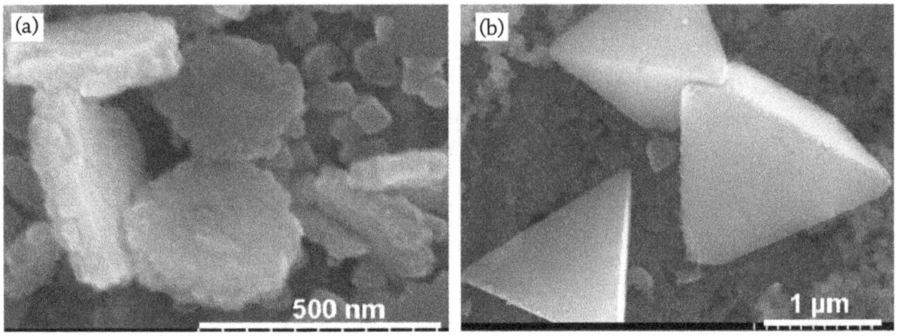

FIGURE 7.21 SEM images of aggregates of hematite obtained by fast nucleation at 100°C from $FeCl_3$ acidic solutions. (a) Discs formed after 48 hours of aging at 100°C and evaporation of 75% of the solvent (initial concentrations $FeCl_3$ 2.10^{-3} $mol.l^{-1}$, HCl 2.10^{-3} $mol.l^{-1}$); (b) tetrahedra obtained after 16 hours of aging at 100°C and evaporation of 50% of the solvent (initial concentrations $FeCl_3$ 10^{-2} $mol.l^{-1}$, HCl 2.10^{-3} $mol.l^{-1}$) (reproduced with permission from [104] Copyright 2008 American Chemical Society).

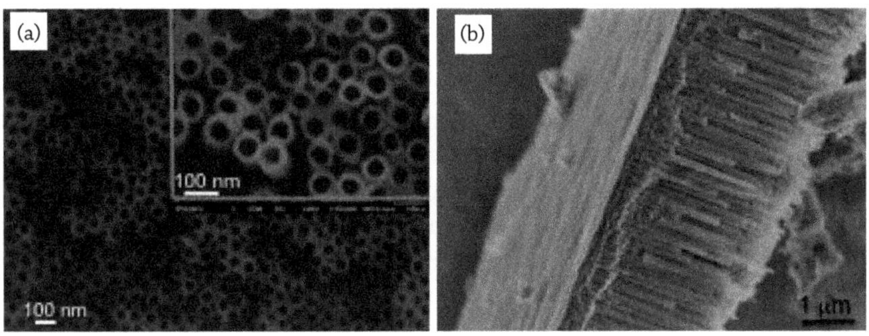

FIGURE 7.22 FESEM top view (a) and cross-sectional (b) images of iron oxide nanotube arrays synthesized by anodization of an iron foil at a potential of 50 V in ethylene glycol containing 0.3% wt NH_4F and 3.0% water for 180 s at 60°C (reproduced with permission from [105] Copyright 2008 American Chemical Society).

electrolyte, and (3) localized chemical dissolution of the oxide by fluoride ions, which creates the tubular structure (§ 8.2.3). These processes are sensitive to temperature because it influences the mobility of the soluble species. Crystallization into hematite is achieved through annealing of the material at 400°C in air. Such hematite nanotubes arrays could have interesting catalytic properties for water photodissociation [106].

c) Hydrothermal Conditions: Hematite

Heating between 120 and 200°C of dilute ferric chloride alone or in the presence of urea with or without cetyltrimethylammonium bromide (CTAB) forms hematite particles with prismatic shapes rather similar to those obtained in thermohydrolysis at 100°C [107–109] (Fig. 7.23). The reaction duration is of course

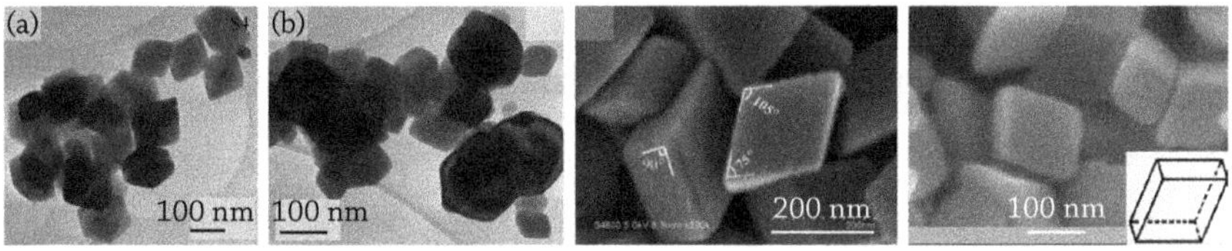

FIGURE 7.23 TEM images of hematite particles synthesized in hydrothermal conditions for 2 hours from $FeCl_3$ solutions at 160°C (a) and 200°C (b) (reproduced with permission from [107] Copyright 2009 American Chemical Society). SEM images of prismatic hematite particles obtained (c) by heating at 120°C for 15 hours of $FeCl_3$ in the presence of urea (the particles exhibit {012}, {01-4} and {-210} faces) (reproduced from [108] with permission of The Royal Society of Chemistry) and (d) by heating at 170°C for 24 hours of $FeCl_3$ in the presence of urea and CTAB (reproduced from [109] with permission of Elsevier).

shortened (around 1 hour at 200°C) in comparison to that required in thermolysis of solutions (a few days at 100°C). One may conclude that the growth mechanisms are the same in both cases.

By contrast, various additives have a marked effect on the shape of particles produced in hydrothermal conditions, as shown by the following examples. Polyhedral platelets (Fig. 7.24a) result from hydrothermal treatment at 180°C for 30 hours of a suspension of goethite obtained by alkalinization at pH 11 of ferrous sulfate in the presence of CTAB [110]. In this case, the morphology resulting from the dissolution–crystallization mechanism involved in the structural change appears to be highly influenced by CTAB.

Nanotubes (Fig. 7.24b) are formed by hydrothermal treatment at 220°C for 48 hours of acidic $FeCl_3$ solutions containing phosphate (pH = 1.8, $PO_4/Fe = 0.04$) [111]. The mechanism involved in the formation of these tubes is somewhat amazing: after 2 hours at 220°C, cigar- or spindle-shaped particles oriented along the (001) direction and similar to those produced by thermolysis (Fig. 7.19d) are obtained. Then dissolution progressively erodes the rod from the tips until a completely hollow tube is formed (Fig. 7.24c). Dissolution is caused both by the acidity of the medium and by complexation of ferric ions by phosphate. The temperature appears to play a critical role in the morphology: at 180°C, only cigar-shaped particles are formed, and at 200°C, a few nanotubes are obtained. Acidity and phosphate concentration are also critical parameters in the synthesis of tubes. An interesting point is the possibility of forming in a topotactic way nanotubes of maghemite, γ Fe_2O_3, by reduction with hydrogen at 360°C of these dried hematite tubes and then reoxidation in air at 240°C [111].

Nanocubes (Fig. 7.24d) homogeneous in size and limited by faces of {012}-type are obtained by hydrothermal treatment at 180°C for 20 hours of mixtures of $FeCl_3$, oleate, and oleic acid in hydroalcoholic medium [112]. The particle shape control is likely due to oleate adsorption, the oleate/Fe molar ratio having to be at least equal to 3 to obtain homogeneous-sized cubic particles. However, study of the morphological evolution during hydrothermal treatment shows that the formation of pseudocubes results from primary particle aggregation.

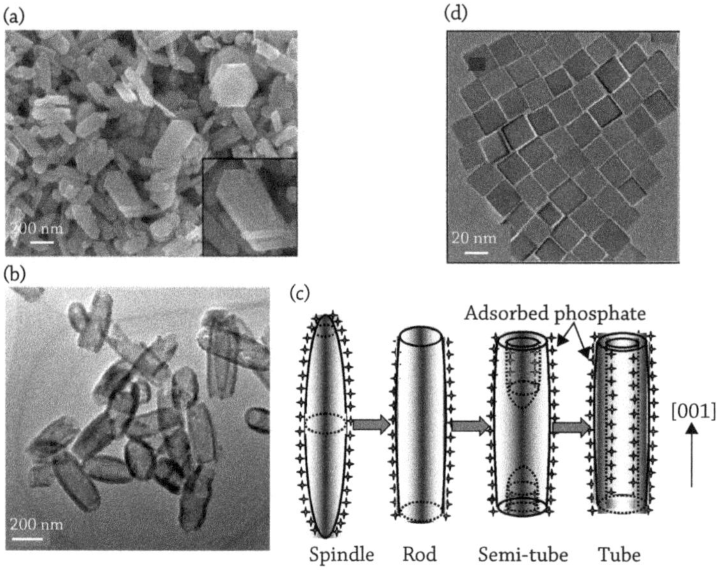

FIGURE 7.24 SEM and TEM images of hematite particles obtained in hydrothermal conditions from $FeCl_3$ in the presence of different additives (see text): (a) platelets formed from goethite (adapted from [109]), (b) nanotubes with scheme of their formation mechanism (c) (courtesy of C.J. Jia, Shandong Univ., L.D. Sun and C.H. Yan, Peking Univ., China). (d) Pseudocubes (reproduced with permission from [112] Copyright 2009 American Chemical Society).

Truncated hexagonal bipyramid hematite particles (Fig. 7.25) [113] are obtained by hydrothermal treatment at 160°C for 8 hours of potassium hexacyanoferrate $K_3[Fe(CN)_6]$ in the presence of hydrazine and of sodium carboxymethyl cellulose, a charged polymer. Adsorption of polymer and the presence of hydrazine are required to promote the dispersion of particles and the growth of {001} and {101} faces.

Moniliforme-shaped straight chains of truncated bipyramidal hematite nanoparticles result from hydrothermal treatment at 180°C for 12 hours of the mixture of $FeCl_3$, sodium oleate, oleylamine, and oleic acid in a mixed solvent made of water, ethanol, *n*-hexane, and *n*-octane [114]. The chains are made of more than a dozen uniformly shaped and sized particles (Fig. 7.26), assembled according to an ordered aggregation process. Contiguous particles inside the chains are in quasi-perfect contact by {006} faces perpendicular to the [001] direction of the rhombohedral crystal structure of hematite. This aggregation mechanism is confirmed by electron microscopy observations of sample during growth showing individual particles and sections with the same diameter as that of the final chains. It is therefore likely that surfactants, as in the previous example, lead to the formation of bipyramids having 12 faces of {113}-type and that the basal truncatures of {006}-type, which may be less protected by ligands and of highest energy, are the place for ordered aggregation and even for recrystallization of individual particles.

An interesting phenomenon leading also to the formation of straight strings of hematite nanoparticles occurs during the treatment of rhombohedral nanoparticles of hematite by hydrazine under heat in strongly alkaline medium

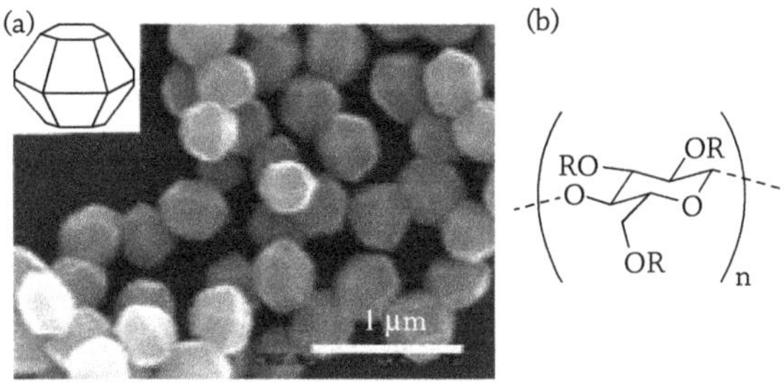

FIGURE 7.25 (a) SEM image of truncated hexagonal bipyramid-shaped hematite particles exhibiting {101} lateral faces and {001} basal faces (reproduced with permission from [113] Copyright 2012 American Chemical Society). (b) Scheme of the base unit of the carboxymethyl cellulose polymer with $R = CH_2CO_2^-$.

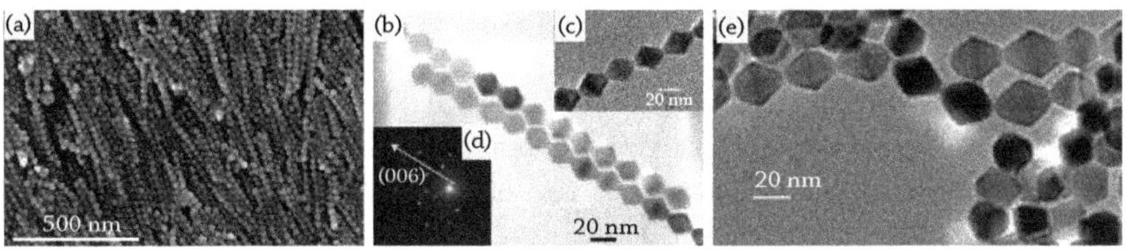

FIGURE 7.26 (a) SEM image of chains of hematite nanoparticles; (b and c) TEM images of straight chains and view of individual truncated bipyramid-shaped particles in chains; (d) electron diffraction pattern showing the common orientation of the {006}-type faces in chains; (e) TEM image showing individual particles and sections of chains being formed (reproduced with permission from [114] Copyright 2010 American Chemical Society).

[115]. The nanoparticles are synthesized by thermolysis at 95°C of ferric nitrate in acidic solution (§ 7.2.2b) and then are treated for 2 days in soda (50% wt water, 50% wt soda) at 110°C in the presence of hydrazine. Electron microscopy shows a change in morphology and ordered aggregation of particles along the direction perpendicular to the faces of {001}-type (Fig. 7.27). Without hydrazine, the same treatment does not significantly modify the hematite particles. Hydrazine N_2H_4 is a Lewis base able to be coordinated to metal cations [116], and it is also a relatively powerful reducing agent (E° = −1.16 V at pH 14). In the reaction with hematite nanoparticles, hydrazine leads first to their change in shape, which is likely induced by complexation on the exposed {104}-type faces. Hydrazine adsorbed can then play its reducing role by electron injection into particles, giving rise to reductive dissolution in producing ferrous complexes from the corners of particles that are the most reactive zones. This phenomenon has been described in the study of ferrous ions adsorption onto hematite (§ 5.4.3). Thus, dissolution forms {001}-type faces, faces by which oriented aggregation occurs to form chains (Fig. 7.28). After prolonged treatment at 110°C, hematite is first reduced to magnetite and then to metallic iron [115].

Nanodiscs have been obtained by hydrothermal treatment of $FeCl_3$ in the presence of phosphate [117] or silicate [118]. Heating solutions containing phosphate at 200°C for 12 hours forms discs of around 150 nm in diameter and 15 nm in

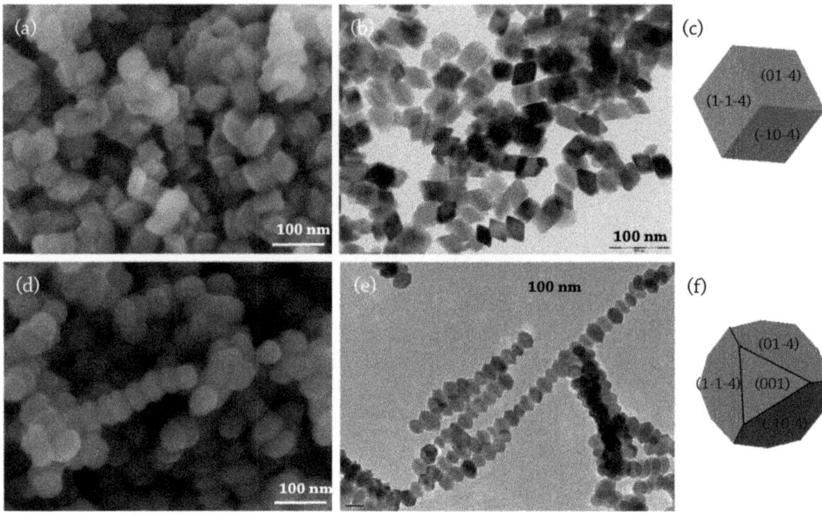

FIGURE 7.27 Hematite nanoparticles obtained by thermolysis in acidic medium of ferric nitrate (top) and treated in alkaline medium in the presence of hydrazine at 110°C (below): (a) and (d) SEM images; (b) and (e) TEM images; (c) and (f) schematized shapes of both types of particles [115].

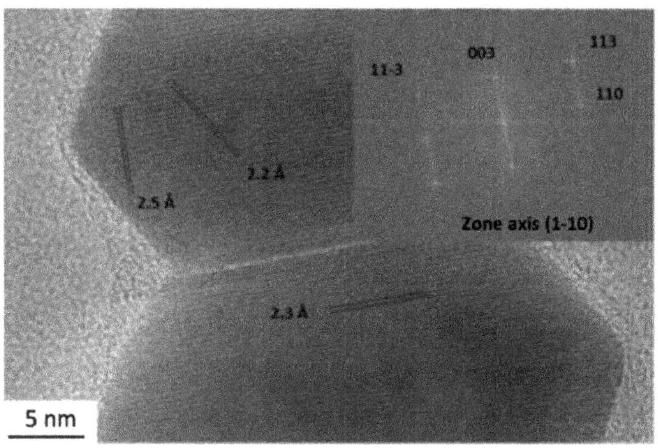

FIGURE 7.28 HRTEM image showing the contact area of particles inside a chain after treatment of hematite in alkaline medium in the presence of hydrazine (insert: electron diffraction image) [115].

thickness. Addition of ethylene glycol decreases the diameter of the discs. In the presence of silicate, discs 150 nm in diameter and 5 nm in thickness are stacked to form piles around 50 nm in length and containing silicate. In both cases, the particle shape results from phosphate or silicate on the {001}-type basal faces.

Dendritic structures of hematite have been formed by hydrothermal treatment of hexacyanoferrate $K_3[Fe(CN)_6]$ at 170–180°C [119] (Fig. 7.29). The morphology of particles appears to be highly dependent on the concentration of ferricyanide. At low concentration ($\leq 3.8 \times 10^{-3}$ mol.l^{-1}), the particles are snowflake-like

(Fig. 7.29a), while concentrations of 10^{-2} mol.l^{-1} or higher result in feather-like (Fig. 7.29b) or leaf-like (Fig. 7.29c) morphologies.

In all cases, the early stage of precipitation yields amorphous material in which energy-dispersive X-ray microanalysis (EDX) shows K, Fe, C, N, and O coexisting. High-resolution transmission electron microscopy (HRTEM) imagery shows that nucleation of hematites takes place at the surface of the early amorphous particles, in the form of thin plates with diameter up to around 100 nm. At longer reaction times, branches develop rapidly from the central crystal, giving a snowflake or feather or leaf morphology, depending on the concentration. HRTEM image analysis shows perfect continuity of lattice fringes across the interfaces, indicating that the crystal growth is epitaxial in all cases.

At room temperature, ferricyanide is a stable complex in solution, but hydrothermal conditions allow substitution of the cyanide ligand by the water molecule. That leads to partial hydroxylation and condensation of mixed cyanohydroxo ferric complexes, followed by precipitation of an amorphous solid of undefined composition according to the following scheme:

$$\left[Fe(CN)_6\right]^{3-}, 3K^+ + x\,H_2O \rightarrow \left[Fe(CN)_{6-x}(OH)_x\right]^{3-} + x\,HCN \rightarrow$$
$$\left[Fe(CN)_{6-x}O_{x/2}\right]\cdot K_3 + x/2\,H_2O$$

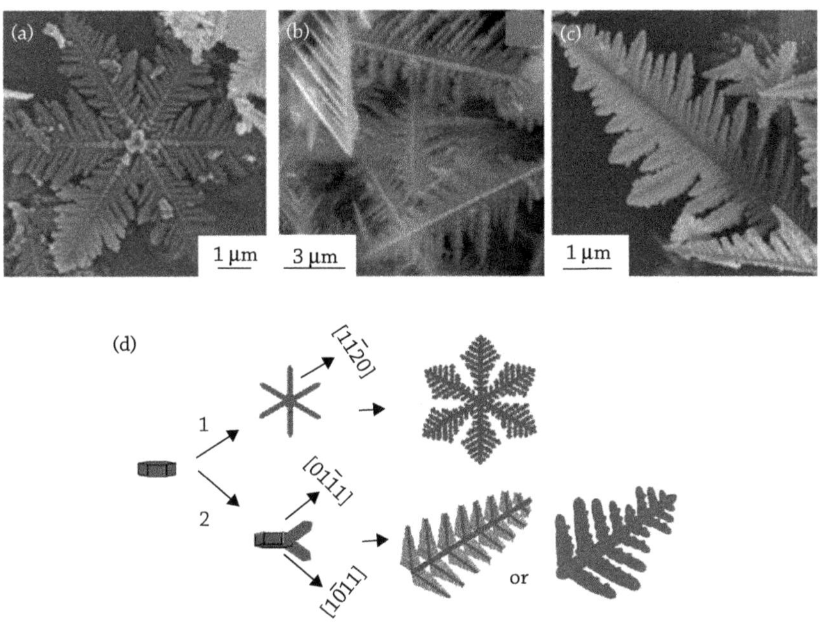

FIGURE 7.29 SEM images of hematite crystals obtained by hydrothermal treatment at 170°C of $K_3[Fe(CN)_6]$: (a) 0.0038 mol.l^{-1}, 2 hours; (b) 0.010 mol.l^{-1}, 48 hours; (c) 0.5 mol.l^{-1}, 90 minutes. (d) Scheme of crystal growth of hematite during thermohydrolysis of ferricyanide, with changes in morphology as induced by changes in concentration: 1, 0.0038 mol.l^{-1}, 2, 0.01 and 0.50 mol.l^{-1} (reproduced with permission from [119] Copyright 2017 American Chemical Society).

Then, with a low concentration of soluble precursors such as mixed cyanohydroxo, the dissolution–crystallization mechanism allows heterogeneous nucleation of hematite on the surface of the early solid in the form of small platelets. They appear to be monocrystals exhibiting facets on which further dendritic crystal growth is observed to occur quickly [119]. During the process, the solution has a more or less high concentration of free cyanide ligand, with two main consequences. On the one hand, the acidity of the medium is fixed by the equilibrium:

$$CN^- + H_2O \Leftrightarrow HCN + HO^-$$

with a pK value that is probably quite different from that at room temperature (§2.1.4). Of course, the pH value is dependent on the concentration of cyanide, and consequently, the electrostatic surface charge of the facets of hematite particles is changed with concentration. On the other hand, cyanide ions can be adsorbed as inner- or outer-sphere complexes onto different facets of hematite. Variation in acidity can obviously change the type of crystal face in which adsorption occurs, which, like many ligands, orients the crystal growth on available and adequate other faces. So, at low concentration, the branches of hematite start to grow at pH of around 9 in the (11–20) crystal direction to form snowflakes, while at high concentrations, the crystal growth starts at around pH 6 and ends at around pH 8 in (10–11) and (01–11) crystal directions for feather- and leaf-like particles [119] (Fig 7.29d).

This example is somewhat similar to that of formation of chromium oxide from chromium sulfate thermolysis (§ 4.2.3, Fig. 4.45) or that of formation of hematite from thermolysis of ferric chloride and of akaganeite suspensions (§ 7.2.2b). They involve the transformation of an unstable phase into a more stable material through a dissolution–crystallization mechanism allowing a controllable crystal growth. Here, the strong affinity of the cyanide ligand toward ferric ions and oxide surfaces induces a high selectivity in crystal growth.

Despite expectations for a stable, compact but strongly stressed crystal structure in which face-sharing ferric octahedra are highly distorted, hematite exhibits an amazing morphological versatility when it is elaborated in various physicochemical conditions. This short review highlights the influence of the kind of precursor and of the presence of various anions, ligands and surfactants in the control of the morphology of hematite nanoparticles.

7.2.3 MIXED FERRIC-FERROUS PHASES: GREEN RUSTS AND MAGNETITE

Precipitation from mixtures of ferric and ferrous ions forms two kinds of mixed-valent compounds: (1) *green rusts*, which are hydroxides with hydrotalcite or pyroaurite crystal structure (belonging to the wide family of layered double hydroxides, § 4.1.2) and (2) *magnetite*, an oxide of the spinel-type crystal structure. These phases are formed through different composition ranges, defined by the ratio $x = [Fe^{3+}/(Fe^{2+}+Fe^{3+})]$ (Fig. 7.2). For green rusts, x is close to 0.33, whereas x is equal to 0.66 for magnetite. A special character of the ferric-ferrous spinel compounds is the electron mobility at room temperature within the framework, while that is not the case within the bidirectional framework of green rusts. Green rusts are very reactive phases in aqueous corrosion of iron. They are also abundant in reductomorphic soils where they assume an important role in bacterial activity and the biogeochemical cycle of iron [2, 120, 121]. Magnetite and its completely oxidized form, maghemite, are ferrimagnetic materials (see below and § 1.1.1) used in many technological applications [3], especially as nanoparticles for medical diagnosis such as magnetic resonance imaging and therapy, for instance, hyperthermia, and vectorization of various drugs with active targeting [122, 123].

a) Green Rusts

For $x \leq 0.33$ and $OH^-/Fe_{total} = 2$ (pH ≈ 8), a bluish-green phase with a layered brucite-type crystal structure derived from that of $Fe(OH)_2$ is obtained (Figs. 7.30a and 7.33). The Fe^{3+} ions in octahedral sites give a positive charge to the brucite-like sheets because the $Fe(OH)_2$ sheets are electrostatically neutral. The balance of charges is carried out by intercalation of anions between the sheets; that is why these compounds are so-called anionic clays as opposed to aluminosilicates in which the sheets are negatively charged (§ 6.3.2). Green rusts (GR) form two families, depending on the nature of the inserted anions. The structural prototype is the carbonated phase, hydrotalcite, $[Mg_6Al_2(OH)_{16}]CO_3{\cdot}4H_2O$ or pyroaurite, its ferric homologous compound. Carbonate and chloride anions form green rusts of type I (GR1) in which the Fe^{II}/Fe^{III} ratio varies from $x = 0.25$ to 0.33, while tetrahedral anions such as sulfate or arsenate form green rusts of type II (GR2) with a fixed composition equal to $x = 0.33$ [124–127]. All these structures are similar but have different stacking sequences of the layers of cations, hydroxyl ions, and anion interlayers [127].

Study of progressive alkalinization of Fe^{2+} and Fe^{3+} ion mixtures ($x = 0.33$) in the presence of sulfate, which corresponds to the $Fe^{II}_4Fe^{III}_2(OH)_{12}SO_4{\cdot}8H_2O$ phase, yields interesting information on its formation mechanism [128–130]. This phase

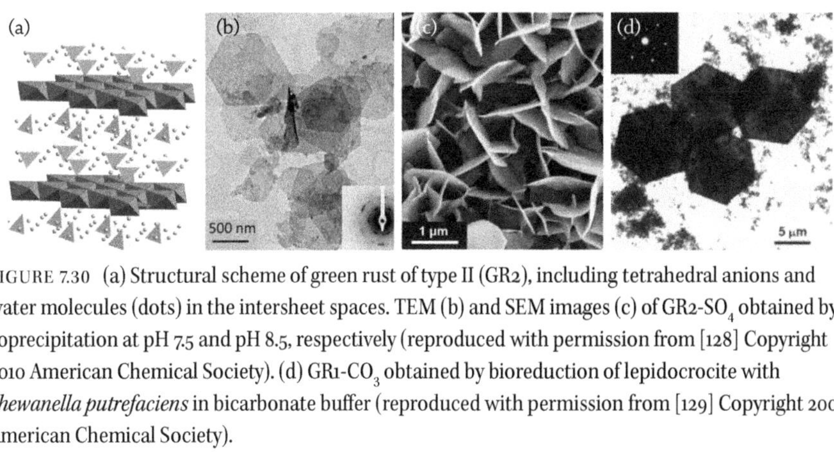

FIGURE 7.30 (a) Structural scheme of green rust of type II (GR2), including tetrahedral anions and water molecules (dots) in the intersheet spaces. TEM (b) and SEM images (c) of GR2-SO_4 obtained by coprecipitation at pH 7.5 and pH 8.5, respectively (reproduced with permission from [128] Copyright 2010 American Chemical Society). (d) GR1-CO_3 obtained by bioreduction of lepidocrocite with *Shewanella putrefaciens* in bicarbonate buffer (reproduced with permission from [129] Copyright 2002 American Chemical Society).

forms according to a heterogeneous crystallization process because ferric and ferrous ions precipitate successively. Up to pH 4, while the ferrous ions remain in solution, alkalinization forms schwertmannite, a purely ferric hydroxysulfate [3]. Its formation is marked by an equivalent point located at $OH^-/Fe \approx 2.7$ on the titration curve of ferric species alone (Fig. 7.31). At higher pH, the ferrous ions in turn are hydroxylated and adsorb onto the ferric phase, inducing the formation of GR. No independent ferrous phase is formed. Crystallization likely proceeds by dissolution–crystallization from the early formed schwertmannite, allowing the incorporation of sulfate anions through the interlayer spaces of the structure [130]. Surprisingly, the end of formation of the GR at $OH^-/Fe = 2$, at around pH 8 (Fig. 7.31), does not correspond to an equivalent point on the titration curve. Nevertheless, it is at this stage that all iron present is included in the GR (green rust) phase, as indicated by Mössbauer spectroscopy and electron microscopy [129]. GR is a reactive metastable phase which, through the subsequent addition of base, forms a mixture of ferrous hydroxide and magnetite. This transformation is marked by an equivalent point around pH 10 (Fig. 7.31). The end of the transformation should correspond to $OH^-/Fe = 2.33$, but, under titration conditions, the reaction is kinetically limited. The instability of GR against hydroxylation could be expected, as indicated by the scheme shown in Figure 7.32, analogous to a disproportionation reaction.

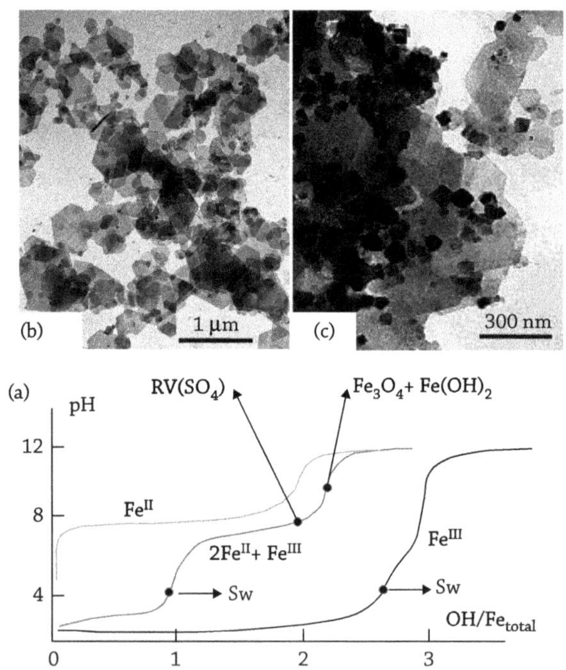

FIGURE 7.31 (a) Protometric titration curves of Fe^{II} ions, Fe^{III} ions, and the mixture of $2Fe^{II} + Fe^{III}$ (all titrations are carried out in the presence of sulfate ions). The formation of schwertmannite (Sw) is indicated on the curves. TEM images of the solid obtained (b) by alkalinization of the $2Fe^{II}+Fe^{III}$ mixture up to $OH^-/Fe_{total} = 2$ and (c) at the equivalent point located at pH 10.

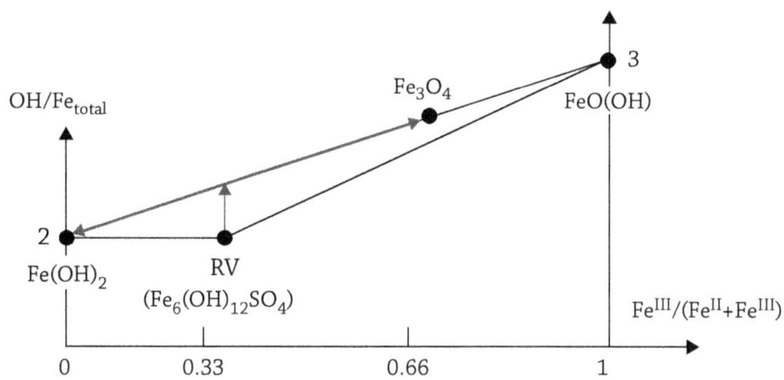

FIGURE 7.32 Scheme of the transformation of GR2-SO_4 into ferrous hydroxide and magnetite for a hydroxylation ratio $OH^-/Fe > 2$.

In spite of the coexistence of ferrous and ferric ions within the sheets of the structure, no electron mobility or intervalence transfer takes place between the cations in GRs, unlike what is observed with the spinel framework of magnetite (see later in this chapter). Indeed, the Mössbauer spectrum of GR2-SO_4 has only two components: one relative to ferrous ions and the other relative to ferric ones, with hyperfine parameters, isomeric shift, and quadrupole splitting, characteristic of these two species [131]. The high-electron concentration due to a majority of ferrous ions in the structure may cause strong electron repulsions between cationic sites, which forces electron localization. In support of this hypothesis, the progressive substitution of ferric ions by Al^{3+} ions shows an ordering at long distance in the distribution of ferrous cations within the sheets of the structure [131], likely caused by electron repulsion between the ferrous ions.

It is interesting to observe the structural relationship between ferrous hydroxide and GR containing similar brucite-type sheets made of edge-sharing octahedra. This structural type exists as long as ferrous ions are the most numerous cations in the compound ($x \leq 0.33$). When the proportion of ferrous ions decreases ($x > 0.33$) up to values of x close to 0.9, crystallization of spinel occurs. As seen earlier, precipitation of purely ferric systems, $x = 1$, forms goethite or hematite, depending on acid–base conditions. Although there are many topotacticity relationships among the transformations of these different phases [44, 132], it is tempting to think that as ferrous ions become the minority, Fe^{II}-Fe^{III} electron transfers can occur from the first steps of the solid formation, inducing crystallization of the spinel phase (see later in this discussion). It is also useful to remember that feroxyhyte, the completely oxidized brucite-like phase ($x = 1$), is obtained only by fast and complete oxidation of ferrous hydroxide, namely, when the brucite structure is already formed. Feroxyhyte thus appears as a kinetical phase. It exists a phase containing only trivalent cations, a chromium oxyhydroxide, CrOOH, of CdI_2 structural type containing sheets of octahedra that are similar to those of feroxyhyte. This phase, obtained by thermolysis at 90°C of acid solutions of chromium nitrate (§ 4.1.3), seems to be stable, but its formation likely results from the chemical inertness of Cr^{3+} ions regarding substitution reactions. For this cation, olation and oxolation reactions involved in crystallization are kinetically competitive and, instead to form the double chains characteristic of the goethite framework, the bidirectional growth of brucite-like sheets occurs (Fig. 7.11). Moreover, it seems that there is no CrOOH phase isotypic with goethite. Consequently, it appears that, because of too high reactivity toward condensation, the precipitation in solution of ferric ions never leads to δ-FeOOH of brucite type.

Green rusts can be formed according to different procedures: coprecipitation of ferric and ferrous ions, as described earlier [129, 130]; partial oxidation of ferrous hydroxide [133]; and chemical, electrochemical, or biological reduction of lepidocrocite [121, 134, 135]. GRs appear as a turntable in many redox processes involved in the corrosion of iron and steel [124, 136], as well as in the transformation of iron in the environment [120, 125, 126, 137–139]. Fougerite is a phase that has only recently been recognized as a mineral [126, 140]; it is so-called because it was first sampled in a forest near Fougères, Ille-et-Vilaine, in France. It is a hydroxycarbonated GR resulting from bioreduction of iron oxyhydroxides in anoxic conditions (Fig. 7.33). The material is partially deprotonated and corresponds to $[Fe^{II}_{6(1-x)}Fe^{III}_{6x}(OH)_{4(4-3x)}(OOH)_{2(3x-1)}]CO_3{\cdot}{\sim}3H_2O$, with x varying from 1/3 to 2/3, and with partial substitutions of Fe^{2+} and Fe^{3+} ions by Mg^{2+} and Al^{3+} ions, respectively [126, 140, 141]. These phases are extremely reactive to oxidation and have been finely characterized thanks to in situ Mössbauer spectroscopy measurements in the place of their formation [126].

b) Magnetite

Magnetite, Fe_3O_4, precipitates quasi-instantaneously by alkalinization of Fe^{3+} and Fe^{2+} mixtures with $x = Fe^{III}/(Fe^{III}+Fe^{II}) = 0.66$ [142]. In the inverse-spinel crystal structure of this oxide, the ferric cations are distributed in one quarter of the octahedral holes (O_h) and in one-eighth of the tetrahedral holes (T_d) of the face-centered cubic, close-packed stacking of oxide ions, when ferrous ions occupy one quarter of octahedral holes, according to the formula $[Fe^{3+}]_{Td}[Fe^{3+}Fe^{2+}]_{Oh}O_4$ [12] (Fig. 7.34).

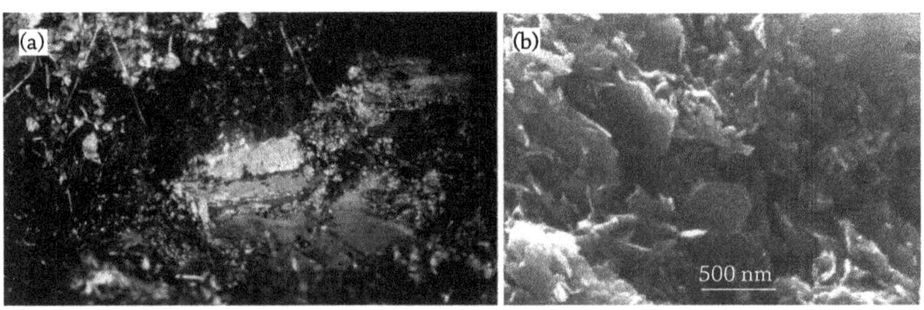

FIGURE 7.33 (a) Soil profile in a forest near Fougères (France). Under the surface overlying an organo-mineral horizon, the bluish-green fougerite stays in a reductomorphic zone (15–30 cm). Fougerite quickly turns to ochre upon contact with oxygen, after opening the profile; (b) SEM image of fougerite particles showing hexagonal platelets 300 to 500 nm in diameter (courtesy of Dr F. Trolard, INRA, Avignon, France).

Magnetite is a semimetal with a narrow band gap (Eg ≈ 0.1 eV) in which fast electron delocalization takes place at RT between ferrous and ferric cations within the octahedral sublattice. This electron mobility is indicated in the Mössbauer spectrum by a component characteristic of octahedral iron ions with a +2.5 electron charge (Fig. 7.34). The electron mobility explains the black color of the solid because the electron transfers are excited by absorption of the whole of the visible light. Other ferric oxy(hydroxi)des, including maghemite, the oxidized derivative from magnetite, are semiconducting with a much larger band gap, from 1.9 to 2.2 eV, which corresponds to absorption of wavelengths shorter than around 600 nm. This energy excites oxygen to metal charge transfers, allowing the transmission of red to orange radiations, which produces the color of the ferric phases.

The presence of ferrous ions and the occurrence of electron mobility are crucial factors in the crystallization of spinel iron oxide. Maghemite, the completely oxidized form of magnetite, $Fe_{2.67}O_4$ or γ-Fe_2O_3, $[Fe^{3+}]_{Td}[Fe^{3+}{}_{5/3}L_{1/3}]_{Oh}O_4$ (L stands for cation vacancies in the spinel framework), does not form from precipitation of ferric ions alone. Instead, the presence of ferrous ions in a poorly ordered solid such as 2- or 6-line ferrihydrite, even in weak proportion ($x \approx 0.90$), induces crystallization of the whole of this solid into spinel iron oxide [143–145]. Whatever the value of x between 0.90 and 0.66 in the early solid resulting from coprecipitation or from ferrous ion adsorption on preexistent ferrihydrite, all Fe^{2+} ions are incorporated in the solid, forming a mixed-valent material with an ordering at short distance and exhibiting an electron hopping as indicated by Mössbauer spectroscopy: the purely ferrous component existing in the spectrum recorded at low temperature below 100 K transforms into a component characteristic of $Fe^{+2.5}$ above 100 K (Fig. 7.35). Also, studies involving adsorption of $^{56}Fe^{II}$ ions onto ^{57}Fe-ferrihydrite, combined with adsorption of $^{57}Fe^{II}$ ions onto ^{56}Fe-ferrihydrite, have shown the occurrence of electron transfers at the ferrihydrite–solution interface [145]: as only ^{57}Fe isotope is detectable in Mössbauer spectroscopy, the existence of ferric ions in the adsorbed layer and the presence of ferrous ions in the core of particles prove the occurrence of electron transfer rather than ionic diffusion (§ 5.4.3).

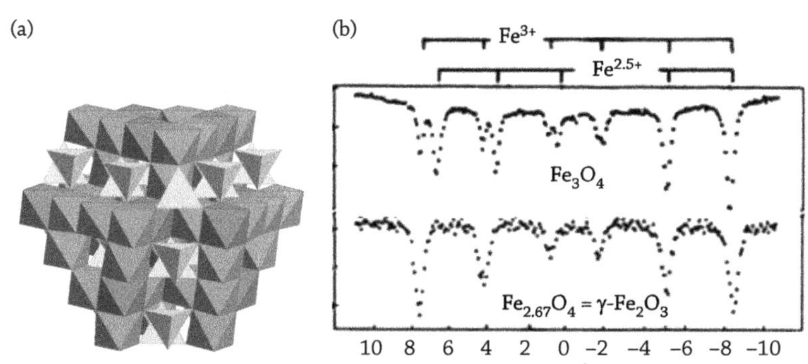

FIGURE 7.34 (a) Scheme of the spinel structure of magnetite and (b) Mössbauer spectra of magnetite and maghemite recorded at room temperature.

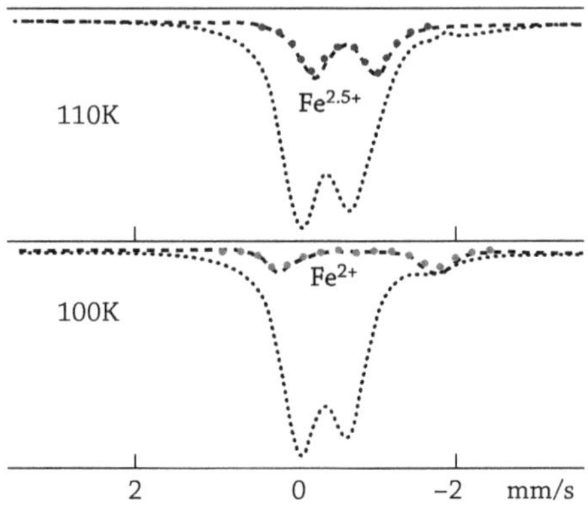

FIGURE 7.35 Mössbauer spectra of the early solid resulting from coprecipitation at pH 11 of the mixture Fe^{III} + 0.15 Fe^{II}, recorded at 100 and 110 K. The calculated Fe^{2+} and $Fe^{2.5+}$ components are drawn as dotted lines.

The injection of electrons provided by ferrous ions into ferrihydrite provokes local structural rearrangements and formation of the spinel ordering in small particles in order to extend the area of electron delocalization [144]. In parallel with this topotactic mechanism, spinel ordering also proceeds by dissolution–crystallization. Thus, these two mechanisms lead to two families of nonstoichiometric spinel particles, ($[Fe^{III}]_{Td}[Fe^{III}_{1+2z/3}Fe^{II}_{1-z}L_{z/3}]_{Oh}O_4$), with different sizes and compositions (Fig. 7.36). The relative importance of these two pathways is dependent on the Fe^{II} content: the more important it is, the more the dissolution–crystallization process prevails and the faster the crystallization occurs (quasi-instantaneous for $x = 0.66$). For $0.74 \geq x \geq 0.66$, a monophased system corresponding to only one particle family forms [143].

Coprecipitation of ferric ions with divalent cations other than Fe^{2+} highlights the role of electron mobility in crystallization of the spinel phase. With cations such as Co^{2+}, Mn^{2+}, and Ni^{2+}, intervalence transfers at room temperature with ferric ions do not noticeably occur. Consequently, spinel ferrites MFe_2O_4 form slowly by coprecipitation involving only a dissolution–crystallization mechanism. The transformation is accelerated, however, and the particle size is increased when the pH of the suspension is high, in correlation with the increase in solubility (Fig. 7.37) [147, 148].

Under strictly anaerobic conditions, formation of stoichiometric magnetite ($x = 0.66$) yields particles that are spheroidal in shape, with a mean size that can be finely adjusted in the range 2–12 nm in carefully controlling pH and the ionic strength of the suspension, without addition of surfactants [149–151] (Fig. 7.38). This result, owing to the influence of electrostatic surface charge density on surface energy, is described elsewhere in this book (§ 5.4.4).

Magnetite can be synthesized under various nanomorphologies through many processes. Most of them proceed by hydrothermal treatment from ferric or ferrous ion solutions in the presence of a reductant or an oxidant, respectively, the more

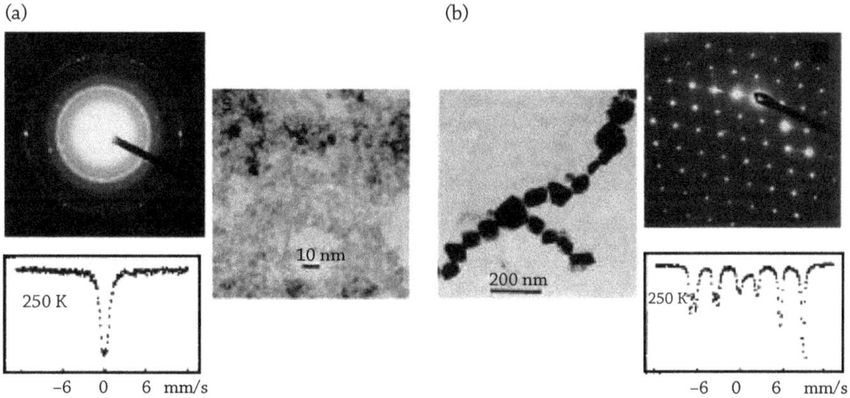

FIGURE 7.36 Electron diffraction patterns, Mössbauer spectra, and TEM images of the two families of spinel particles formed after 1 day aging at room temperature from the mixture Fe^{III} + 0.15 Fe^{II} precipitated at pH 11. The small particles (a) of 4 nm in mean size and with the composition Fe^{II}/Fe^{III} = 0.07 are formed by topotactic structuration of the early ferrihydrite, while the large, more or less prismatic particles (b) formed by dissolution–crystallization are 150 nm in mean size with the composition Fe^{II}/Fe^{III} = 0.33.

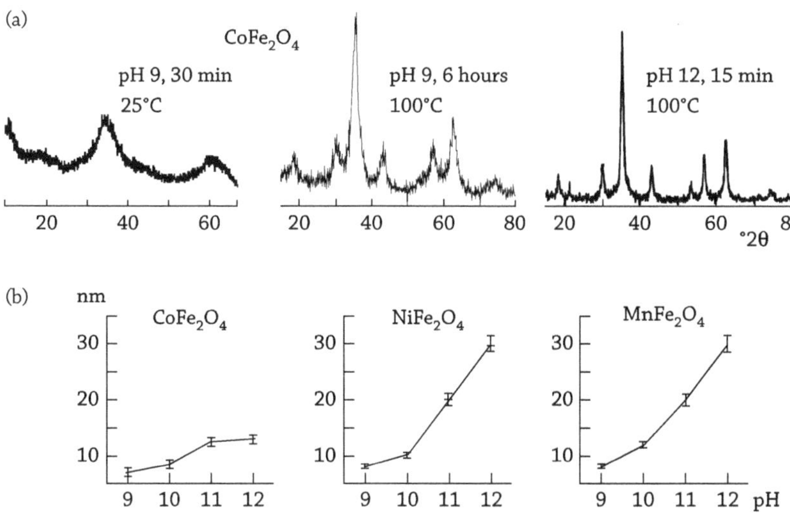

FIGURE 7.37 (a) X-ray diffraction patterns of the solid in suspension in which the ferrite $CoFe_2O_4$ is formed from coprecipitation of Co^{2+} and Fe^{3+} nitrates (Co/Fe = 1/2). The pattern of the solid precipitated at pH 9 and aged 30 mn at 25°C is characteristic of 2-line ferrihydrite, but the infrared absorption spectrum of the solid indicates the coexistence of $Co(OH)_2$. After heating at 100°C for 6 hours of the suspension at pH 9, around 35% of the solid are converted into spinel ferrite. In the suspension formed at pH 12 and heated 15 mn at 100°C, around 70% of the solid are converted into spinel ferrite. (b) Particle mean diameter of $CoFe_2O_4$, Mn Fe_2O_4 and Ni Fe_2O_4 aged 24 hours at 100°C in suspension at various pH [152].

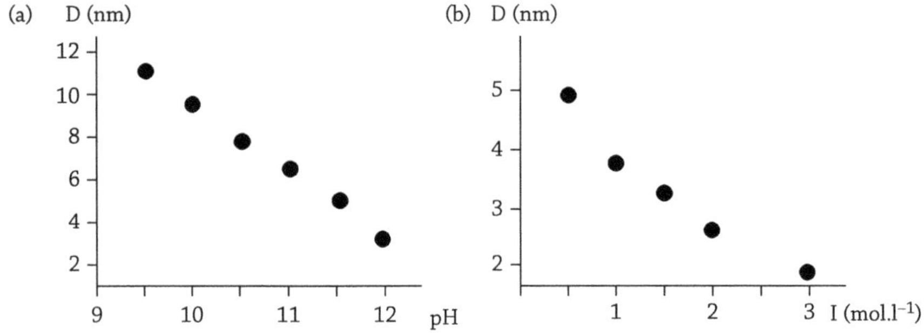

FIGURE 7.38 Mean particle diameter of magnetite Fe_3O_4 aged 8 days at 25°C in suspensions at (a) different pH ($NaNO_3$ 1 mol l^{-1}) and (b) different ionic strength I (pH = 12).

often hydrazine is oxidized into dinitrogen or reduced into ammonia. Uniform octahedra are produced by hydrothermal treatment at 160°C of ferric ions in aqueous solutions in the presence of hydrazine and EDTA in strongly basic medium (Fig. 7.39a) [152]. Also in hydrothermal conditions (180°C, 24 hours), rhombic dodecahedra are formed from aqueous solutions of ferrous sulfate in the presence of hydrazine and polyvinylpyrrolidone at pH 9 (Fig. 7.39b) [153]. Microwave heating (240W, 10 mn) of ferrous sulfate solutions containing citrate and phosphite in basic medium yields hexagonal platelets (Fig. 7.39c) [154]. Nanorods are obtained from a solution of ferrous acetate under an intense exposure to ultrasound in the presence of cyclodextrine, which acts to stabilize and disperse the particles [155]. Hydrothermal treatment between 110 and 160°C in a strongly alkaline medium (NaOH > 1.5 mol.l^{-1}) of these magnetite nanorods produces nanowires with a diameter of 15 to 50 nm and several micrometers in length, depending on the concentration of soda (Fig. 7.39d) [156]. Magnetite nanowires can also be obtained from hydrothermal treatment at 130°C under a magnetic field (0.35 T) of ferrous chloride in the presence of hydrazine in alkaline medium (Fig. 7.39) [157]. Nanotubes 50 to 60 nm in diameter and 15 to 20 nm in thickness are obtained by hydrothermal treatment at 140°C of ferric sulfate solutions in the presence of hydrazine and egg albumin (Fig. 7.39e) [158]. The protein, on which ferric ions are adsorbed, seems to act as a template in inducing the rolling up of the oxide sheet. Finally, astonishing dendritic crystals of magnetite have been synthesized by hydrothermal treatment at 160°C for 48 hours of the mixture of potassium ferricyanide $K_3[Fe(CN)_6]$ with CTAB and hydrazine [159] (Fig. 7.39f–i). The morphologies of dendrites, depending on concentrations in iron and CTAB, are all formed of a trunk oriented along the (110) direction, with four branches along the (111) direction.

Magnetite nanoparticles are sensitive to oxidation and transform *in fine* into maghemite, γ-Fe_2O_3. It is not clear whether the Fe_3O_4 stoichiometry is exact in the above-mentioned examples because the synthesis of stoichiometric magnetite by coprecipitation requires extreme caution to avoid oxidation. It is nevertheless important to consider the effective composition of spinel iron oxide particles because there is a continuous solid solution, from magnetite ($x = 0.33$) to maghemite ($x = 1$), and many properties of the material such as chemical reactivity, redox

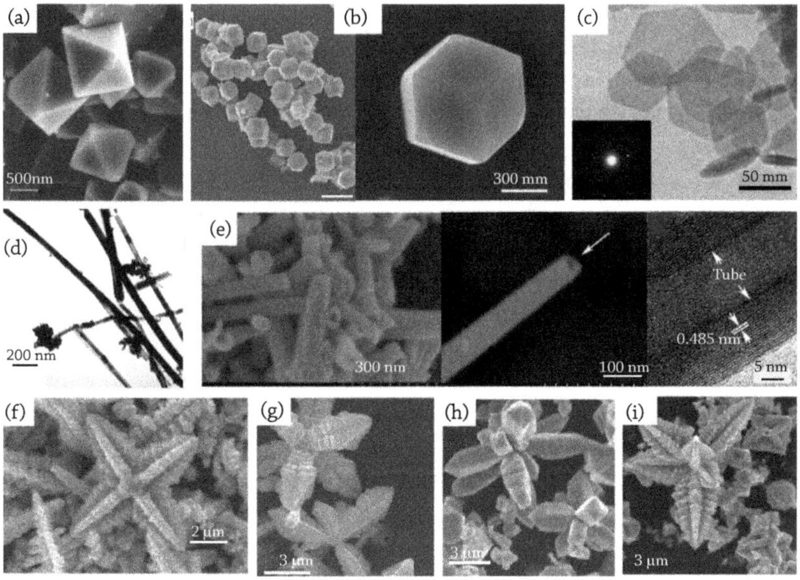

FIGURE 7.39 TEM and SEM images of different morphologies of magnetite obtained under various conditions (see text). (a) Octahedra (reproduced with permission from [152] Copyright 2007 American Chemical Society); (b) rhombic dodecahedra (reproduced with permission from [153] Copyright 2008 American Chemical Society); (c) hexagonal platelets (adapted from [154]); (d) nanowires (reproduced from [156] with permission of Elsevier 2004); (e) nanotubes (reproduced from [158] with permission of The Royal Society of Chemistry); (f–i) dendrites (reproduced with permission from [159] Copyright 2009 American Chemical Society).

behavior, adsorption capacity, and magnetism, are closely dependent on its stoichiometry. Knowledge of the real composition of particles, which is often neglected or ignored in many studies, can, however, be easily determined from three main techniques: (1) chemical analysis involving dissolution of oxide in acidic medium and titration of ferrous ions by potassium bichromate on a hand and then in the same sample, titration of the total amount in iron by bichromate after reduction of ferric ions by tin (II) [143]; (2) Mössbauer spectroscopy [160]; and (3) sometimes with more difficulty, X-ray diffraction [144, 160].

c) Transformation of Magnetite Induced by Adsorption

The high reactivity of magnetite nanoparticles is of course due to their high surface-to-volume ratio and, because of the high electron mobility within the solid, they give rise to an interesting surface chemistry.

Magnetite nanoparticles can be converted into maghemite not only by oxidation with oxygen, but also by involving various ion and/or electron transfers through the solid–solution interface. Basically, the mechanism of the transformation implies, in all cases, the elimination of an electron from a surface ferrous ion with formation of a ferric ion and a cationic vacancy in the octahedral sublattice to maintain electrical neutrality. This process can go to completion within the particles and form maghemite because the electron mobility within the solid allows renewal of the ferrous sites in surface. Simultaneously, the electrical neutrality is preserved in the whole of particles by migration of ferric ions toward the surface in creating cationic vacancies. Of course, this ion diffusion within the framework is the limiting step of reaction rate.

In neutral to basic conditions in suspension (or in solid state), oxidation of magnetite by oxygen proceeds by reduction of dioxygen on the surface of particles. The oxide ions produced are coordinated to the formed surface ferric ions. This transformation gives rise to an increase in particle size because of the adduction of oxygen layers. In acidic medium, around pH 2 and in strictly anoxic conditions, the transformation proceeds according to a completely different mechanism in which surface Fe^{2+} ions are desorbed as hexaaquo complexes in solution involving interfacial electron and ion transfer according to the balance:

$$\left[Fe^{3+}\right]_{Td}\left[Fe^{2.5+}{}_{2}\right]_{Oh}O_4 + 2\,H^+ \rightarrow$$
$$0.75\left[Fe^{3+}\right]_{Td}\left[Fe^{3+}{}_{5/3}V_{1/3}\right]_{Oh}O_4 + \left[Fe\right]^{2+}{}_{solvated} + H_2O$$

V stands for a cationic vacancy.

It formally corresponds to a $Fe^{2+}/2H^+$ net exchange between the solid and solution [161]. The process is illustrated in Scheme 7.3. Because of nanometric particle size and elimination of ferrous ions, magnetite oxidation in acidic conditions (pH $\approx$ 2) does not cause significant variation in particle size. From the kinetics standpoint, around 50% of the reaction is quasi-instantaneous. Complete oxidation needs around 1 month at room temperature for particles around 8 nm in diameter. Of course, the smaller the particles are, the faster the reaction becomes. In similar acidic conditions, cobalt ferrite $CoFe_2O_4$ nanoparticles, in which there is no Co^{II}-Fe^{III} intervalence transfer, give rise to desorption of only 10% of Co^{2+} ions, which corresponds to a fraction of the surface cations [161]. Such a result highlights the role of electron mobility in oxidation and the reactivity of magnetite nanoparticles. The treatment of magnetite particles in acidic medium and separation from the released ferrous ions lead to the formation of spontaneous dispersions as stable sols because the nanoparticles of maghemite bear a high positive electrostatic surface charge (§ 5.2.1).

This reaction of magnetite oxidation is seemingly reversible. Indeed, the ferrous adsorption on maghemite nanoparticles occurs by increasing the pH of the dispersion to reach the magnetite composition (x = 0.66). Adsorption takes place at around pH 6, at a pH where ferrous ions alone do not precipitate and where the hydroxylated surface behaves as a ligand able to coordinate the ferrous ions (§ 5.4.2). All the iron present in the system appears in Mössbauer spectroscopy as spinel-type iron [162–164]. However, as structural investigations by X-ray diffraction and Mössbauer spectroscopy show that the core of particle remains deficient in cations, iron ions do not migrate inside the particles. Electrons, and likely also protons to keep the charge balance, are injected into the core of particles from the more or less hydroxylated ferrous surface layer. As ferric ions are formed in the presence of ferrous ions in this surface layer, simultaneously it epitactically

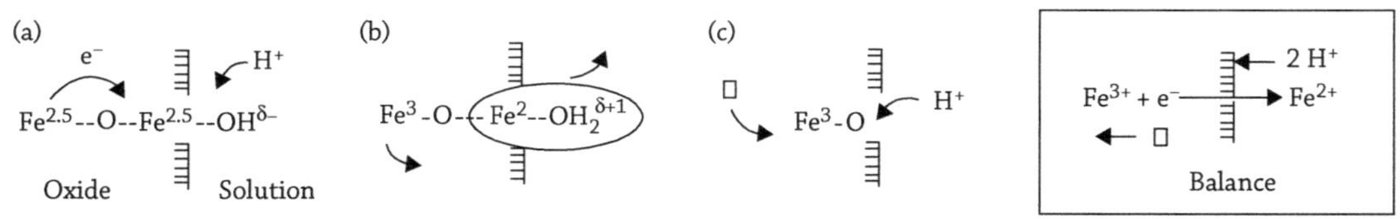

SCHEME 7.3 Sketches of the conversion of magnetite nanoparticles into maghemite in acidic suspension (pH ≈ 2): (a) the protonation of a surface oxygen atom provokes the localization of a mobile electron in forming a surface octahedral ferrous ion; (b) the ferrous ion is released in solution as a hexaaquo complex in leaving an excess of positive charge inside the particle; (c) the charge excess is balanced by the displacement of an internal ferric ion toward the surface in creating a cationic vacancy inside the framework and by fixation of a second proton. The net balance of the Fe(II) ion elimination is the consumption of two protons per Fe^{2+}, the diffusion of cationic vacancies toward the core of particles, and finally the oxidation of the material.

crystallizes into spinel, as previously described, and the reaction occurs up to the equipopulation in Fe^{3+} and Fe^{2+} ions within the octahedral sublattice of particles. The end of this phenomenon is particularly well evidenced in the protometric titration curves of maghemite suspensions containing variable amounts of ferrous ions (Fig. 7.40).

A similar result has been observed in the study of $^{57}Fe^{2+}$ ion adsorption onto ^{56}Fe-synthesized magnetite and, inversely, in $^{56}Fe^{2+}$ ion adsorption onto ^{57}Fe-synthesized nonstoichiometric magnetite [165]. The study clearly shows that the adsorption of ferrous ions occurs with the growth of a layer of spinel oxide and, simultaneously, the reduction of ferric ions in the core of particles. The study also shows that the amount of ferrous ions so adsorbed increases as the departure from stoichiometry of particles is large. Also, the adsorption of Co^{2+} or Ni^{2+} ions, which do not give rise to intervalence transfer with ferric ions in corresponding ferrite spinels, is limited, pointing to the fundamental role of electron mobility in the interfacial chemistry of the spinel iron oxides.

Adsorption of ferric ions onto magnetite involves similar interfacial electron transfer [166]. At pH ≈ 8, the ferric ions immediately form a layer of ferrihydrite on the magnetite particles, and this layer, partially reduced by desorption of mobile electrons, crystallizes in epitaxy on the oxidized particles. The magnetite oxidation proceeds in these conditions, with an increase in particle size corresponding approximately to one to two spinel unit cells in thickness. The reaction stops at this stage because the rate of the process is limited by the diffusion of cations from the core of particles toward the exterior through the adsorbed layer. The adsorption of metal cations that are easily reducible, such as silver, platinum, and palladium, also allows oxidation of nanometric magnetite [167]. The reaction occurs without iron ions desorption but with surface hydroxylation and sudden acidification of the suspension (Scheme 7.4). It leads to a mixture of maghemite and metallic nanoparticles of silver, platinum, or palladium (Fig. 7.41). The set of

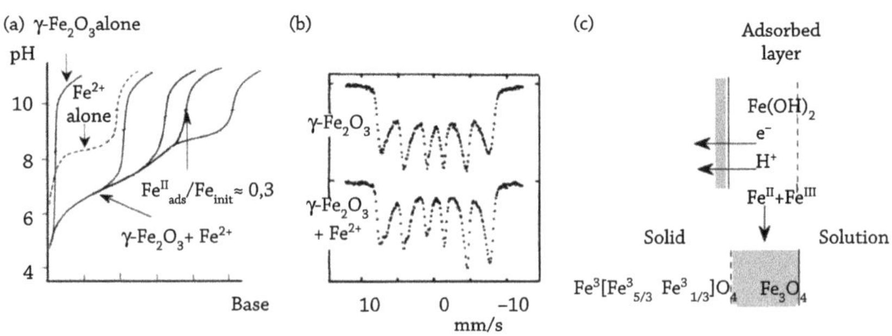

FIGURE 7.40 (a) Protometric titration curves of maghemite nanoparticles in the presence of variable amounts of ferrous ions (reproduced from [163] with permission of The Royal Society of Chemistry); (b) Mössbauer spectra of maghemite nanoparticles before and after ferrous ions adsorption (reproduced from [164] with permission of Springer); (c) schematized balance of ferrous ion adsorption onto maghemite.

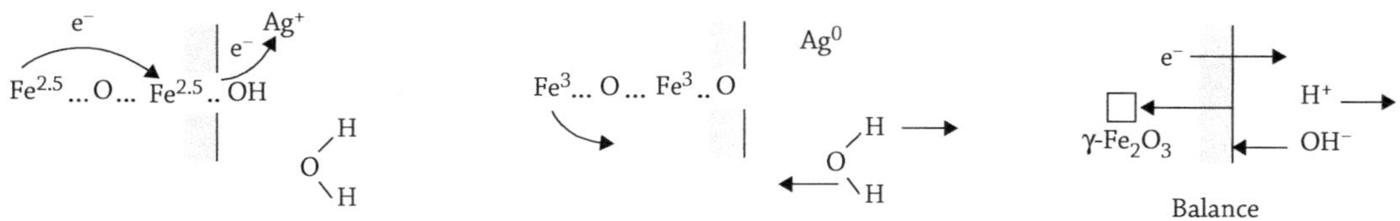

SCHEME 7.4 Adsorption of an Ag^+ ion on magnetite provokes, as protonation (Scheme 7.3), the localization of a mobile electron from the framework on a surface site and reduction of silver. The charge balance in the core of particles occurs by migration of a Fe^{3+} ion toward the surface. The solvation and deprotonation of a water molecule help balance locally the charge of the newly formed surface ferric ion and account for the acidification of the medium. The global balance corresponds to the release of an electron from magnetite and migration of cationic vacancies toward the core of the particle, in association with the fixation of a hydroxo ligand on the surface and the release of a proton in solution.

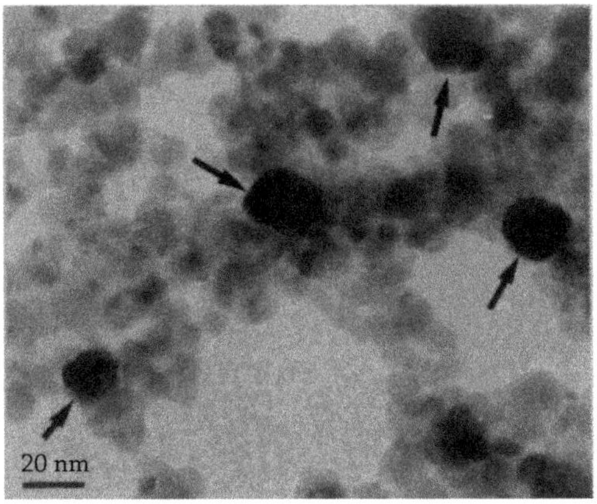

FIGURE 7.41 TEM image of silver nanoparticles (labeled by arrows) formed by nitrate silver reduction in a suspension of magnetite nanoparticles at initial pH 8 (final pH 4) (reproduced from [167] with permission of Elsevier).

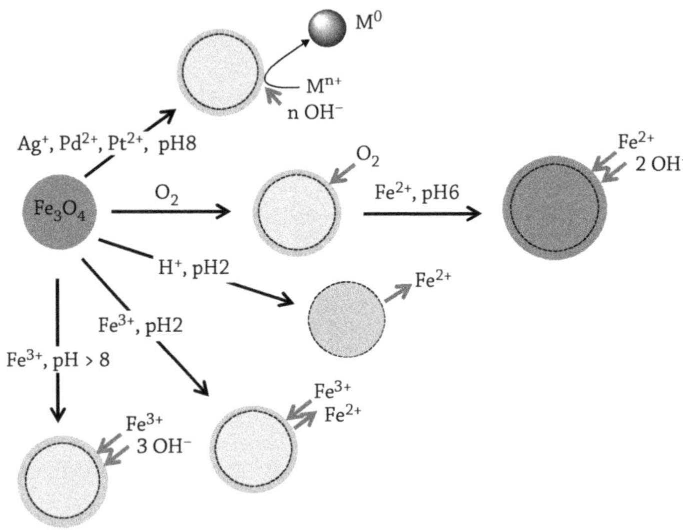

SCHEME 7.5 Sketch of transformations of magnetite induced by adsorption. Each of the steps is described in the text. The dotted circles stand for the initial size of particles in these transformations.

these reactions (Scheme 7.5) shows that magnetite nanoparticles behave like rechargeable batteries.

d) Dispersions of Magnetic Particles

Maghemite particles resulting from magnetite oxidation form easily concentrated dispersions that are stable against aggregation and sedimentation. In acidic medium at low ionic strength, the nanoparticles bear a high surface positive charge density and spontaneously disperse in water in forming sols that are practically free from aggregation (§ 5.3) [142, 168, 169]. Stable dispersions can also be obtained by adsorption of various species, charged or not (e.g., citrate, phosphonate, polymers, polyelectrolytes), depending on the nature of the chosen dispersion medium, aqueous or not aqueous, and on the function of nanoparticles, for instance, for medical uses with introduction in the human body [122, 123] or applications in mechanics such as magnetic liquids or ferrofluids [124, 170–174]. The formation of such ferrofluids is often carried out by oleic acid adsorption on particles that are then easily dispersed in oil [175].

Dispersions of the magnetic particles in solid matrixes, organic (polyvinylic alcohol, epoxy resin) or inorganic (silica glasses), are obtained by gelification of an aqueous sol in which the precursor of the matrix is introduced and is more or less adsorbed on particles. By simple drying, stiff magnetic composites with adjusted concentrations in particles (i.e., with adjusted distances between particles) are obtained; they are well adapted for physical studies of the magnetism of nanoparticles and various measurements on unique particles. Such samples have been successfully used in the study of interparticle magnetic coupling and the magnetic relaxation involved in superparamagnetism [176, 177].

Silica-based magnetic composites have been synthesized by condensation of silicic acid, $Si(OH)_4$, or alkoxides $Si(OR)_4$ (e.g., R = CH_3–, C_2H_5–) introduced into the acidic aqueous sol of maghemite nanoparticles [178]. Hydrolysis of alkoxo groups and acid-catalyzed oxolation of silanol functions (§ 4.1.4a) occurred in situ in forming a transparent gel and then a monolithic glass by drying at room temperature (Fig. 7.42). In the silica matrix, the nanoparticles are indeed embedded

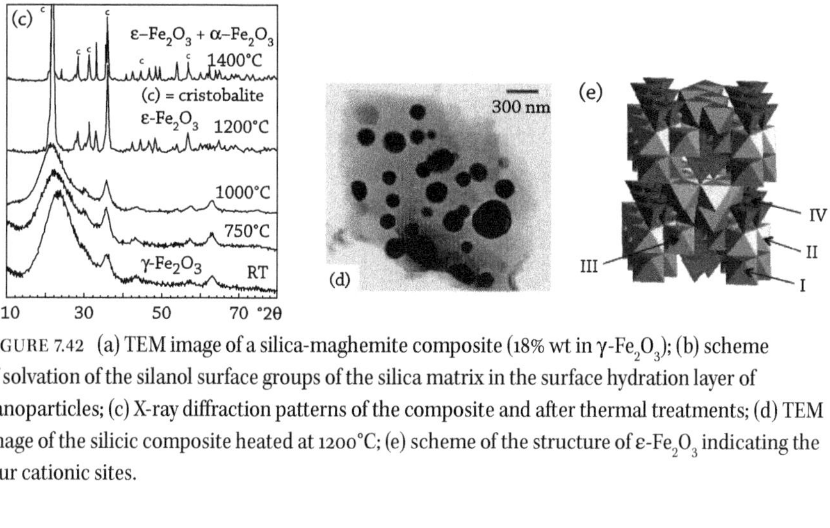

FIGURE 7.42 (a) TEM image of a silica-maghemite composite (18% wt in γ-Fe_2O_3); (b) scheme of solvation of the silanol surface groups of the silica matrix in the surface hydration layer of nanoparticles; (c) X-ray diffraction patterns of the composite and after thermal treatments; (d) TEM image of the silicic composite heated at 1200°C; (e) scheme of the structure of ε-Fe_2O_3 indicating the four cationic sites.

with their hydration water shell without direct chemical interaction between silica and particles. Infrared, near-infrared, and Mössbauer spectroscopies show that there is no formation of Si–O–Fe bonds. In addition, thermogravimetric and infrared studies show that the particle hydration layers are still present in the dried material.

The silica matrix behaves as an antisintering agent during the thermal treatment that stabilizes the maghemite nanoparticles against their transformation into hematite. Indeed, Mössbauer spectroscopy shows that for powders of nanoparticles around 10 nm in diameter, not protected by adsorbed species, and dried at room temperature, the $\gamma \rightarrow \alpha$-$Fe_2O_3$ transformation begins around 300°C and ends around 400 to 500°C. For nanoparticles that have adsorbed sulfate or phosphate, which are thermally stable entities, the crystal transformation is pushed away to around 800°C [179]. In composites containing low enough concentrations of nanoparticles, any transformation occurs above around 1000°C, a temperature at which the silica glass begins to soften, allowing particles to move and to diffuse within the matrix. At around 1200°C, partial crystallization of the silica matrix into cristobalite occurs. That allows limited aggregation of the nanoparticles and the transformation of aggregates into a rarely observed crystal phase, ε-Fe_2O_3 (§ 1.1.3a). The particles have a volume corresponding approximately to a few tens of initial maghemite nanoparticles [180] (Fig. 7.42). ε-Fe_2O_3, isotypical with kappa alumina [180, 181], is ferrimagnetic with a very high coercivity and is likely piezoelectric [182]. Figure 7.43 illustrates the magnetic behavior (magnetization and coercivity) of nanosized phases of maghemite γ-Fe_2O_3, ε-Fe_2O_3 and hematite α-Fe_2O_3. Maghemite behaves as a soft magnetic material (high saturation magnetization but very low coercivity), ε-Fe_2O_3 behaves as a hard magnetic material (high coercivity), and hematite is a weakly ferromagnetic material.

At 1400°C, the matrix is in its whole crystallized as cristobalite and iron oxide, transformed into hematite as crystals that are larger by at least one order of magnitude than the ε-Fe_2O_3 particles. All observations show that the ε-Fe_2O_3 phase results from sintering a restricted number of γ-Fe_2O_3 particles, while in powders of unprotected nanoparticles, the thermal transformation leads directly to large hematite particles, which is the thermodynamically more stable phase. In powders, the positive surface component of the free enthalpy of transformation of iron oxide particles can be lowered by growing, so that the free enthalpy of the system is minimized by direct formation of hematite, which is the most stable phase [49]. In composites, in which the particle diffusion and consequently sintering are restricted, the decrease in surface area leads to lowering the free enthalpy of the system and to allowing a crystal transformation. However, the surface component remains important so that the minimum in free enthalpy cannot be reached, whatever the kinetical constraints might be. It results only in the formation of a metastable phase, if it exists. With careful control of the thermal treatment, it should be possible to observe a series of intermediate phases between γ-Fe_2O_3 and α-Fe_2O_3 (§ 1.1.3a), as for transition aluminas [12].

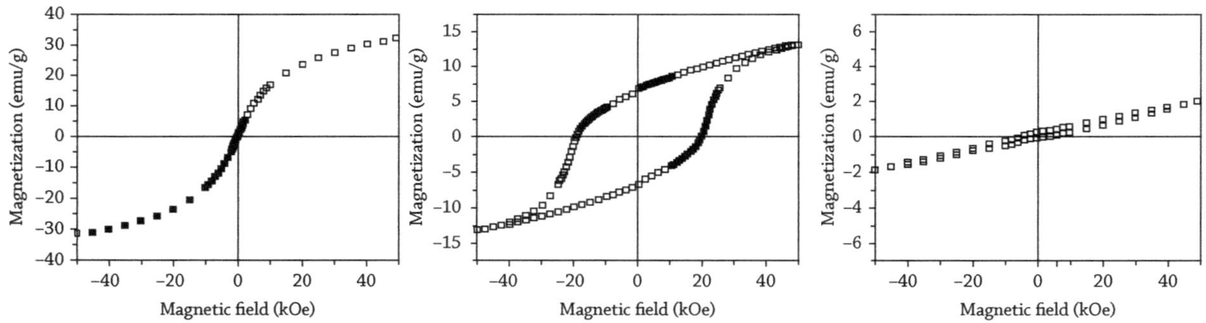

FIGURE 7.43 Magnetization (*M*) versus applied magnetic field (*H*) plots at room temperature for nanoparticles of (a) γ-Fe_2O_3 (6 nm), (b) ε-Fe_2O_3, and (c) α-Fe_2O_3. The unit of magnetization (ume/g) is a calibrated value for a given Fe_2O_3 weight (courtesy of S.I. Ohkoshi, School of Science, Univ. of Tokyo, Japan).

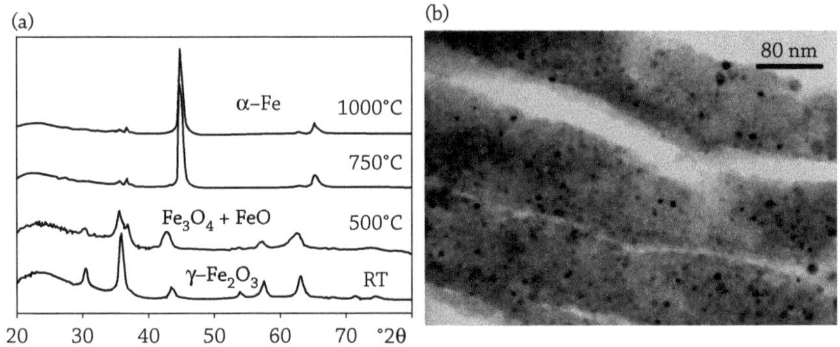

FIGURE 7.44 (a) X-ray diffraction patterns of silica-maghemite composites formed from triethoxysilane (18% wt in γ-Fe_2O_3) and after thermal treatment under argon flux. (b) TEM image of α-Fe particles formed by heating the composite at 1000°C under argon flux (the white bands are due to variations in thickness of the sample slices obtained by microtomy).

Reactive silica matrixes can be synthesized by adding in the aqueous sol of nanoparticles an alkoxysilane bearing a reactive function, such as triethoxysilane $SiH(OC_2H_5)_3$ [178]. The Si–H bond resists hydrolysis and condensation, and after drying the suspension at room temperature, a silica material with a composition close to $SiH_{0.8}O_{1.5}(OH)_{0.2}$, and seemingly similar by its stiffness and transparency to that formed with silicic acid, is obtained. The thermal behavior of this material is very different, however. By heating the composite under argon atmosphere, the Si–H bonds are broken in releasing hydrogen. That causes the reduction of iron oxide into a mixture of magnetite and ferrous oxide around 500°C and then in metal iron from 750°C, while the matrix is oxidized into silica [183, 184] (Fig. 7.44).

7.2.4 POLYMETALLIC FERRITES: SPINELS, HEXAFERRITES, GARNETS

The synthesis of oxides containing several metal ions is often faced to phase separation (§ 4.1.6). The following examples highlight some strategies that can help circumvent the difficulty in the synthesis of some important iron-based magnetic oxides.

a) Cubic Ferrites

The crystal structure of spinel-type ferrites $M^{II}M'^{III}_2O_4$, like magnetite, is formed from a face-centered cubic framework of O^{2-} ions, with one-half of the octahedral holes and one-eighth of the tetrahedral holes occupied by tri- and divalent cations, respectively. These ferrites are magnetically soft, (of low magnetic coercivity), and they are used in inductances and in electrical transformers. They can be synthesized by the classical processes of solid chemistry (milling and heating of powders) for many combinations of M^{II} and M'^{III} cations [185], but they do not act the same with coprecipitation in solution. With many divalent ions such as Mg, Cd, Co, Ni, Mn, Zn, and Pb, for instance, coprecipitation at pH 10 at room temperature of the mixture M^{2+} + 2 Fe^{3+} yields a mixture of hydroxide $M(OH)_2$ and ferrihydrite (§ 7.2.2a). $M(OH)_2$ hydroxide is a stable phase, and iron precipitates aside. For chromium ferrite, $Fe^{II}(Fe^{III}_{1-x}Cr^{III}_{x})_2O_4$, coprecipitation of Fe^{2+}, Fe^{3+}, and Cr^{3+} ions also forms a mixture of different phases because the chemical reactivity of iron ions is much higher than that of chromium and the Cr^{3+} ions cannot be directly incorporated into the iron oxide phase, while crystallization of the purely ferric-ferrous phase (magnetite) is quasi-immediate because of the fast intervalence transfer between iron cations (§ 7.2.3b). The synthesis of such ferrites is possible, however, in soft conditions, and various methods can be used.

Coprecipitation of the cations M^{2+} + 2 Fe^{3+} (M = Co, Ni, Mn) at room temperature immediately forms a mixture of ferrihydrite and $M(OH)_2$ hydroxide. By thermolysis of the suspension at 100°C or even at a lower temperature, the spinel phases are formed according to a dissolution–crystallization mechanism (Fig. 7.37) [147, 186, 187]. The acidity of the medium is the main parameter of the synthesis in determining the particle size because the solubility of the cations, strongly depending on pH, directly influences the dissolution–crystallization process.

Such spinel phases can also be formed by coprecipitation of the divalent cation and iron as ferrous form. The solid is then oxidized either with a soluble reactive, nitrate for instance, or with air. The formation of ferric ions results in spinel oxide crystallization. This is typically the case for chromium ferrite $Cr_{0.42}Fe_{2.56}O_4$ [188] and titanium ferrite Fe_2TiO_4 [189]. This method, however, often forms micronic-sized particles that contain heterogeneities in composition due to concentration gradients of the divalent cation (Co, Mn, Ni) between the core and the surface of the particles [190].

It is also possible to form the spinel phase from mixed oxalic complexes. So, nickel ferrite $Ni_{0.6}Fe_{2.4}O_4$ is obtained by heating the complex $Ni_{0.2}Fe_{0.8}C_2O_4{\cdot}2H_2O$ at 300°C precipitated at RT from a solution of the same Ni/Fe composition containing a slight excess of oxalic acid. By heating the solid between 600 and 1000°C, the solid evolves toward formation of the stoichiometric ferrite $NiFe_2O_4$, with an excess of hematite Fe_2O_3 [191]. Similarly, spinel ferrites $FeAl_{2-2z}Cr_{2z}O_4$ ($0 \le z \le 1$) can be obtained by heating at 700°C ternary oxalic complexes $(NH_4)[Fe_{1-(x+y)}Cr_xAl_y(C_2O_4)_3]{\cdot}3H_2O$ with (x + y = 2/3) [192]. The thermal treatment unavoidably provokes a more or less important sintering, with an increase in particle size, but the technique allows formation of the spinel phases at a temperature that is at least 300° lower than that involved in diffusion in solid state.

b) Barium Hexaferrite $BaFe_{12}O_{19}$

Hexagonal ferrites, known as hexaferrites, are part of a large group of magnetic materials that have an important technological interest as permanent magnets and materials for magnetic recording and data storage, as well as components in electric devices operating at microwave frequencies [193]. The complex crystal structures of such compounds are based on different stackings of spinel slices.

Barium hexaferrite, $BaFe_{12}O_{19}$, which is likely the best known compound among the numerous hexaferrites, belongs to the magnetoplombite ($PbFe_{12}O_{19}$) structural type. It is built from the stacking of half spinel unit cells (S blocks), connected to junction blocks made of three layers of oxide ions in hexagonal compact packing (R blocks) (Fig. 7.45). Iron ions exhibit 4- and 6-coordinations in the spinel groups and 5- and 6-coordinations in the R blocks. The compound is ferrimagnetic with a uniaxial magnetization parallel to the *c* axis of the crystal cell. Because of a very high magnetocrystalline anisotropy, the material has a high remanent magnetization and a high coercivity (hard ferrite) and exhibits an important resistance to demagnetization. Such materials are used for fabrication of permanent magnets [193].

The conventional synthesis of $BaFe_{12}O_{19}$ involves heating at high temperature (1100°C) the mixture of iron oxide and barium carbonate powders. The casting of the material is carried out after milling and sintering at 1200°C. This elaboration process induces many structural defects, mainly involving nonstoichiometry and stacking, and local precipitation of Fe_2O_3 that lowers the coercitive forces of the material [193, 194]. To avoid such drawbacks, synthesis in softer conditions from solutions is possible.

Coprecipitation in a soda excess of ferric chloride in the presence of barium carbonate forms a mixture of mixed hydroxides and basic carbonates that transforms into barium hexaferrite by calcination at 710°C. The magnetic properties are optimized by annealing at 950°C. The method yields platelet-shaped particles 0.3 μm in mean size [195, 196]. Another technique consists of oxidizing by hydrogen peroxide the ferrous oxalate in solution in the presence

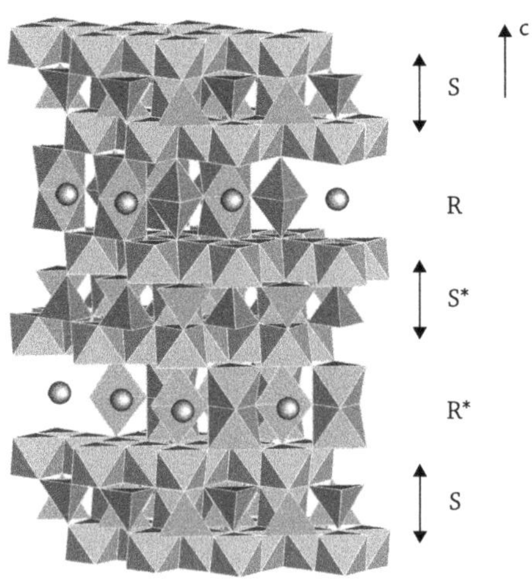

FIGURE 7.45 Scheme of the crystal structure of barium hexaferrite, $BaFe_{12}O_{19}$. The S blocks (spinel) and R blocks labeled with an asterisk have turned 180° around the *c* axis of the crystal cell (vertical axis on the figure). The octahedra represent the FeO_6 units, tetrahedra the FeO_4 units, and bipyramids the FeO_5 units. The spheres represent the Ba^{2+} ions. In the S blocks, the octa- and tetrahedral Fe ions are antiferromagnetically coupled, as the octa- and pentahedral Fe ions in the R blocks.

of an excess of oxalic acid. Addition of barium carbonate (Ba/Fe = 1/12) forms mixed complexes of Fe^{3+} and Ba^{2+} ions. These complexes are destroyed in alkaline medium (carbonate buffer) and form a gel of mixed hydroxide. Heated at 700°C, the gel crystallizes into hexaferrite as circular platelets [196]. The easy magnetization axis of these platelets is perpendicular to basal faces, and the particles of micrometric size (from around 0.1 to 0.3 μm) allow the fabrication of materials with a high storage density using the vertical component of the magnetic field of the recording head. For longitudinal recording, acicular particles with the easy magnetization axis along the particle's major axis are best suited, but this morphology for hexaferrite is unusual. An indirect technique, however, does permit reaching this goal, starting from acicular particles of goethite, α-FeOOH. A gel of these particles is treated by barium ethoxide, $Ba(C_2H_5O)_2$, in alcoholic medium. The reaction of the alkoxide with the surface hydroxylated groups of goethite provokes the adsorption of barium and the elimination of alcohol. A thermal treatment allows the diffusion of barium ions within the particles, and crystallization of hexaferrite occurs from 750°C in preserving the initial shape of goethite particles [201].

Nanoparticles of $BaFe_{12}O_{19}$ can also be obtained from thermal decomposition of the citrate complex obtained in dissolving ferric and barium nitrates in the stoichiometric ratio (Ba/Fe = 1/12) in a solution of citric acid containing (or not) ethylene glycol [197]. The precipitate is washed carefully in order to eliminate nitrate ions (nitrate compete with citrate to complex barium). Heated at 900°C, the complex yields exclusively the magnetoplombite phase of hexaferrite, as particles 100 to 200 nm in mean diameter.

Many other techniques, as well as hydrothermal processes, have been used to form barium hexaferrites. An interesting review is given in the article by Pullar [193].

c) Yttrium–Iron Garnet $Y_3Fe_5O_{12}$

Garnets are silicate minerals $A^{II}_3B^{III}_2(SiO_4)_3$ with A = Mg, Fe, Mn or Ca and B = Al, Fe, Cr. The complex pseudocubic structure is built from SiO_4 tetrahedra, BO_6 octahedra, and distorted AO_8 cubes. Compounds of iron and of some lanthanide ions (from Gd to Y) have the same structural type. The best known compound is $Y_3Fe_5O_{12}$ (Fig. 7.46), so-called YIG for yttrium iron garnet, in which the Y^{3+} ions are octacoordinated. The structure is strongly constrained because the polyhedra, corner sharing, and edge sharing, including tetrahedra, are highly distorted. These ferrimagnetic compounds are soft ferrites, as spinel ferrites. YIG has many uses and applications in hyperfrequency transmission, such as radar and telecommunications by satellites. It has been used largely for magnetic bubble devices.

The classical ceramic route used to synthesize YIG from Fe_2O_3 and Y_2O_3 powders involves a very high temperature (1500°C). This major drawback can be avoided by using softer procedures involving solution chemistry.

A process based on Pechini's technique consists of forming a mixed complex with a polycarboxylate, citrate, or malonate, within a polyester organic matrix [198]

FIGURE 7.46 Crystal structure of garnet $Y_3Fe_5O_{12}$. Octahedra and tetrahedra represent FeO_6 and FeO_4 units, respectively. The 8-coordinated polyhedra represent YO_8 units.

(§ 4.1.6). Two reactions are involved in the process: esterification between the organic acid and glycol to form the matrix, and complexation of cations by carboxylate. The role of the polyester matrix is to minimize the segregation of cations inside metal chelates during the subsequent thermal treatment, allowing elimination of the organic components and formation of the oxide. For that purpose, iron and yttrium nitrates are dissolved with stoichiometric proportions (Y/Fe 3/5) in glycerol, and then malonic acid is added in the solution. Esterification of malonic acid produces a gel that can be easily spread on a substrate before thermal treatment. The different steps of the treatment are identified by infrared absorption spectroscopy [199]. At 330°C, the polyester is decomposed, and the organic matrix contains only metal malonate complexes in which the cations are bicoordinated. From 400 to 600°C, the carboxylates have disappeared and are replaced by metal carbonates. At 800°C, the infrared spectrum exhibits only the metal–oxygen vibrations characteristic of the YIG framework, but X-ray diffraction shows that the YIG phase is the main product, together with small amounts of Fe_2O_3 and $YFeO_3$, which are likely formed after thermal decomposition of the organic matrix.

It seems that an efficient complexation of the cations in the starting solution is required to avoid the formation of phase mixture. The adjustment of pH at around 2 of the starting aqueous solution containing iron and yttrium nitrates (Y/Fe = 3/5) and concentrated citric acid, without glycol addition, allows the simultaneous complexation of iron and yttrium cations. Heating this solution at 80°C forms a gel. After drying the gel at 110°, the thermal treatment allows crystallization of YIG from 650°C. At 700°C, YIG is the only phase identified. The particle size varies from 30 to 500 nm, depending on the duration (2–25 hours) and temperature (700–900°C) of the thermal treatment [200, 201]. It also seems that the presence of ethylene glycol in the reaction medium favors the persistence of parasitic phases such as Fe_2O_3 and/or $YFeO_3$. The same procedure appears to be suitable for the synthesis of nanoparticles of the solid solution $Y_3Fe_{5-x}Al_xO_{12}$ ($0 \leq x \leq 2$) [202].

The YIG phase can also be formed by hydrothermal treatment at 225°C for 6 hours of the iron and yttrium (Y/Fe = 0.84) coprecipitate at pH 10.5 [203]. Why such a slight excess of yttrium is required is not known (the stoichiometric ratio is Y/Fe = 0.60). The pH value is critical in the formation of YIG: at pH 7.5; only Fe_2O_3 is obtained. Yttrium remains in solution, while at pH 9.5, Y_2O_3 and other unidentified phases coexist with YIG. In addition, the pH must be adjusted with ammonia. With soda, YIG does not form, likely because the presence of Na^+ ions and hydrothermal treatment yield only a mixture of $Y(OH)_3$ and Fe_2O_3. The YIG particles obtained under optimal conditions are 1.5 µm in mean size. This important particle size demonstrates that a dissolution–crystallization mechanism is involved in the growth of YIG. Hydrothermal treatment at a temperature above 225°C decomposes the YIG phase into $YFeO_3$ and Fe_2O_3.

References

[1] Stumm W., Sulzberger B. The cycling of iron in natural environments: Considerations based on laboratory studies of heterogeneous redox processes. *Geochim. Cosmochim. Acta* 1992, 56, 3233–3257.

[2] Pédro G.G. Cycles biogéochimiques et ecosystèmes continentaux. Rapport sur la Science et la Technologie n° 27, Académie des Sciences, EDP Sciences, Les Ulis, France, 2007.

[3] Cornell R. M., Schwertmann U. *The Iron Oxides, Structure, Properties, Reactions, Occurrence and Uses*. Weinheim, Germany: Wiley-VCH, 2003.

[4] Schmidt C., Behrens S., Kappler A. Ecosystem functioning from a geomicrobiological perspective—a conceptual framework for biogeochemical iron cycling. *Environ. Chem.* 2010, 7, 399–405.

[5] Borch T., Kretzschmar R., Kappler A., Van Cappellen Ph., Ginder-Vegel M., Voegelin A., Campbell K. Biogeochemical redox processes and their impact on contaminant dynamics. *Environ. Sci. Technol.* 2010, 44, 15–23.

[6] Chapman S.K., Daft S., Munro A.W. Heme: The most versatile redox centre in biology? *Struct. Bond.* 1997, 88, 39–70.

[7] Chasteen N.D., Harrison P.M. Mineralization in ferritin: an efficient means of iron storage. *J. Struct. Biol.* 1999, 126, 182–194.

[8] McClean R.G., Schofield M.A., Kean W.F., Sommer C.V., Robertson D.P., Toht D., Gajdardziska-Josifovska M. Botanical iron minerals: correlation between nanocrystal structure and modes of biological self-assembly. *Eur. J. Mineral.* 2001, 13, 1235–1242.

[9] Faivre D., Godec T.U. From bacteria to mollusks: The principles underlying the biomineralization of iron oxide materials. *Angew. Chem. Int. Ed.* 2015, 54, 4728–4747.

[10] Mann S., Webb J., Williams R. J. P. *Biomineralization, Chemical and Biochemical Perspectives*. Weinheim, Germany: Wiley-VCH, 1989.

[11] Herbillon A. Les hydroxydes ferrosiques, les rouilles vertes et la fougérite dans le cycle biogéochimique du fer. *C. R. Geoscience* 2006, 338, 393–401.

[12] Wells A. F. *Structural Inorganic Chemistry*, 5th ed. Oxford: Oxford University Press, 1991.

[13] Blesa M.A., Matijevic E. Phase transformations of iron oxides, oxohydroxides, and hydrous oxides in aqueous media. *Adv. Colloid Interface Sci.* 1989, 29, 173–221.

[14] Schwertmann U., Cornell R. M. *Iron Oxides in the Laboratory, Preparation and Characterization*. Weinheim, Germany: Wiley-VCH, 2000.

[15] Baes C.F., Mesmer R.E. *The Hydrolysis of Cations*. New York: Wiley, 1976.

[16] Bottero J. Y., Manceau A., Villieras F., Tchoubar D. Structure and mechanisms of formation of FeOOH(Cl) polymers. *Langmuir* 1994, 10, 316–319.

[17] Zhu M., Puls B.W., Frandsen C., Kubicki J.D., Zhang H., Waychunas G.A. In situ structural characterization of ferric iron dimers in aqueous solutions: identification of μ-oxo species. *Inorg. Chem.* 2013, 52, 6788–6797.

[18] Zhu M., Frandsen C., Wallace A.F., Legg B., Khalid S., Zhang H., MØrup S., Banfield J.F., Waychunas G.A. Precipitation pathways for ferrihydrite formation in acidic solutions. *Geochim. Cosmochim. Acta* 2016, 172, 247–264.

[19] Hou Y. Rodriguez M.A., Nyman M. Cation-bonding and protonation of the Fe_4-square cluster. *Cryst. Growth Des.* 2012, 12, 1422–1431.

[20] Lippard S. J. Oxo-bridged polyiron centers in biology and chemistry. *Angew. Chem., Int. Ed. Engl.* 1988, 27, 344–361.

[21] Hagen K.S. Synthetic models for iron-oxygen aggregation and biomineralization. *Angew. Chem. Int. Ed. Engl.* 1992, 31, 1010–1012.

[22] Kurtz D.M. Jr. Oxo- and hydroxo-bridged diiron complexes: a chemical perspective on a biological unit. *Chem. Rev.* 1990, 90, 585–606.

[23] Schmitt W., Murugesu M., Goodwin J.C., Hill J.P., Mandel A., Bhalla R., Anson C.E., Heath S.L., Powell A.K. Strategies for producing cluster-based magnetic arrays. *Polyhedron* 2001, 20, 1687–1697.

[24] Murugesu M., Clérac R., Wernsdorfer W., Anson C.E., Powell A.K. Hierarchical assembly of $\{Fe_{13}\}$ oxygen-bridged clusters into a close-packed superstructure. *Angew. Chem. Int. Ed.* 2005, 44, 6678–6682.

[25] Bino A., Ardon M., Lee D., Springler B., Lippard S.J. Synthesis and structure of $[Fe_{13}O_4F_{24}(OMe)_{12}]^{5-}$: The first open-shell Keggin ion. *J. Am. Chem. Soc.* 2002, 124, 4578–4579.

[26] Brachtel G., Hoppe R. Neue oxoferrate(III). Zur kenntnis von Na_5FeO_4. *Z. Anorg. Allg. Chem.* 1978, 446, 77–86.

[27] Beverskog B., Puigdomenech I. Revised Pourbaix diagrams for iron at 23–300°C. *Corrosion Sci.* 1996, 38, 2121–2135.

[28] Weller M.T., Hector A.L. The structure of the $Fe^{IV}O_4^{4-}$ ion. Angew. *Chem. Int. Ed.* 2000, 39, 22, 4162–4163.

[29] Hoppe R., Mader K. Zur konstitution von $K_3[FeO_4]$. *Z. Anorg. Allg. Chem.* 1990, 586, 115–124.

[30] Delattre J.L., Stacy A.M., Young Jr., V.G., Long G.L., Hermann R., Grandjean F. Study of the structural, electronic and magnetic properties of the barium-rich iron(IV) oxides, Ba_2FeO_4 and Ba_3FeO_5. *Inorg. Chem.* 2002, 41, 2834–2838.

[31] Licht S., Yu X. Recent advances in Fe(VI) synthesis. *ACS Symposium Series*, Vol. 985, 2–51, 2008.

[32] Denvir, A., Pletch, D. Electrochemical generation of ferrate. Part I: Dissolution of an iron wool bed anode. *J Appl. Electrochem.* 1996, 26, 815–822.

[33] Lapicque F., Valentin G. Direct electrochemical preparation of solid potassium ferrate. *Electrochem. Comm.* 2002, 4, 764–766.

[34] Machala L., Zboril R., Sharma V.K., Filip J., Schneeweiss O., Madarász J., Homonnay Z., Pokol G., Yngard R. Thermal stability of solid ferrates(VI): a review. *ACS Symposium Series,* Vol. 985, 124–144, 2008.
[35] Rush J.D., Bielski B. H. J. Decay of ferrate(V) in neutral and acidic solutions. A premix pulse radiolysis study. *Inorg. Chem.* 1994, 33, 5499–5502.
[36] Sharma V.K., O'Connor D.B., Cabelli D.E. Oxidation of thiocyanate by iron(V) in alkaline medium. *Inorg. Chim. Acta* 2004, 357, 4587–4591.
[37] Sharma V.K., Burnett C.R., Yngard R.A., Cabelli D.E. Iron(VI) and iron(V) oxidation of copper(I) cyanide. *Environ. Sci. Technol.* 2005, 39, 3849–3854.
[38] Sharma V.K. Oxidation of inorganic compounds by ferrate(VI) and ferrate(V): one-electron and two-electron transfer steps. *Environ. Sci. Technol.* 2010, 44, 5148–5152.
[39] Jiang J. Q., Lloyd B. Progress in the development and use of ferrate(VI) salt as an oxidant and coagulant for water and wastewater treatment. *Water Research* 2002, 36, 1397–1408.
[40] Patrat G., Debergevin F., Perner M., Joubert J.C. Structure locale de δ-FeOOH. *Acta Cryst. B* 1983, 39, 165–170.
[41] Drits V.A., Sakharov B.A., Manceau A. Structure of feroxyhite as determinated by simulation of X-ray diffraction curves. *Clay Miner.* 1993, 28, 209–221.
[42] Manceau A., Drits V.A. Local structure of ferrihydrite and ferroxyhite by EXAFS spectroscopy. *Clay Miner.* 1993, 28, 165–184.
[43] Cudennec Y., Lecerf A. Etude du type structural de γ-FeOOH et comparaison avec la structure de $Cu(OH)_2$. *C.R. Chimie* 2001, 4, 885–891.
[44] Cudennec Y., Lecerf A. Etude des mécanismes de formation des oxy-hydroxydes de fer; hypothèses de transformations topotactiques. *C. R. Chimie* 2003, 6, 437–444.
[45] Cudennec, Y., Lecerf, A. Topotactic transformations of goethite and lepidocrocite into hematite and maghemite. *Sol. State Sci.* 2005, 7, 520–529.
[46] Majzlan J., Lang B.E., Stevens R., Navrotsky A., Woodfield B.F., Boerio-goates J. Thermodynamics of Fe oxides: Part I. Entropy at standard temperature and pressure and heat capacity of goethite (α-FeOOH), lepidocrocite (γ-FeOOH), and maghemite (γ-Fe_2O_3). *Amer. Miner.* 2003, 88, 846–854.
[47] Majzlan J., Grevel K.D., Navrotsky A. Thermodynamics of Fe oxides: Part II. Enthalpies of formation and relative stability of goethite (α-FeOOH), lepidocrocite (γ-FeOOH), and maghemite (γ-Fe_2O_3). *Amer. Miner.* 2003, 88, 855–859.
[48] Majzlan J., Navrotsky A., Schwertman U. Thermodynamics of iron oxides: Part III. Enthalpies of formation and stability of ferrihydrite (~$Fe(OH)_3$), schwertmannite (~$FeO(OH)_{3/4}(SO_4)_{1/8}$), and ε-$Fe_2O_3$. *Geochim. Cosmochim. Acta* 2004, 68, 1049–1059.
[49] Navrotsky A., Mazeina L., Majzlan J. Size-driven structural and thermodynamic complexity in iron oxides. *Science* 2008, 319, 1635–1638.
[50] Jambor L., Dutrizac J. E. Occurrence and constitution of natural and synthetic ferrihydrite, a widespread iron oxyhydroxide. *Chem. Rev.* 1998, 98, 2549–2585.
[51] Schwertmann U., Friedl J., Stanjek H. From Fe(III) ions to ferrihydrite and then to haematite. *J. Colloid Interface Sci.* 1999, 209, 215–223.
[52] Dixit S., Hering J.G. Comparison of arsenic(V) and arsenic(III) sorption onto iron oxide minerals: Implications for arsenic mobility. *Environ. Sci. Technol.* 2003, 37, 4182–4189.

[53] Polizotto M.L., Harvey C.F., Sutton S.R., Fendorf S. Processes conducive to the release and transport of arsenic into aquifers of Bangladesh. *Proc. Natl. Acad. Sci. USA* 2005, 102, 18819–18823.
[54] Theil E.C. Ferritin: structure, gene regulation, and cellular functions in animals, plants, and microorganisms. *Ann. Rev. Biochem.* 1987, 56, 289–315.
[55] Towe K.M., Bradley W.F. Mineralogical constitution of colloidal hydrous ferric oxides. *J. Colloid Interface Sci.* 1967, 24, 384–392.
[56] Combes J. M., Manceau A., Calas G., Bottero J. Y. Formation of ferric oxides from aqueous solutions: A polyhedral approach by X-ray absorption spectroscopy: I. Hydrolysis and formation of ferric gels. *Geochim. Cosmochim. Acta* 1989, 53, 583–594.
[57] Drits V.A., Sakharov B.A., Salyn A.L., Manceau A. Structural model for ferrihydrite. *Clay Miner.* 1993, 28, 185–207.
[58] Manceau A., Skanthakumar S., Soderholm L. PDF analysis of ferrihydrite: Critical assessment of the under-constrained akdalaite model. *Amer. Miner.* 2014, 99, 102–108.
[59] Eggleton R.A., Fitzpatrick R.W. New data and a revised structural model for ferrihydrite. *Clays Clay Miner.* 1988, 36, 11–124.
[60] Michel F.M., Ehm L., Antao S.M., Lee P.L., Chupas P.J., Liu G., Strongin D.R., Schoonen M.A.A., Phillips B.L., Parise J.B. The structure of ferrihydrite, a nanocrystalline material. *Science* 2007, 316, 1726–1729.
[61] Maillot F., Morin G., Wang Y., Bonnin D., Ildefonse P., Chanéac C., Calas G. New insight into the structure of nanocrystalline ferrihydrite: EXAFS evidence for tetrahedrally coordinated iron(III). *Geochim. Cosmochim. Acta* 2011, 75, 2708–2720.
[62] Harrington R., Hausner D.B., Xu W., Bhandari N., Michel F.M., Brown Jr. G.E., Strongin D.R., Parise J.B. Neutron pair distribution function study of two-line ferrihydrite. *Environ. Sci. Technol.* 2011, 45, 9883–9890.
[63] Gilbert B., Erbs J.J., Penn R.L., Petkov V., Spagnoli D., Waychunas G.A. A disordered nanoparticle model for 6-line ferrihydrite. *Amer. Miner.* 2013, 98, 1465–1476.
[64] Manceau A. Critical evaluation of the revised akdalaite model for ferrihydrite. *Amer. Miner.* 2011, 96, 521–533.
[65] Manceau A., Gates W.P. Incorporation of Al in iron oxyhydroxides: implications for the structure of ferrihydrite. *Clay Miner.* 2013, 48, 481–489.
[66] Paktung D., Manceau A., Dutrizac J. Incorporation of Ge in ferrihydrite: Implications for the structure of ferrihydrite. *Amer. Miner.* 2013, 98, 848–858.
[67] Combes J. M., Manceau A., Calas G. Formation of ferric oxides from aqueous solutions: A polyhedral approach by X-ray Absorption Spectroscopy: II. Hematite formation from ferric gels. *Geochim. Cosmochim. Acta* 1990, 54, 1083–1091.
[68] Birch W. D., Pring A., Reller A., Schmalle H. W. Bernalite, $Fe(OH)_3$, a new mineral from Broken Hill, New South Wales: Description and structure. *Amer. Miner.* 1993, 78, 827–834.
[69] Lemaire B.J., Davidson P., Ferré J., Jamet J.P., Petermann D., Panine P., Dozov I., Jolivet J.P. Physical properties of aqueous suspensions of goethite (α-FeOOH) nanorods Part I: In the isotropic phase. *Eur. Phys. J. E* 2004, 13, 291–308; Lemaire B.J., Davidson P., Petermann D., Panine P., Dozov I., Stoenescu D., Jolivet J.P. Physical properties of aqueous suspensions of goethite (α-FeOOH) nanorods Part II: In the nematic phase. *Eur. Phys. J. E* 2004, 13, 309–319.
[70] Chemin N., Rozes L., Chanéac C., Cassaignon S., Le Bourhis E., Jolivet J.P., Spalla O., Barthel E., Sanchez C. Structure and mechanical properties of mesostructured

functional hybrid coatings based on anisotropic nanoparticles dispersed in poly(hydroxylethyl methacrylate). *Chem. Mater.* 2008, 20, 4602–4611.
[71] Parfitt, R. L., van der Gast, S. J., Childs, C. W. A structural model for natural siliceous ferrihydrite. *Clays Clay Miner.* 1992, 40, 675–681.
[72] Rose J., Manceau A., Masion A., Bottero J. Y., Garcia F. Nucleation and growth mechanisms of Fe oxyhydroxide in the presence of PO_4 ions. 1. Fe K-Edge EXAFS study. *Langmuir* 1996, 12, 6701–6707.
[73] Rose J., Flanck A. M., Masion A., Bottero J. Y., Elmerich P. Nucleation and growth mechanisms of Fe oxyhydroxide in the presence of PO_4 ions. 2. P K-Edge EXAFS study. *Langmuir* 1997, 13, 1827–1834.
[74] Doelsch E., Rose J., Masion A., Bottero J. Y., Nahon D., Bertsch P. M. Speciation and crystal chemistry of iron(III) chloride hydrolyzed in the presence of SiO_4 ligands. 1. An Fe K-edge EXAFS study. *Langmuir* 2000, 16, 4726–4731.
[75] Pokrovski G.S., Schott J., Farges F., Hazemann J.L. Iron (III)-silica interactions in aqueous solution: Insights from X-ray absorption fine structure spectroscopy. *Geochim. Cosmochim. Acta* 2003, 67, 3559–3573.
[76] Chan M. A., Parry W. T. Rainbow of rocks, mysteries of sandstone colors and concretions in Colorado plateau canyon country. Public Information Series 77, Utah Geological Survey 1–17, 2002.
[77] Matijevic E., Scheiner P. Ferric hydrous oxide sols. III. Preparation of uniform particles by hydrolysis of Fe(III)-chloride, -nitrate, and –perchlorate solutions. *J. Colloid Interface Sci.* 1978, 63, 509–524.
[78] Matijevic E. Colloidal chemical aspects of the corrosion of metals. *Pure Appl. Chem.* 1980, 52, 1179–1193.
[79] Hernandez J. Synthèse de nanoparticules d'oxydes de fer et d'aluminium pour l'étude de l'adsorption d'entités inorganiques polycondensées. Thèse, Université Pierre et Marie Curie, Paris, France, 1998.
[80] Rodriguez R. D., Demaille D., Lacaze E., Jupille J., Chanéac C., Jolivet J.P. Rhombohedral shape of hematite nanocrystals synthesized via thermolysis of an additive-free ferric chloride solution. *J. Phys. Chem. C* 2007, 111, 16866–16870.
[81] Morales M. P., Gonzales-Carreno T., Serna C. J. The formation of α-Fe_2O_3 monodispersed particles in solution. *J. Mater. Res.* 1992, 7, 2538–2545.
[82] Bottero J. Y., Manceau A., Villieras F., Tchoubar D. Structure and mechanisms of formation of FeOOH(Cl) polymers. *Langmuir* 1994, 10, 316–319.
[83] Paterson R., Rahman H. The ion exchange properties of crystalline inorganic oxide-hydroxides. Part I. βFeOOH: A variable capacity anion exchanger. *J. Colloid Interface Sci.* 1983, 94, 60–69.
[84] Paterson R., Rahman H. The ion exchange properties of crystalline inorganic oxide-hydroxides. Part II. Exclusion of perchlorate from β-FeOOH by an ion sieve mechanism. *J. Colloid Interface Sci.* 1984, 97, 423–427.
[85] Cai J., Liu J., Gao Z., Navrotsky A., Suib S.L. Synthesis and anion exchange of tunnel structure akaganeite. *Chem. Mater.* 2001, 13, 4595–4602.
[86] Post J.E., Heaney P.J., Von Dreele R.B., Hanson J. Neutron and temperature-resolved synchrotron X-ray powder diffraction study of akaganeite. *Amer. Miner.* 2003, 88, 782–788.
[87] Sillen L.G. Stability constants of metal-ion complexes. The Chemical Society. London, Special Publication n°17, 1964; Special Publication n°25, 1971.

[88] Holt S.H. *Inorganic Syntheses*. New York: Wiley, 1983.

[89] Maeda H., Maeda Y. Atomic force microscopy studies for investigating the smectic structures of colloidal crystals of β-FeOOH. *Langmuir* 1996, 12, 1446–1452.

[90] Zboril R., Mashlan M., Petridis D. Iron(III) oxides from thermal processes-synthesis, structural and magnetic properties, Mössbauer spectroscopy characterization, and applications. *Chem. Mater.* 2002, 14, 969–982.

[91] Machala L., Tucek J., Zboril R. Polymorphous transformations of nanometric iron(III) oxide: a review. *Chem. Mater.* 2011, 23, 3255–3272.

[92] Rahman M.M., Jamal A., Khan S.B., Faisal M. Characterization and applications of as-grown β-Fe_2O_3 nanoparticles prepared by hydrothermal method. *J. Nanopart. Res.* 2011, 13, 3789–3799.

[93] Sugimoto T., Muramatsu A., Sakata K., Shindo D. Characterization of hematite particles of different shapes. *J. Colloid Interface Sci.* 1993, 158, 420–428.

[94] Bailey J.K., Brinker C.J., Mecartney M.L. Growth mechanisms of iron oxide particles of differing morphologies from the forced hydrolysis of ferric chloride solutions. *J. Colloid Interface Sci.* 1993, 157, 1–13.

[95] Sugimoto T., Waki S., Itoh H., Muramatsu A. Preparation of monodisperse platelet-type hematite particles from a highly condensed β-FeOOH suspension. *Colloids Surf., A* 1996, 109, 155–165.

[96] Sugimoto T., Muramatsu A. Formation mechanism of monodispersed α-Fe_2O_3 particles in dilute $FeCl_3$ solutions. *J. Colloid Interface Sci.* 1996, 184, 626–638.

[97] Almeida T.P., Fay M. W., Zhu Y., Brown P.D. A valve-assisted snapshot approach to understand the hydrothermal synthesis of α-Fe_2O_3 nanorods. *CrystEngComm.* 2010, 12, 1700–1704.

[98] Kandori K., Kawashima Y., Ishikawa T. Effects of citrate ions on the formation of monodispersed cubic hematite particles. *J. Colloid Interface Sci.* 1992, 152, 284–288.

[99] Kandori K., Okamoto N., Ishikawa T. Preparation of nanoporous micrometer-scale hematite particles by a forced hydrolysis reaction in the presence of polyethylene glycol. *Langmuir* 2002, 18, 2895–2900.

[100] Kandori K., Sakai M., Inoue I., Ishikawa T. Effects of amino acids on the formation of hematite particles in a forced hydrolysis reaction. *J. Colloid Interface Sci.* 2006, 293, 108–115.

[101] Aschauer U., Jones F., Richmond W.R., Bowen P., Rohl A.L., Parkinson G.M., Hofmann H. Growth modification of hematite by phosphonate additives. *J. Crystal Growth* 2008, 310, 688–698.

[102] Chang H.C., Matijevic E. Interactions of metal hydrous oxides with chelating agents IV. Dissolution of hematite. *J. Coll. Interface Sci.* 1983, 92, 479–488.

[103] Zhang Y., Kallay N., Matijevic E. Interactions of metal hydrous oxides with chelating agents. 7. Hematite-oxalic acid and citric acid systems. *Langmuir* 1985, 1, 201–206.

[104] Wang W., Howe J.Y., Gu B. Structure and morphology evolution of hematite (α-Fe_2O_3) nanoparticles in forced hydrolysis of ferric chloride. *J. Phys. Chem. C* 2008, 112, 9203–9208.

[105] LaTempa T.J., Feng X., Paulose M., Grimes C.A. Temperature-dependent growth of self-assembled hematite (α-Fe_2O_3) nanotube arrays: rapid electrochemical synthesis and photoelectrochemical properties. *J. Phys. Chem. C* 2009, 113, 16293–16298.

[106] Mohapatra S.K., John S.E., Banerjee S., Misra M. Water photooxidation by smooth and ultrathin α-Fe_2O_3 nanotube arrays. *Chem. Mater.* 2009, 21, 3048–3055.

[107] Almeida T.P., Fay M. W., Zhu Y., Brown P.D. Process map for the hydrothermal synthesis of α-Fe_2O_3 nanorods. *J. Phys. Chem. C* 2009, 113, 18689–18698.

[108] Li X., Wei W., Wang S., Kuai L., Geng B. Single-crystalline α-Fe_2O_3 oblique nanoparallelepipeds: high-yield synthesis, growth mechanism and structure enhanced gas-sensing properties. *Nanoscale* 2011, 3, 718–724.

[109] Song C., Zhang Y., Lin Y., Wang D. Preparation and magnetic property of Fe_2O_3 parallelepiped nanocrystals. *Mater. Lett.* 2011, 65, 3195–3198.

[110] Jia B., Gao L., Sun J. Synthesis of single crystalline hematite polyhedral nanorods via a facile hydrothermal process. *J. Am. Ceram. Soc.* 2007, 90, 1315–1318.

[111] Jia C.J., Sun L.D., Yan Z.G., You L.P., Luo F., Han X.D., Pang Y.C., Zhang Z., Yan C.H. Single-crystalline iron oxide nanotubes. *Angew. Chem. Int. Ed.* 2005, 44, 4328–4333.

[112] Wang L., Gao L. Morphology-controlled synthesis and magnetic property of pseudocubic iron oxide nanoparticles. *J. Phys. Chem. C* 2009, 113, 15914–15920.

[113] Van T.K., Cha H.G., Nguyen C.K., Kim S.W., Jung M.H., Kang Y.S. Nanocystals of hematite with unconventional shape-truncated hexagonal bipyramid and its optical and magnetic properties. *Cryst. Growth Des.* 2012, 12, 862–868.

[114] Meng L., Chen W., Chen C., Zhou H., Peng Q., Li Y. Uniform α-Fe_2O_3 nanocrystal moniliforme-shape straight-chains. *Cryst. Growth Des.* 2010, 10, 479–482.

[115] Duchateau A. Réduction par électrolyse de nanoparticules d'oxydes de fer en milieu alcalin à 110°C. Thèse, Université Pierre et Marie Curie, Paris, France, 2013.

[116] Manimekalai R. Transition metal complexes of phenylacetates and substituted phenyacetates with neutral hydrazine as a ligand. *Intern. J. Applied Biol. Pharm. Techn.* 2011, 2, 251–262.

[117] Bao L., Yang H., Wang X., Zhang F., Shi R., Liu B., Wang L., Zhao H. Synthesis and size-dependent magnetic properties of single-crystalline hematite nanodiscs. *J. Cryst. Growth* 2011, 328, 62–69.

[118] Qu J., Yu Y., Cao C.Y., Song W.G. α-Fe_2O_3 Nanodisks: layered structure, growth mechanism, and enhanced photocatalytic property. *Chem. Eur. J.* 2013, 19, 11172–11177.

[119] Green A.E., Chiang C.Y., Greer H.F., Waller A., Ruszin A., Webster J., Niu Z., Self K., Zhou W. Growth mechanism of dendritic hematite via hydrolysis of ferricyanide. *Cryst. Growth Des.* 2017, 17, 800–808.

[120] Bourrié G., Trolard F., Génin J.-M. R., Jaffrezic A., Maître V., Abdelmoula M. Iron control by equilibria between hydroxy-green rusts and solutions in hydromorphic soils. *Geochim. Cosmochim. Acta.* 1999, 63, 3417–3427.

[121] Berthelin J., Ona-Nguema G., Stemmler S., Quantin C., Abdelmoula M., Jorand F. Bioreduction of ferric species and biogenesis of green rusts in soils. *C. R. Geoscience* 2006, 338, 447–455.

[122] Mornet S., Vasseur S., Grasset F., Duguet E. Magnetic nanoparticle design for medical diagnosis and therapy. *J. Mater. Chem.* 2004, 14, 2161–2175.

[123] Laurent S., Forge D., Port M., Roch A., Robic C., Vander Elst L., Muller R.N. Magnetic iron oxide nanoparticles: synthesis, stabilization, vectorization, physicochemical characterizations, and biological applications. *Chem. Rev.* 2008, 108, 2064–2110.

[124] Refait Ph., Abdelmoula M., Génin J.-M.R. Mechanisms of formation and structure of green rust one in aqueous corrosion of iron in the presence of chloride ions. *Corros. Sci.* 1998, 40, 1547–1560.

[125] Géhin A., Ruby C., Abdelmoula M., Benali O., Ghanbaja J., Refait Ph., Génin J.-M.R. Synthesis of Fe(II-III) hydroxysulphate green rust by coprecipitation. *Sol. State Sci.* 2002, 4, 61–66.
[126] Génin J.-M.R., Rabha A., Géhin A., Abdelmoula M., Benali O., Ernstsen V., Ona-Nguema G., Upadhyay C., Ruby C. Fougerite and Fe^{II-III} hydroxycarbonate green rust; ordering, deprotonation and/or cation substitution; structure of hydrotalcite-like compounds and mythic ferrosic hydroxide $Fe(OH)_{(2+x)}$. *Sol. State Sci.* 2005, 7, 545–572.
[127] Génin J.-M.R., Abdelmoula M., Ruby C., Upadhyay C. Speciation of iron; characterisation and structure of green rusts and FeII–III oxyhydroxycarbonate fougerite. *C. R. Geoscience* 2006, 338, 402–419.
[128] Ahmed I.A.M., Benning L.G., Kakonyi G., Sumoondur A.D., Terrill N.J., Shaw S. Formation of green rust sulfate: a combined in situ time-resolved X-ray scattering and electrochemical study. *Langmuir* 2010, 26, 6593–6603.
[129] Ona-Nguema G., Abdelmoula M., Jorand F., Benali O., Géhin, A. Block J.-C., Génin J.-M. R. Iron (II, III) Hydroxy-carbonate green rust formation and stabilization from lepidocrocite bioreduction. *Environ. Sci. Technol.* 2002, 36, 16–20.
[130] Ruby C., Géhin A., Abdelmoula M., Génin J.-M. R., Jolivet J.P. Coprecipitation of Fe(II) and Fe(III) cations in sulphated aqueous medium and formation of hydroxysulphate green rust. *Sol. State Sci.* 2003, 5, 1055–1062.
[131] Ruby C., Aïssa R., Géhin A., Cortot J., Abdelmoula M., Génin J.-M.R. Green rusts synthesis by coprecipitation of FeII –FeIII ions and mass-balance diagram. *C. R. Geoscience* 2006, 338, 420–432.
[132] Ruby C., Abdelmoula M., Aissa R., Medjahdi G., Brunelli M., François M. Aluminium substitution in iron (II–III)-layered double hydroxides: Formation and cationic order. *J. Solid State Chem.* 2008, 181, 2285–2291.
[133] Cudennec, Y., Lecerf, A. Topotactic transformations of goethite and lepidocrocite into hematite and maghemite. *Sol. State Sci.* 2005, 7, 520–529.
[134] Génin J.-M.R., Ruby C., Géhin A., Refait Ph. Synthesis of green rusts by oxidation of $Fe(OH)_2$, their products of oxidation and reduction of ferric oxyhydroxides; Eh–pH Pourbaix diagrams. *C. R. Geoscience* 2006, 338, 433–446.
[135] Zegeye A., Ona-Nguema G., Carteret C., Huguet L., Abdelmoula M., Jorand F. Formation of hydroxysulphate green rust 2 as a single iron(II-III) mineral in microbial culture. *Geomicrobiol. J.* 2005, 22, 389–399.
[136] Refait Ph., Abdelmoula M., Génin J.-M.R., Sabot R. Green rusts in electrochemical and microbially influenced corrosion of steel. *C. R. Geoscience* 2006, 338, 476–487.
[137] Trolard F., Génin J.-M. R., Abdelmoula M., Bourrié G., Humbert B., Herbillon A. Identification of a green rust mineral in a reductomorphic soil by Mossbauer and Raman spectroscopies. *Geochim. Cosmochim. Acta* 1997, 61, 1107–1111.
[138] Trolard F., Bourrié G. Découverte d'un minéral recherché depuis plus de 20 ans, la fougérite, une rouille verte. *INRA mensuel* 1997, 92, 2–3.
[139] Génin J.-M.R., Bourrié G., Trolard F., Abdelmoula M., Jaffrezic A., Refait Ph., Maître V., Humbert, B., Herbillon A. Thermodynamic equilibria in aqueous suspensions of synthetic and natural Fe(II)-Fe(III) green rusts: occurrences of the mineral in hydromorphic soils. *Environ. Sci. Technol.* 1998, 32, 1058–1068.
[140] Trolard F., Bourrié G., Abdelmoula M., Refait Ph., Feder F. Fougerite, a new mineral of the pyroaurite-iowaite group: description and crystal structure. *Clays Clay Miner.* 2007, 55, 323–334.

[141] Ruby C., Upadhyay C., Géhin A., Ona-Nguema G., Génin J.-M.R. In situ redox flexibility of FeII-III oxyhydroxy-carbonate green rust and fougerite. *Environ. Sci. Technol.* 2006, 40, 4696–4702.
[142] Jolivet J.P., Massart R., Fruchart J. M. Synthèse et étude physicochimique de colloïdes magnétiques non surfactés en milieu aqueux. *Nouv. J. Chim.* 1983, 7, 325–331.
[143] Jolivet J.P., Belleville P., Tronc E., Livage J. Influence of Fe(II) ions on the formation of spinel iron oxide in aqueous medium. *Clays Clay Miner.* 1992, 40, 531–540.
[144] Tronc E., Belleville P., Jolivet J. P., Livage J. Transformation of ferric hydroxide into spinel by Fe(II) adsorption. *Langmuir* 1992, 8, 313–319.
[145] Yang L., Steefel C.I., Marcus M.A., Bargar J.R. Kinetics of Fe(II)-catalyzed transformation of 6-line ferrihydrite under anaerobic flow conditions. *Environ. Sci. Technol.* 2010, 44, 5469–5475.
[146] Williams A.G.B., Scherer M.M. Spectroscopic evidence for Fe(II)-Fe(III) electron transfer at the iron oxide-water interface. *Environ. Sci. Technol.* 2004, 38, 4782–4790.
[147] Pottier A. Précipitation de cations métalliques en solution aqueuse. Influence de l'acidité sur les caractéristiques des particules d'oxyde. Thèse, Université Pierre et Marie Curie, Paris, France, 1999.
[148] Jolivet J.P., Tronc E., Chanéac C. Synthesis of iron oxide-based magnetic nanomaterials and composites. *C. R. Chimie* 2003, 5, 659–664.
[149] Vayssières L., Chanéac C., Tronc E., Jolivet J. P. Size tailoring of magnetite particles formed by aqueous precipitation an exemple of thermodynamic stability of nanometric oxide particles. *J. Colloid Interface Sci.* 1998, 205, 205–212.
[150] Jolivet J.P., Vayssières L., Chanéac C., Tronc E. Precipitation of spinel iron oxide: nanoparticle size control. *Mater. Res. Symp. Proc.* 1997, 432, 145–150.
[151] Jolivet J.P., Froidefond C., Pottier A., Chanéac C., Cassaignon S., Tronc E., Euzen P. Size tailoring of oxide nanoparticles by precipitation in aqueous medium. A semi-quantitative modeling. *J. Mater. Chem.* 2004, 14, 3281–3288.
[152] Zhang D.E., Zhang X.J., Ni X.M., Song J.M., Zheng H.G. Fabrication and characterization of Fe_3O_4 octahedrons via an EDTA-assisted route. *Cryst. Growth Des.* 2007, 7, 2117–2119.
[153] Geng B.Y., Ma J.Z., Hou J.H. Controllable synthesis of single-crystalline Fe_3O_4 polyhedra possessing the active basal facets. *Cryst. Growth Des.* 2008, 8, 1443–1447.
[154] Zhou H., Yi R., Li J., Su Y., Liu X. Microwave-assisted synthesis and characterization of hexagonal Fe_3O_4 nanoplates. *Sol. State Sci.* 2010, 12, 99–104.
[155] Kumar R.V., Koltypin Y., Xu X.N. Yeshurun Y., Gedanken A., Felner I., Fabrication of magnetite nanorods by ultrasound irradiation. *J. Appl. Phys.* 2001, 89, 6324–6328.
[156] Lian S., Wang E., Gao E.L., Kang Z., Wu D., Lan Y., Xu L. Growth of single-crystal magnetite nanowires from Fe_3O_4 nanoparticles in a surfactant-free hydrothermal process. *Solid State Comm.* 2004, 132, 375–378.
[157] Wang J., Chen Q., Zeng C., Hou B. Magnetic-field-induced growth of single-crystalline Fe_3O_4 nanowires. *Adv. Mater.* 2004, 16, 137–140.
[158] Geng B.Y., Zhan F., Jiang H., Guo Y., Xing Z. Egg albumin as a nanoreactor for growing single-crystalline Fe_3O_4 nanotubes with high yields. *Chem. Commun.* 2008, 5773–5775.
[159] Hu M., Jiang J.S., Li X. Surfactant-assisted hydrothermal synthesis of dendritic magnetite microcrystals. *Cryst. Growth Des.* 2009. 9, 820–824.

[160] Gorski C.A., Scherer M.M. Determination of nanoparticulate magnetite stoichiometry by Mössbauer spectroscopy, acidic dissolution, and powder X-ray diffraction: A critical review. *Amer. Miner.* 2010, 95, 1017–1026.

[161] Jolivet J.P., Tronc E. Interfacial electron transfer in colloidal spinel iron oxide. Conversion of Fe_3O_4 into γ-Fe_2O_3 in aqueous medium. *J. Colloid Interface Sci.* 1988, 125, 688–701.

[162] Tronc E., Jolivet J. P. Exchange and redox reactions at the interface of spinel-like iron oxide colloids in solution: Fe(II) adsorption. *Adsorption Sci. Techn.* 1984, 1, 247–251.

[163] Tronc E., Jolivet J. P., Lefebvre J., Massart R. Ion adsorption and electron transfer in spinel-like iron oxide colloids. *J. Chem. Soc. Faraday Trans* I 1984, 80, 2619–2629.

[164] Tronc E., Jolivet J. P., Belleville P., Livage J. Redox phenomena in spinel iron oxide colloids induced by adsorption. *Hyperf. Interact.* 1989, 46, 635–643.

[165] Gorski C.A., Scherer M.M. Influence of magnetite stoichiometry on FeII uptake and nitrobenzene reduction. *Environ. Sci. Technol.* 2009, 43, 3675–3680.

[166] Belleville Ph., Jolivet J. P., Tronc E., Livage J. Crystallisation of ferric hydroxide into spinel by adsorption on colloidal magnetite. *J. Colloid Interface Sci.* 1992, 150, 453–460.

[167] Jolivet J.P., Tronc E., Barbé C., Livage J. Interfacial electron transfer in colloidal spinel iron oxide. Silver ion reduction in aqueous medium. *J. Colloid Interface Sci.* 1990, 138, 465–472.

[168] Jolivet J.P., Chanéac C., Vayssières L., Tronc E. Wet chemistry of spinel iron oxide nanoparticles. *J. Phys. IV France* 1997, C1, 573–576.

[169] Prené Ph., Tronc E., Jolivet J.P., Livage J., Cherkaoui J.R., Noguès M., Dormann J.L. Magnetic properties of isolated γ-Fe_2O_3 particles. *IEEE Trans. Mag.* 1993, 29, 2658–2660.

[170] Bacri J. C., Perzinski R., Salin D., Cabuil V., Massart R. Ionic ferrofluids: A crossing of chemistry and physics. *J. Magn. Magn. Mater.* 1990, 85, 27–32.

[171] Dabadie J. C., Fabre P., Veyssié M., Cabuil V., Massart R. Magnetic lyotropic phases. *J. Phys. Condens. Matter* 1990, 2 Supplement SA291-SA294.

[172] Fabre P., Casagrande C., Veyssié M., Cabuil V., Massart R. Ferrosmectics: a new magnetic and mesomorphic phase. *Phys. Rev. Lett.* 1990, 64, 539–542.

[173] Lefebure S., Dubois E., Cabuil V., Neveu S., Massart R. Monodisperse magnetic nanoparticles: preparation and dispersion in water and oils. *J. Mater. Res.* 1998, 13, 2975–2981.

[174] Cousin F., Dubois E., Cabuil V. Tuning the interactions of a magnetic colloidal suspension. *Phys. Rev. E* 2003, 68, 21405–21414.

[175] Rosensweig R.E. Magnetic fluids. *Ann. Rev. Fluid Mech.* 1987, 19, 437–463.

[176] Dormann J.L., Fiorani D., Tronc E. *Magnetic Relaxation in Fine Particle Systems. Advances in Chemical Physics*. New York: Wiley, 1997.

[177] Tronc E., Fiorani D., Noguès, Testa A. M., Lucari F., D'Orazio F., Grenèche J. M., Wernsdorfer W., Galvez N., Chanéac C., Mailly D., Jolivet J. P. Surface effects in non-interacting and interacting γ-Fe_2O_3 nanoparticles. *J. Magn. Magn. Mater.* 2003, 262, 6–14.

[178] Chanéac C., Tronc E., Jolivet J.P. Magnetic iron oxide-silica nanocomposites. Synthesis and characterization. *J. Mater. Chem.* 1996, 6, 1905–1911.

[179] Tronc E., Jolivet J.P., Livage J. Mössbauer investigation of the $\gamma \rightarrow \alpha$-$Fe_2O_3$ transformation in small particles. *Hyperf. Interact.* 1990, 54, 737–740.

[180] Tronc E., Chanéac C., Jolivet J.P. Structural and magnetic characterization of ε-Fe_2O_3. *J. Solid State Chem.* 1998, 139, 93–104.
[181] Machala L., Tucek J., Zboril R. Polymorphous transformations of nanometric iron(III) oxide: a review. *Chem. Mater.* 2011, 23, 3255–3272.
[182] Jin J., Ohkoshi S., Hashimoto K. Giant coercive field of nanometer-sized iron oxide. *Adv. Mater.* 2004, 16, 48–51.
[183] Jolivet J.P., Tronc E., Chanéac C. Synthesis of iron oxide- and metal-based nanomaterial. *Eur. Phys. J. AP* 2000, 10,167–172.
[184] Chanéac C., Tronc E., Jolivet J.P. Thermal behavior of spinel iron oxide-silica composites. *NanoStruct. Mater.* 1995, 6, 715–718.
[185] Rousset A., Chassagneux F; Gillot B. Etude de spinelles à valences mixtes du fer. *Ann. Chim. Fr.* 1983, 8, 227–236.
[186] Sato T., Kuroda C., Saito M. Preparation and magnetic characteristics of ultra-fine spinel ferrites. *Ferrites: Proc. Intern. Conf.* July 1970, Japan 72–74.
[187] Tang Z.X., Sorensen C.M., Klabunde K.J., Hadjipanayis G.C. Preparation of manganese ferrite fine particles from aqueous solution. *J. Coll. Interface Sci.* 1991, 146, 38–52.
[188] Tamaura Y., Rasyid U., Katsura T. Formation of chromium-bearing ferrite $Cr_{0.42}Fe_{2.56}O_{4.00}$ in aqueous suspension by nitrate oxidation. *J. Chem. Soc. Dalton Trans.* 1980, 2125–2128.
[189] Katsura T., Tamaura Y., Chyo G.S. The formation of the oxidized Fe_3O_4-Fe_2TiO_4 solid solution by the air oxidation of the aqueous suspension. *Bull. Chem. Soc. Japan* 1979, 52, 96–100.
[190] Utsunomiya T., Hoshino Y., Show K. Uniformity of the composition of spinel ferrite particles prepared by air oxidation of hydroxide precursors in suspension. *Mat. Res. Bull.* 1985, 20, 85–92.
[191] Gallagher P.K., O'Bryan H.M. Jr, Schrey F., Monforte F.R. Preparation of a nickel ferrite from coprecipitated $Ni_{0.2}Fe_{0.8}C_2O_4.2H_2O$. *Ceramic Bull.* 1969, 48, 1053–1059.
[192] Chassagneux F., Rousset A. Préparation et étude structurale des spinelles $(FeAl_{2-2z}Cr_z)O_4$. *J. Solid State Chem.* 1976, 16, 161–166.
[193] Pullar R.C. Hexagonal ferrites: a review of the synthesis, properties and applications of hexaferrite ceramics. *Prog. Mat. Sci.* 2012, 57, 7, 1191–1334.
[194] Bernier J.C., Poix P. *Ann. Chim. Fr.* 1979, 4, 460.
[195] Haneda K., Miyakawa C., Kojima H. Preparation of high-coercivity $BaFe_{12}O_{19.}$ *J. Amer. Ceram. Soc.* 1974, 57, 354–357.
[196] Bernier J.C., Poix P., Najmi M. Sol-Gel processing of acicular particles of barium ferrite. In *Ultrastructure Processing of Advanced Ceramics*, J.D. Mackenzie, D.R. Ulrich, Eds. New York: Wiley, 1988, 557–567.
[197] Zhong W., Ding W., Jiang Y., Zhang N., Zhang J., Du Y., Yan Q. Preparation and magnetic properties of barium hexaferrite nanoparticles produced by the citrate process. *J. Am. Ceram. Soc.* 1997, 80, 3258–3262.
[198] Pechini M.P. Method of preparing lead and alkaline earth titanates and niobates and coating method using the same to form a capacitor. US Patent Office n°3, 330, 697, 11 July 1967.
[199] Veitch C.D. Synthesis of polycrystalline yttrium iron garnet and yttrium aluminium garnet from organic precursors. *J. Mater. Sci.* 1991, 26, 6527–6532.

[200] Vaqueiro P., Crosnier-Lopez M.P., Lopez-Quintela M.A. Synthesis and characterization of yttrium-iron garnet nanoparticles. *J. Solid State Chem.* 1996, 126, 161–168.
[201] Vaqueiro P., Lopez-Quintela M.A. Influence of complexing agents and pH on yttrium-iron garnet synthesized by the sol-gel method. *Chem. Mater.* 1997, 9, 2836–2841.
[202] Grasset F., Mornet S., Demourgues A., Portier J., Bonnet J., Vekris A., Duguet E. Synthesis, magnetic properties, surface modification and cytotoxicity evaluation of $Y_3Fe_{5-x}Al_xO_{12}$ ($0 \leq x \leq 2$) garnet submicron particles for biomedical applications. *J. Magn. Magn. Mat.* 2001, 234, 409–418.
[203] Cho Y.S., Burdick V.L., Amarakoon V.R.W. Hydrothermal preparation and morphology characteristics of $Y_3Fe_5O_{12}$. *J. Am. Ceram. Soc.* 1997, 80, 1605–1608.

8 }

Titanium, Manganese, and Zirconium Dioxides

The dioxides of titanium (TiO_2), manganese (MnO_2), and zirconium (ZrO_2) are important materials because of their technological uses. TiO_2 is used mainly as white pigment. Because of its semiconducting properties, TiO_2, in its nanomaterial form, is also used as an active component of photocells and photocatalysis for self-cleaning glasses and cements [1]. MnO_2 is used primarily in electrode materials. ZrO_2 is used in refractory ceramics, abrasive materials, and stabilized zirconia as ionic conductive materials stable at high temperature. Many of these properties are, of course, dependent on particle size and shape (§ Chap. 1). Dioxides of other tetravalent elements with interesting properties have been studied elsewhere in this book, especially VO_2, which exhibits a metal–isolator transition at 68°C, used, for instance, in optoelectronics (§ 4.1.5), and silica, SiO_2 (§ 4.1.4), which is likely the most ubiquitous solid for many applications and uses.

Aqueous chemistry is of major interest in synthesizing these oxides in the form of nanoparticles from inorganic salts and under simple, cheap, and environmental friendly conditions. However, as the tetravalent elements have restricted solubility in water (§ 2.2), metal–organic compounds such as titanium and zirconium alkoxides are frequently used in alcoholic solution as precursors for the synthesis of TiO_2 and ZrO_2 nanoparticles. An overview of the conversion of alkoxides into oxides is indicated about silica formation (§ 4.1.4), and since well-documented works have already been published [2–4], these compounds are not considered here.

The crystal structures of most MO_2 dioxides are of TiO_2 rutile type for hexacoordinated cations (e.g., Ti, V, Cr, Mn, Mo, W, Sn, Pb) and CaF_2 fluorite type for octacoordinated, larger cations (e.g., Zr, Ce), but polymorphism is common [5]. Some dioxides of elements such as chromium and tin form only one crystalline phase. So, hydrolysis of $SnCl_4$ [6, 7] or acidification of stannate $[Sn(OH)_6]^{2-}$ [8] leads both to the same rutile-type phase, cassiterite, SnO_2. Many other dioxides are polymorphic, especially TiO_2, which exists in three main crystal phases: anatase, brookite, and rutile; and MnO_2, which gives rise to a largely diversified crystal chemistry. ZrO_2 is also polymorphic, with a variable coordination number of the Zr^{4+} cation—7 in the monoclinic phase and 8 in quadratic and cubic phases.

8.1 Speciation of Ti^{IV}, Mn^{IV}, and Zr^{IV} in Solution

The tetravalent elements of the *d*-block of classification are strong acids in aqueous solution because the aquo ligands are strongly polarized by highly charged cations. Consequently, purely aquocomplexes do not exist. Hydroxylation and condensation occur from strongly acidic media in which polycations form and precipitation begins simultaneously.

Ti^{IV} ion is soluble only at pH ≤ 0 as the monomeric hexacoordinated species $[Ti(OH)_2(OH_2)_4]^{2+}$, $[Ti(OH)_3(OH_2)_3]^+$, and the polycation $[Ti_8O_{12}(OH_2)_{24}]^{8+}$ [9, 10] (Fig. 8.1a). The structure of the polycation has been determined by X-ray diffraction from the compound $[Ti_8O_{12}(OH_2)_{24}]Cl_8{\cdot}7H_2O$ [11, 12], obtained in highly concentrated acidic medium or by reaction of $TiCl_4$ with moist air. Surprisingly, this polycation, in which the Ti–O bonds approximately form a cube, contains only oxo and aquo ligands. Formally, the polycation results from condensation of the $h = 3$ complex, $[Ti(OH)_3(OH_2)_3]^+$. This trifunctional complex should, in principle, form by olation an infinite tridirectional system containing hydroxo and aquo ligands. Instead, its condensation is soon restricted because bridging oxo ligands transforms the other hydroxo ligands into water molecules, leading to stabilization of a cubic-shaped octamer, with no possibility of subsequent condensation. In the polycation, only very short oxo bridges (mean Ti–O bond length of 1.820 Å) and terminal aquo ligands with long $Ti{-}OH_2$ bond length (2.129 Å) exist. In comparison, Ti–O bond lengths in different TiO_2 crystal varieties are 1.946 and 1.983 Å in rutile; 1.937 and 1.966 Å in anatase; and 1.865, 1.919, 1.945, 1.993, and 2.04 Å in brookite. In addition, hydroxide $Ti(OH)_4$ does not exist (see later in this chapter), thereby showing that the high polarizing power of the Ti^{IV} cation destabilizes the hydroxo bridges. This explains why, in a hypothetical hydroxo Ti_8 polycation, deprotonation of bridging hydroxo ligands allows reinforcement of the Ti–O bond and simultaneously the increase in the basicity of terminal hydroxo ligands to form aquo ligands.

Zr^{IV} ion, which is larger than Ti^{IV} ion ($r_{Zr4+} = 0.84$ Å vs. $r_{Ti4+} = 0.60$ Å), adopts in solution the square-based antiprismatic 8-coordination. Consequently, zirconium is far less acidic than titanium. The $[Zr(OH_2)_8]^{4+}$ complex (pK ≈ −0.3), however, remains a strong acid that spontaneously hydrolyses to form $[Zr(OH)(OH_2)_7]^{3+}$ and $[Zr(OH)_2(OH_2)_6]^{2+}$ species. This last complex condenses by olation (δ(OH) = −0.07) and, in the absence of complexing anions, forms a cyclic tetramer $[Zr_4(OH)_8(OH_2)_{16}]^{8+}$ containing double hydroxo bridges. The tetramer is a stable species well characterized in solution by various techniques [13–16]. In the tetramer in both solution and solid state, zirconium preserves its antiprismatic 8-coordination [17, 18] (Fig. 8.1b). Unlike the Ti_8 polycation, the hydroxo bridges, which are more numerous and less polarized in the Zr_4 entities, remain stable.

Mn^{IV} ion is smaller than Ti^{IV} ($r_{Mn4+} = 0.53$Å) and does not exist in solution as an aquohydroxo species. Potentially even more acidic than titanium, it precipitates MnO_2 over all the accessible range of acidity, likely from the virtual $[Mn(OH)_4(OH_2)_2]^0$ species. Mn^{IV} ion results from redox reactions involving

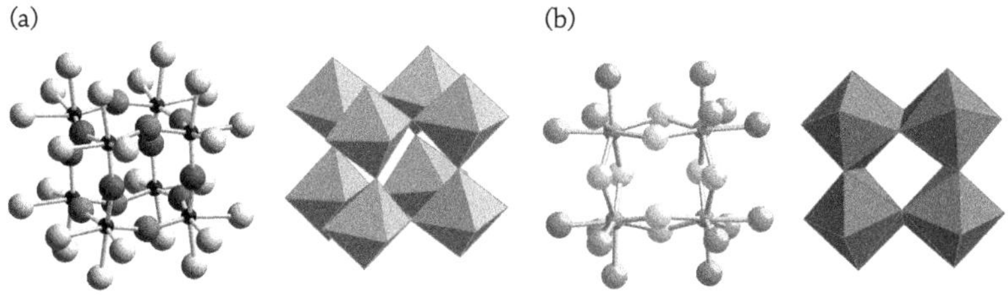

FIGURE 8.1 Structures of (a) $[Ti_8O_{12}(OH_2)_{24}]^{8+}$ and (b) $[Zr_4(OH)_8(OH_2)_{16}]^{8+}$ polycations.

soluble forms of Mn^{II}, Mn^{III}, or Mn^{VII}. However, in using strongly complexing polydentate ligands, it is possible to stabilize soluble forms of Mn^{IV} such as the $[Mn^{IV}(biguanide)_3]^{4+}$ chelate obtained by reduction of permanganate by biguanide sulfate in alkaline medium, the complex being then crystallized in strongly acidic medium [19]. There are also some other polynuclear oxobridged complexes of Mn^{IV} stabilized by various bridging and chelating ligands (Fig. 8.2).

8.2 Titanium Oxide

TiO_2 exists under three main crystalline varieties: rutile, anatase, and brookite [5] (Fig. 8.3). The difference in morphology of natural crystals reflects their differences in crystal structure—octahedra for anatase, platelets for brookite, and rods for rutile. In these structures, the TiO_6 octahedra are linked by edge sharing in three, two, and one directions, respectively, indicating the preferred directions of crystal growth. In principle, rutile is the thermodynamically most stable phase, although the relative stabilities of the three phases are highly dependent on particle size in the nanometric range (§ 1.1.1a). The formation enthalpies of bulk solids indicate increasing stabilities from anatase, brookite to rutile ($\Delta H^{\circ}_f = -939$, -941, and -944 kJ.mol^{-1}, respectively) [20], but the values are very close and the ranking may be modified by small differences in surface energy. Thus, anatase is the thermodynamically most stable phase for particles smaller than 11 nm in diameter; brookite becomes more stable for particles from 11 to 35 nm in diameter; and for bigger particles, rutile is effectively the most stable phase [21]. Phase transformations can thus occur at variable temperatures (for instance, anatase–rutile between 400° and 1200°C) and according to sequences that may be conflicting, depending on the mean particle size—for example, A→R, A→B→R or B→A→R. The presence of adsorbed species, anions or cations, is also an important factor determining the structural orientation during synthesis and subsequent transformations.

Nanoparticles of TiO_2 are more often synthesized in aqueous solution from tetrachloride $TiCl_4$, and the three polymorphs can be selectively formed according to the conditions of the precipitation. The dilution of $TiCl_4$ in water strongly acidifies the solution because the formation of the hydroxylated complexes $[Ti(OH)_2(OH_2)_4]^{2+}$ and $[Ti(OH)_3(OH_2)_3]^{+}$ releases 2 to 3 moles of protons per mole of titanium. The precipitation of TiO_2 can then take place by addition of a

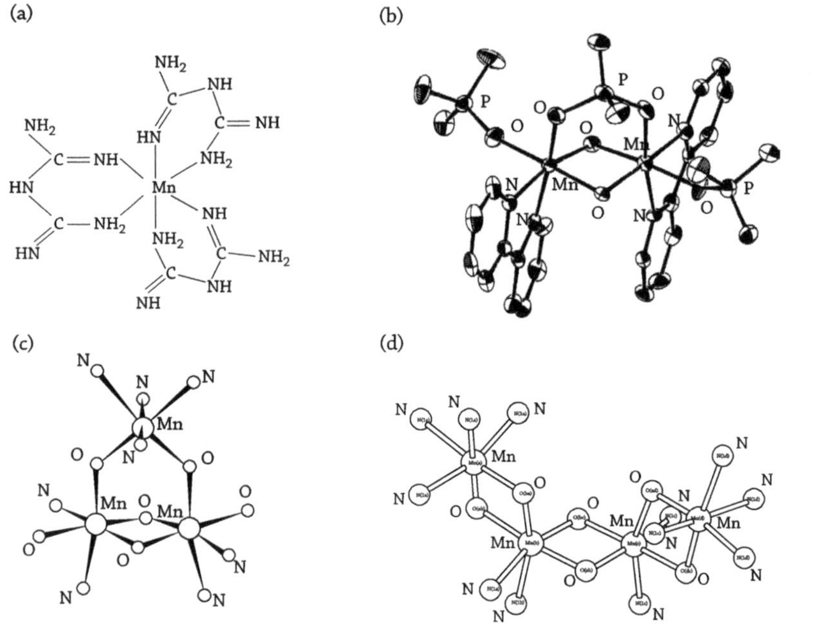

FIGURE 8.2 Complexes of Mn^{IV} stable in aqueous solution: (a) $[Mn(biguanide)_3]^{4+}$ [19]; (b) $[Mn_2O_2(\mu\text{-}HPO_4)(bpy)_2(H_2PO_4)_2]$ (reproduced with permission from [Sarneski J. E., Didiuk M., Thorp H. H., Brudvig G. W., Crabtree R. H., Faller J. W., Schulte G. K., A high-valent oxomanganese dimer containing bridging and terminal inorganic phosphate ligands, Inorg. Chem. 1991, 30, 2833–2835] Copyright 1991 American Chemical Society); (c) $[Mn_3O_4(bpy)_4(OH_2)_2]^{4+}$ (reproduced with permission from [Sarneski J. E., Thorp H. H., Brudvig G. W., Crabtree R. H., Schulte G. K., Assembly of high-valent oxomanganese clusters in aqueous solution. Redox equilibrium of water-stable $Mn_3O_4^{4+}$ and $Mn_2O_2^{3+}$ complexes, J. Am. Chem. Soc. 1990, 112, 7255–7260] Copyright 1990 American Chemical Society); (d) $[Mn_4O_6(bpy)_6]^{4+}$ (reproduced with permission from [Philouze C., Blondin G., Girerd J. J., Guilhem J., Pascand C., Lexa D., Aqueous chemistry of high-valent manganese. Structure, magnetic, and redox properties of a new type of Mn-oxo cluster, $[Mn(IV)_4O_6(bpy)_6]^{4+}$: relevance to the oxygen evolving center in plants, J. Am. Chem. Soc. 1994, 116, 8557–8565] Copyright 1994 American Chemical Society). (In (c) and (d), only the atoms directly coordinated to Mn^{4+} ions are shown.)

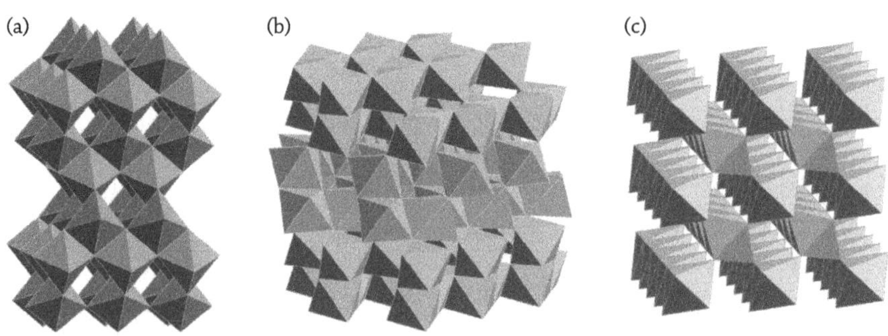

FIGURE 8.3 Crystal structures of the three main polymorphs of titanium dioxide (a) anatase, (b) brookite, and (c) rutile.

base and/or by heating at low temperature, the thermolysis being able to cause hydroxylation of tetravalent elements up to formation of the noncharged complex. The use of various ligands permits selecting the crystalline variety, the particle size and shape, and even the composition of titanium dioxide. Other synthesis methods are based on transformation of titanates obtained from hydrothermal treatment of various precursors in strongly alkaline medium. Lastly, a third type of protocol involves oxidation processes in solution, chemical oxidation of the trichloride $TiCl_3$, or anodic oxidation of the metal. These different procedures for synthesizing the TiO_2 polymorphs are examined in the following sections.

8.2.1 PRECIPITATION OF TI^{4+} IONS IN ACIDIC TO NEUTRAL MEDIA

a) Anatase

The addition of a base to a solution of $TiCl_4$ up to $2 \leq pH \leq 7$ immediately forms an amorphous precipitate. After aging for 5 hours at room temperature, the mean composition of the still amorphous solid in suspension is around $TiO_{0.3}(OH)_{3.4}$ [22]. That indicates that the hydroxide $Ti(OH)_4$ does not exist, likely because of the high polarizing power of Ti^{4+} ions, as previously noted in the formation of the Ti_8 polycation. Crystallization of the amorphous solid into anatase occurs after aging at 60°C for around 1 day, without any significant evolution in particle size. One may conclude that the crystallization proceeds topotactically by dehydration and ordering in solid state (Fig. 8.4).

The particles of anatase are approximately spherical and have a mean size depending on the pH of precipitation and the length of aging in suspension. From pH 2 to pH 7, the mean diameter varies from 5 nm to 11 nm. With regard to this particle size range, the anatase phase is effectively the stable phase, as indicated earlier [21]. This change in particle size induces shifting and widening of the Raman absorption peak located at around 150 cm^{-1} on the spectrum of the solid. The phenomenon is due to the confinement effect of phonons [22], which is typical of nanometric systems (§ 1.1.3c). The change in particle size results from the change in surface charge density and finally in surface energy with pH (§ 5.3.4). At pH > 7, aging of the precipitate leads to alkaline titanates (§ 8.2.2).

Under similar conditions (pH of precipitation of $TiCl_4$ between 1 and 6, aging at 60°C) but in the presence of amino acid, anatase nanoparticles with a morphology dependent on the amino acid are obtained [23]. The most active acids are glutamic, aspartic, and serine (Scheme 8.1). They permit obtaining anatase exclusively on the complete acidity range ($1 \leq pH \leq 6$), with a mean particle size between 4 and 8 nm.

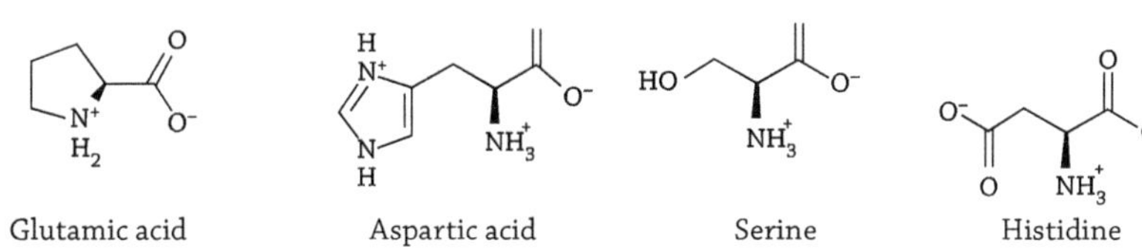

SCHEME 8.1

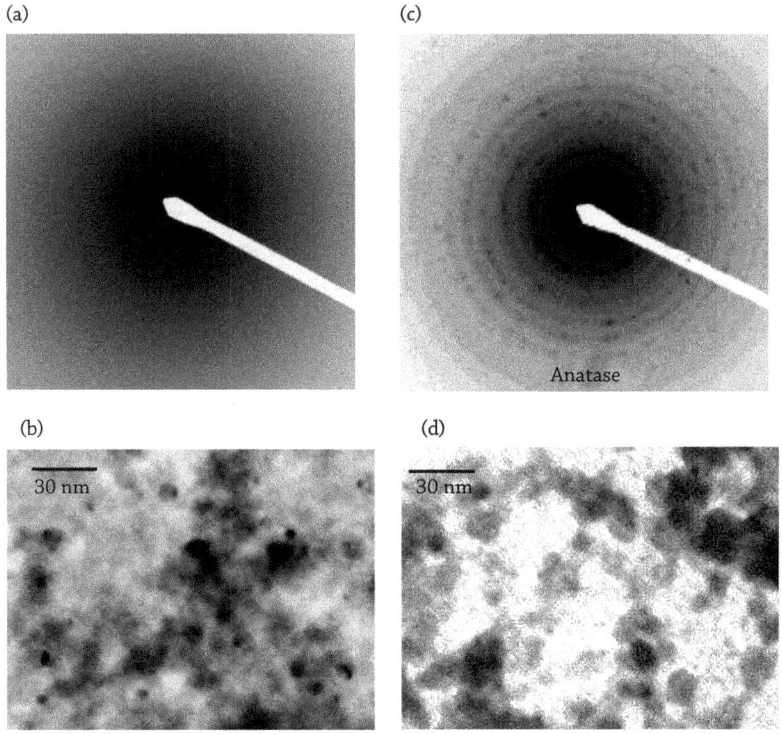

FIGURE 8.4 Electron diffraction patterns and TEM images of the solid obtained by precipitation of $TiCl_4$ at pH 4 and aged for 5 hours at 25°C (a, b) and for 24 hours at 60°C (c, d) (reproduced from [22] with permission of The Royal Society of Chemistry).

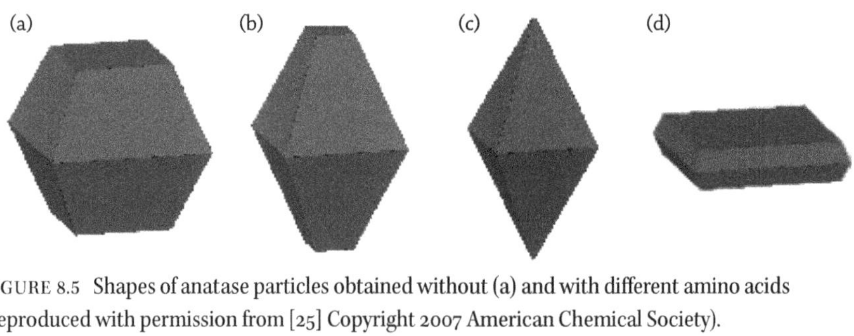

FIGURE 8.5 Shapes of anatase particles obtained without (a) and with different amino acids (reproduced with permission from [25] Copyright 2007 American Chemical Society).

The particle shapes are determined from the crystal lengths along different directions of X-ray diffraction and change from a truncated bipyramid that is roughly spherical (Fig. 8.5a) except in acidic medium (pH 1) where the {101} faces are preferentially developed (Fig. 8.5b). Glutamic and aspartic acids induce formation of octahedra without truncation and exhibit only {101} faces (Fig. 8.5c). With histidine at pH 6, the anatase particles are platelets with the predominance of {001}

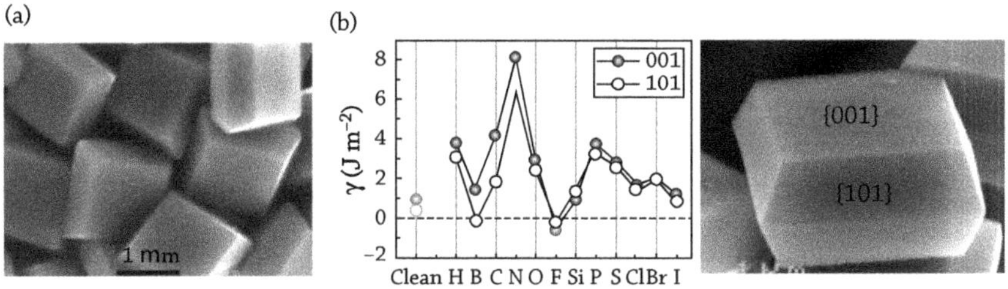

FIGURE 8.6 (a) SEM images of anatase particles obtained by hydrothermal treatment (180°C, 14 hours) of an aqueous solution of TiF_4 ($5.33.10^{-3}$ mol.l^{-1}); (b) variation in surface energy calculated for adsorption of various atoms on the {101} and {001} faces of anatase (reproduced from [25] with permission of the Nature Publishing Group).

faces (Fig. 8.5d). The effect of amino acids seems to be correlated to their adsorption ability, allowing selective lowering of the surface energy of particles (§ 5.4.4). The specific role of glutamic and aspartic acids among the others seems to be due to the presence of two carboxylate groups at the ends of the molecule, allowing an efficient coordination with surface Ti^{4+} ions [23].

Fluoride ion also appears to be an active agent in controlling the morphology of anatase particles, especially in developing the {001} faces of monocrystals. It has been calculated that fluoride adsorption strongly lowers the energy of {001} and {101} faces compared to other atoms—typically, hydrogen, natrium, phosphorus, sulfur, chlorine, bromine, and iodine [24, 25]. In the presence of fluoride, the {001} faces become even more stable than the {101} faces (Fig. 8.6). That is effectively observed in the synthesis of micrometric anatase particles by hydrothermal treatment of TiF_4 in the presence of hydrofluoric acid, HF, in yielding truncated square-based bipyramid monocrystals (Fig. 8.6). This process produces particles of much bigger size than is seen in thermolysis because of the thermal effect, which permits an easier jump of activation barriers for reactions in solution and at the solid–solution interface. The thermal effect also favors the redissolution–crystallization processes involved in the decrease in the surface area of the system. The increase in concentration of titanium fluoride or of hydrofluoric acid causes the decrease in particle size, with a truncation degree increasing with the TiF_4/HF ratio [26]. Similar micrometric truncated bipyramids (Fig. 8.7) have been obtained by microwave heating (210°C, 90 minutes) of aqueous solutions of titanium fluoride in the presence of ionic liquid (1-methyl-imidazolium tetrafluoroborate) [27]. Microwave heating considerably shortens the synthesis duration (§ 4.2.4). In addition, the {001}/{101} face extension ratio can be continuously adjusted from 27 to 50% in increasing the concentration of the ionic liquid. This is likely due to tetrafluoroborate adsorption favoring {001} face extension. EDTA (ethylenediaminetetraacetate) is also a strong complexing species for titanium, which, from TiF_4 in similar hydrothermal conditions, allows the formation of particles of anatase with prevalent {001} faces [28].

FIGURE 8.7 FESEM images of an anatase particle as viewed under two angles, obtained by microwave heating (210°C, 90 minutes) of an aqueous hydrochloric acid solution of TiF_4 (0.04 mol.l^{-1}) and 1-methyl-imidazolium tetrafluoroborate (reproduced with permission from [27] Copyright 2010 American Chemical Society).

FIGURE 8.8 FESEM images of an anatase particle obtained by hydrothermal treatment (180°C, 10h) of a mixture of Ti powder 0.01 g, water 27 ml, H_2O_2 3 ml and HCl 40% 0.1 ml (a {110} face is indicated by the circle) (reproduced from [29] with permission of The Royal Society of Chemistry).

The hydrothermal treatment (180°C, 10 hours) of an aqueous suspension of powdered metallic titanium with hydrogen peroxide and hydrofluoric acid also produces truncated bipyramids of anatase, with additional truncations ({110} faces) appearing at the corners of particles, between the {101} faces [29] (Fig. 8.8).

Titanium peroxocomplexes heated in solution around 100°C generally form anatase particles, without specific character owing to this precursor, in comparison to the procedures involving titanium sulfate or chloride [30–32].

Decomposition by hydrothermal treatment at 200°C of ammonium lactate $[Ti_4O_4(C_3H_4O_3)_8](NH_4)_8{\cdot}4H_2O$ [33], which is a tetrameric titanium complex stable in aqueous solution (Fig. 8.9c), forms pure anatase particles 5 nm in diameter [34]. However, under similar conditions but in the presence of urea, heating (160°C, 24 hours) of the same titanium complex forms either anatase (0.1 mol.l^{-1} urea) or brookite (6 mol.l^{-1} urea) in the form of rods (25 × 150 nm) [34] (Fig. 8.9b). In this process, urea apparently acts more as an ammonium ion producer than as a ligand, but the anatase-to-brookite ratio is dependent on the urea concentration. The increase in urea concentration probably raises the basicity of the medium,

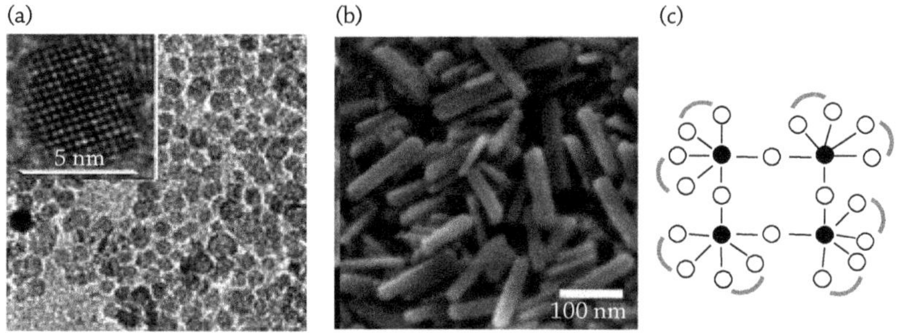

FIGURE 8.9 (a) TEM image of anatase particles resulting from heating (200°C, 72 hours) of an aqueous solution of the titanium dihydroxo bis-ammonium lactato complex (C_{Ti} = 2.08 mol.l^{-1}) (reproduced from [34] with permission of The Royal Society of Chemistry). (b) SEM image of rods of brookite obtained by heating (160°C, 24 hours) the same titanium complex in the presence of 6 mol.l^{-1} urea (reproduced with permission from [Kandiel T.A., Feldhoff A., Robben L., Dillert R., Bahnemann D. W., Tailored titanium dioxide nanomaterials: anatase nanoparticles and brookite nanorods as highly active photocatalysts, Chem. Mater. 2010, 22, 2050–2060] Copyright 2010 American Chemical Society). (c) Scheme of the structure of titanium lactate $[Ti_4O_4(C_3H_4O_3)_8]^{8-}$.

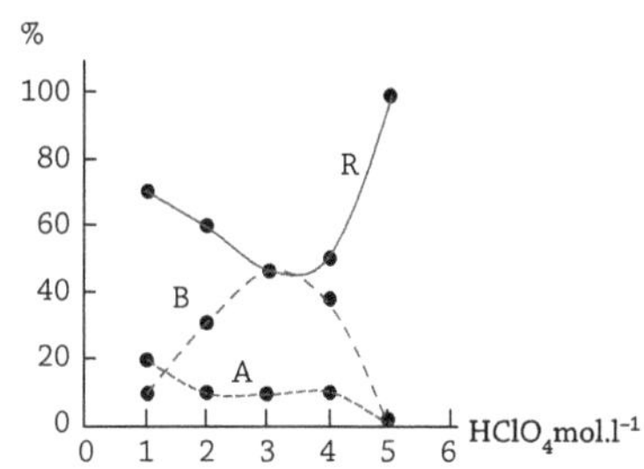

FIGURE 8.10 Proportions of the TiO_2 phases (R rutile, B brookite, A anatase) obtained by thermolysis (100°C, 24 hours) of $TiCl_4$ (0.15 mol.l^{-1}) in perchloric acid of various concentrations.

favoring the kinetics of titanium lactate hydrolysis, increase in particle size, and change in the crystal structure of solid.

b) Rutile

In a strong acidic and noncomplexing medium, which is typically concentrated perchloric acid, 1 to 5 mol.l^{-1}, thermolysis at 100°C for some days of $TiCl_4$ solutions yields rutile. (*Caution: Use of hot and concentrated perchloric acid can be hazardous, resulting in explosion.*) By heating for 24 hours, precipitation of titanium is quasi-complete along the whole acidity range, but the solid obtained is a mixture of anatase, brookite, and rutile, with proportions determined from X-ray diffraction patterns, are dependent on the acidity (Fig. 8.10) [35. 36]. Aging at 100°C for 1 month allows the transformation into rutile to be complete, but the particle size depends largely on acidity. The phase transformation occurs according to a dissolution–crystallization process, as indicated by particle growth and anisotropy increasing with acidity, when the solubility

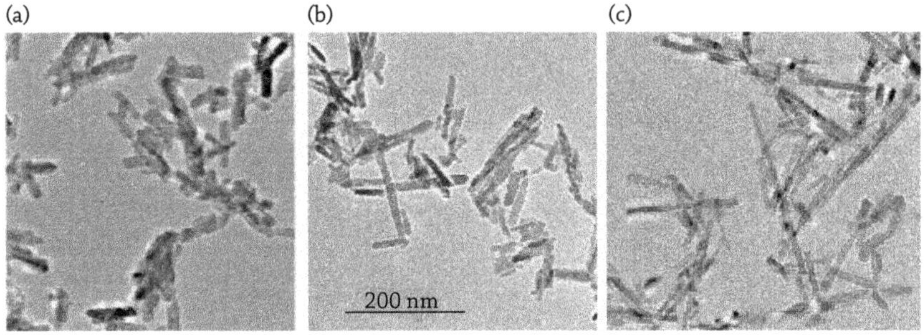

FIGURE 8.11 TEM images of rods of rutile obtained by thermolysis (100°C, 2 weeks) of $TiCl_4$ in $HClO_4$ (a) 1, (b) 2, and (c) 4 mol.l^{-1}. The mean aspect ratio is 6, 10, and 13.5, respectively [Da Costa S., Chanéac C., Jolivet J.P., Unpublished results, 2003].

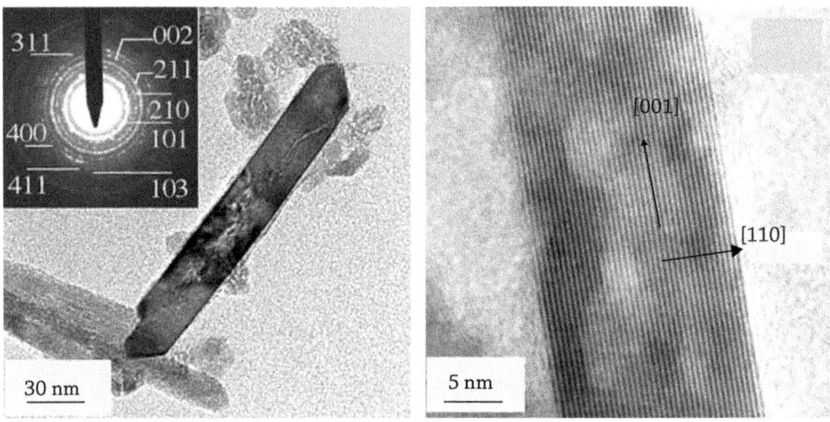

FIGURE 8.12 TEM images of a rod of rutile synthesized by thermolysis (95°C, 1 week) of $TiCl_4$ 0.15 mol.l^{-1} in HNO_3 1 mol.l^{-1} (reproduced from [37] with permission of Springer).

increases and the amount of early metastable phases to transform is important (Fig. 8.11). Rutile rods are also obtained from thermolysis of $TiCl_4$ in HNO_3 1 mol.l^{-1} [37] (Fig. 8.12), but the mixture of rutile (50%), anatase (25%), and brookite (25%) cannot be resolved by subsequent thermolysis.

As for other polymorphic compounds, the precipitation of TiO_2 occurs with the initial and temporary formation of the less stable phases. Previous study of the nucleation-growth steps of the solid (§ 4.2) showed that the activation energy of the process decreases when supersaturation is high and the solid–solution interfacial tension is low. Among polymorphs, that is always the less thermodynamically stable phase that has the lowest volumic mass and interfacial tension. These two parameters are linked, however, as indicated by the empirical relationship $\gamma = (2E/d)(a/\pi^2)$, where E is the Young modulus of the solid, d is the interplane distance normal to the exposed face, and a is the atomic radius. As the distance d is in inverse proportion to the volumic mass, the interfacial tension is in proportion to the cubic root of volumic mass. As it is 3.84, 4.17, and 4.26 g.cm^{-3} for anatase,

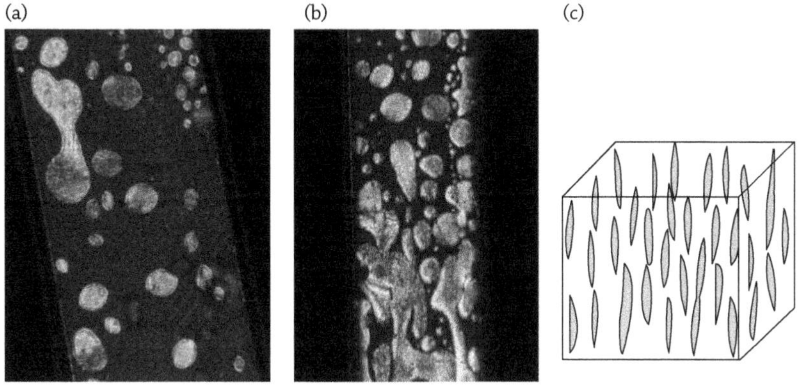

FIGURE 8.13 Images in optical microscopy between cross-polarizing crystals of suspensions of TiO_2 rutile particles (mean size 200 × 11 – 15 nm, L/D = 15, pH 2) with volumic fraction ≈ 6% (a), ≈ 10% (b). The colored zones are nematic domains (c) spontaneously formed in the isotropic phase (reproduced with permission from [38] Copyright 2007 American Chemical Society).

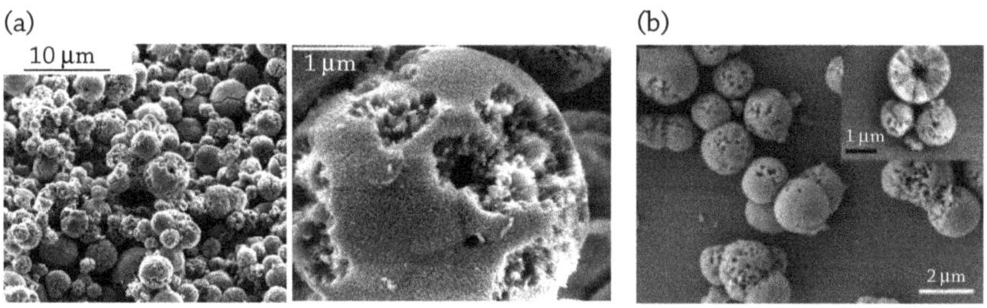

FIGURE 8.14 Mesostructures resulting from assembling nanorods of rutile: (a) after heating at 100°C of $TiCl_4$ 1 mol.l^{-1} in HCl 4 mol.l^{-1} [Da Costa S., Chanéac C., Jolivet J.P., Unpublished results, 2003]; (b) hydrothermal heating at 130°C of $TiCl_4$ 3 mol.l^{-1} in the presence of Na_2CO_3, L-serine, and *N, N'*-dicyclohexylcarbodiimide (reproduced with permission from [Liu S.J., Gong J.Y., Hu B., Yu S.H., Mesocrystals of rutile TiO_2: mesoscale transformation, crystallization, and growth by a biologic molecules-assisted hydrothermal process, Cryst. Growth Des. 2009, 9, 203–209] Copyright 2009 American Chemical Society).

brookite, and rutile, respectively, one may deduce that the interfacial tension of the three polymorphs varies in the same way and that anatase is thus the kinetically favored phase.

In concentrated aqueous suspensions, the rod-like rutile particles tend to spontaneously align along a common direction in forming liquid crystals exhibiting a strong birefringence [38] (Fig. 8.13).

In the presence of complexing species and at high concentration, the rod-like rutile particles aggregate to form hollow spherical mesostructures (Fig. 8.14). The ligands can be chloride ions or molecules of biological interest such as *N, N'*-dicyclohexylcarbodiimide and L-serine. However, the effective mechanism involved in such processes and the role of the ligands have not yet been clearly established.

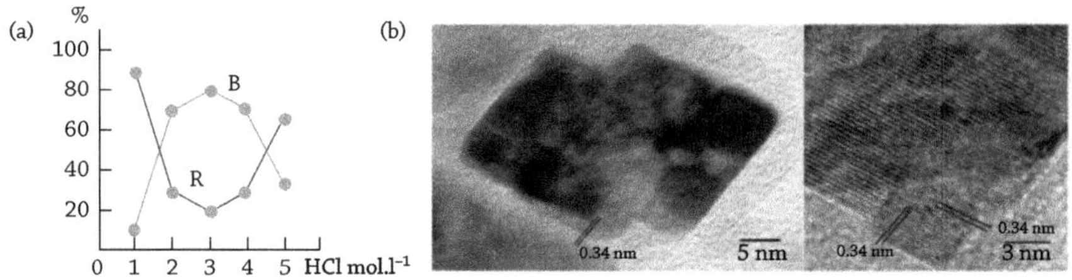

FIGURE 8.15 (a) Relative proportions of brookite (B) and rutile (R) formed by thermolysis (100°C, 48 hours) of $TiCl_4$ (0.15 mol.l^{-1}) with various HCl concentrations; (b) TEM images of brookite particles obtained in HCl 3 mol.l^{-1} (reproduced from [36] with permission of The Royal Society of Chemistry).

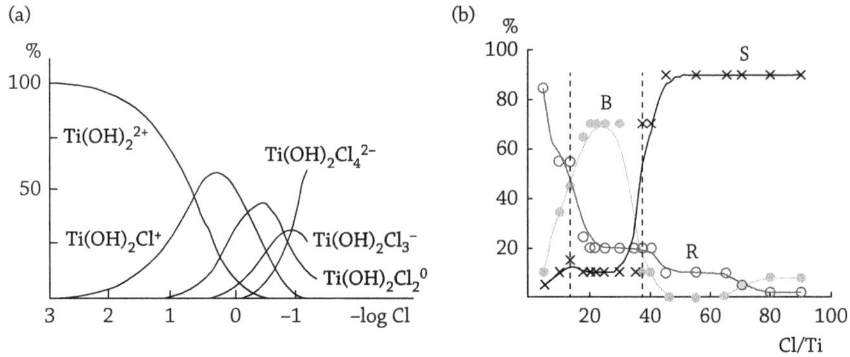

FIGURE 8.16 (a) Distribution of hydroxochloro titanium complexes as a function of chloride concentration [from 40]. (b) Distribution of titanium in solid form (brookite (B), rutile (R)) and in soluble species (S) as a function of the Cl/Ti molar ratio after 24 hours of thermolysis at 100°C of $TiCl_4$ and HCl mixtures (reproduced from [36] with permission of The Royal Society of Chemistry).

c) Brookite

In concentrated aqueous hydrochloric acid solutions, thermolysis of $TiCl_4$ forms mixtures of brookite and rutile, with proportions depending on acid concentration (Fig. 8.15a). In these conditions, brookite in the form of diamond-shaped nanoparticles is a stable phase that does not transform during aging at 100°C (Fig. 8.15b). Chloride ions play a fundamental role in stabilizing the brookite phase and, in such acidic media, in forming a series of hydroxochloroaquo mixed complexes of titanium (Fig. 8.16a) [39–41]. Among them, the noncharged species, $[Ti(OH)_2Cl_2(OH_2)_2]^0$, seems to be the precursor of the solid. Indeed, when the amount of brookite is expressed as a function of the Cl/Ti ratio (Fig. 8.16b), instead of acidity, it appears that the maximum amount of brookite corresponds exactly to the Cl/Ti composition range in which the noncharged $[Ti(OH)_2Cl_2(OH_2)_2]^0$ complex is the predominant species.

It is thus possible to propose a pathway in which the chloride ligand orients the condensation toward formation of brookite (Fig. 8.17) [36]. The first step of the process is olation between $[Ti(OH)_2Cl_2(OH_2)_2]^0$ complexes in which elimination of coordinated water forms a edge-sharing dimer. These dimers can themselves

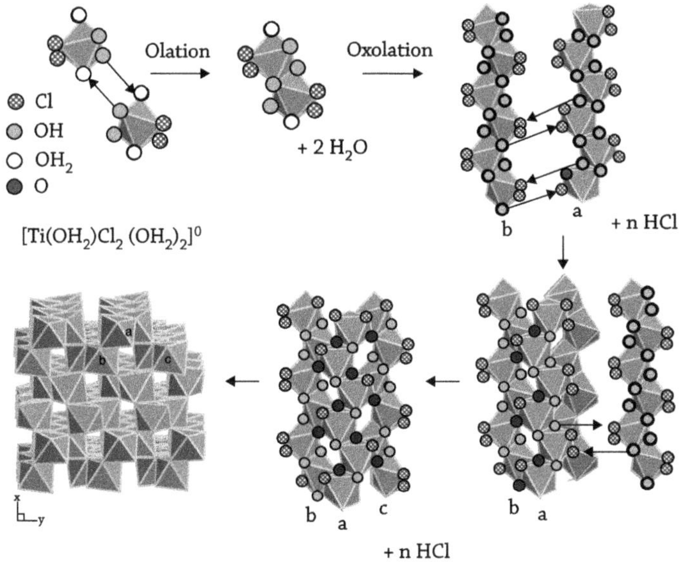

FIGURE 8.17 Possible pathway for the formation of brookite from the $Ti(OH)_2Cl_2(OH_2)_2$ complex (reproduced from [36] with permission of The Royal Society of Chemistry).

condense, still by olation, to form chaining from which all coordinated water is eliminated, containing only terminal hydroxo and chloride ligands. At this stage, oxolation with HCl elimination can occur between small chains to form the local structure of brookite when all chloride ligands are removed. Thus, chloride ions are temporarily active, able to orient the brookite crystallization process. Although they are not present in the solid, their presence is essential for crystallization and stabilization in suspension of brookite. Indeed, if brookite particles formed in HCl 3 $mol.l^{-1}$ are transferred and aged in $HClO_4$ 3 $mol.l^{-1}$ at 100°C for 1 month, the transformation into rutile is complete, whereas if aging occurs in the presence of chloride ($HClO_4$ 3 $mol.l^{-1}$ + NaCl 3 $mol.l^{-1}$), no transformation of brookite is observed.

A rather similar effect is observed in concentrated nitric acid medium. On the one hand, nitrate ions in strong acidic medium can become ligands, playing a role similar to that of chloride ions [37, 42, 43]. On the other hand, if the thermolysis of $TiCl_4$ is carried out in sulfuric acid medium, anatase TiO_2 is obtained. Sulfate ions are strongly complexing of titanium (Fig. 8.18a) as bidentate ligand and not monodentate as chloride and nitrate, so that condensation of titanium from the noncharged precursor, $Ti(OH)_2(SO_4)(OH_2)_2$, occurs in a different way. By olation, this complex forms the dimer $Ti_2(OH)_4(SO_4)_2(OH_2)_2$. Through subsequent condensation, still by olation, it leads merely to fragments of double chains, able to assemble in removing sulfate and prefiguring the anatase crystal structure (Fig. 8.18b). Small sulfate amounts, however, remain trapped in the solid, being likely at the origin of the stabilization of anatase during the aging of suspensions.

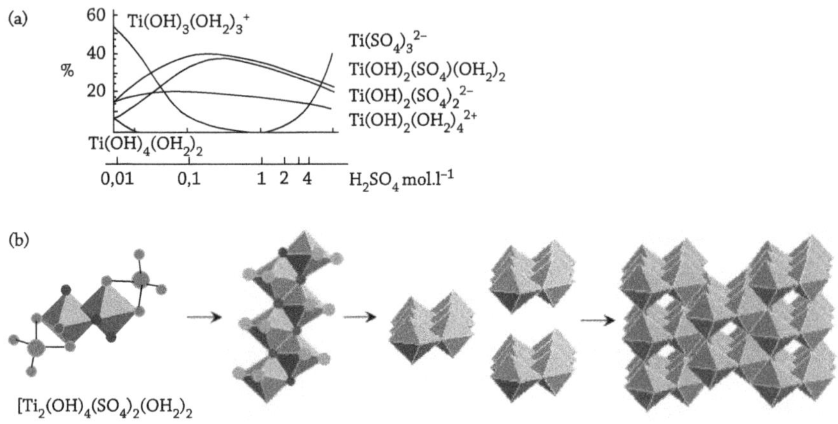

FIGURE 8.18 (a) Distribution of titanium hydroxysulfate complexes as a function of sulfuric acid concentration [from Babko A.K., Mazurenko E.A., Nabivanets B.I., The sulphato complexes of titanium(IV) ion solution, Russ. J. Inorg. Chem. 1969, 14, 1091–1093]]. (b) Schematic pathway of anatase formation from the noncharged titanium hydroxoaquo-sulfato complex.

An interesting pathway to brookite is offered by the titanium oxalato-oxocomplex, $Ti_2O_3(H_2O)_2(C_2O_4)\cdot H_2O$, obtained by heating at 90°C a titanium oxysulfate solution containing lithium oxalate (Ox/Ti = 1) [44, 45]. The crystal structure of the complex in solid state is made of layers containing zigzag chains of corner-shared, strongly distorted TiO_6 octahedra [46]. The chains are connected by corners of octahedra, and the layers are linked between them by oxalato ligands (Fig. 8.19 b, c). The octahedral coordination of titanium thus results from bonding with a bidentate oxalato ligand (O1 and O2 Fig. 8.19b), one water molecule (O6), two oxygen atoms that connect the octahedra in the chains (O3 and O4), and a third (O5) that ensures the interchain's bond. The egg-shaped particles transform by thermal treatment at 300°C into brookite in preserving their morphology (Fig. 8.19d, e). The material exhibits an important mesoporosity (255 m^2/g in a specific surface area measured by Brunauer, Emmett, and Teller adsorption method (BET), with a mean pore size of 3.4 nm), which is likely due to the evacuation of water and CO_2 during the thermal transformation of the material. Taking into account the large width of X-ray diffraction peaks, the material seems to be nanotextured, made of nanocrystals of brookite forming micrometric-sized particles. Such a porous brookite so obtained is stable up to 600°C, a temperature at which the rutile phase appears. This example shows once again that the formation of brookite is obtained from a complexed form of titanium.

8.2.2 TRANSFORMATION OF LAYERED TITANATES

The alkaline titanates of the general formula $M_{2-y}H_yTi_nO_{2n+1}$ form a large family of compounds with a lamellar crystal structure. The anionic charge of the lattice is

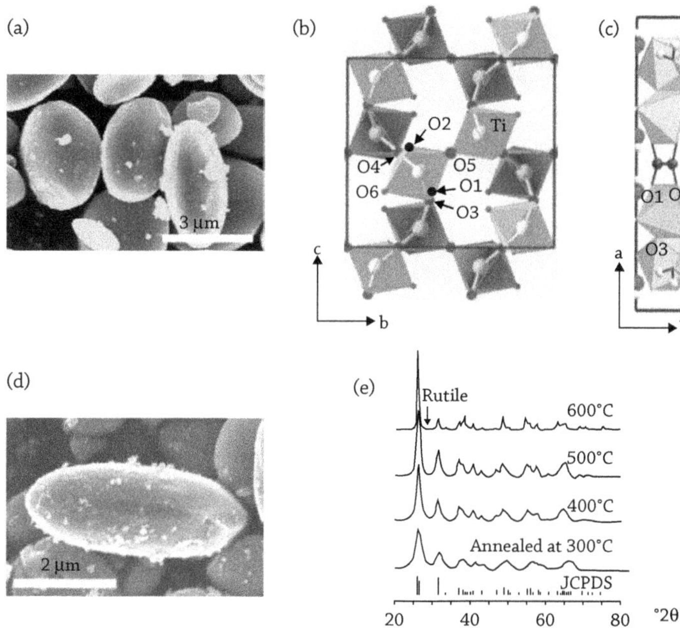

FIGURE 8.19 (a) Egg-shaped particles of titanium oxo-oxalato hydrated complex $Ti_2O_3(H_2O)_2(C_2O_4)\cdot H_2O$ (reproduced with permission from [52] Copyright 2010 American Chemical Society); (b) layers in the crystal structure containing chains of octahedra linked by corners (octahedra of different color are located at different heights); (c) connection of the chains by oxalato ligands (the spheres stand for crystallization water molecules) (adapted from [46]); (c) particle of brookite obtained by heating at 300°C the titanium oxalato complex; (d) X-ray diffraction patterns of titanium oxo-oxalato complex heated at various temperatures (reproduced with permission from [44] Copyright 2010 American Chemical Society).

balanced by that of alkaline cations or protons located within the interlayer spaces. In their protonated form, these materials may be interesting precursors of dioxide TiO_2 because they are chemically reactive solids able to transform in aqueous suspension at low temperatures in preserving their morphology or, at least, their low directionality.

The crystal structure of titanates may be considered as being formed of strips of edge-shared octahedra, the strips being linked by corner sharing in forming corrugated layers. Alkaline cations and/or protons are located in the spaces between the strips [47] (Fig. 8.20).

Alkaline titanates can be synthesized either by reaction in solid state of TiO_2 with an alkaline nitrate $K(Na)NO_3$ or by hydrothermal treatment in a strongly basic medium of a titanium salt or the polymorphs of TiO_2 [48]. Typically, the heating of sulfate $TiOSO_4\cdot xH_2O$ in NaOH $\approx$ 7 mol.l^{-1} forms the titanate $Na_2Ti_3O_7\cdot xH_2O$ [49]. The solid is made of fibers (Fig. 8.21a), with length depending on the synthesis temperature: 30–60 nm in length and 3–5 nm in width at 100°C, 100–200 nm × 8–10 nm at 150°C, 30 μm × 40–80 nm at 200°C and an increasing quality in

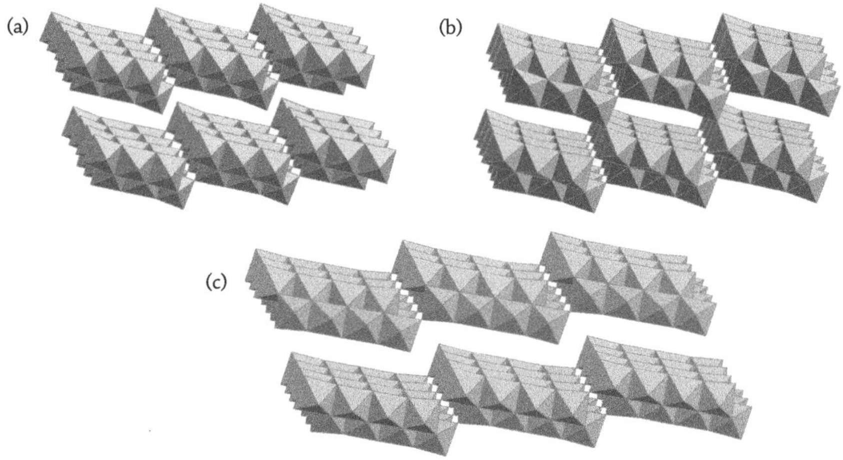

FIGURE 8.20 Schematic crystal structure of some titanates: $Na_2Ti_3O_7$ (a), $Na_2Ti_6O_{13}$ (b), $Na_2Ti_4O_9$ (c).

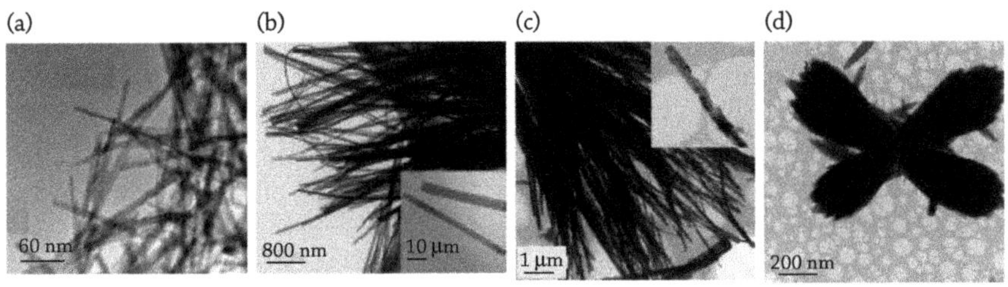

FIGURE 8.21 TEM images (a) of H-titanate formed in hydrothermal conditions at 155°C and (b) at 200°C; (c) anatase particles obtained from long H-titanate fibers shown in (b) by treatment at 120°C, HNO_3 5.10^{-2} $mol.l^{-1}$; (d) rutile particles obtained from the same long H-titanate fibers shown in (b) by treatment at 80°C, HNO_3 2.65 $mol.l^{-1}$ (reproduced with permission from [50] Copyright 2005 American Chemical Society).

crystallization as the temperature is high [50]. The treatment of sodium titanate with HCl at pH 7 yields, by ion exchange, the protonated form $H_2Ti_3O_7{\cdot}xH_2O$ without morphological change. Because of titanate's layered structure, various morphologies of nanoparticles can be obtained: small strip, tube, rod, wire, or fiber [51]. The experimental conditions of the hydrothermal synthesis (e.g., temperature, duration of heating, rate of cooling, basicity of the medium) have an important impact on the particle shape, even if some of them, especially nanotubes, are kinetically stabilized [52].

Layered titanates are interesting precursors of titanium dioxide because they are reactive, and through treatment in solution at low temperature of their Na- or H-form, anatase and rutile are easily obtained in diluted (5.10^{-2} $mol.l^{-1}$) and concentrated (≥ 2 $mol.l^{-1}$) nitric acid, respectively. The transformation is easy, all the more so as the titanate fibers are short—that is, the temperature of synthesis is low. On the one hand, the particle shape of anatase obtained in weakly acidic medium is closely related to that of the parent titanate particles: short titanate fibers give anatase nanoparticles of about 10 nm, while the longest fibers of titanate formed at 200°C transform into fibrils consisting of aggregates of anatase

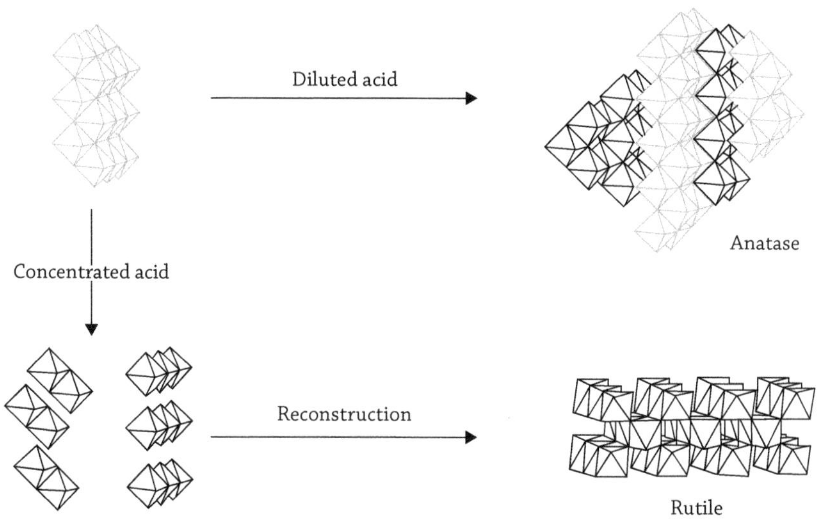

FIGURE 8.22 Scheme of the transformation of titanate $H_2Ti_3O_7{\cdot}xH_2O$ into anatase or rutile according to the acidity (reproduced with permission from [50] Copyright 2005 American Chemical Society).

nanoparticles (Fig. 8.21b). On the other hand, in strongly acidic medium, there is no shape filiation between the initial titanate fibers and the obtained needles of rutile (Fig. 8.21c). These results suggest that the mechanisms involved in these two transformations are completely different. The preservation of morphology observed in transforming titanate into anatase may be explained by a surface reaction requiring only a structural rearrangement, because there is a close structural relationship between titanate and anatase in the organization of the edge-shared TiO_6 octahedra within the corrugated strips (Fig 8.22). For rutile, however, the reaction occurs in strongly acidic solution, allowing solubilization of titanate and reconstruction of the solid in the thermodynamically most stable phase.

Acidity and concentration of sodium ions govern the formation and transformation of titanate. For instance, the precipitate obtained by addition of soda to titanium sulfate and filtration to eliminate the sodium sulfate, and then treated in hydrothermal conditions (220°C, 48 hours), forms pure anatase at pH 0.8, titanate at pH 7, and pure brookite at pH 12.5 [53]. Brookite so formed is obtained as an aggregate in the form of flowers or sand roses (Fig. 8.23). During the progress of the reaction at high pH, on the one hand, titanate forms first and then transforms into brookite. On the other hand, if alkalinization is carried out with LiOH or KOH, mixtures of anatase and rutile are formed, showing that the Na^+ ions have a crucial role in the titanate-to-brookite transformation.

The treatment of anatase particles (120°C, 20 hours) in sodium hydroxide 10 $mol.l^{-1}$ yields the titanate $Na_2Ti_3O_7$ in the form of nanotubes (Fig. 8.24a, b) [54, 55]. Heating this titanate at 200°C for 20 hours at pH ≈ 11 transforms it into brookite in preserving the particle shape [56] (Fig. 8.24c, d). In both cases, multiwall nanotubes with an internal diameter of around 5 nm are obtained. This suggests that, as for the formation of flower-shaped brookite, as well as for the transformation of

FIGURE 8.23 SEM images of brookite formed from hydrothermal treatment (220°C, 48 hours) at pH 12.5 of the titanate resulting from precipitation of $TiOSO_4$ by soda (reproduced with permission from [53] Copyright 2009 American Chemical Society).

FIGURE 8.24 TEM images of titanate nanotubes, (a), (b), formed in alkaline medium by hydrothermal treatment of anatase. Nanotubes of brookite, (c), (d), resulting from the hydrothermal transformation of the titanate nanotubes (reproduced from [56] with permission of The Royal Society of Chemistry).

titanate into anatase in weakly acidic medium, the structural transformation of titanate proceeds by rearrangement in solid state. With brookite, some structural relationship with the early titanate also exists.

As shown earlier, the treatment of anatase at 150°C for 72 hours in NaOH 10 mol.l^{-1} forms nanotubes, while at higher temperatures and more concentrated soda (170°C, NaOH 15 mol.l^{-1}), nanowires of titanate $Na_2Ti_3O_7$ are obtained [57]. Washing this material in dilute acidic medium causes the Na^+/H^+ exchange, and

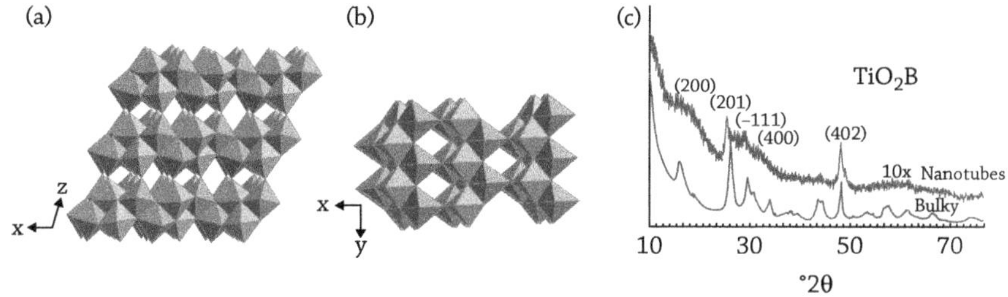

FIGURE 8.25 (a) Crystal structure of $TiO_2(B)$ viewed under two orientations; (b) X-ray diffraction patterns of bulky and nanotubes of $TiO_2(B)$ (adapted from [59]).

heating at 400°C converts, with preservation of the morphology, the nanowires into $TiO_2(B)$, a form of dioxide with a crystal structure analogous to that of bronzes Na_xTiO_2 ($Na_xTi^{III}{}_xTi^{IV}{}_{1-x}O_2$) [58] (Fig. 8.25). It is noteworthy that hydrothermal treatment (150°C, 10 hours, NaOH 10 mol.l^{-1}) of crystals of the Ti_8 polycation $[Ti_8O_{12}(OH_2)_{24}]Cl_8{\cdot}7H_2O$ (§8.1) forms $TiO_2(B)$ nanotubes, 5–8 nm in outer diameter and 3–6 nm in inner diameter [59]. Because of the relatively open structure of this titanium dioxide form, lithium intercalation gives this material an interesting application as electrode material [60].

8.2.3 OXIDATION OF TI^{III} AND TI^0 IN ACIDIC TO NEUTRAL MEDIUM

Reduced forms of titanium can be hydroxylated and oxidized to form dioxide. The main interest of such procedures lies in the original morphologies obtained.

a) Ti^{III} Oxidation

The trichloride $TiCl_3$ is an interesting precursor because the Ti^{3+} ions, which are much less acidic than Ti^{4+} ions, are easier to manipulate in solution. These ions are soluble, hydroxylable, and precipitable in conditions that are easily manageable, and their oxidation, which occurs in solution as well as in solid state, forms the dioxide TiO_2. Hydroxylation of $TiCl_3$ starts from around pH 2. At pH 3, the purple solution becomes darker and turns black. As the titration is carried out in the presence of air, this means that some polycationic entities underwent a partial oxidation, allowing intervalent Ti^{III}-Ti^{IV} electron transfers that cause the black color. The precipitation of a blue–black solid begins beyond pH 3 up to pH 9. It corresponds to a hydroxide or an oxyhydroxide $Ti(OH)_3$/TiOOH, which, in air, is rather rapidly oxidized in suspension or after drying. These blue-colored mixed phases of Magneli's type, $[Ti^{III}{}_{2-x}Ti^{IV}{}_xO_{3+x/2}]$, are formed transitorily. Then, the completely oxidized white compound, TiO_2, is obtained (Fig. 8.26) [61].

In comparison with thermolysis of $TiCl_4$ in perchloric medium which forms mixtures of rutile and anatase at pH < 1, pure anatase at 1 < pH < 6, and precursors of titanate at pH > 6 [25], the thermolysis at 60°C under air for 24 hours of $TiCl_3$ yields quasi-selectively different polymorphs in well-delimited ranges of

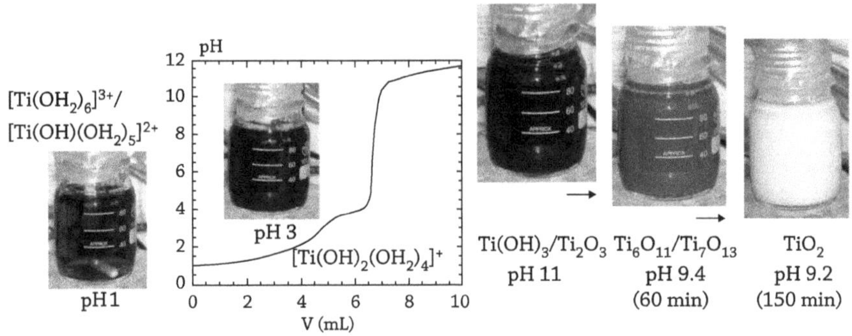

FIGURE 8.26 Titration curve by NaOH of a $TiCl_3$/HCl solution. From the left to right, images and compositions of solutions and suspensions at pH 1, 3, and 11 and suspensions aged at room temperature for 60 and 150 minutes in which the pH is decreasing with oxidation [42].

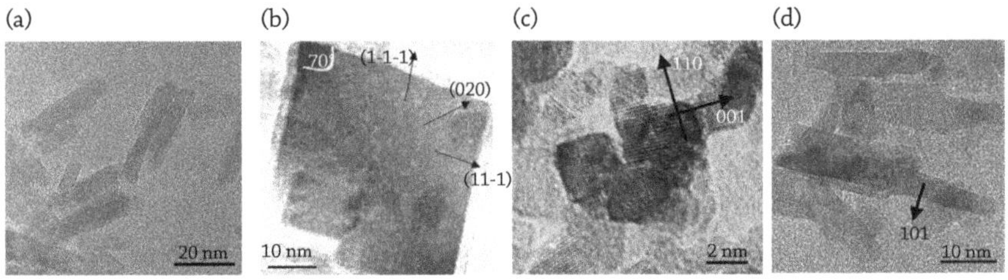

FIGURE 8.27 TEM images of particles formed by thermolysis at 60°C for 24 hours of $TiCl_3$ 0.15 mol.l^{-1} at various pHs: (a) rutile pH 0.5, (b) brookite pH 2, (c) rutile pH 4, (d) anatase pH 7 (reproduced from [61] with permission of Elsevier).

acidity: rutile at pH < 1, brookite at pH ≈ 1, pseudo-rutile with a very distorted unit cell at pH 3.5–4, and anatase at pH 6.5–7 [61] (Fig. 8.27). These crystalline varieties are obtained under unusual nanomorphologies, compared to that obtained from $TiCl_4$. It seems that during the thermolysis at pH ≤ 3, oxidation occurs before precipitation, so that Ti^{IV} species precipitate, while at higher pH, precipitation proceeds before oxidation. It occurs, then, in situ in the solid state, and the final morphology is likely imposed by that of the initial particles of Ti^{III} oxyhydroxide.

The use of urea during heating the $TiCl_3$ solutions seems to favor the formation of brookite. Thus, heating at refluxing for 7 hours at 100°C of $TiCl_3$ 0.015 mol.l^{-1} in the presence of urea 0.5 mol.l^{-1} forms spherical particles 10–20 nm in size [62]. Heating the same mixture at 70°C for 2 hours yields aggregates of platelets around 25 nm in size [63].

Hydrothermal treatment with microwave heating at 200°C for 1 minute of the commercial $TiCl_3$ solution (15% in HCl 10%) as single reagent system yields amazing mesocrystals made by ordered assembling of rutile nanorods 7.5 nm in diameter and 2 μm in length (Fig. 8.28) [64]. The growth of the rods is oriented along the (001) direction. Heating between 200 and 700°C of $TiCl_3$ in HCl solution at pH 1 for 3 days yields mixtures of phases in which brookite is the most abundant

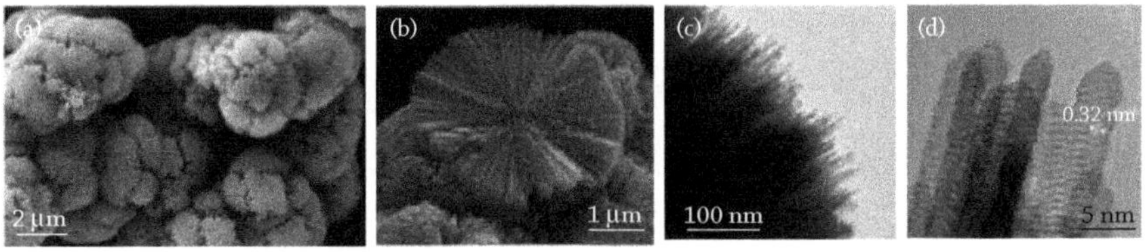

FIGURE 8.28 SEM (a,b) and TEM images (c,d) of rutile mesocrystals obtained by microwave heating of $TiCl_3$ (reproduced from [64] with permission of The Royal Society of Chemistry).

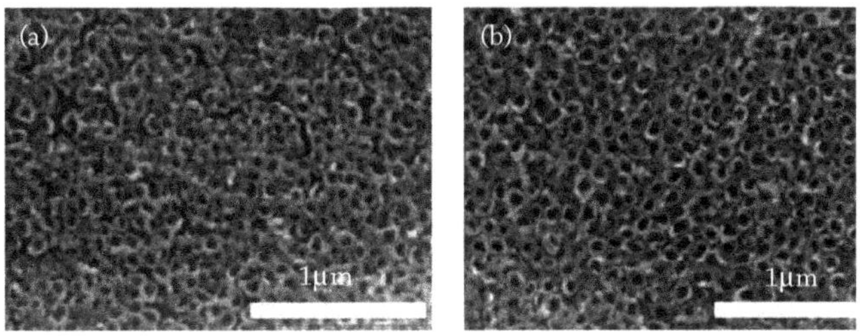

FIGURE 8.29 SEM images of TiO_2 nanotube arrays obtained from titanium foil anodization in aqueous HF 0.5 wt %, 20V, 1 hour (a) at room temperature, (b) in an ice bath. The nanotubes formed at RT have an average inner diameter of 50 nm and an outer diameter of 110 nm. The average inner diameter of the nanotubes formed in an ice bath is 60 nm with an outer diameter of 110 nm (reproduced with permission from [70] Copyright 2009 American Chemical Society.)

(65 and 53% at 200 and 700°C, respectively), with anatase around 30% and rutile around 6%. The particles of brookite are 25 nm in mean size [65]. Intriguingly, an unknown Ti_2O_3 phase of corundum crystal-type is formed beyond 400°C (around 20%), which does not oxidize by heating under air.

Apart from the fact that the most acidic media favor the formation of rutile, the formation of the other crystal phases of TiO_2 from oxidation of Ti^{III} in solution is difficult to understand. The main cause seems to be the conditions of the oxidation, which are never really controlled: for example, a more or less high oxidation rate according to the temperature and acidity of the medium; the presence of air in various amounts or special reagents in the reactor; or the simultaneity or not of hydroxylation and oxidation reactions. The example of ferric and ferric-ferrous mixed phases formation from ferrous hydroxide (§ 7.2.1) highlights the influence of the oxidation reaction conditions on the crystal structure and the characteristics of the products obtained.

b) Ti^0 Oxidation

Anodic oxidation of metallic titanium is used to produce TiO_2 nanotubes [66, 67]. The tubes are a few tens of nanometers in diameter and up to about 100 micrometers in length. The synthesis is based on electrochemical anodization of a titanium foil as anode and a thin platinum foil as counterelectrode in a two-electrode cell containing an electrolyte—typically, fluoride (or chloride) ions in aqueous solution (Fig. 8.29) or in nonaqueous

FIGURE 8.30 FESEM images of TiO_2 nanotube-array sample (134 μm long) grown at 60 V for 17 hours in ethylene glycol containing 0.25 wt % NH_4F (a) bottom surface, (b) top surface views; TiO_2 nanotube-array samples (45 μm long) grown at 40 V in dimethylsulfoxide containing 2% HF for 69 hours, (c) cross-sectional (d) lateral and (e) top surface views (reproduced with permission from [68] Copyright 2006 American Chemical Society.)

medium such as glycerol, ethylene glycol, and dimethylsulfoxide (Fig. 8.30). Most often the tubes are longer and more uniform when the anodization is performed in a nonaqueous medium [68–70], but many other parameters influence the characteristics of these nanostructures, for example, anodization potential, temperature, nature and concentration of the electrolyte, and acidity of the medium.

The anodic anodization with a potential of 20 to 80 volts releases Ti^{4+} ions, which are immediately hydroxylated at the contact of the electrolyte and precipitate in forming a more or less dense oxide layer on the surface of the metal. The growth of the layer results from ionic migration within the oxide under the electric field: diffusion of Ti^{4+} ions toward the solution and diffusion of oxygenated species toward the metal. The newly formed oxide grows at the metal–oxide or at the oxide–electrolyte interfaces, according to the relative mobilities of the ionic species. Under a constant voltage, the growth stops when the effect of the electric field, that is, the driving force of the process, disappears for a given thickness of the oxide film. A homogeneous, compact, and more or less thick oxide layer is thus formed, which constitutes the passivation layer of the metal. In the presence of complexing species, typically fluorine, the titanium ions are solubilized as anionic fluorined complexes when the fluoride ions cross over the oxide layer and also by dissolution of oxide at the oxide–electrolyte interface. There are thus three regimes, depending on the F^- ion concentration: (1) at very low concentration (typically, 0.05 wt %), not any effect occurs, and a compact oxide layer forms; (2) at high concentration (around 1%), no oxide is formed, and all produced Ti^{4+} ions are complexed, that is, the electropolishing of the metal; and (3) at intermediate fluoride ion concentration, the formation of oxide competes with the solubilization of Ti^{4+} ions, leading to a porous oxide layer or to formation of nanotubes.

The mechanism involved in tube formation can be described as follows [71]. After the initial oxidation step which forms a compact oxide layer at around 2 minutes (the current in the cell decreases) (Fig. 8.31a), the presence of fluoride causes the formation of rather irregular nanometric pores that penetrate the oxide layer (the current increases), and then the growth of an even layer of nanopores or nanotubes occurs under a stationary regime (the current in the cell decreases again). Under the electric field, the fluoride ions hold at the oxide–solution interface a rather thin oxide layer by complexation of the Ti^{4+} ions; as fluoride ions diffuse faster than the oxygenated species within the oxide layer, a zone rich in fluoride of around 20 nm exists at the metal–oxide interface (Fig. 8.31b). The formation of long, independent tubes can be explained by considering that the formation of oxide generates a plastic flow of the layer due to compressive stresses induced by the volume expansion and electrostrictive forces generated during the oxide growth. The viscous oxide is then pushed up the pore walls and thus extends along the tube length, with accumulation of fluoride and titanium oxyfluoride at the pores' boundaries (Fig. 8.31c). Dissolution leads finally to formation of independent tubes that remain attached to the barrier layer that fills the bottom. After elimination of the oxide layer, the surface of metal appears to be marked by small craters. The longer length of the nanotubes formed in glycerol is due to the lower mobility of fluoride ions and less dissolution of oxide in glycerol than in water.

The dimensional and morphological characteristics of the nanotubes are dependent on anodization conditions such as voltage, duration, temperature, and electrolyte composition [71]. In all cases, nanotubes stemming from metal anodization are amorphous. Crystallization, more often into anatase, is obtained by annealing under air at a temperature around 450–500°C.

8.2.4 SYNTHESIS OF BARIUM TITANATE $BaTiO_3$

Barium titanate $BaTiO_3$ belongs to the large family of perovskite-like oxides ($BaTiO_3$, $PbTiO_3$, $PbZr_{1-x}Ti_xO_3$, $PbFe_{0.5}Nb_{0.5}O_3$, etc.) (Fig. 8.32), which are interesting for their ferro- or piezoelectrical properties [72]. $BaTiO_3$ is especially used as nanomaterial in multilayer ceramic capacitances. Some of these oxides, such as $PbZr_{0.5}Ti_{0.5}O_3$ and $(Pb_{1-x}La_xZr_yTi_{1-y})_{1-x/4}O_3$ exhibit electro-optic properties, allowing modulation of light by an electric field. $LiNbO_3$ is used for nonlinear optics. The synthesis of such materials in the form of nanoparticles presents a considerable technological challenge.

The conventional synthesis of $BaTiO_3$ and of many other titanates is carried out by calcination beyond 1000°C of the crushed mixture of titanium oxide and

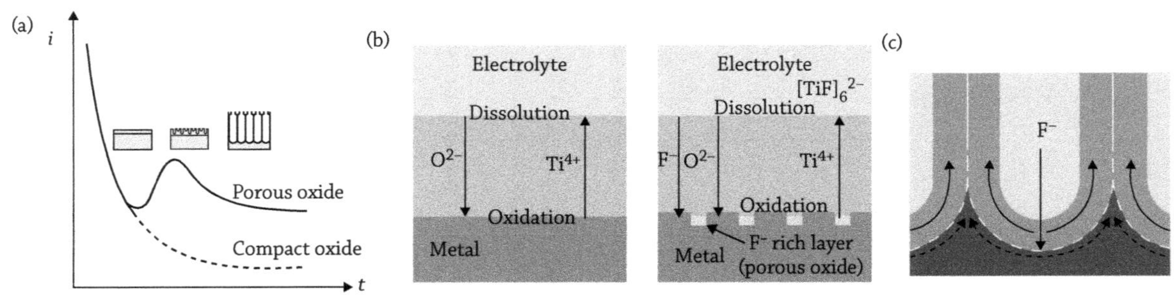

FIGURE 8.31 (a) Scheme of a typical current-time curve relative to anodic oxidation of titanium in the absence (dashed line) and the presence (continuous line) of fluoride ions in the electrolyte. Either a compact (fluoride free) or porous/tubular (containing fluoride) oxide layer is formed by different morphological stages. (b) Schematic drawing showing field-aided transport of mobile ionic species through the oxide layers in the absence or presence of fluoride ions. The rapid fluoride migration leads to accumulation at the metal–oxide interface. (c) Scheme showing the flow mechanism pushing oxide and fluoride layer up the cell walls by viscous flow (adapted from [71]).

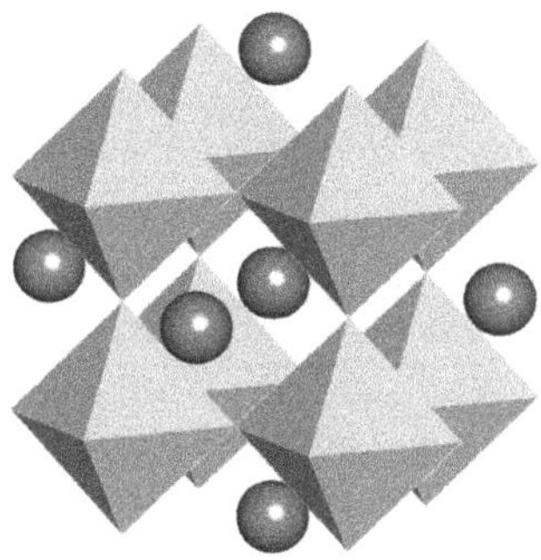

FIGURE 8.32 Scheme of idealized perovskite-like $BaTiO_3$ crystal structure. The 12-coordinated Ba^{2+} ions represented by spheres are trapped between TiO_6 corner-shared octahedra.

barium carbonate. This technique involves many drawbacks: pollution of the material because of grinding, competing formation of the stable pyrochlore-type crystal phase that lowers the dielectric performance of the material, and cost due to duration and high temperature of calcination. Other methods, based on solution chemistry and coprecipitation of titanium and barium, have allowed the perovskite-type phase at low temperatures with a higher purity and in the form of small-sized particles. An advantage of such a process is the ability to form thin films of this material for purposes of fabricating miniaturized capacitors: a suspension or a gel of small particles is deposited on a metallic electrode and then heated at a temperature low enough to allow the sintering of particles to be compatible with cheap electrodes in nickel or copper. Various synthesis methods can thus be used.

One of these methods is used to form a mixed complex of titanium and barium with oxalate or citrate [73, 74]. The mixed citrate $BaTi(C_6H_5O_7)_3{\cdot}6H_2O$ crystallizes by acidification at pH ≤ 2.6 of the mixture of solutions of these two elements, regardless of the stoichiometry. Thermal decomposition of the complex leads in several stages to the formation of $BaTiO_6$ at temperatures between 600°C and 700°C: from 360°C, barium carbonate forms, and at around 500°C, the mixture $BaCO_3$ and TiO_2 can be identified by X-ray diffraction. Thus, barium titanate is not directly formed, but it crystallizes at around 700°C by reaction in solid state between barium carbonate and titanium oxide. The perovskite phase is therefore formed at a temperature that is largely lower than that required with the use of coarse powders of carbonate and oxide and without pyrochlore phase as a by-product. The crystallization temperature decreases as a result of the high reactivity of smaller particles, allowing ion diffusion on shorter distances to facilitate crystallization of the mixed oxide. Pechini's technique (§ 3.1.6) also helps form the perovskite phase of $BaTiO_3$ from mixed titanium barium citric complexes in ethylene glycol [75] in the same temperature range.

A less energetic process for synthesizing $BaTiO_3$ consists of heating at 90°C aqueous suspensions of $Ba(OH)_2$ and TiO_2 [76, 77]. The reaction occurs only with finer powders of anatase (7 m^2/g) or TiO_2 B (72 m^2/g). TiO_2 B (Fig. 8.25) is obtained from treatment of titanate $K_2Ti_4O_9$ in acidic medium and heating at 500°C. The conversion of TiO_2 B into barium titanate is complete after the suspension is heated for 24 hours at 90°C, forming particles of around 50 nm in diameter. The conversion is around 50% with anatase and 15% with rutile under the same conditions. Similar results are obtained by hydrothermal treatment (250°C, 3.5 MPa). Under these conditions, crystallization of barium titanate occurs by a surface reaction of barium on the TiO_2 particles involving the diffusion of Ba^{2+} ions within the surface titanate layer. The reaction is of course the fastest and the most complete when TiO_2 has a less stable crystal structure (anatase instead rutile) and a small size.

In even softer conditions, smaller titanate particles can be obtained from barium and titanium metal-organic compounds. The stoichiometric mixture of barium and titanium isopropoxides, $M(O^iPr)_4$, (^iPr = $-CH(CH_3)_2$), is first heated at refluxing in isopropylic alcohol (^iPrOH) to form, through ether $^iPr_2O$ elimination, the cluster $Ba_4Ti_4O_4(O^iPr)_{16}{\cdot}x^iPrOH$ (Fig. 8.33), an oxoalkoxide in which Ba–O–Ti bonds stable against hydrolysis already exist [78]. (The crystal contains two types of complexes that differ mainly in the number of solvating alcohol molecules, $x = 3$ and 4.) Then the reaction of the cluster with water at room temperature eliminates the organic ligands at the cluster periphery; this allows the subsequent condensation of the clusters and the quantitative precipitation of crystalline $BaTiO_3$ particles with a size of around 5 to 15 nm in diameter.

A different procedure involves hydrolysis of titanium, zirconium, or niobium alkoxides by an aqueous suspension of barium or strontium hydroxide [79]. The low reactivity of the divalent metal alkoxides is thus avoided [80]. In the case

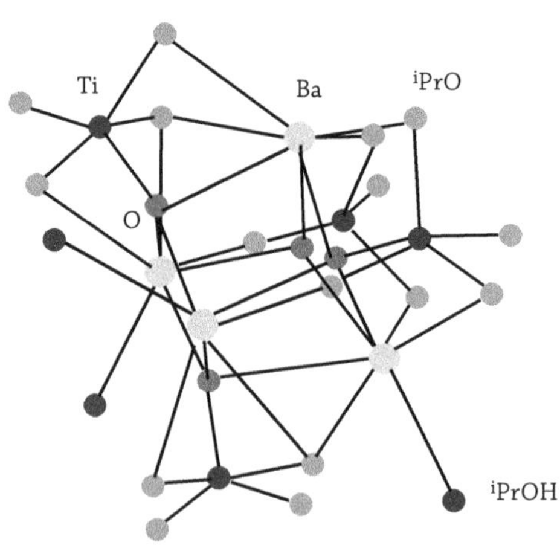

FIGURE 8.33 Crystal structure of the $Ba_4Ti_4O_4(O^iPr)_{16}{\cdot}3^iPrOH$ cluster (^iPr = $-CH(CH_3)_2$).

of barium titanate, the drying of the solid formed at 60°C for 1 hour permits obtaining the perovskite phase. Through this process, different compositions can be easily synthesized, and doping of various phases, for instance, compounds of $Ba_{1-x}Sr_x(Ti_{1-y}Zr_y)O_3$ type, can be carried out.

8.3 Manganese Oxides

Manganese is an abundant element, the 12th (in weight) in the Earth's crust. Manganese oxides are thus widespread. They have a fundamental role in (bio) geochemistry and in the environment, in oceans as well as in soils. The manganese dioxides, which are among the strongest oxidants in the natural medium, play a major role in regulating the redox cycles of most elements [81, 82]. Manganese and its oxides also have an important role in human organisms and in plants, where manganese ions are the active centers in metalloenzymes, especially in photosynthetic systems. They also have the ability to form solid phases in some vegetables [83]. In natural waters, manganese oxides have an important action because of their high adsorption capacity of many metallic elements, thus controlling their distribution and biodisponibility [84]. Thus, the manganese dioxides are also largely used as molecular sieves or ion exchangers to fix metals from industrial effluents. Frequently nonstoichiometric, manganese oxide nanoparticles exhibit interesting electrochemical properties that allow their use as catalysts and as cathodic materials for dry batteries [85]. Most of these dioxides are black in color because Mn^{III} centers resulting from their nonstoichiometry give rise to visible light-induced Mn^{III}-Mn^{IV} intervalence transfers. Because of their black color, manganese oxides were used as pigment in prehistoric decorated caves, such as the Lascaux and Pech-Merle caves in France. From antiquity to modern times, manganese dioxide, which is strongly oxidizing, has been used in the glass industry as a bleaching agent known as the "soap of glassworkers." It allowed oxidation of ferrous ions into ferric ions, causing the green color of glass to become practically colorless in forming Mn^{2+} ions, which also are almost colorless. For these ions having the d^5 electron configuration, the *d-d* transitions are doubly forbidden by selection rules, which avoids light absorption on the visible range.

8.3.1 THE MAIN CRYSTAL PHASES OF MnO_2 DIOXIDE

The crystal chemistry of manganese oxides, especially that of MnO_2, is complicated by a large polymorphism, a high variability in composition, and the important phenomenon of nonstoichiometry. The presence of Mn^{3+} and/or Mn^{2+} ions when substituted for Mn^{4+} ions causes various charge-compensating cations within the structure [5, 86]. This leads to a variability in the names of natural phases and synthetic compounds [87]. The main structural characteristics of manganese oxides and oxyhydroxides that often occur in syntheses in the laboratory are briefly examined in this section.

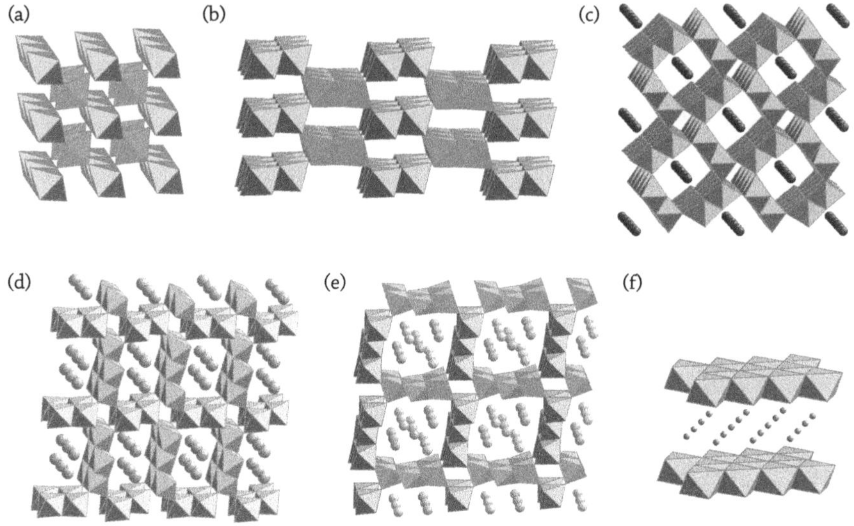

FIGURE 8.34 Crystal structures of MnO_2 polymorphs. (a) pyrolusite, β-MnO_2; (b) ramsdellite, γ-MnO_2; (c) cryptomelane, α-MnO_2; (d) romanechite; (e) todokorite; (f) birnessite, δ-MnO_2; the spheres in channels and between planes represent intercalated cations and/or water molecules.

The most stable phase thermodynamically is pyrolusite (β-MnO_2), and it has a rutile-like crystal structure. Single chains of edge-sharing MnO_6 octahedra are linked together by corners (Fig. 8.34a).

Ramsdellite (γ-MnO_2) contains double chains of MnO_6 octahedra connected by the sharing of corners, as do goethite, α-FeOOH, and diaspore, α-AlOOH (Fig. 8.34b). This phase, which often is not ordered at long distance, contains many defects, essentially ramsdellite/pyrolusite intergrowths and twinning. A typical example is that of nsutite, a natural phase in which octahedra double chains contain single chains that are statistically dispersed. Its X-ray diffraction pattern, similar to that of ramsdellite, does not indicate a superlattice [88]. Nsutite is characterized by the presence of manganous ions. Its formula is $Mn^{4+}{}_{1-x}Mn^{2+}{}_{x}O_{2-2x}(OH)_{2x}$, with $x = 0.06–0.07$ [89]. Also attached to the ramsdellite structural type is the ε-MnO_2 phase, which is characterized by a low rate of intergrowth and a high concentration of microtwinnings, leading to an order of cations along the *a* direction of the crystal and complete disorder in the (*bc*) plane.

Cryptomelane, α-MnO_2, contains the same octahedra double chains as ramsdellite, but they are arranged to form square-section channels, as is found in akaganeite β-FeOOH (Fig. 8.34c), where K^+ ions and water molecules are located. These cations balance out the presence of Mn^{3+} ions (around 20%) in substitution of Mn^{4+} ions. This phase is named hollandite when the compensating cations are mainly Ba^{2+}. There is a large variety of other mineralogical phases of the same type, which are also nonstoichiometric and contain various compensating cations. For example, there is romanechite ($Ba_{0.13}Mn^{IV}{}_{0.74}Mn^{III}{}_{0.26}O_2$) with a crystal structure formed by association of double and triple chains of octahedra (Fig. 8.34d) and todorokite formed by association of triple chains of MnO_6 octahedra (Fig. 8.34e). Its composition is complex and variable. It contains Mn^{IV}, Mn^{III} ions, and Mg^{II}, Ni^{II}, or Al^{III} ions in substitution [86].

Birnessite (or vernadite), δ-MnO_2 ($Na_{0.24}H_2O_{0.72}Mn^{IV}{}_{0.94}O_2$), is a layered structure resulting from the stacking of MnO_6 octahedra sheets (Fig. 8.34f). Various cations (e.g., alkaline, alkaline earth, and ammonium) compensate for the negative charge of the sheets resulting from nonstoichiometry due to vacancies of Mn^{4+} ions and to the presence of Mn^{3+} ions. The name "birnessite" includes a set of phases that are of biogenic origin or synthetic, which, in the nanometric state, exhibit structural modifications due to differences in composition and in stacking of the MnO_6 octahedra sheets [87, 90].

Under the +III oxidation state, manganese forms MnOOH oxyhydroxides. Manganite, γ-MnOOH, is the most stable phase. It is isostructural with pyrolusite, but all the cations are Mn^{3+} and half of the oxygen atoms are hydroxyl ions. A second polymorph is groutite, α-MnOOH, which is isostructural with ramsdellite. A third variety is feitknechtite, β-MnOOH, whose crystal structure has not yet been precisely established (see later in this chapter). These three structures contain Mn^{3+} ions of d^4 electron configuration. The Jahn–Teller effect causes a strong distortion of the MnO_6 octahedra. Haussmanite, Mn_3O_4 ($Mn^{II}Mn^{III}{}_2O_4$), is a mixed Mn^{II}-Mn^{III} compound with a distorted spinel structure. Finally, the hydroxide $Mn(OH)_2$, pyrochroite, adopts the brucite-type structure (§ 4.1.1). It is extremely sensitive to oxidation.

8.3.2 PRECIPITATION OF MANGANESE OXIDES

Because of the lack of soluble aquohydroxo complexes of manganese under the +III and +IV oxidation degrees (§ 8.1), the synthesis of oxides and oxyhydroxides from solutions at relatively low temperature (≤ 95°C) has to proceed by redox reactions involving soluble and stable species of manganese, such as

- oxidation of Mn^{II} in various conditions, abiotic [91–94] or biotic [81, 95];
- reduction of Mn^{VII} with various reactants such as sugars [96], hydrogen peroxide [97], and thiosulfate [98]; and
- mediamutation between Mn^{II} and Mn^{VII}, which appear to be rich in possibilities [99]. The use of hydrothermal conditions diversifies even more the products of all these redox reactions [99–101].

Because of the numerous oxidation states of manganese and the structural variety of its oxides, the dimensional, morphological, and structural characteristics of the solid are closely dependent on the conditions in which the reaction is carried out, especially acidity, composition (Mn^{VII}/Mn^{II} ratio in the case of mediamutation), temperature, and duration of aging in suspension. The nature of the countercations can also have a strong influence on the crystal structure of the oxide formed.

a) Mediamutation

The reaction in solution is formally written as

$$2\left[Mn^{VII}O_4\right]^- + 3\left[Mn^{II}(OH_2)_6\right]^{2+} + 4\,H_2O \rightarrow 5\left[Mn^{IV}(OH)_4(OH_2)_2\right]^0 + 4\,H^+$$

The Mn^{IV} complex, noncharged on the whole acidity range, condenses by olation and oxolation to form MnO_2 oxide (§ 4.1.4) according to

$$\left[Mn(OH)_4(OH_2)_2\right]^0 \rightarrow MnO_2 + 4\,H_2O$$

So, the addition of potassium permanganate in a solution of manganous sulfate at room temperature, in stoichiometric proportions or in excess ($Mn^{VII}/Mn^{II} \geq 2/3$, total concentration in Mn 0.2 mol.l^{-1}), almost immediately triggers the precipitation of a black solid. The acidity of the medium, adjusted by acid or base addition, is a key factor on which the characteristics of this solid depend [99].

1. When the medium is initially acidic (pH ≤ 2), the solid obtained is amorphous and the average oxidation degree of manganese is close to 4. Heating at 95°C for 7 days (the final pH is close to 1) forms cryptomelane, α-MnO_2, in the form of sticks around 30 nm in diameter and 500 nm in length (Fig. 8.35 a). The average oxidation degree of manganese is then close to 3.7, which means that the solid contains K^+ ions (the K/Mn molar ratio is about 0.25) and water molecules ($H_2O/Mn \approx 0.3$). The mechanism involved in crystallization is a combination of a dissolution–crystallization process forming primary nanosticks about 10 nm in diameter with a longitudinal ordered aggregation leading to the final particles (Fig. 8.35a, insert). This aggregation process is clearly evidenced by a comparison of X-ray diffraction and transmission electron

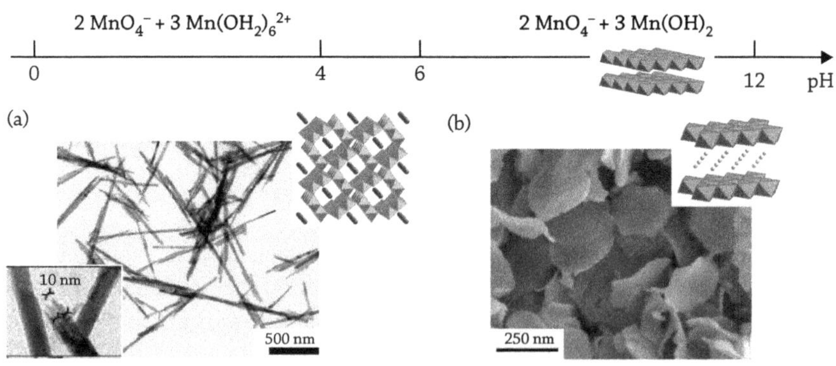

FIGURE 8.35 TEM images of sticks of cryptomelane α-MnO_2 (a) and of platelets of birnessite $K_{0.3}MnO_2$ (b) obtained by thermolysis of the precipitate resulting from mediamutation in acidic and basic media, respectively [from Portehault D., Synthèse par chimie douce en milieu aqueux d'oxydes de manganèse nano-structurés. Thèse, Université Pierre et Marie Curie, Paris, France, 2008].

microscopy data [102]. In the absence of K^+ ions, when the reaction is carried out using a large excess of sodium or lithium permanganate (obtained from $KMnO_4$ solution by ion exchange on a Li-cationic resin), heating forms pyrolusite, β-MnO_2. The Na^+ and Li^+ ions are too small to stabilize the α-MnO_2 phase, and in pyrolusite, the channels of the structure are empty. The particles of pyrolusite are hollow rods with a rather square section (Fig. 8.36), which suggests that their formation results from an ordered aggregation of primary rods [99].

2. In a basic precipitation medium (pH ≥ 8), the black solid immediately formed is crystallized with a birnessite, δ-MnO_2-type structure and with the average composition $K_{0.3}Mn^{IV}{}_{0.7}Mn^{III}{}_{0.3}O_2$. The particles are hexagonal platelets about 300 nm in diameter and 15 to 40 nm in thickness (Fig. 8.35b). Aging under heat (95°C) does not lead to change. In these conditions, the Mn^{2+} ions are initially precipitated as pyrochroite, $Mn(OH)_2$, so that their oxidation by permanganate occurs within the solid phase. The structural and morphological relationship between the two phases, pyrochroite and birnessite, would suggest that oxidation proceeds in situ in the solid state, coupled with the diffusion of K^+ ions in the interlayer space. Na^+ and Li^+ ions do not influence structural control, as they can indeed just as easily as K^+ ions insert between the sheets of the structure [102].
3. In the presence of a large excess of permanganate (Mn^{VII}/Mn^{II} = 20/3, total concentration in Mn 0.2 mol.l^{-1}), two kinetic steps arise successively during the reaction. In strongly acidic medium (with initial pH ≤ 1) as at pH 11, the rapid and nearly complete oxidation of Mn^{II} forms, as previously seen, amorphous spheroidal aggregates and

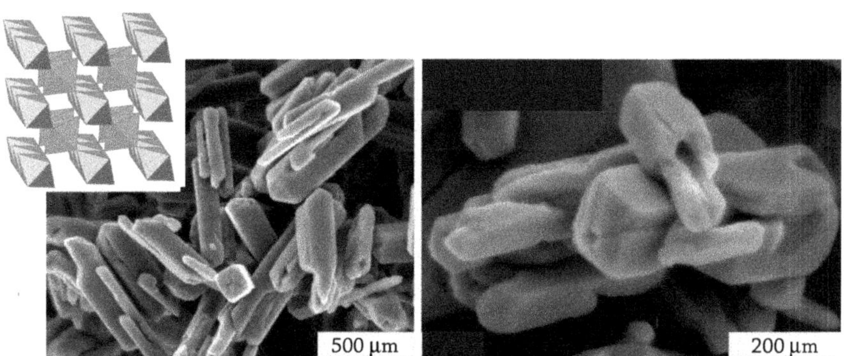

FIGURE 8.36 FESEM images of hollow rods of pyrolusite β-MnO_2 formed by thermolysis of the solid obtained at pH 1.3 from mediamutation reaction of $LiMnO_4$ with $MnSO_4$ [Portehault D., Synthèse par chimie douce en milieu aqueux d'oxydes de manganèse nano-structurés. Thèse, Université Pierre et Marie Curie, Paris, France, 2008].

FIGURE 8.37 FESEM images of nanoparticles immediately formed by oxidation of Mn^{2+} ions with an excess of potassium permanganate (Mn^{VII}/Mn^{II} = 20/3) at room temperature (a) in acidic medium (initial pH 2) and (b) in basic medium (initial pH 11); (c) and (d): particles obtained after heating at 95°C by reaction of the excess of permanganate in the presence of the particles formed in acidic medium (a) and in basic medium (b), respectively [Portehault D., Synthèse].

hexagonal birnessite platelets, respectively (Fig. 8.37 a,b). In a second step occurring during heating at 95°C for a week, the suspensions with an initial pH of 1 still form nanosticks of cryptomelane, while at pH ≥ 2, nanometric objects similar to sponges due to a wrinkled veil are obtained (Fig. 8.37 c,d). In both cases, structural analysis indicates the presence of birnessite [88, 103, 104]. In this stage, all introduced Mn^{2+} ions having been rapidly oxidized, the excess of permanganate reacts with water during the heating of suspensions according to

$$4\left[MnO_4\right]^- + 18\ H_2O \rightarrow 4\left[Mn(OH)_4(OH_2)_2\right]^0 + 3\ O_2 + 4\ OH^-$$

This reaction is slow and causes alkalinization of the medium: for an initial pH of 2, the final pH after heating for 7 days is 10. Consequently, birnessite is the solid phase favorably formed [105–107]. In both cases, the decoration of particles resulting from initial precipitation in acidic (Fig. 8.38c) or basic (Fig. 8.38d) media proceeds by heterogeneous nucleation on existing particles acting as germs. So, the growth of the basal faces of birnessite forms wrinkled small strips about 10 nm thick.

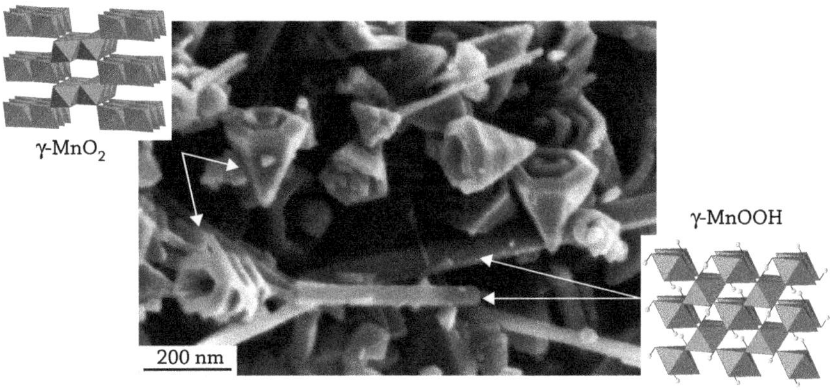

FIGURE 8.38 FESEM image of sticks of manganite, γ-MnOOH, and cones of ramsdellite, γ-MnO_2 [Portehault D., Synthèse].

A similar heteroepitaxy mechanism can be involved in synthesizing composite materials made of different manganese oxides. The elaboration of heterogeneous nanostructures that associate materials different in composition and/or in crystal structure is today an active part of nanotechnology because such composite materials may have original properties. In the case of manganese oxides, the {001} interplanar spacing of birnessite, $d_{001\delta}$ = 0.7 nm, is similar to the {110} d-spacing of cryptomelane, $d_{110\alpha}$ = 0.69 nm (misfit of about 0.1%). Thus, the formation of composites made by epitaxy between a tunnel-based structure (cryptomelane α-MnO_2) and a layered structure (birnessite δ-MnO_2) can be expected. For that, square section sticks of cryptomelane are formed first, as previously indicated, by mediamutation in acidic medium in stoichiometric proportions (Mn^{VII}/Mn^{II} = 2/3, Mn concentration 0.2 mol.l^{-1}, pH = 2, 60°C, 7 days). The particles are then dispersed in water at pH 2, a solution of permanganate is added to the suspension, and the mixture is heated at 95°C for a day. The X-ray diffraction pattern of the particles that are obtained indicates the coexistence of birnessite and cryptomelane, and electron microscopy shows that the growth of sheets of birnessite occurred from the four lateral faces of the cryptomelane sticks, in perfect epitaxy on lengths greater than 500 nm (§ 4.2.3, Fig. 4.53) [108]. Because of the tetragonal symmetry of cryptomelane, four equivalent {110} planes perpendicular between them allow the growth of four platelets of birnessite, with a shape similar to that of an arrow feather.

4. With less potassium permanganate, and especially for the ratio Mn^{VII}/Mn^{II} = 1/4, the redox potential of the system is lower and the mediamutation reaction yields oxyhydroxides corresponding to the +III oxidation state of manganese. The formal reaction in basic medium is written as

$$[MnO_4]^- + 4\,Mn(OH)_2 \rightarrow 5\,MnO(OH) + OH^- + H_2O$$

and in acidic medium, the reaction is

$$[MnO_4]^- + 4\ Mn(OH_2)_6^{2+} \rightarrow 5\ MnO(OH) + 7\ H^+ + 18\ H_2O$$

The acidity has an important role in the nature of the obtained solid phases. The pH is controlled by the addition of potassium hydroxide throughout introduction of reactants, before heating the suspension at 95°C [109, 110].

On all the pH ranges after heating for one day, mixtures are obtained in which manganite, γ-MnOOH, is always present. When the pH is maintained close to 10, feitknechtite, β-MnOOH, is obtained in the form of twinned triangle-shaped particles (§ 4.2.3, Fig. 4.55) [98]. Between final pH 5 and 6, groutite, α-MnOOH, is formed. In acidic medium, at around final pH ≈ 2, manganite mixed with ramsdellite γ-MnO_2 is formed. Manganite is in the form of sticks with a variable length from 400 nm to some micrometers, and ramsdellite forms hexagonal-based cones (Fig. 8.38). These cones result from ordered aggregation of rods by their {110} lateral faces, according to a mechanism described in § 4.2.3, Fig. 4.57.

5. With even less permanganate (ratio $Mn^{VII}/Mn^{II} = 2/30$), heating at 95°C for 1 day forms in alkaline medium (final pH ≈ 11) rhombohedral particles about 30 nm in diameter of hausmanite, Mn_3O_4 ($Mn^{II}Mn^{III}{}_2O_4$), while in acidic medium (final pH ≤ 2), nanorods of pyrolusite, β-MnO_2, are obtained (Fig. 8.39) [99]. The amount of permanganate is not enough to oxidize the two-thirds of Mn^{II} in alkaline medium, and the oxidation of $Mn(OH)_2$ has to occur by air during aging. In acidic medium, the oxidation of Mn^{2+} ions is complete. The adsorption of Mn^{2+} ion on the solid early formed from

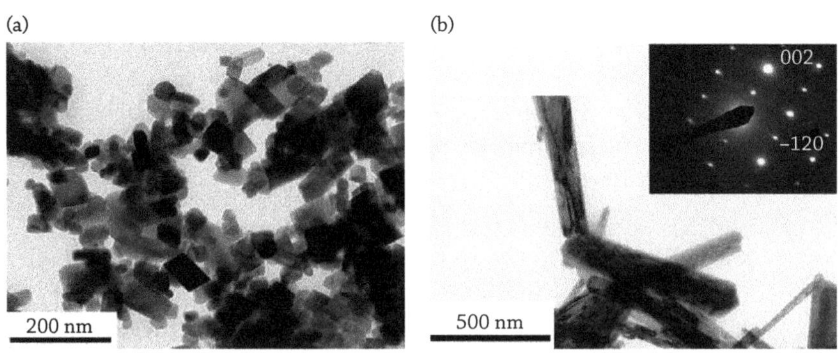

FIGURE 8.39 TEM images of (a) haussmanite Mn_3O_4 nanoparticles formed by heating the Mn^{VII}-Mn^{II} mixture ($Mn^{VII}/Mn^{II} = 2/30$, pH 11) and (b) rods of pyrolysite obtained at pH = 1 [Portehault D., Synthèse].

the mixing of reactant likely increases their reducing power because of the strong stabilization of the oxidized form, which favors oxidation by water or dissolved oxygen during heating. The Mn^{3+} and/or Mn^{4+} ions resulting from oxidation has an ability to hydroxylate and to form a solid far greater than Mn^{2+} ions. This means that they are also much more able to be structurally integrated into the solid in which they are adsorbed. The oxidation of manganous ions can then occur either by a reactant in solution (water molecule or dioxygen) or by direct electron injection in the supporting solid (§ 5.4.3). Such a phenomenon has previously been considered to explain the strong reducing power of ferrous ions adsorbed onto goethite and other iron oxides [111–113]. A quite analogous mechanism is also involved in the transformation of a synthetic birnessite by Mn^{2+} ion adsorption in solution [114]. After aging for some days at room temperature at different pH values, birnessite transforms into different phases: ramsdellite (pH 2.4), cryptomelane (pH 4), groutite (pH 6), and manganite (pH 8). The transformation is speeded up when the concentration in Mn^{2+} ions increases, but as the experiments were not carried out under anaerobic conditions, it is difficult to know the real mechanism involved, either surface-catalyzed oxidation by oxygen or electron injection into the solid. The fact that adsorbed Mn^{2+} ions yield Mn^{IV} in acidic medium and Mn^{III} at $pH > 4$ could mean that dissolved oxygen is the electron receiver, but it is also possible that the structural reconstruction results from electron injection in the early solid.

b) Mn^{II} Oxidation

The mediamutation almost instantaneously forms an ill-organized solid. Its structuration occurs during heating in suspension. To have better control of the formation of the solid, it may be interesting to use a powerful oxidant of Mn^{2+} ions involving slow reactions. The precipitation can be carried out with many reactants, either in acidic solution with the $[Mn(OH_2)_6]^{2+}$ complex or, at $pH \geq 8$, in heterogeneous medium with the hydroxide $Mn(OH)_2$. The $[Mn(OH_2)_6]^{2+}$ complex is stable in solution, while the hydroxide, white when formed, burnishes rapidly in suspension and oxidizes into hausmanite, Mn_3O_4, under air. Depending on the power of the oxidizer and the composition of the system, oxidation is more or less complete and powerful, allowing diverse phases to form.

The addition of hydrogen peroxide to a solution of $MnCl_2$ ($H_2O_2/Mn = 1$) at pH 7 does not cause precipitation, but at pH 9, the white solid immediately formed turns rapidly to brown. After heating for 2 days at 95°C, hausmanite, Mn_3O_4, is obtained. However, 25% of manganese remain in solution as Mn^{2+} ions, which means that, in these conditions, only 50% of the initial amount of manganese have been oxidized into Mn^{III}. With a bigger amount of hydrogen peroxide ($H_2O_2/Mn = 10$), oxidation is quite quantitative in forming manganite, γ-MnOOH [115].

With a more powerful oxidizer such as persulfate, $S_2O_8^{2-}$, and at 60–95°C, the reaction quantitatively yields Mn^{IV} on a large acidity range. The countercation plays an important role in the crystal structure of the solid. As for mediamutation, oxidation by persulfate leads to strong acidification of the suspension:

$$Mn^{2+} + S_2O_8^{2-} + 2\,H_2O \rightarrow MnO_2 + 2\,SO_4^{2-} + 4\,H^+$$

In basic medium (initial pH ≥ 8) with potassium persulfate, $Mn(OH)_2$ formed at pH 10 rapidly oxidizes into birnessite if the pH is maintained at 10 by addition of potassium hydroxide during heating. If not controlled, the pH drops to 2 after one day at 95°C, and cryptomelane α-MnO_2 is obtained [116]. At initial pH < 8, the reaction is slower. A black precipitate containing a mixture of ramsdellite γ-MnO_2 and cryptomelane appears after 1 hour of heating at 60°C [116]. As previously seen, the presence of the potassium countercation favors the formation of cryptomelane, which always contains about 25% of K^+ ions. With ammonium persulfate, ramsdellite is exclusively formed. It transforms into pyrolusite, β-MnO_2, after 7 days of heating at 95°C [124]. Mn^{2+} ion oxidation by ammonium persulfate in hydrothermal conditions (120°C, 12 hours) also forms rods of pyrolusite 40 to 100 nm in diameter and 2.5 to 4 μm in length [117, 118]. On the other hand, under these conditions, addition of an excess of ammonium sulfate allows cryptomelane to stabilize in the form of needles 5 to 20 nm in diameter and 5 to 10 μm in length. It is likely that the influence of ammonium ions on the crystal orientation of the MnO_2 phase is closely dependent on temperature and also on acidity.

Catalysis with silver allows oxidation of Mn^{2+} ions by persulfate at room temperature. The mechanism of catalysis seems to involve SO_4^- radicals arising from persulfate decomposition that could form Ag^{3+} ions, oxidizing in turn Mn^{2+} into Mn^{4+} ions [119]. Special morphologies of cryptomelane α-MnO_2 can in this way be obtained at room temperature [120, 121]. The addition of silver nitrate to the equimolar mixture of manganous sulfate and ammonium persulfate forms urchin-like structures about 2 μm in diameter (Fig. 8.40a). In the presence of silver foil instead of silver nitrate, spherical structures interconnected by nanowires and forming a spherical network are obtained (Fig. 8.40b), while periodical short applications of ultrasound form a network of entangled nanowires of cryptomelane (Fig. 8.40c). Without silver ions, no solid is formed at room temperature. Thus, the catalysis effect acts not only on the reaction rate, but also on the morphology of the obtained objects. In the case of urchin-like structures, two steps are clearly identified during their growth: after 3 hours, spherical aggregates 600 to 800 nm in diameter made of small nanoparticles are formed, which after 6 hours transform into spheres around 1.5 μm in diameter completely covered by small rods. It seems that when the reaction starts, the high concentration of reactants and the catalyst leads to the fast growth of small particles aggregated into spherical clusters. When the reactants deplete, the concentration of silver may also decrease, for instance, in being trapped in the large channels of cryptomelane, and the reaction slows down. Thus, heterogeneous nucleation occurs at the surface of the aggregates, leading to a radial growth of rods that form the spines observed on

FIGURE 8.40 FESEM images at different magnifications of various morphologies of cryptomelane α-MnO_2 obtained at room temperature from Mn^{2+} ion oxidation by ammonium persulfate (a) in the presence of Ag^+ ions after 2 days; (b) in the presence of a silver foil after 3 days; (c) in the presence of a silver foil after 5 days and periodical ultrasonic treatment (reproduced with permission from [118] Copyright 2005 American Chemical Society); (d) in the presence of Ag^+ ions in sulfuric acid (reproduced from [120] with permission of The Royal Society of Chemistry).

the final particles. When silver foil is used as a catalyst, the Ag^+ ions are released at lower concentration than previously, thus favoring the growth of rods that form in the first spherical dense aggregates. As Ag^+ ions are continuously supplied, long monodimensional structures can be formed, allowing connections between adjacent spherical aggregates (Fig. 8.40b). The epitaxial growth of such long fibers requires a calm environment and cannot occur in stirred medium. However, short and periodical ultrasonic treatment does not disrupt the growth, but rather allows these long fibers to disperse and promotes the formation of loosely packed spherical aggregates connected by entangled long fibers (Fig. 8.40c). When the oxidation of Mn^{2+} ions by ammonium persulfate catalyzed by silver nitrate is carried out in acidic medium, a cryptomelane core-shell structure somewhat different from the urchin-like structure is obtained (Fig. 8.40d) [120]. Spheres about 2 µm in diameter resulting from dense aggregation of nanoparticles are covered with a spherical shell made of nanorods 30 to 40 nm in length, aligned in a compact and ordered manner. The difference in morphology with urchin-like particles seems to be due to the slowdown of oxidation in acidic medium. The catalytic mechanism could be somewhat different under these conditions, involving the transient formation of Mn^{3+} ions to which the disproportion into Mn^{4+} and Mn^{2+} ions could be slower as the acidity increases [119, 121].

Spiny spherical architectures, α-MnO_2 cryptomelane-based (Fig. 8.41), are also obtained from Mn^{2+} ion oxidation by potassium bichromate in acidic medium under hydrothermal conditions [122]. At 120°C, clusters about 4 µm in diameter result from assembling square-section rods 200 nm thick and 2 µm long (Fig. 8.41a), while at 180°C, clusters 5 µm in diameter in the form of the dandelion flower with hairs 40 nm thick and 2.5 µm long are obtained (Fig. 8.41b). In both cases, the growth direction of the rods is the (001) axis, which is also that of the channels of the structure. In contrast to the structures made of a core formed by

FIGURE 8.41 SEM images at different magnifications of clusters obtained by oxidation of manganous sulfate by potassium bichromate in sulfuric acid in hydrothermal conditions (a) at 120°C (12 hours) and (b) at 180°C (12 hours) (reproduced with permission from [122] Copyright 2005 American Chemical Society).

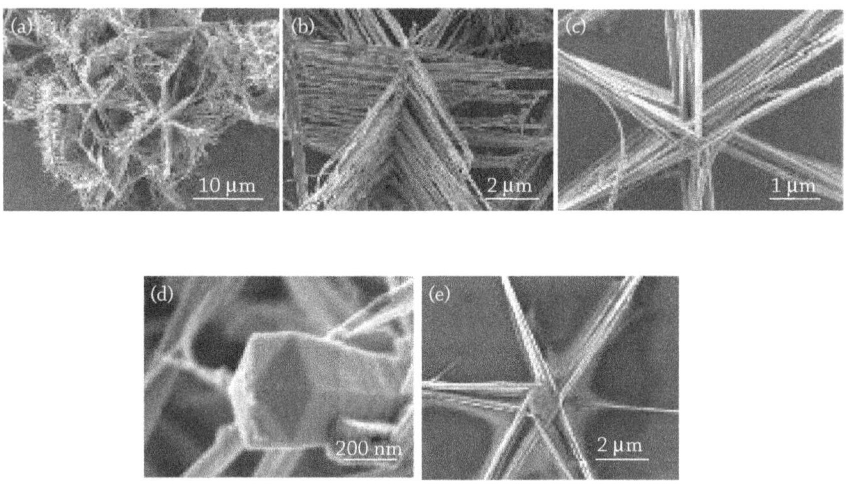

FIGURE 8.42 MEB images of star-like ε-MnO_2 structures obtained by oxidation of Mn^{2+} ions by $[ClO_4]^-$ ions at 160°C for 24 hours (a), (b) and (c); (d) image of the central axis of the structures; (e) image of the branches formed after heating for 4 hours (courtesy of S.L. Suib, Univ. of Connecticut, USA).

dense agglomerates of nanoparticles (Fig. 8.40), the rods forming these clusters are derived from the same center. Such a morphology requires accurate control of nucleation and the growth steps of the solid. This control is likely provided by the softness of the redox reaction because of the small difference in potential of the two redox couples (Cr^{VI}/Cr^{III} 1.33V, Mn^{IV}/Mn^{II} 1.23V).

The oxidation of Mn^{2+} ions by perchlorate in hydrothermal conditions at 160°C yields particles of ε-MnO_2, a phase in which the crystal structure is made of pyrolusite and ramsdellite intergrowths (single and double chains of MnO_6 octahedra, Fig. 8.34) [101]. Surprisingly, the particles consist of the stacking of six-armed stars on an hexagonal central axis from which are derived six ribbon-shaped branches (Fig. 8.42). The branches are 50 to 300 nm wide, 10 nm thick, and several µm long. Between them is an angle very close to 60°. The central axis and the branches have the same ε-MnO_2 crystal structure, and the whole forms an object 5 to 10 µm long.

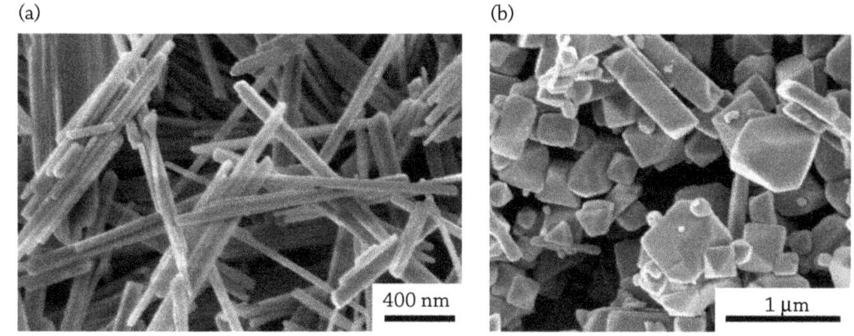

FIGURE 8.43 FESEM images of (a) manganite and (b) hausmanite nanoparticles obtained from reduction of permanganate by thiosulfate at pH 11 at 60°C and 95°C, respectively [Portehault D., Synthèse].

The formation of such an architecture results from at least two nucleation steps. After heating the solution at 160°C for 2 hours, a small amount of solid is obtained, formed of nanopyramids about 500 nm to 1 μm in size similar to the central axis of the products obtained after 24 hours of hydrothermal treatment (Fig. 8.42d). After 4 hours, the star-like structures are formed (Fig. 8.42e). The growth of the branches results from a heterogeneous nucleation process carried out on the facets of the pyramids. Clearly, the perchlorate ions have a specific effect (but one that has not yet been identified) on the oxidation of Mn^{2+} ions, yielding the unique type of crystal growth.

c) Mn^{VII} Reduction

Because of its strong oxidizing power, permanganate is reducible by many reactants in solution, especially organic compounds such as glucose [96]. Apparently, no peculiarities are obtained in this system. Notably, however, the reduction of permanganate by sulfite ions in alkaline medium (pH ≈ 11) leads to birnessite δ-MnO_2 at 60°C and to hausmanite Mn_3O_4 at 95°C [116]. Under similar conditions, the reduction by thiosulfate at 60°C leads to manganite (Fig. 8.43a) with, transiently, formation of feitknechtite (§ 4.2.3b). At 95°C, hausmanite is also the ultimate stage of the reduction (Fig. 8.43b).

The reduction of permanganate by hydrogen peroxide in nitric acid or in acetic buffer (4.7 ≥ pH ≥ 2) yields, after refluxing, sticks or rods of cryptomelane (α-MnO_2). The particles are tens of nanometers in width and hundreds of nanometers to tens of micrometers in length [97]. The rise in concentration of hydrogen peroxide (between 0.25 and 5% in volume) increases the reaction rate and decreases the particle size. At higher concentration, manganite (γ-MnOOH) becomes the main solid phase.

The reduction of permanganate by water in hydrothermal conditions (240°C, 4 days) in the presence of different countercations, such as magnesium, or sodium, provides a good example of structural control of MnO_2 oxide [123]. Magnesium leads to the formation of platelets of a todorokite phase (Fig. 8.44a), with the composition

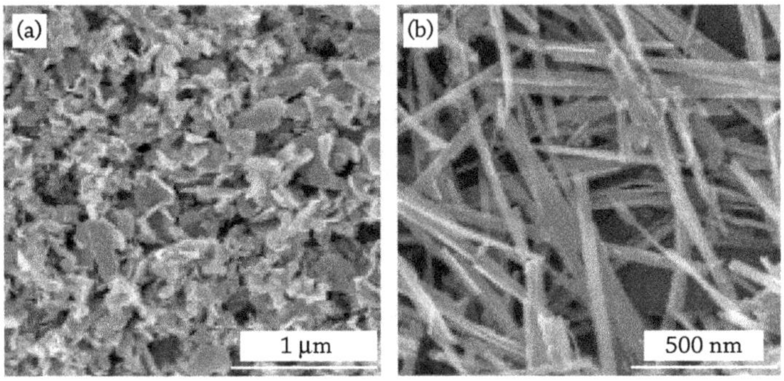

FIGURE 8.44 SEM images of particles obtained from hydrothermal treatment at 240°C for 4 days of aqueous solutions of (a) magnesium permanganate and (b) sodium permanganate (reproduced from [123] with permission of The Royal Society of Chemistry).

$Mg_{3.17}Mn_{5.05}O_{12}$, 4.52 H_2O containing square-section channels formed by triple MnO_6 octahedra chains. With sodium, a phase containing rectangular-section channels formed by double and quadruple MnO_6 octahedra chains are obtained in the form of lamellae 40 nm in width and some μm in length (Fig. 8.44b). Two factors are critical in the structural control: the temperature of the treatment and the nature of the cation. It has been previously seen that the reduction of permanganate causes an important increase in the pH of the medium:

$$4[MnO_4]^- + 18\ H_2O \rightarrow 4\left[Mn(OH)_4(OH_2)_2\right]^0 + 3O_2 + 4\ OH^-$$

which leads to the formation of a birnessite-type solid. When the temperature of the hydrothermal treatment is about 180–200°C, this layered phase is stable with either Mg^{2+} or Na^+ cations and exhibits the same X-ray diffraction pattern. At 240°C, however, this phase transforms. If the reaction is carried out with magnesium permanganate, the Mg^{2+} ions, in forming the $Mg(OH)_2$ hydroxide, allows a decrease in the pH around neutrality while with sodium permanganate, the Na^+ ions do not have the ability to regulate the pH, which remains high—close to 13—during the length of the treatment. Together with acidity, the size and charge of the cations act as a template specifically influencing the structuration of the oxide. It seems that other permanganate salts (Rb^+, Cs^+ and Cu^{2+} or Zn^{2+} ions) lead to similar structures of manganese oxides, with channels of different sizes [123].

8.4 Zirconium Oxides

Zirconium is an element of the titanium group in the classification. It forms zirconia, ZrO_2, which is used for a wide variety of applications. In particular, zirconia is the main component of technical ceramics because of its chemical inertness and its very high thermal stability. Many efforts have been devoted to the synthesis of zirconia by precipitation because this process yields, at the difference of the solid chemistry, small-sized particles, whether crystallized or not. These particles are interesting precursors to the elaboration and casting of ceramic materials.

8.4.1 THE CRYSTAL VARIETIES OF ZIRCONIA

Under atmospheric pressure, zirconia exists in three crystalline phases. Monoclinic zirconia, the low-temperature phase, is thermodynamically stable up to 1100°C. Zr^{4+} ions adopt a 7-coordination in an irregular polyhedron, with Zr-O distances varying from 2.04 to 2.26 Å and variable O-Zr-O angles (Fig. 8.45). Tetragonal zirconia is stable from 1100°C to 2380°C, and at higher temperatures, the cubic phase persists up to melting at 2710°C. The cubic phase has the fluorite CaF_2 structural type in which the Zr^{4+} ions adopt the cubic 8-coordination and the O^{2-} ions the tetrahedral 4-coordination. This structure can be simply described as a cubic stacking of the O^{2-} ions in which the Zr^{4+} ions occupy half of the cubic sites. Finally, the tetragonal phase can be considered a distortion of the cubic structure, with a slight extension along the *c* axis involving an alternative move apart and together of oxide ions with regard to zirconium ions (Fig. 8.45). The nanometric size of particles and hydration may, however, modify the relative stabilities of these different crystal phases (§ 1.1.3).

Alloying or doping elements such as Mg^{2+}, Ca^{2+}, Y^{3+}, and lanthanides Ln^{3+} allow modification of the relative stabilities of these phases. With transition temperatures strongly lowered, the tetragonal and cubic phases can become stable at room temperature. The origin of stabilization lies in the oxygen vacancies resulting from the charge balance and allowing the framework relaxation by atomic displacement. It is typically the case for the "stabilized" cubic zirconia that Zr^{4+} ions are substituted by Y^{3+} ions ($Zr_{1-x}Y_xO_{2-x/2}$) or Ca^{2+} ions ($Zr_{1-x}Ca_xO_{2-x}$). The oxygen vacancies allow mobility of oxide ions within the framework, giving an important ionic conductivity to these phases at high temperature.

8.4.2 PRECIPITATION OF ZIRCONIA

Addition of a base to a solution of a zirconium salt (oxychloride, oxynitrate) at room temperature causes the immediate precipitation of a nonstructured, gelatinous, and highly hydrated solid that is not the hydroxide $Zr(OH)_4$ (§ 4.1). That is an oxyhydroxide of ill-defined composition $ZrO_{2-x}(OH)_{2x}$,yH_2O, often named

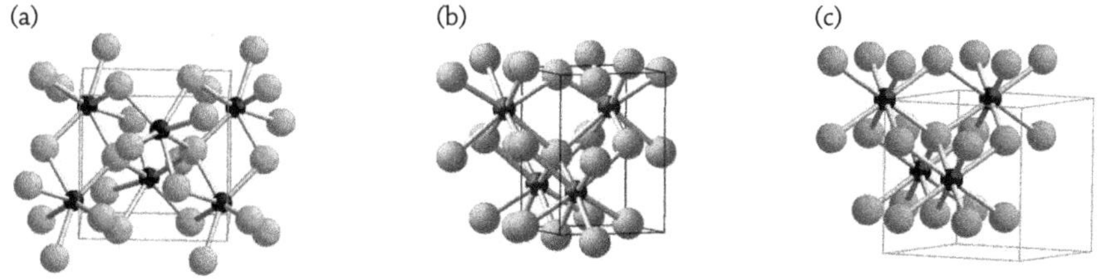

FIGURE 8.45 Crystal structure of (a) monoclinic, (b) tetragonal, and (c) cubic zirconia.

hydrated zirconia. Depending on the pH of precipitation and the nature of the base (ammoniac, sodium, or potassium hydroxides), the solid contains variable amounts of adsorbed anions or cations to balance the surface charge of particles. Chloride and nitrate can be eliminated by washing at controlled pH [124]. On the other hand, sulfate is complexing enough to remain in the coordination sphere of zirconium within the hydroxylated phases, forming basic salts $Zr_a(OH)_b(SO_4)_c(OH_2)_d$ with diversified composition and crystal structures. For this reason, zirconium oxysulfate is not currently used as a zirconia precursor.

It is sometimes assumed that amorphous zirconia could result from rapid condensation by olation of the neutralized tetramer $[Zr_4(OH)_{16}(OH_2)_8]^0$ deriving from the polycation that preexists in acidic solution (§ 8.1). X-ray and neutron diffractions suggest, however, that the ordering at short distance is characteristic of that existing in tetragonal zirconia [125], in which the coordination of zirconium is cubic, while it is antiprismatic in the tetramer. It is therefore doubtful that the structure of the tetramer is preserved within the solid.

Crystallization of the amorphous solid is induced by thermal treatment. It can occur by calcination of the dried precipitate between 400 and 600°C, leading to an increase in the particle size by sintering. Crystallization may also happen by thermohydrolysis or by hydrothermal treatment of suspensions. The obtained crystal structure is highly dependent on the acid–base conditions: monoclinic zirconia in acidic medium, tetragonal and/or cubic zirconia in neutral and alkaline medium.

Refluxing of a molar solution of oxychloride at pH between 1 and 2.5 causes important acidification: the pH lowers from 2.5 to 0.4 after heating for 2 days. This corresponds to hydroxylation of zirconium and leads to precipitation after about 20 hours of monoclinic zirconia [126]. Depending on conditions, nonaggregated spheroidal nanoparticles about 5 nm in mean size [127] or small rods forming cubic aggregates around 50 nm in size [128] are obtained (Fig. 8.46). A similar result is obtained by heating the suspension of amorphous zirconia, provided that the solid

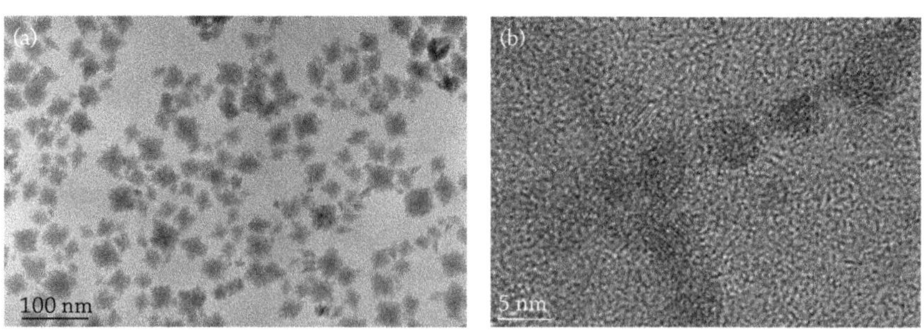

FIGURE 8.46 TEM images of monoclinic zirconia nanoparticles obtained by thermolysis of zirconium oxychloride (a) at 100°C for 3 days (0.075 mol.l^{-1}) (reproduced from [Hu M.Z.C., Harris M.T., Byers C.H., Nucleation and growth for synthesis of nanometric zirconia particles by forced hydrolysis, J. Colloid Interface Sci. 1998, 198, 87–99] with permission of Elsevier); (b) at 95°C for 2 days (0.5 mol.l^{-1}, pH 0.85) [127].

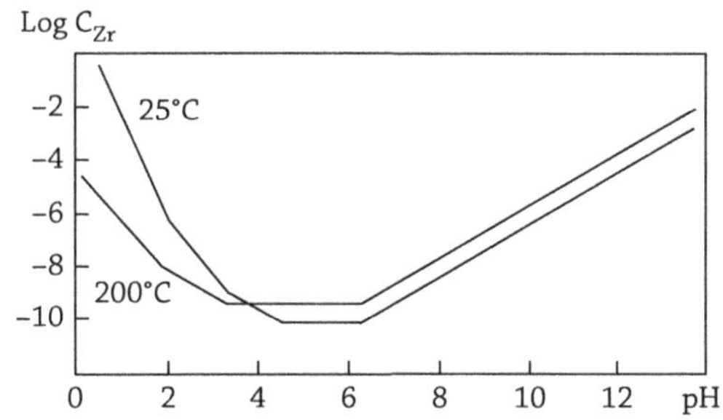

FIGURE 8.47 Theoretical solubility of amorphous zirconia at 25°C and of the monoclinic phase at 200°C as a function of pH.

was not previously washed and the chloride ions were eliminated. Indeed, when they are eliminated, the refluxing no longer causes either acidification of the suspension or zirconia crystallization [126]. That shows, on the one hand, that the chloride ions present within the amorphous solid effectively balance a positive charge, which is annihilated by hydroxylation in acidifying the medium during heating. On the other hand, it shows that the structural transformation, which requires acidification, has to proceed according to a dissolution–crystallization mechanism. Hydrothermal heating at 110°C of acidic solutions (pH 0.5) of oxynitrate similarly yields nanoparticles of monoclinic zirconia after about 20 hours [129].

At higher pH, the precipitation at the same temperature is much more rapid. Hydrothermal treatment of oxynitrate at 110°C forms tetragonal zirconia after 1 to 2 hours at pH 3.6 as well as pH 12 [129]. This phase represents all the zirconium amount present in the reactor, but prolonged heating causes its transformation into the thermodynamically stable monoclinic phase. The transformation is accelerated when the pH is raised. That shows the tetragonal phase nucleates more easily than the thermodynamically stable phase, and the t-ZrO_2 → m-ZrO_2 transformation occurs according to dissolution–crystallization. The formation of the monoclinic phase is favored, as the solubility of zirconia is important (Fig. 8.47).

A globally similar behavior is observed in hydrothermal conditions at a lower temperature (95°C) [130], but the marked kinetic delay of crystallization causes the coexistence of diverse phases, which complicates interpretation of the results. At higher temperature (240°C), hydrothermal heating of amorphous zirconia in a solution of soda forms, after 24 hours, rods of tetragonal zirconia 100 nm in length and 50 nm in thickness (Fig. 8.48). They transform after 72 hours into rods of monoclinic zirconia of the same thickness but twice as long [131]. When the hydrothermal liquid is neutral water, the t-ZrO_2 → m-ZrO_2 transformation is strongly restricted, providing evidence of the requirement of a noticeable solubility so that the dissolution–crystallization occurs. At 300°C, similar results are obtained with the formation of pure monoclinic zirconia with a mean particle size of 40 nm in concentrated soda (30 wt. %). When the reaction is achieved in neutral water, monoclinic and tetragonal zirconias coexist, with particle sizes of 17 and 15 nm, respectively [132].

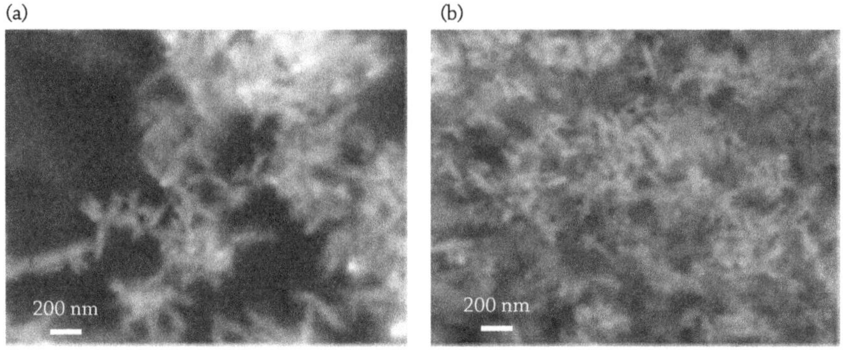

FIGURE 8.48 SEM images (a) of tetragonal zirconia nanoparticles obtained by hydrothermal treatment for 24 hours at 240°C of amorphous zirconia precipitated from $ZrOCl_2$ in NaOH 0.5 mol.l^{-1}; (b) nanoparticles of monoclinic zirconia formed after heating the suspension for 72 hours.

The formation of the tetragonal phase occurs only if amorphous zirconia is initially present in the medium hydrothermally treated, and the am-ZrO_2 → t-ZrO_2 transformation seems to occur in situ, topotactically in the solid state [132]. That is in accordance with the fact that the ordering at short distance in amorphous zirconia deduced from X-ray and neutron diffraction is characteristic of that existing in tetragonal zirconia [125] and that the morphology of the amorphous precursor is roughly preserved following the transformation into tetragonal zirconia [129].

In strongly alkaline medium (NaOH or KOH 20 wt. %), at refluxing or in hydrothermal conditions at 110°C, amorphous zirconia suspension forms the cubic phase [126, 133]. In the presence of $CaCl_2$, heating of oxychloride at pH 10 for 2 hours at 220°C also yields pure cubic zirconia in the form of nanoparticles 7 nm in mean size [134]. It is not a stabilized zirconia (§ 8.4.3) because the medium is insufficiently alkaline. The molar calcium content of the solid is lower than 1%, and the crystal–cell parameter (a = 0.526 nm), which does not vary with the amount of calcium introduced for the synthesis, is very close to that of the pure cubic ZrO_2 phase (a = 0.512 nm). The cubic phase can also be easily formed by hydrothermal treatment at 180°C for 18 hours in strongly alkaline medium (NaOH 10 mol.l^{-1}) of an amorphous zirconia obtained from hydrolysis of zirconium isopropoxide ($Zr[OCH(CH_3)_2]_4$) [135]. The particles, 5–6 nm in size (Fig. 8.49), are exempt of sodium. In NaOH 3 mol.l^{-1}, particles of the monoclinic phase are obtained, and in NaOH 5 mol.l^{-1}, a mixture of the monoclinic (80%) and cubic (20%) phases is formed.

In summary, hydroxylation of zirconium in solution rapidly forms an amorphous zirconia. It crystallizes by heating or hydrothermal treatment, according to mechanisms depending on acidity, that is, the solubility of the solid: dissolution–crystallization, to form the thermodynamically stable monoclinic phase; or in situ in the solid-state transformation in the minimum of solubility, to form the tetragonal zirconia, the phase of lowest activation energy. In turn, dissolution–crystallization can transform this phase into the monoclinic phase, the mechanism being promoted by an increase in the duration of thermal treatment, the temperature, and the alkalinity of the medium. Apart from the differences in crystal

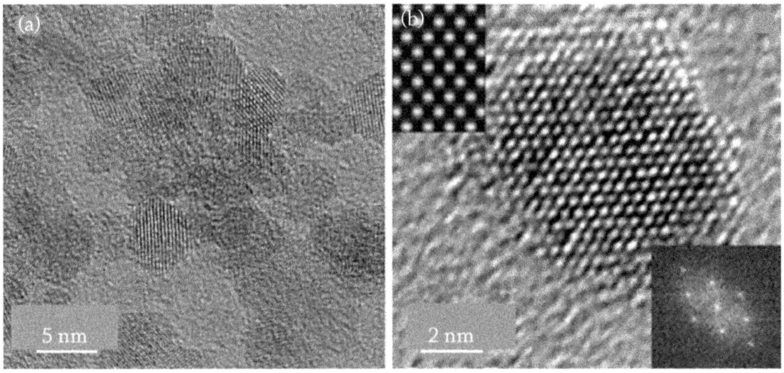

FIGURE 8.49 TEM images of (a) cubic zirconia nanoparticles formed at 180°C in NaOH 10 mol.l^{-1}; (b) a monocrystalline nanoparticle along [101] axis projection (upper left, simulated HRTEM image; bottom right Fourier transform) (adapted from [135]).

structures, this system behaves similarly to others such as iron (III), aluminum, and titanium, of which the high reactivity in solution toward hydroxylation and precipitation more frequently leads to an initial amorphous solid phase.

An interesting process helps form stable aqueous dispersions of tetragonal zirconia nanoparticles [136]. It consists of heating at 110°C for 24 hours a solution of zirconium oxychloride with methacrylic acid, MA (MA = CH_3CH_2CCOOH, 1 mmole MA per gram of ZrO_2) and then neutralizing this solution by the addition of concentrated soda up to pH 10 in an ammonium bicarbonate–triethanolamine buffer (TEA, 1 mmole per gram ZrO_2). The mixture, aged for 18 hours at RT, forms a transparent grayish sol, which is flocculated at pH 5 by the addition of hydrochloric acid. The washed solid is resuspended at pH 11.5 and hydrothermally heated at 110°C for 15 hours. The process yields pure tetragonal zirconia as nanoparticles 11 nm in mean size. The initial complexation of the precursor by methacrylic acid is essential to obtain nonaggregated amorphous zirconia particles. During the hydrothermal treatment, methacrylate is substituted by TEA, which remains linked in avoiding the growth of particles. They can then easily be dispersed in water by the addition of citric or tartaric acid.

8.4.3 SYNTHESIS OF STABILIZED ZIRCONIA

Pure zirconia undergoes phase transitions and changes in volume during thermal cyclings that cause cracks in ceramic materials. Zirconia containing as much as 8% in yttrium oxide or 20% in calcium oxide forms a solid solution of fluorite crystal structure stable from RT up to melting, beyond 2000°C. With a strong chemical inertness and a high conductivity by oxygen ions, this so-called stabilized zirconia is a high-performance ceramic material, used in particular as solid electrolyte in fuel cells and as electric resistance in high-temperature ovens. However, the synthesis of nanoparticles allowing the elaboration of ceramic raises questions about the synthesis of mixed oxides that are still encountered in the synthesis of ferrites (§ 7.2.4) and barium titanate (§ 8.2.4).

Pechini's process (§ 4.1.6), which consists of copolymerized polyethylene glycol and citric acid by polyesterification in the solution of cations and then calcinating the resin, has been successfully used in the elaboration of yttrium-stabilized zirconia in the form of nanoparticles of tetragonal (400°C) and cubic (700°C) crystal structures [137].

Hydrothermal treatment at 110°C for a week of the amorphous zirconia suspension, freshly precipitated and containing commercial powder of yttrium oxide (6 to 10% in molar ratio) in the presence of sodium carbonate Na_2CO_3 in high concentration (3 $mol.l^{-1}$), yields the cubic solid solution as particles 5 nm in mean size [138]. With soda, the mixed phase does not form, and hydrothermal treatment yields monoclinic or tetragonal zirconia in the presence of Y_2O_3. Sodium carbonate allows both a high alkalinity and a sufficient solubility of yttrium oxide to form the mixed phase. The soda and sodium carbonate mixture, on the other hand, accelerates crystallization of the stabilized zirconia.

Coprecipitation of zirconium and yttrium inevitably causes their segregation in distinct solid phases because of the distance between their pHs of precipitation (pH ≥ 0 for zirconium, pH ≥ 7 for yttrium) and their difference in solubility (Fig. 8.50). A marked heterogeneity of the mixed crystalline phases obtained from thermal treatment of the coprecipitate can therefore be expected. The attempt should then be made to adjust the hydroxylation of the two cations in using a common complexing ligand. Oxalate is inappropriate because on the whole acidity range, at low concentration in oxalate, the cations precipitate mixtures of oxalic complexes and oxyhydroxides, while at higher concentration in oxalate at pH < 8, zirconium is solubilized as anionic oxalato complexes and yttrium precipitates as an oxalato complex [139]. With acetylacetone (acacH), zirconium is efficiently complexed between pH 2 and 6. At higher pH, it precipitates an oxyhydroxide, while the trivalent cations form insoluble acetylacetonates. In an unexpected way, acetylacetonate (acac) forms with the mixture of Zr^{IV}, Y^{III}, Nd^{III} cations ([Zr] = 0.1

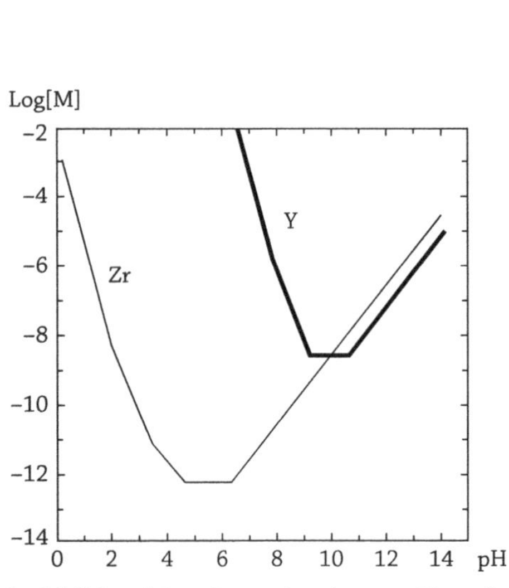

FIGURE 8.50 Theoretical solubilities of zirconium and yttrium at 25°C as a function of pH.

$mol.l^{-1}$, [Y] = [Nd] = 0.025 $mol.l^{-1}$, [acac] = 0.1 $mol.l^{-1}$) at $6 \leq pH \leq 7$, a sol that is stable during a few days, made of nanoparticles about 4–5 nm in mean diameter and containing nearly 100% of the cations. The acac:Zr ratio is critical because with an insufficient amount of acac, the sol is unstable, and with an excess amount, precipitation of the complex $Zr(acac)_4$ occurs.

A careful physicochemical analysis of this system [139, 140] shows that the sol contains nanoparticles of amorphous zirconium oxyhydroxide stabilized against aggregation by acac adsorption. Acac forms a surface layer allowing steric protection toward the van der Waals attractive forces between particles [141]. The trivalent cations are themselves adsorbed on these particles, thanks to the ligand exchange with surface acac. They thus reinforce the stability of the dispersion of nanoparticles in bringing positive charges to their surface. Because of the very small particle size, the trivalent cations can be adsorbed in totality. After some weeks of aging, the nanoparticles aggregate and form a homogeneous gel. Its thermal treatment at 600°C for 3 hours leads to crystallization of the cubic zirconia (Fig. 8.51).

The diffraction pattern of the calcined gel exhibits finer and more intense peaks than those of the calcined solid of the same composition resulting from coprecipitation without acac. It may be deduced that the reactivity of the nanoparticles in the gel and the homogeneity in the cation distribution within the solid favor the formation and growth of cubic zirconia. For this composition, a treatment at 1000°C attenuates the difference in texture of the two solids. The easy formation of this polymetallic crystalline phase containing neodyme should make possible use of stabilized zirconia as a host for actinides, especially americium, which can be used for treatment by transmutation in nuclear reactors, with the ultimate goal of obtaining nuclear fuel [139].

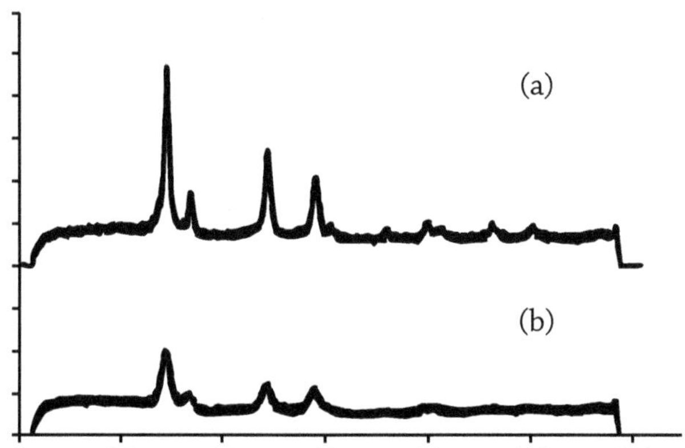

FIGURE 8.51 X-ray diffraction patterns of (a) the gel and (b) the coprecipitate of composition $Zr_{0.663}Y_{0.176}Nd_{0.161}O_{1.831}$ calcinated for 3 hours at 600°C (reproduced from [140] with permission of The Royal Society of Chemistry).

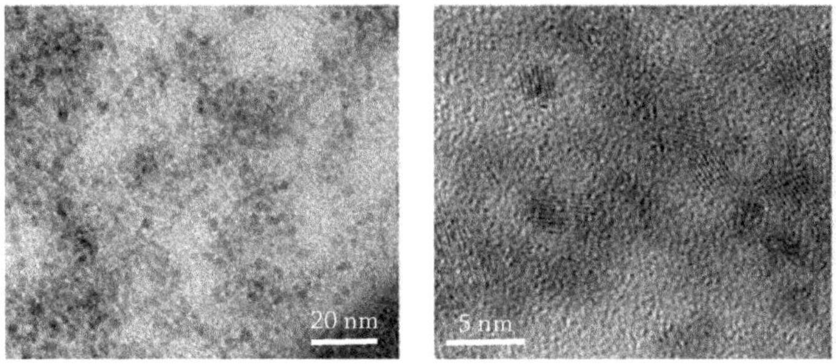

FIGURE 8.52 TEM images at different magnifications of yttria-stabilized zirconia nanoparticles $Zr_{0.83}Y_{0.17}O_{1.915}$ obtained by hydrothermal synthesis at 160° from sols of zirconium oxynitrate, yttrium nitrate, and acetylacetone (reproduced with permission from [142] Copyright 2009 American Chemical Society).

The hydrothermal treatment, at 160°C for 3 days, of the sol obtained with the acac addition to the solution of zirconium and yttrium ([Zr] = 0.1 mol.l^{-1}, [Y] = 0.05 mol.l^{-1}, [Acac] = 0.1 mol.l^{-1}, pH = 7 adjusted with ammonia) allows dispersion of yttria-stabilized zirconia [142]. It is known that 100% of zirconium and 40% of yttrium are reproducibly associated in the crystallized mixed phase, forming the compound $Zr_{0.83}Y_{0.17}O_{1.915}$ (ZrO_2 – 9% Y_2O_3), as spherical nanoparticles 5 nm in mean size (Fig. 8.52). The advantage of this procedure is that it achieves the synthesis in one step, avoiding prior precipitation of amorphous zirconia. Another advantage is that it yields a gel that is easily isolated by syneresis from the liquid containing the excess of reactants. This gel produces, by mechanical or ultrasonic stirring, particles that are perfectly dispersible in water.

Cubic zirconia, including other trivalent cations in solid solution, especially aluminum, can also be synthesized [143]. The coprecipitation of zirconium and aluminum is achieved by the addition of ammonia in the 5°C cooled solution of $ZrOCl_2$ and $AlCl_3$ (Zr/Al = 4) up to pH 8. The mixture lowers the pH to 3.5 and forms a transparent gel. It contains a solid that is amorphous to X-ray diffraction. By heating the gel at 200°C for a few hours, cubic zirconia is obtained in the form of nanoparticles 6 to 8 nm in mean size. The cubic phase containing 5% of aluminum is stable up to 800°C. It forms tetragonal zirconia at 1000°C and monoclinic zirconia at 1200°C in the form of particles about 25 nm in size.

Cubic zirconia can also be stabilized with calcium in ZrO_2-CaO solid solutions. They can be formed by hydrothermal treatment (110°C, 7 days) of suspensions of amorphous zirconia in the presence of a mixture of NaOH (or LiOH) and lime (17 to 25 molar %). Lime alone does not allow the crystallization of the mixed phase [144]. The increase in soda concentration, from 0.25 to 3 mol.l^{-1}, causes the decrease in particle size. Apparently, the soda concentration favors solubilization of the reactants in the crystallization process, allowing the nucleation rate to increase.

As for the synthesis of ZrO_2 nanoparticles, the hydrothermal treatment assisted by microwave heating, strikingly accelerate the formation of nanoparticles of the

ZrO_2-CaO and ZrO_2-Y_2O_3 solid solutions, decreasing the duration of syntheses up to about 60 minutes [145, 146].

References

[1] Henderson M.A. A surface science perspective on TiO_2 photocatalysis. *Surf. Sci. Reports* 2011, 66, 185–297.

[2] Brinker C.J., Scherer G.W. *Sol-Gel Science. The Physics and Chemistry of Sol-Gel Processing*. San Diego: Academic Press, 1990.

[3] Turova N.Y., Turevskaya E.P., Kessler V.G., Yanoskaya M.I. *The Chemistry of Metal Alkoxides*. Boston: Kluwer, 2002.

[4] Sanchez C., Julian B., Belleville P., Popall M. Applications of hybrid organic-inorganic nanocomposites. *J. Mater. Chem*. 15, 2005, 3559–3592.

[5] Wells A. F. *Structural Inorganic Chemistry*, 5th ed. Oxford: Oxford University Press, 1991.

[6] Ocana M., Matijevic E. Well-defined colloidal tin(IV) oxide particles. *J. Mater. Res*. 1990, 5, 1083–1091.

[7] Hiratsuka R.S., Pulcinelli S.H., Santilli C.V. Formation of SnO_2 gels from solutions. *J. Non Cryst. Solids* 1990, 121, 76–83.

[8] Donalson J.D., Fuller M.J. Ion exchange properties of tin(IV) materials. *J. Inorg. Nucl. Chem*. 1968, 30, 1083–1092.

[9] Baes C.F., Mesmer R.E. *The Hydrolysis of Cations*. New York: Wiley, 1976.

[10] Einaga H. Hydrolysis of titanium (IV) in aqueous (Na,H)Cl solution. *J. Chem. Soc. Dalton* 1979, 1917–1919.

[11] Reichmann M.G., Hollander F.J., Bell A.T. Structure of $[Ti_8O_{12}(H_2O)_{24}]Cl_8.HCl.7H_2O$. *Acta Cryst C* 1987, 43, 1681–1683.

[12] Liu C.E., Rouet A., Sutrisno H., Puzenat E., Terrisse H., Brohan L., Richard-Plouet M. Low temperature synthesis of nanocrystallized titanium oxides with layered or tridimensional frameworks, from $[Ti_8O_{12}(H_2O)_{24}]Cl_8{\cdot}HCl{\cdot}7H_2O$ hydrolysis. *Chem. Mater*. 2008, 20, 4739–4748.

[13] Johnson J.S., Kraus K.A. Hydrolytic behavior of metal ions. VI, Ultracentrifugation of zirconium(IV) and hafnium(IV); effect of acidity on the degree of polymerization. *J. Am. Chem. Soc*. 1956, 78, 3937–3943.

[14] Muha G.M., Vaughan P.A. Structure of the complex ion in aqueous solutions of zirconyl and hafnyl oxyhalides. *J. Chem. Phys*. 1960, 33, 194–199.

[15] Åberg M., Glaser J. ^{17}O and ^{1}H NMR study of the tetranuclear hydroxo zirconium complex in aqueous solution. *Inorg. Chim. Acta* 1993, 206, 53–61.

[16] Singhal A., Toth L.M., Lin J. S., Affholter K. Zirconium(IV) tetramer/octamer hydrolysis equilibrium in aqueous hydrochloric acid solution. *J. Am. Chem. Soc*. 1996, 118, 11529–11534.

[17] Clearfield A., Vaughan P.A. The crystal structure of zirconyl chloride octahydrate and zirconyl bromide octahydrate. *Acta Cryst*. 1956, 9, 555–558.

[18] Mak T.C.W. Refinement of the crystal structure of zirconyl chloride octahydrate. *Can. J. Chem*. 1968, 46, 3491–3493.

[19] Das G., Bharadwaj P.K., Ghosh D., Chaudhuri B., Banerjee R. Synthesis and structure of the $[Mn^{4+}(biguanide)_3]^{4+}$ ion: the simplest source for water-stable manganese (IV). *Chem. Commun.* 2001, 323–324.
[20] Ride D.R., Ed. *Handbook of Chemistry and Physics*, 85th ed. Boca Raton, FL: CRC Press, 2004.
[21] Zhang H., Banfield J.F. Understanding polymorphic phase transformation behavior during growth of nanocrystalline aggregates: Insights from TiO_2. *J. Phys. Chem. B* 2000, 104, 3481–3487.
[22] Pottier A., Cassaignon S., Chanéac C., Villain F., Tronc E., Jolivet J.P. Size tailoring of TiO_2 anatase nanoparticles in aqueous medium and synthesis of nanocomposites. Characterization by Raman spectroscopy. *J. Mater. Chem.* 2003, 13, 877–882.
[23] Durupthy O., Bill J., Aldinger F. Bioinspired synthesis of crystalline TiO_2: effect of amino acids on nanoparticles structure and shape. *Cryst Growth Des.* 2007, 7, 2696–2704.
[24] Yang H.G., Sun G.H., Qiao S.Z., Zou J., Liu G., Campbell Smith S., Cheng H.M., Lu G.Q. Anatase TiO_2 single crystals with a large percentage of reactive facets. *Nature* 2008, 453, 638–642.
[25] Yang H.G., Liu G., Qiao S.Z., Sun G.H., Jin Y.G., Campbell Smith S., Zou J., Cheng H.M., Lu G.Q. Solvothermal synthesis and photoreactivity of anatase TiO_2 nanosheets with dominant {001} facets. *J. Am. Chem. Soc.* 2009, 131, 4078–4083.
[26] Nguyen C.K., Cha H.G., Kang Y.S. Axis-oriented anatase TiO_2 single crystals with dominant {001} and {100} facets. *Cryst. Growth Des.* 2011, 11, 3947–3953.
[27] Zhang D., Li G., Wang H., Chan K.M., Yu J.C. Biocompatible anatase single-crystal photocatalysts with tunable percentage of reactive facets. *Cryst. Growth Des.* 2010, 10, 1130–1137.
[28] Ma X.Y., Chen Z.G., Hartono S.B., Jiang H.B., Zou J., Qiao S.Z. Yang H.G. Fabrication of uniform anatase TiO_2 particles exposed by {001} facets. *Chem. Commun.* 2010, 46, 6608–6610.
[29] Liu M., Piao L., Zhao L., Ju S., Yan Z., He T., Zhou C., Wang W. Anatase TiO_2 single crystals with exposed {001} and {110} facets: facile synthesis and enhanced photocatalysis. *Chem. Commun.* 2010, 46, 1664–1666.
[30] Zhang Y., Wu L., Zeng Q., Zhi J. An approach for controllable synthesis of different phase titanium dioxide nanocomposites with peroxotitanium complex as precursor. *J. Phys. Chem. C* 2008, 112, 16457–16462.
[31] Liao J., Shi L., Yuan S., Zhao Y., Fang J. Solvothermal synthesis of TiO_2 nanocrystal colloids from peroxotitanate complex solution and their photocatalytic activities. *J. Phys. Chem. C* 2009, 113, 18778–18783.
[32] Seok S.I., Vithal M., Chang J.A. Colloidal TiO_2 nanocrystals prepared from peroxotitanium complex solutions: Phase evolution from different precursors. *J. Colloid Interface Sci.* 2010, 346, 66–71.
[33] Groenke N., Seisenbaeva G.A., Kaminskyy V., Zhivotovsky B., Kost B., Kessler V.G. Structural characterization, solution stability and potential health and environmental effects of the nano-TiO_2 bioencapsulation matrix and the model product of its biodegradation. *TiBALDH. RSC Adv.* 2012, 2, 4228–4235.
[34] Möckel H., Giersig M., Willig F. Formation of uniform size anatase nanocrystals from bis(ammoniumlactato)titanium dihydroxide by thermohydrolysis. *J. Mater. Chem.* 1999, 9, 3051–3056.

[35] Pottier A. Précipitation de cations métalliques en solution aqueuse. Influence de l'acidité sur les caractéristiques des particules d'oxyde. Thèse, Université Pierre et Marie Curie, Paris, France, 1999.
[36] Pottier A., Chanéac C., Tronc E., Mazerolles L., Jolivet J. P. Synthesis of brookite TiO_2 nanoparticles by thermolysis of $TiCl_4$ in strongly acidic aqueous medium. *J. Mater. Chem.* 2001, 11, 1116–1121.
[37] Cassaignon S., Koelsch M., Jolivet J.P. Selective synthesis of brookite, anatase and rutile nanoparticles: thermolysis of $TiCl_4$ in aqueous nitric acid. *J. Mater. Sci.* 2007, 42, 6689–6695.
[38] Dessombz A., Chiche D., Davidson P., Panine P., Chanéac C. Jolivet J.P. Design of liquid-crystalline aqueous suspensions of rutile nanorods: evidence of anisotropic photocatalytic properties. *J. Am. Chem. Soc.* 2007, 129, 5904–5909.
[39] Nabivanets B.I. Adsorption of titanium(IV) ions on ion-exchange resins. Russ. *J. Inorg. Chem.* 1962, 7, 212–215.
[40] Nabivanets B.I., Kudritskaya L.N. Determination of the composition and dissociation constants of titanyl chloro complexes. *Russ. J. Inorg. Chem.* 1967, 12, 789–792.
[41] Nabivanets B.I., Kudritskaya L.N. A study on the polymerization of titanium (IV) in hydrochloric acid solutions. *Russ. J. Inorg. Chem.* 1967, 12, 616–620.
[42] Koelsch M. Nanoparticules de TiO_2: contrôle structural, morphologique, dimensionnel et propriétés électrochimiques. Thèse, Université Pierre et Marie Curie, Paris, France, 2004.
[43] Koelsch M., Cassaignon S., Jolivet J.P. Synthesis of nanometric TiO_2 in aqueous solution by soft chemistry: obtaining of anatase, brookite and rutile with controlled shapes. *Mat. Res. Soc. Symp. Proc.* 2004, 822, S5.3.1-S5.3.6.
[44] Dambournet D., Belharouak I., Amine K. Tailored preparation methods of TiO_2 anatase, rutile, brookite: Mechanism of formation and electrochemical properties. *Chem. Mater.* 2010, 22, 1173–1179.
[45] Dambournet D., Belharouak I., Ma J., Amine K. Toward high surface area TiO_2 brookite with morphology control. *J. Mater. Chem.* 2011, 21, 3085–3090.
[46] Boudaren C., Bataille T., Auffrédic J.P., Louër D. Synthesis, structure determination from powder diffraction data and thermal behaviour of titanium(IV) oxalate $[Ti_2O_3(H_2O)_2](C_2O_4)\cdot H_2O$. *Sol. State Sci.* 2003, 5, 175–182.
[47] Andersson S., Wadsley A.D. The crystal structure of $Na_2Ti_3O_7$. *Acta Cryst.* 1961, 14, 1245–1249.
[48] Zhang T., Chen Q., Peng L.M. Hydrothermal reaction mechanism and pathway for the formation of $K_2Ti_6O_{13}$ nanowires. *Adv. Funct. Mater.* 2008, 18, 3018–3025.
[49] Zhu H., Gao X., Lan Y., Song D., Xi Y., Zhao J. Hydrogen titanate nanofibers covered with anatase nanocrystals: a delicate structure achieved by the wet chemistry reaction of the titanate nanofibers. *J. Am. Chem. Soc.* 2004, 126, 8380–8381.
[50] Zhu H.Y., Lan Y., Gao X.P., Ringer S.P., Zheng Z.F., Song D.Y., Zhao J.C. Phase transition between nanostructures of titanate and titanium dioxides via simple wet-chemical reactions. *J. Am. Chem. Soc.* 2005, 127, 6730–6736.
[51] Wu D., Liu J., Zhao X., Li A., Chen Y., Ming N. Sequence of events for the formation of titanate nanotubes, nanofibers, nanowires, and nanobelts. *Chem. Mater.* 2006, 18, 547–553.
[52] Huang J., Cao Y., Huang Q., He H., Liu Y., Guo W. High-temperature formation of titanate nanotubes and the transformation mechanism of nanotubes into nanowires. *Cryst. Growth Des.* 2009, 9, 3632–3637.

[53] Hu W., Li L., Li G., Tang C., Sun L. High-quality brookite TiO_2 flowers: synthesis, characterization and dielectric performance. *Cryst. Growth Des.* 2009, 9, 3676–3682.
[54] Du G.H., Chen Q., Che R. C., Yuan Z. Y., Peng L.-M. Preparation and structure analysis of titanium oxide nanotubes. *Appl. Phys. Lett.* 2001, 79, 3702–3704.
[55] Chen Q., Du G.H., Zhang S., Peng L.-M. The structure of trititanate nanotubes. *Acta Cryst. B* 2002, 58, 587–593.
[56] Deng Q., Wei M., Ding X., Jiang L., Ye B., Wei K. Brookite-type TiO_2 nanotubes. *Chem. Commun.* 2008, 3657–3659.
[57] Armstrong A.R., Armstrong G., Canales J., Bruce B.G. TiO_2-B nanowires. *Angew. Chem. Int. Ed.* 2004, 43, 2286–2288.
[58] Marchand R., Brohan L., Tournoux M. TiO_2(B) A new form of titanium dioxide and the potassium octatitanate $K_2Ti_8O_{17}$. *Mat. Res. Bull.* 1980, 15, 1129–1133.
[59] Sutrisno H. Synthesis and characterization of TiO_2(B) nanotubes prepared by hydrothermal method using $[Ti_8O_{12}(OH_2)_{24}]Cl_8{\cdot}HCl{\cdot}7H_2O$ as precursor. *Makara J. Sci.* 2010, 14, 27–31.
[60] Zukalova M., Kalba M., Kavan L., Exnar I., Haeger A., Graetzel M. Electrochemical and gas-phase photocatalytic performance of nanostructured TiO_2(B) prepared by novel synthetic route. *Progr. Solid State Chem.* 2005, 33, 253–261.
[61] Cassaignon S., Koelsch M., Jolivet J.P. From $TiCl_3$ to TiO_2 nanoparticles (anatase, brookite and rutile): Thermohydrolysis and oxidation in aqueous medium. *J. Phys. Chem. Solids* 2007, 68, 695–700.
[62] Lee D.H., Park J.G., Choi K.J., Choi H.J., Kim D.W. Preparation of brookite-type TiO_2/carbon nanocomposite electrodes for application to Li ion batteries. *Eur. J. Inorg. Chem.* 2008, 878–882.
[63] Li J.G., Tang C., Li D., Haneda H., Ishigaki T. Monodispersed spherical particles of brookite-type TiO_2: synthesis, characterization and photocatalytic property. *J. Am. Ceram. Soc.* 2004, 87, 1358–1361.
[64] Zhang D., Li G., Wang F., Yu J.C. Green synthesis of a self-assembled rutile mesocrystalline photocatalyst. *Cryst. Eng. Comm.* 2010, 12, 1759–1763.
[65] Bokhimi X., Pedraza F. Characterization of brookite and a new corundum-like titania phase synthesized under hydrothermal conditions. *J. Solid State Chem.* 2004, 177, 2456–2463.
[66] Zwilling V., Aucouturier M., Darque-Ceretti E. Anodic oxidation of titanium and TA6V alloy in chromic media. An electrochemical approach. *Electrochim. Acta* 1999, 45, 921–929.
[67] Gong D., Grimes C.A., Varghese O.K., Hu W., Singh R.S. Titanium oxide nanotube arrays prepared by anodic oxidation. *J. Mater. Res.* 2001, 16, 3331–3334.
[68] Paulose M., Shankar K., Yoriya S., Prakasam H.E., Varghese O.K., Mor G.K., Latempa T.A., Fitzgerald A., Grimes C.A. Anodic growth of highly ordered TiO_2 nanotube arrays to 134 μm in length. *J. Phys. Chem. B Lett.* 2006, 110, 16179–16184.
[69] Fahim N.F., Sekino T. A novel method for synthesis of titania nanotube powders using rapid breakdown anodization. *Chem. Mater.* 2009, 21, 1967–1979.
[70] Wang J., Lin Z. Anodic formation of ordered TiO_2 nanotube arrays: effects of electrolyte temperature and anodization potential. *J. Phys. Chem. C* 2009, 113, 4026–4030.
[71] Roy P., Berger S., Schmuki P. TiO_2 nanotubes: synthesis and applications. *Angew. Chem. Int. Ed.* 2011, 50, 2904–2939.

[72] Lines M.E., Glass A.M. *Principles and Applications of Ferroelectrics and Related Materials*. Oxford: Oxford University Press, 2001.
[73] Mulder B.J. Preparation of $BaTiO_3$ and other ceramic powders by coprecipitation of citrates in an alcohol. *Ceram. Bull.* 1970, 49, 990–993.
[74] Hennings D., Mayr W. Thermal decomposition of (BaTi)citrates into barium titanate. *J. Solid State Chem.* 1978, 26, 329–338.
[75] Kakihana M., Arima M., Nakamura Y., Yashima M., Yoshimura M. Spectroscopic characterization of precursors used in the Pechini-type polymerizable complex processing of barium titanate. *Chem. Mater.* 1999, 11, 438–450.
[76] Hertl W. Kinetics of barium titanate synthesis. *J. Am. Ceram. Soc.* 1988, 71, 879–883.
[77] Pfaff G. $BaTiO_3$ preparation by reaction of TiO_2 with $Ba(OH)_2$. *J. European Ceram. Soc.* 1991, 8, 35–39.
[78] Yanovsky A., Yanovskaya M., Limar V.K., Kessler V.G., Turovac N.Y., Struchkov Y.T. Synthesis and crystal structure of the double barium-titanium isopropoxide $[Ba_4Ti_4(\mu_4\text{-}O)_4(\mu_3\text{-}OR)_2(\mu\text{-}OR)_8(OR)_6(ROH)_4]$ $[Ba_4Ti_4(\mu_4\text{-}O)_4(\mu_3\text{-}OR)_2(\mu\text{-}OR)_8(OR)_6(ROH)_3]$. *J. Chem. Soc., Chem. Commun.* 1991, 1605–1606.
[79] Chaput F., Boilot J.P. Chemical processing of $(Ba, Sr)(Ti, Zr)O_3$ perovskite ceramics. *J. Mater. Sci. Lett.* 1987, 6, 1110–1112.
[80] Bradley D.C., Mehrotra R.C., Gaur D.P. *Metal Alkoxides*. London: Academic Press, 1978.
[81] Tebo B.M., Bargar J.R., Clement B.G., Dick G.J., Murray K.J., Parker D., Verity R., Webb S.M. Biogenic manganese oxides: properties and mechanisms of formation. *Annu. Rev. Earth Planet. Sci.* 2004, 32, 287–328.
[82] Borch T., Kretzschmar R., Kappler A., Van Cappellen Ph., Ginder-Vegel M., Voegelin A., Campbell K. Biogeochemical redox processes and their impact on contaminant dynamics. *Environ. Sci. Technol.* 2010, 44, 15–23.
[83] Lanson B., Marcus M.A., Fakra S., Panfili F., Geoffroy N., Manceau A. Formation of Zn–Ca phyllomanganate nanoparticles in grass roots. *Geochim. Cosmochim. Acta* 2008, 72, 2478–2490.
[84] McKenzie R.M. The manganese oxides in soils—A review. *Zeit. Pflanzenernährung und Bodenkunde* 1972, 131, 221–242.
[85] Suib S.L. Porous manganese oxide octahedral molecular sieves and octahedral layered materials. *Acc. Chem. Res.* 2008, 41, 479–487.
[86] Post J.E. Manganese oxide minerals: Crystal structures and economic and environmental significance. *Proc. Natl. Acad. Sci. USA* 1999, 96, 3447–3454.
[87] Villalobos M., Toner B., Bargar J., Sposito G. Characterization of the manganese oxide produced by *Pseudomonas putida* strain MnB1. *Geochim. Cosmochim. Acta* 2003, 67, 2649–2662.
[88] Turner S., Buseck P.R. Defects in nsutite (γ-MnO_2) and dry-cell battery efficiency. *Nature* 1983, 304, 143–146.
[89] Faulring G.M. Unit cell determination and thermal transformations of nsutite. *Amer. Miner.* 1965, 50, 170–179.
[90] Villalobos M., Lanson B., Manceau A., Toner B., Sposito G. Structural model for the biogenic Mn oxide produced by *Pseudomonas putida*. *Amer. Miner.* 2006, 91, 489–502.

[91] Yuan J., Li W.N., Gomez S., Suib S.L. Shape-controlled synthesis of manganese oxide octahedral molecular sieve three-dimensional nanostructures. *J. Am. Chem. Soc.* 2005, 127, 14184–14185.

[92] Li Y., Ding Y., Xiong Y., Xie Y. One-step solution-based catalytic route to fabricate novel α-MnO_2 hierarchical structures on a large scale. *Chem. Commun.* 2005, 918–920.

[93] Pike J., Hanson J., Zhang L., Chan S.W. Synthesis and redox behavior of nanocrystalline hausmanite (Mn_3O_4). *Chem. Mater.* 2007, 19, 5609–5616.

[94] Hill J.P., Alam S., Ariga K., Anson C.E., Powell A.K. Nanostructured microspheres of MnO_2 formed by room temperature solution processing. *Chem. Commun.* 2008, 383–385.

[95] Saratovsky I., Wightman P.G., Pasten P.A., Gaillard J.F., Poeppelmeier K.R. Manganese oxides: parallels between abiotic and biotic structures. *J. Am. Chem. Soc.* 2006, 128, 11188–11198.

[96] Brock S.L., Duan N., Tian Z.R., Giraldo O., Zhou H., Suib S.L. A review of porous manganese oxide materials. *Chem. Mater.* 1998, 10, 2619–2628.

[97] Villegas J.C., Garces L.J., Gomez S., Durand J.P., Suib S.L. Particle size control of cryptomelane nanomaterials by use of H_2O_2 in acidic conditions. *Chem. Mater.* 2005, 17, 1910–1918.

[98] Portehault D., Cassaignon S., Baudrin E., Jolivet J.P. Evolution of nanostructured manganese (oxyhydr)oxides in water through MnO_4^- reduction. *Cryst. Growth Des.* 2010, 10, 2168–2173.

[99] Portehault D., Cassaignon S., Baudrin E., Jolivet J.P. Structural and morphological control of manganese oxide nanoparticles upon soft aqueous precipitation through MnO_4^-/Mn^{2+} reaction. *J. Mater. Chem.* 2009, 19, 2407–2416.

[100] Shen X.F., Ding Y.S., Jiu J., Cai J., Laubernds K., Zerger R.P., Vasiliev A., Aindow M., Suib S.L. Control of nanometer-scale tunnel sizes of porous manganese oxide octahedral molecular sieve nanomaterials. *Adv. Mater.* 2005, 17, 805–809.

[101] Ding Y.S., Shen X.F., Gomez S., Luo H., Aindow M., Suib S.L. Hydrothermal growth of manganese dioxide into three-dimensional hierarchical nanoarchitectures. *Adv. Funct. Mater.* 2006, 16, 549–555.

[102] Portehault D., Cassaignon S., Baudrin E., Jolivet J.P. Morphology control of cryptomelane type MnO_2 nanowires by soft chemistry. Growth mechanisms in aqueous medium. *Chem. Mater.* 2007, 19, 5410–5417.

[103] Portehault D., Cassaignon S., Nassif N., Baudrin E., Jolivet J.P. A core–corona hierarchical manganese oxide and its formation by an aqueous soft chemistry mechanism. *Angew. Chem. Int. Ed.* 2008, 47, 6441–6444.

[104] Portehault D., Cassaignon S., Baudrin E., Jolivet J.P. Design of hierarchical core-corona architectures of layered manganese oxides by aqueous precipitation. *Chem. Mater.* 2008, 20, 6140–6147.

[105] Luo J., Huang A., Park S.H., Suib S.L., O'Young C.L. Crystallization of sodium-birnessite and accompanied phase transformation. *Chem. Mater.* 1998, 10, 1561–1568.

[106] Ma Y., Luo J., Suib S.L. Syntheses of birnessites using alcohols as reducing reagents: effects of synthesis parameters on the formation of birnessites. *Chem. Mater.* 1999, 11, 1972–1979.

[107] Liu Z., Ma R., Ebina Y., Takada K., Sasak T. Synthesis and delamination of layered manganese oxide nanobelts. *Chem. Mater.* 2007, 19, 6504–6512.

[108] Portehault D., Cassaignon S., Baudrin E., Jolivet J.P. Synthesis of a manganese oxide nanocomposite through heteroepitaxy in aqueous medium. *Chem. Commun.* 2009, 674–676.

[109] Portehault D., Cassaignon S., Baudrin E., Jolivet J.P. Twinning driven growth of manganese oxide hollow cones through self-assembly of nanorods in water. *Cryst. Growth Des.* 2009, 9, 2562–2565.

[110] Portehault D., Cassaignon S., Baudrin E., Jolivet J.P. Selective heterogeneous oriented attachment of manganese oxide nanorods in water: toward 3D nanoarchitectures. *J. Mater. Chem.* 2009,19, 7947–7954.

[111] Stumm W., Sulzberger B. The cycling of iron in natural environments: Considerations based on laboratory studies of heterogeneous redox processes. *Geochim. Cosmochim. Acta* 1992, 56, 3233–3257.

[112] Silvester E., Charlet L., Tournassat C., Géhin A., Grenèche J.M., Liger E. Redox potential measurements and Mössbauer spectrometry of FeII adsorbed onto FeIII (oxyhydr)oxides. *Geochim. Cosmochim. Acta* 2005, 69, 4801–4815.

[113] Gorski C.A., Scherer M.M. Fe^{2+} Sorption at the Fe oxide-water interface: a revised conceptual framework, 315–343. In *Aquatic Redox Chemistry*, Tratnyek P. et al., Eds. ACS Symposium Series, American Chemical Society, Washington, DC, 2011.

[114] Tu S., Racz G.J., Goh T.B. Transformations of synthetic birnessite as affected by pH and manganese concentration. *Clays Clay Miner.* 1994, 42, 321–330.

[115] Cassaignon S., Koelsch M., Jolivet J.P. Controlled synthesis of manganese (oxihydr) oxides in aqueous solution by thermolysis. *Mat. Res. Soc. Symp. Proc.* 2004, 822, S.3.8.1-S.3.8.6.

[116] Portehault D. Synthèse par chimie douce en milieu aqueux d'oxydes de manganèse nano-structurés. Thèse, Université Pierre et Marie Curie, Paris, France, 2008.

[117] Wang X., Li Y. Selected-control hydrothermal synthesis of α- and β-MnO_2 single crystal nanowires. *J. Am. Chem. Soc.* 2002, 124, 2880–2881.

[118] Wang X., Li Y. Synthesis and formation mechanism of manganese dioxide nanowires/nanorods. *Chem. Eur. J.* 9, 2003, 300–306.

[119] House D.A. Kinetics and mechanism of oxidations by peroxodisulfate. *Chem. Rev.* 1962, 62, 185–203.

[120] Li Y., Ding Y., Xiong Y., Xie Y. One-step solution-based catalytic route to fabricate novel α-MnO_2 hierarchical structures on a large scale. *Chem. Commun.* 2005, 918–920.

[121] Li Y., Ding Y., Xiong Y., Xie Y. Rational growth of various α-MnO_2 hierarchical structures and β-MnO_2 nanorods via a homogeneous catalytic route. *Cryst. Growth Des.* 2005, 5, 1953–1958.

[122] Yuan J., Li W.N., Gomez S., Suib S.L. Shape-controlled synthesis of manganese oxide octahedral molecular sieve three-dimensional nanostructures. *J. Am. Chem. Soc.* 2005, 127, 14184–14185.

[123] Huang H., Chen C.H., Xu L., Genuino H., Garcia-Martinez J., Garces H.F., Jin L., Kithongo C.K., Suib S.L. Single-step synthesis of manganese oxide octahedral molecular sieves with large pore sizes. *Chem. Commun.* 2010, 46, 5945–5947.

[124] Clearfield A. Structural aspects of zirconium chemistry. *Rev. Pure Appl. Chem.* 1964, 14, 91–108.

[125] Livage J., Doi K., Mazières C. Nature and thermal evolution of amorphous hydrated zirconium oxide. *J. Amer. Ceram. Soc.* 1968, 51, 349–353.

[126] Clearfield A. Crystalline hydrous zirconia. *Inorg. Chem.* 1964, 3, 146–148.
[127] Kamp H. Couches fonctionnelles inorganiques par voie sol-gel aqueuse. Thèse, Université Pierre et Marie Curie, Paris, France, 2010.
[128] Hu M.Z.C, Hunt R.D., Payzant E.A., Hubbard C.R. Nanocrystallization and phase transformation in monodispersed ultrafine zirconia particles from various homogeneous precipitation methods. *J. Am. Ceram. Soc.* 1999, 82, 2313- 2320.
[129] Denkewicz R.P., Jr., TenHuisen K.S., Adair J.H. Hydrothermal crystallization kinetics of m-ZrO_2 and t-ZrO_2. *J. Mater. Res.* 1990, 5, 2698–2705.
[130] Stefanic G., Popovic S., Music S. Influence of pH on the hydrothermal crystallization kinetics and crystal structure of ZrO_2. *Thermochim. Acta* 1997, 303, 31–39.
[131] Bugrov A.N., Almjasheva O.V. Effect of hydrothermal synthesis conditions on the morphology of ZrO_2 nanoparticles. *Nanosystems: Phys. Chem. Math.* 2013, 4, 810–815.
[132] Tani E., Yoshimura M., Somiya S. Formation of ultrafine tetragonal ZrO_2 powder under hydrothermal conditions. *J. Amer. Ceram. Soc.* 1983, 66, 11–14.
[133] Dell'Agli G., Colantuono A., Mascolo G. The effect of mineralizers on the crystallization of zirconia gel under hydrothermal conditions. *Solid State Ionics* 1999, 123, 87–94.
[134] Cheng H., Wu L., Ma J., Zhao Z., Qi L. Hydrothermal preparation of nanosized cubic ZrO_2 powders. *J. Mater. Sci. Lett.* 1996, 15, 895–897.
[135] Tahir M.N., Gorgishvili L., Li J., Gorelik T., Kolb U., Nasdala L., Tremel W. Facile synthesis and characterization of monocrystalline cubic ZrO_2 nanoparticles. *Sol. State Sci.* 2007, 9, 1105–1109.
[136] Chen C.W., Yang X.S., Chiang A.S.T. An aqueous process for the production of fully dispersible t-ZrO_2 nanocrystals. *J. Taiwan Institute of Chemical Engineers* 2009, 40, 296–301.
[137] Laberty-Robert C., Ansart F., Deloget C., Gaudon M., Rousset A. Powder synthesis of nanocrystalline ZrO_2–8%Y_2O_3 via a polymerization route. *Mat. Res. Bull.* 2001, 36, 2083–2101.
[138] Dell'Agli G., Mascolo G. Hydrothermal synthesis of ZrO_2-Y_2O_3 solid solutions at low temperature. *J. Eur. Ceram. Soc.* 2000, 20, 139–145.
[139] Lemonnier S. Synthèse par voie douce d'oxydes polymétalliques incluant des actinides. Réactivité et structure de la solution au solide. Thèse, Université Pierre et Marie Curie, Paris, France, 2007.
[140] Lemonnier S., Grandjean S., Robisson A.C., Jolivet J.P. Synthesis of zirconia sol stabilized by trivalent cations (yttrium and neodymium or americium): a precursor for Am-bearing cubic stabilized zirconia. *Dalton Trans.* 2010, 39, 2254–2262.
[141] Peyre V., Spalla O., Belloni L., Nabavi M. Stability of a nanometric zirconia colloidal dispersion under compression. Effect of surface complexation by acetylacetone. *J. Colloid Interface Sci.* 1997, 187, 184–200.
[142] Guiot C., Grandjean S., Lemonnier S., Jolivet J.P., Batail P. Nano single crystals of yttria-stabilized zirconia. *Cryst. Growth Des.* 2009, 9, 3548–3550.
[143] Mondal A., Ram S. Al^{3+}-stabilized c-ZrO_2 nanoparticles at low temperature by forced hydrolysis of dispersed metal cations in water. *Solid State Ionics* 2003, 160, 169–181.
[144] Dell'Agli G., Mascolo G. Low temperature hydrothermal synthesis of ZrO_2-CaO solid solutions. *J. Mater. Sci.* 2000, 35, 661–665.

[145] Rizzuti A., Corradi A., Leonelli C., Rosa R., Pielaszek R., Lojkowski W. Microwave technique applied to the hydrothermal synthesis and sintering of calcia stabilized zirconia nanoparticles. *J. Nanopart. Res.* 2010, 12, 327–335.
[146] Vernieuwe K., Lommens P., Martins J.C., Van Den Broeck F., Van Driessche I., De Buysser K. Aqueous ZrO_2 and YSZ colloidal systems through microwave aassisted hydrothermal synthesis. *Materials* 2013, 6, 4082–4095.

Conclusion

Metal oxide nanostructures are of major interest in technology. It is therefore essential to have a full understanding of the phenomena involved in the aqueous synthesis of nanoparticles, so that their properties can be adjusted to a desired application. Understanding these phenomena is also important in other fields, for instance, in geology and environmental sciences, enabling us to explain the presence and formation of a given mineral.

The precipitation of metal oxy(hydroxi)des is a complex phenomenon initiated by hydroxylation of the cations in solution and resulting from condensation of the hydroxylated species. Therefore, the acidity of the cations is the main characteristic of their reactivity. Three main parameters are essential in predicting and rationalizing the behavior of metal cations in water: the formal charge (the oxidation degree), size, and electronegativity, which determine the degree of polarization of the oxygenated ligands. One may thus define five classes of cations:

1. The too weakly polarizing cations that form only aquocomplexes unable to condense and to precipitate; for instance, the alkaline cations M^+,
2. The cations that condense by olation and form polycations and hydroxides, typically, the divalent cations and also Al^{3+}.
3. The cations that condense by olation and oxolation and form oxyhydroxides and oxides (such as Cr^{3+}, Fe^{3+}, and Mn^{3+}).
4. The cations that condense essentially by oxolation and form oxides that are more or less hydrated (Ti^{4+}, Mn^{4+}, V^{5+}).
5. The cations that form anionic oxocomplexes and exhibit no trend toward condensation, typically, Mn^{VII}.

This series thus includes cations of increasing polarizing power, that is cations of increasing oxidation degree and electronegativity .

Precipitation usually generates nanosized particles. In a system that is not continuously fed, in which a limited amount of matter is available in the reactor, the nucleation step is always sudden and easy enough, allowing lower supersaturation and creating nuclei that have stopped growing because of the too low concentration in soluble precursor. That does not, however, exclude an intense dynamics of dissolution–crystallization because of the evolution of the criticality of the particle size during the decrease in supersaturation. Other phenomena may also occur during the evolution of suspensions, such as changes in crystal structure and/or in the morphology of particles.

The physical properties and reactivity of nanomaterials are closely related to the size, shape, and crystal structure of nanoparticles. For many oxides, these characteristics are adjustable over a more or less wide range, as long as the formation mechanism of the solid, the kinetics of involved processes, and the surface chemistry of the material are well understood and controlled. Many examples show the importance of the surface in the control of the size and morphology of nanoparticles.

Among the set of parameters with which it is possible to act during the synthesis, for instance, acidity, temperature, ionic strength, concentration, and nature of possible ligands, the crucial factor is really the acidity of the medium. It indeed plays simultaneously on many phenomena. In particular, the acidity is the main factor

- controlling the formation of the solid in determining the concentration of the noncharged precursor;
- determining the protonation/deprotonation equilibria and consequently the efficiency of complexing ligands;
- adjusting the surface charge of particles on which their aggregation or their dispersion depend; and
- determining the mechanism of crystallization of the solid, such as dissolution–crystallization, in situ crystallization, and possibly ordered aggregation.

Consequently, it is important to control the acidity of the medium as well as the process of hydroxylation of the cations: base addition into the solution of cations or, inversely, addition of cations into the basic solution that may avoid the formation of unwanted reaction intermediates, thermohydrolysis, or hydrothermal conditions. These different techniques may be useful in forming various polymorphic forms and synthesizing polymetallic phases.

Various wisely chosen ligands may allow control of the morphology of nanoparticles. The preferred growth of some crystal faces involves a subtle balance between the complexing power and structural adequacy of the ligand toward the surface.

Finally, the synthesis of complex architectures may be based on differences in the kinetics of parallel or consecutive reactions and/or also the ordered aggregation phenomenon as exemplified by titanium and manganese oxides.

A great deal of knowledge has already been acquired on the controlled precipitation of metal oxides in aqueous medium. Even so, much remains to be done in gaining an understanding of the mechanisms of formation of nanoparticles and improving their characterization. The intensity of the work provided in this field at the present time holds promise of success in the near future.

INDEX

www.ingramcontent.com/pod-product-compliance
Ingram Content Group UK Ltd.
Pitfield, Milton Keynes, MK11 3LW, UK
UKHW021133160926
475595UK00020B/29

9 780190 928117